Exercises and Solutions in Statistical Theory

CHAPMAN & HALL/CRC
Texts in Statistical Science Series

Series Editors
Francesca Dominici, *Harvard School of Public Health, USA*
Julian J. Faraway, *University of Bath, UK*
Martin Tanner, *Northwestern University, USA*
Jim Zidek, *University of British Columbia, Canada*

Analysis of Failure and Survival Data
P. J. Smith

The Analysis of Time Series — An Introduction, Sixth Edition
C. Chatfield

Applied Bayesian Forecasting and Time Series Analysis
A. Pole, M. West, and J. Harrison

Applied Categorical and Count Data Analysis
W. Tang, H. He, and X.M. Tu

Applied Nonparametric Statistical Methods, Fourth Edition
P. Sprent and N.C. Smeeton

Applied Statistics — Handbook of GENSTAT Analysis
E.J. Snell and H. Simpson

Applied Statistics — Principles and Examples
D.R. Cox and E.J. Snell

Applied Stochastic Modelling, Second Edition
B.J.T. Morgan

Bayesian Data Analysis, Second Edition
A. Gelman, J.B. Carlin, H.S. Stern, and D.B. Rubin

Bayesian Ideas and Data Analysis: An Introduction for Scientists and Statisticians
R. Christensen, W. Johnson, A. Branscum, and T.E. Hanson

Bayesian Methods for Data Analysis, Third Edition
B.P. Carlin and T.A. Louis

Beyond ANOVA — Basics of Applied Statistics
R.G. Miller, Jr.

The BUGS Book: A Practical Introduction to Bayesian Analysis
D. Lunn, C. Jackson, N. Best, A. Thomas, and D. Spiegelhalter

A Course in Categorical Data Analysis
T. Leonard

A Course in Large Sample Theory
T.S. Ferguson

Data Driven Statistical Methods
P. Sprent

Decision Analysis — A Bayesian Approach
J.Q. Smith

Design and Analysis of Experiments with SAS
J. Lawson

Elementary Applications of Probability Theory, Second Edition
H.C. Tuckwell

Elements of Simulation
B.J.T. Morgan

Epidemiology — Study Design and Data Analysis, Second Edition
M. Woodward

Essential Statistics, Fourth Edition
D.A.G. Rees

Exercises and Solutions in Statistical Theory
L.L. Kupper, B.H. Neelon, and S.M. O'Brien

Exercises and Solutions in Biostatistical Theory
L.L. Kupper, B.H. Neelon, and S.M. O'Brien

Extending the Linear Model with R — Generalized Linear, Mixed Effects and Nonparametric Regression Models
J.J. Faraway

A First Course in Linear Model Theory
N. Ravishanker and D.K. Dey

Generalized Additive Models: An Introduction with R
S. Wood

Generalized Linear Mixed Models: Modern Concepts, Methods and Applications
W. W. Stroup

Graphics for Statistics and Data Analysis with R
K.J. Keen

Interpreting Data — A First Course in Statistics
A.J.B. Anderson

Introduction to General and Generalized Linear Models
H. Madsen and P. Thyregod

An Introduction to Generalized Linear Models, Third Edition
A.J. Dobson and A.G. Barnett

Introduction to Multivariate Analysis
C. Chatfield and A.J. Collins

Introduction to Optimization Methods and Their Applications in Statistics
B.S. Everitt

Introduction to Probability with R
K. Baclawski

Introduction to Randomized Controlled Clinical Trials, Second Edition
J.N.S. Matthews

Introduction to Statistical Inference and Its Applications with R
M.W. Trosset

Introduction to Statistical Limit Theory
A.M. Polansky

Introduction to Statistical Methods for Clinical Trials
T.D. Cook and D.L. DeMets

Introduction to the Theory of Statistical Inference
H. Liero and S. Zwanzig

Large Sample Methods in Statistics
P.K. Sen and J. da Motta Singer

Large Sample Methods in Statistics
P.K. Sen and J. da Motta Singer

Linear Algebra and Matrix Analysis for Statistics
S. Banerjee and A. Roy

Logistic Regression Models
J.M. Hilbe

Markov Chain Monte Carlo — Stochastic Simulation for Bayesian Inference, Second Edition
D. Gamerman and H.F. Lopes

Mathematical Statistics
K. Knight

Modeling and Analysis of Stochastic Systems, Second Edition
V.G. Kulkarni

Modelling Binary Data, Second Edition
D. Collett

Modelling Survival Data in Medical Research, Second Edition
D. Collett

Multivariate Analysis of Variance and Repeated Measures — A Practical Approach for Behavioural Scientists
D.J. Hand and C.C. Taylor

Multivariate Statistics — A Practical Approach
B. Flury and H. Riedwyl

Multivariate Survival Analysis and Competing Risks
M. Crowder

Pólya Urn Models
H. Mahmoud

Practical Data Analysis for Designed Experiments
B.S. Yandell

Practical Longitudinal Data Analysis
D.J. Hand and M. Crowder

Practical Multivariate Analysis, Fifth Edition
A. Afifi, S. May, and V.A. Clark

Practical Statistics for Medical Research
D.G. Altman

A Primer on Linear Models
J.F. Monahan

Principles of Uncertainty
J.B. Kadane

Probability — Methods and Measurement
A. O'Hagan

Problem Solving — A Statistician's Guide, Second Edition
C. Chatfield

Randomization, Bootstrap and Monte Carlo Methods in Biology, Third Edition
B.F.J. Manly

Readings in Decision Analysis
S. French

Sampling Methodologies with Applications
P.S.R.S. Rao

Stationary Stochastic Processes: Theory and Applications
G. Lindgren

Statistical Analysis of Reliability Data
M.J. Crowder, A.C. Kimber, T.J. Sweeting, and R.L. Smith

Statistical Methods for Spatial Data Analysis
O. Schabenberger and C.A. Gotway

Statistical Methods for SPC and TQM
D. Bissell

Statistical Methods in Agriculture and Experimental Biology, Second Edition
R. Mead, R.N. Curnow, and A.M. Hasted

Statistical Process Control — Theory and Practice, Third Edition
G.B. Wetherill and D.W. Brown

Statistical Theory: A Concise Introduction
F. Abramovich and Y. Ritov

Statistical Theory, Fourth Edition
B.W. Lindgren

Statistics for Accountants
S. Letchford

Statistics for Epidemiology
N.P. Jewell

Statistics for Technology — A Course in Applied Statistics, Third Edition
C. Chatfield

Statistics in Engineering — A Practical Approach
A.V. Metcalfe

Statistics in Research and Development, Second Edition
R. Caulcutt

Stochastic Processes: An Introduction, Second Edition
P.W. Jones and P. Smith

Survival Analysis Using S — Analysis of Time-to-Event Data
M. Tableman and J.S. Kim

The Theory of Linear Models
B. Jørgensen

Time Series Analysis
H. Madsen

Time Series: Modeling, Computation, and Inference
R. Prado and M. West

Understanding Advanced Statistical Methods
P.H. Westfall and K.S.S. Henning

Texts in Statistical Science

Exercises and Solutions in Statistical Theory

Lawrence L. Kupper
Brian H. Neelon
Sean M. O'Brien

CRC Press
Taylor & Francis Group
Boca Raton London New York

CRC Press is an imprint of the
Taylor & Francis Group an **informa** business
A CHAPMAN & HALL BOOK

CRC Press
Taylor & Francis Group
6000 Broken Sound Parkway NW, Suite 300
Boca Raton, FL 33487-2742

Printed on acid-free paper
Version Date: 20130401

International Standard Book Number-13: 978-1-4665-7289-8 (Paperback)

Library of Congress Cataloging-in-Publication Data

Kupper, Lawrence L.
Exercises and solutions in statistical theory / Lawrence L. Kupper, Brian H. Neelon, Sean M. O'Brien.
pages cm -- (Chapman & Hall/CRC texts in statistical science)
ISBN 978-1-4665-7289-8 (pbk.)
1. Probabilities--Problems, exercises, etc. I. Neelon, Brian H. II. O'Brien, Sean M. III. Title.

QA273.25.K87 2013
519.5076--dc23 2013004215

Visit the Taylor & Francis Web site at
http://www.taylorandfrancis.com

and the CRC Press Web site at
http://www.crcpress.com

To Sandy, Mark, and Chieko; and to Dean Smith, a great basketball coach and an even greater mentor.

Lawrence L. Kupper

To Sara and my family.

Brian Neelon

To Sarah, Jacob, and Avery.

Sean M. O'Brien

Contents

Preface

This book contains exercises and selected detailed solutions covering statistical theory (from basic probability theory through the theory of statistical inference) that is typically taught in courses taken by advanced undergraduate students and graduate students in many quantitative disciplines (e.g., statistics, biostatistics, mathematics, engineering, physics, computer science, psychometrics, epidemiology, etc.).

Many exercises in this book deal with real-life scenarios in such important application areas as medicine, epidemiology, clinical trials, actuarial science, social science, psychometrics, parapsychology, engineering, wear testing, lot acceptance sampling, physics, chemistry, biology, environmental health, highway safety research, genetics, gambling, and sports. Several of these exercises illustrate the utility of important statistical areas such as study design strategies, sampling from finite populations, maximum likelihood, asymptotic theory, correlated data analysis, multilevel models, queueing theory, latent class analysis, conditional inference, order statistics, regression analysis, autoregressive models, survival analysis, generalized linear models, Bayesian analysis, measurement and misclassification error, group testing, and interrater agreement methods. Numerous exercises contain references to published books and articles that both students and instructors can use to obtain more information about the particular statistical topics being considered.

The authors strongly believe that the best way to obtain an in-depth understanding of the principles of statistical theory is to work through exercises whose solutions require nontrivial and illustrative utilization of relevant theoretical concepts. The exercises in this book have been prepared with this belief in mind. Mastery of the theoretical statistical strategies needed to solve the exercises in this book will prepare the user for successful study of even higher-level statistical theory.

Chapter 1, Concepts and Notation, contains basic results needed to help solve the exercises that follow. The exercises, and selected detailed solutions, are divided into five chapters: Chapter 2, Basic Probability Theory; Chapter 3, Univariate Distribution Theory; Chapter 4, Multivariate Distribution Theory;

Chapter 5, Estimation Theory; and, Chapter 6, Hypothesis Testing Theory. The chapters are arranged sequentially in the sense that a good understanding of basic probability theory is needed for exercises dealing with univariate distribution theory, and univariate distribution theory provides the basis for extensions to multivariate distribution theory. Mastery of the material in Chapters 2 through 4 is needed for the exercises in Chapters 5 and 6 on statistical inference. The exercises in each chapter vary in level of difficulty from fairly basic to challenging, with more difficult exercises identified with an asterisk. In each chapter, solutions are provided only for the odd-numbered exercises; a solutions manual for the even-numbered exercises can be obtained directly from CRC Press. This solutions manual should be quite useful to instructors who are looking for interesting and challenging lecture examples, homework problems, and examination questions in statistical theory. The book also contains a brief summary of some useful mathematical results (see the Appendix), a detailed index, and a list of useful references.

The main mathematical prerequisite for this book is an excellent working knowledge of multivariable calculus, along with some basic knowledge about matrices (e.g., matrix multiplication, the inverse of a matrix, etc.).

This book is not meant to be used as the primary textbook for a course in statistical theory. Some examples of excellent primary textbooks on statistical theory include Casella and Berger (2002), Hogg, Craig, and McKean (2005), Kalbfleisch (1985), Ross (2006), and Wackerly, Mendenhall III, and Scheaffer (2008). Rather, our book should serve as a supplemental source of a wide variety of exercises and selected detailed solutions both for advanced undergraduate students and graduate students who take such courses and for instructors of such courses. In addition, this book will be useful to individuals who are interested in enhancing or refreshing their own theoretical statistical skills. All solutions are sufficiently detailed so that users of the book can see how the relevant statistical theory is used in a logical manner to address important statistical questions in a wide variety of settings.

Lawrence L. Kupper
Brian Neelon
Sean M. O'Brien

Acknowledgments

Lawrence L. Kupper acknowledges the hundreds of students who have taken his classes in statistical theory. Many of these students have provided valuable feedback on the lectures, homework sets, and examinations that make up some of the material for this book. The authors acknowledge the fact that some exercises may overlap in concept with exercises found in other statistical theory books; such conceptual overlap is unavoidable given the breadth of material being covered. We thank the staff at Chapman Hall/CRC Press for their help with the production of this book. We especially thank our editor Rob Calver for his always insightful and extremely helpful advice, and for his constant encouragement and support during the preparation of both of our exercises and solutions books.

Authors

Lawrence L. Kupper, PhD, is emeritus alumni distinguished professor of biostatistics, School of Public Health, University of North Carolina (UNC), Chapel Hill, North Carolina. Dr. Kupper is a fellow of the American Statistical Association (ASA), and he received a Distinguished Achievement Medal from the ASA's Environmental Statistics Section for his research, teaching, and service contributions. During his 40 academic years at UNC, Dr. Kupper has won several classroom teaching and student mentoring awards. He has sole-authored and co-authored over 170 papers in peer-reviewed journals, and he has published several co-authored book chapters. Dr. Kupper has also co-authored four textbooks, namely, *Epidemiologic Research—Principles and Quantitative Methods*, *Applied Regression Analysis and Other Multivariable Methods* (four editions), *Quantitative Exposure Assessment*, and *Exercises and Solutions in Biostatistical Theory.* The contents of this exercises-and-solutions book come mainly from course materials developed and used by Dr. Kupper for his graduate-level courses in biostatistical theory, taught over a period of more than three decades.

Brian Neelon, PhD, is an assistant professor in the Department of Biostatistics & Bioinformatics at the Duke University School of Medicine. He obtained his doctorate from the University of North Carolina, Chapel Hill, where he received the Kupper Dissertation Award for outstanding dissertation-based publication. Before arriving at Duke University, Dr. Neelon was a postdoctoral research fellow in the Department of Health Care Policy at Harvard University. His research interests include Bayesian methods, longitudinal data analysis, health policy statistics, and environmental health.

Sean M. O'Brien, PhD, is an assistant professor in the Department of Biostatistics & Bioinformatics at the Duke University School of Medicine. He works primarily on studies of cardiovascular interventions using large multicenter clinical registries. He is currently statistical director of the Society of Thoracic Surgeons National Data Warehouse at Duke Clinical Research Institute. His methodological contributions are in the areas of healthcare provider performance evaluation, development of multidimensional composite measures, and clinical risk adjustment. Before joining Duke University, he was a research fellow at the National Institute of Environmental Health Sciences. He received his PhD in biostatistics from the University of North Carolina at Chapel Hill in 2002.

Chapter 1

Concepts and Notation

1.1 Basic Probability Theory

1.1.1 Counting Formulas

1.1.1.1 N-tuples

With sets $\{a_1, a_2, \ldots, a_q\}$ and $\{b_1, b_2, \ldots, b_s\}$ containing q and s distinct items, respectively, it is possible to form qs distinct pairs (or 2-tuples) of the form $(a_i, b_j), i = 1, 2, \ldots, q$ and $j = 1, 2, \ldots, s$. Adding a third set $\{c_1, c_2, \ldots, c_t\}$ containing t distinct items, it is possible to form qst distinct triplets (or 3-tuples) of the form $(a_i, b_j, c_k), i = 1, 2, \ldots, q,\ j = 1, 2, \ldots, s$, and $k = 1, 2, \ldots, t$. Extensions to more than three sets of distinct items are straightforward.

1.1.1.2 Permutations

A *permutation* is defined to be an ordered arrangement of r distinct items. The number of distinct ways of arranging n distinct items using r at a time is denoted P_r^n and is computed as

$$\mathrm{P}_r^n = \frac{n!}{(n-r)!},$$

where $n! = n(n-1)(n-2)\cdots(3)(2)(1)$ and where $0! \equiv 1$. If the n items are not distinct, then the number of distinct permutations is less than P_r^n.

1.1.1.3 Combinations

The number of ways of dividing n distinct items into k distinct groups with the ith group containing n_i items, where $n = \sum_{i=1}^{k} n_i$, is equal to

$$\frac{n!}{n_1!n_2!\cdots n_k!} = \frac{n!}{\left(\prod_{i=1}^{k} n_i!\right)}.$$

The above expression appears in the *multinomial expansion*

$$(x_1 + x_2 + \cdots + x_k)^n = \sum\nolimits^{*} \frac{n!}{\left(\prod_{i=1}^{k} n_i!\right)} x_1^{n_1} x_2^{n_2} \cdots x_k^{n_k},$$

where the summation symbol $\sum^{*}$ indicates summation over all possible values of $n_1, n_2, \ldots, n_k$ with $n_i, i = 1, 2, \ldots, k$, taking the set of possible values $\{0, 1, \ldots, n\}$ subject to the restriction $\sum_{i=1}^{k} n_i = n$.

With $x_1 = x_2 = \cdots = x_k = 1$, it follows that

$$\sum\nolimits^{*} \frac{n!}{\left(\prod_{i=1}^{k} n_i!\right)} = k^n.$$

As an important special case, when $k = 2$, then

$$\frac{n!}{n_1!n_2!} = \frac{n!}{n_1!(n - n_1)!} = \mathrm{C}_{n_1}^{n},$$

which is also the number of ways of selecting *without replacement* n_1 items from a set of n distinct items (i.e., the number of *combinations* of n distinct items selected n_1 at a time).

The above combinational expression appears in the *binomial* ($k = 2$) *expansion*

$$(x_1 + x_2)^n = \sum\nolimits^{*} \frac{n!}{n_1!n_2!} x_1^{n_1} x_2^{n_2} = \sum_{n_1=0}^{n} \mathrm{C}_{n_1}^{n} x_1^{n_1} x_2^{n-n_1}.$$

When $x_1 = x_2 = 1$, it follows that

$$\sum_{n_1=0}^{n} \mathrm{C}_{n_1}^{n} = 2^n.$$

Example. As a simple example using the above counting formulas, if 5 cards are dealt from a well-shuffled standard deck of 52 playing cards, the number of ways in which such a 5-card hand would contain exactly 2 aces is equal to $qs = \mathrm{C}_2^4 \mathrm{C}_3^{48} = 103{,}776$, where $q = \mathrm{C}_2^4 = 6$ is the number of ways of selecting 2 of the 4 aces and where $s = \mathrm{C}_3^{48} = 17{,}296$ is the number of ways of selecting 3 of the remaining 48 cards.

1.1.1.4 Pascal's Identity

$$C_k^n = C_{k-1}^{n-1} + C_k^{n-1}$$

for any positive integers n and k such that $C_k^n \equiv 0$ if $k > n$.

1.1.1.5 Vandermonde's Identity

$$C_r^{m+n} = \sum_{k=0}^{r} C_{r-k}^m C_k^n,$$

where m, n, and r are nonnegative integers satisfying $r \leq \min\{m, n\}$.

1.1.2 Probability Formulas

1.1.2.1 Definitions

Let an *experiment* be any process via which an observation or measurement is made. An experiment can range from a very controlled experimental situation to an uncontrolled observational situation. An example of the former situation would be a laboratory experiment where chosen amounts of different chemicals are mixed together to produce a certain chemical product. An example of the latter situation would be an epidemiological study where subjects are randomly selected and interviewed about their smoking and physical activity habits.

Let $A_1, A_2, \ldots, A_p$ be $p(\geq 2)$ possible events (or outcomes) that could occur when an experiment is conducted. Then:

1. For $i = 1, 2, \ldots, p$, the *complement* of the event A_i, denoted $\overline{A}_i$, is the event that A_i does *not* occur when the experiment is conducted.
2. The *union* of the events $A_1, A_2, \ldots, A_p$, denoted $\cup_{i=1}^p A_i$, is the event that *at least* one of the events $A_1, A_2, \ldots, A_p$ occurs when the experiment is conducted.
3. The *intersection* of the events $A_1, A_2, \ldots, A_p$, denoted $\cap_{i=1}^p A_i$, is the event that *all* of the events $A_1, A_2, \ldots, A_p$ occur when the experiment is conducted.

Given these definitions, we have the following probabilistic results, where $\text{pr}(A_i), 0 \leq \text{pr}(A_i) \leq 1$, denotes the *probability* that event A_i occurs when the experiment is conducted:

(i) $\text{pr}(\overline{\text{A}}_i) = 1 - \text{pr}(\text{A}_i)$. More generally,

$$\text{pr}\left(\overline{\cup_{i=1}^{p}\text{A}_i}\right) = 1 - \text{pr}\left(\cup_{i=1}^{p}\text{A}_i\right) = \text{pr}\left(\cap_{i=1}^{p}\overline{\text{A}}_i\right)$$

and

$$\text{pr}\left(\overline{\cap_{i=1}^{p}\text{A}_i}\right) = 1 - \text{pr}\left(\cap_{i=1}^{p}\text{A}_i\right) = \text{pr}\left(\cup_{i=1}^{p}\overline{\text{A}}_i\right).$$

(ii) The probability of the union of p events is given by:

$$\begin{aligned}\text{pr}\left(\cup_{i=1}^{p}\text{A}_i\right) &= \sum_{i=1}^{p}\text{pr}(\text{A}_i) - \sum_{i=1}^{p-1}\sum_{j=i+1}^{p}\text{pr}(\text{A}_i \cap \text{A}_j) \\ &+ \sum_{i=1}^{p-2}\sum_{j=i+1}^{p-1}\sum_{k=j+1}^{p}\text{pr}(\text{A}_i \cap \text{A}_j \cap \text{A}_k) - \cdots \\ &+ (-1)^{p-1}\text{pr}\left(\cap_{i=1}^{p}\text{A}_i\right).\end{aligned}$$

As important special cases, we have, for $p = 2$,

$$\text{pr}(\text{A}_1 \cup \text{A}_2) = \text{pr}(\text{A}_1) + \text{pr}(\text{A}_2) - \text{pr}(\text{A}_1 \cap \text{A}_2)$$

and, for $p = 3$,

$$\begin{aligned}\text{pr}(\text{A}_1 \cup \text{A}_2 \cup \text{A}_3) &= \text{pr}(\text{A}_1) + \text{pr}(\text{A}_2) + \text{pr}(\text{A}_3) \\ &- \text{pr}(\text{A}_1 \cap \text{A}_2) - \text{pr}(\text{A}_1 \cap \text{A}_3) - \text{pr}(\text{A}_2 \cap \text{A}_3) \\ &+ \text{pr}(\text{A}_1 \cap \text{A}_2 \cap \text{A}_3).\end{aligned}$$

1.1.2.2 *Mutually Exclusive Events*

For $i \neq j$, two events A_i and A_j are said to be *mutually exclusive* if these two events cannot both occur (i.e., cannot occur together) when the experiment is conducted; equivalently, the events A_i and A_j are mutually exclusive when $\text{pr}(\text{A}_i \cap \text{A}_j) = 0$. If the p events $\text{A}_1, \text{A}_2, \ldots, \text{A}_p$ are *pairwise* mutually exclusive, that is, if $\text{pr}(\text{A}_i \cap \text{A}_j) = 0$ for every $i \neq j$, then

$$\text{pr}\left(\cup_{i=1}^{p}\text{A}_i\right) = \sum_{i=1}^{p}\text{pr}(\text{A}_i),$$

since pairwise mutual exclusivity implies that any intersection involving more than two events must necessarily have probability zero of occurring.

1.1.2.3 Conditional Probability

For $i \neq j$, the *conditional* probability that event A_i occurs *given that* (or conditional on the fact that) event A_j occurs when the experiment is conducted, denoted $\mathrm{pr}(\mathrm{A}_i|\mathrm{A}_j)$, is given by the expression

$$\mathrm{pr}(\mathrm{A}_i|\mathrm{A}_j) = \frac{\mathrm{pr}(\mathrm{A}_i \cap \mathrm{A}_j)}{\mathrm{pr}(\mathrm{A}_j)}, \quad \mathrm{pr}(\mathrm{A}_j) > 0.$$

Using the above definition, we then have:

$$\begin{aligned}
\mathrm{pr}\left(\cap_{i=1}^{p}\mathrm{A}_i\right) &= \mathrm{pr}\left(\mathrm{A}_p| \cap_{i=1}^{p-1} \mathrm{A}_i\right) \mathrm{pr}\left(\cap_{i=1}^{p-1}\mathrm{A}_i\right) \\
&= \mathrm{pr}\left(\mathrm{A}_p| \cap_{i=1}^{p-1} \mathrm{A}_i\right) \mathrm{pr}\left(\mathrm{A}_{p-1}| \cap_{i=1}^{p-2} \mathrm{A}_i\right) \mathrm{pr}\left(\cap_{i=1}^{p-2}\mathrm{A}_i\right) \\
&\;\;\vdots \\
&= \mathrm{pr}\left(\mathrm{A}_p| \cap_{i=1}^{p-1} \mathrm{A}_i\right) \mathrm{pr}\left(\mathrm{A}_{p-1}| \cap_{i=1}^{p-2} \mathrm{A}_i\right) \cdots \mathrm{pr}(\mathrm{A}_2|\mathrm{A}_1)\mathrm{pr}(\mathrm{A}_1).
\end{aligned}$$

Note that there would be $p!$ ways of writing the above product of p probabilities. For example, when $p = 3$, we have

$$\begin{aligned}
\mathrm{pr}(\mathrm{A}_1 \cap \mathrm{A}_2 \cap \mathrm{A}_3) &= \mathrm{pr}(\mathrm{A}_3|\mathrm{A}_1 \cap \mathrm{A}_2)\mathrm{pr}(\mathrm{A}_2|\mathrm{A}_1)\mathrm{pr}(\mathrm{A}_1) \\
&= \mathrm{pr}(\mathrm{A}_2|\mathrm{A}_1 \cap \mathrm{A}_3)\mathrm{pr}(\mathrm{A}_1|\mathrm{A}_3)\mathrm{pr}(\mathrm{A}_3) \\
&= \mathrm{pr}(\mathrm{A}_1|\mathrm{A}_2 \cap \mathrm{A}_3)\mathrm{pr}(\mathrm{A}_3|\mathrm{A}_2)\mathrm{pr}(\mathrm{A}_2), \text{ and so on.}
\end{aligned}$$

1.1.2.4 Independence

The events A_i and A_j are said to be *independent* events if and only if the following equivalent probability statements are true:

1. $\mathrm{pr}(\mathrm{A}_i|\mathrm{A}_j) = \mathrm{pr}(\mathrm{A}_i)$;
2. $\mathrm{pr}(\mathrm{A}_j|\mathrm{A}_i) = \mathrm{pr}(\mathrm{A}_j)$;
3. $\mathrm{pr}(\mathrm{A}_i \cap \mathrm{A}_j) = \mathrm{pr}(\mathrm{A}_i)\mathrm{pr}(\mathrm{A}_j)$.

When the events $\mathrm{A}_1, \mathrm{A}_2, \ldots, \mathrm{A}_p$ are *mutually independent*, so that the conditional probability of any event is equal to the unconditional probability of that same event, then

$$\mathrm{pr}(\cap_{i=1}^{p}\mathrm{A}_i) = \prod_{i=1}^{p} \mathrm{pr}(\mathrm{A}_i).$$

1.1.2.5 Partitions and Bayes' Theorem

When $\text{pr}(\cup_{i=1}^{p}\text{A}_i) = 1$, and when the events $\text{A}_1, \text{A}_2, \ldots, \text{A}_p$ are pairwise mutually exclusive, then the events $\text{A}_1, \text{A}_2, \ldots, \text{A}_p$ are said to constitute a *partition* of the experimental outcomes; in other words, when the experiment is conducted, exactly one and only one of the events $\text{A}_1, \text{A}_2, \ldots, \text{A}_p$ must occur. If B is any event and $\text{A}_1, \text{A}_2, \ldots, \text{A}_p$ constitute a partition, it follows that

$$\begin{aligned} \text{pr(B)} &= \text{pr}\left[\text{B} \cap (\cup_{i=1}^{p}\text{A}_i)\right] = \text{pr}\left[\cup_{i=1}^{p}(\text{B} \cap \text{A}_i)\right] \\ &= \sum_{i=1}^{p} \text{pr}(\text{B} \cap \text{A}_i) = \sum_{i=1}^{p} \text{pr}(\text{B}|\text{A}_i)\text{pr}(\text{A}_i). \end{aligned}$$

As an illustration of the use of the above formula, if the events $\text{A}_1, \text{A}_2, \ldots, \text{A}_p$ represent an exhaustive list of all p possible causes of some observed outcome B, where $\text{pr(B)} > 0$, then, given values for $\text{pr}(\text{A}_i)$ and $\text{pr}(\text{B}|\text{A}_i)$ for all $i = 1, 2, \ldots, p$, one can employ *Bayes' Theorem* to compute the probability that A_i was the cause of the observed outcome B, namely,

$$\text{pr}(\text{A}_i|\text{B}) = \frac{\text{pr}(\text{A}_i \cap \text{B})}{\text{pr(B)}} = \frac{\text{pr}(\text{B}|\text{A}_i)\text{pr}(\text{A}_i)}{\sum_{j=1}^{p} \text{pr}(\text{B}|\text{A}_j)\text{pr}(\text{A}_j)}, \quad i = 1, 2, \ldots, p.$$

Note that $\sum_{i=1}^{p} \text{pr}(\text{A}_i|\text{B}) = 1$.

As an important special case, suppose that the events $\text{A}_1, \text{A}_2, \ldots, \text{A}_p$ constituting a partition are *elementary* events in the sense that none of these p events can be further decomposed into smaller events (i.e., for $i = 1, 2, \ldots, p$, the event A_i cannot be written as a union of mutually exclusive events each having a smaller probability than A_i of occurring when the experiment is conducted). Then, any more complex event B (sometimes called a *compound* event) must be able to be represented as the union of two or more of the elementary events $\text{A}_1, \text{A}_2, \ldots, \text{A}_p$. In particular, with $2 \leq m \leq p$, if

$$\text{B} = \cup_{j=1}^{m}\text{A}_{i_j},$$

where the set of positive integers $\{i_1, i_2, \ldots, i_m\}$ is a subset of the set of positive integers $\{1, 2, \ldots, p\}$, then

$$\text{pr(B)} = \sum_{j=1}^{m} \text{pr}(\text{A}_{i_j}).$$

In the very special case when the elementary events $\text{A}_1, \text{A}_2, \ldots, \text{A}_p$ are *equally likely* to occur, so that $\text{pr}(\text{A}_i) = \frac{1}{p}$ for $i = 1, 2, \ldots, p$, then $\text{pr(B)} = \frac{m}{p}$.

Example. To continue an earlier example, there would be $p = \text{C}_5^{52} =$ 2,598,960 possible 5-card hands that could be dealt from a well-shuffled standard deck of 52 playing cards. Thus, each such 5-card hand has probability

$\frac{1}{2{,}598{,}960}$ of occurring. If B is the event that a 5-card hand contains exactly two aces, then

$$\mathrm{pr(B)} = \frac{m}{p} = \frac{103{,}776}{2{,}598{,}960} = 0.0399.$$

1.2 Univariate Distribution Theory

1.2.1 Discrete and Continuous Random Variables

A *discrete* random variable X takes either a finite or a countably infinite number of values. A discrete random variable X is characterized by its probability distribution $\mathrm{p}_X(x) = \mathrm{pr}(X = x)$, which is a formula giving the probability that X takes the (permissible) value x. Hence, a valid discrete probability distribution $\mathrm{p}_X(x)$ has the following two properties:

i. $0 \leq \mathrm{p}_X(x) \leq 1$ for all (permissible) values of x and
ii. $\sum_{\text{all } x} \mathrm{p}_X(x) = 1$.

A *continuous* random variable X can theoretically take all the real (and hence uncountably infinite) numerical values on a line segment of either finite or infinite length. A continuous random variable X is characterized by its density function $\mathrm{f}_X(x)$. A valid density function $\mathrm{f}_X(x)$ has the following properties:

i. $0 \leq \mathrm{f}_X(x) < +\infty$ for all (permissible) values of x;
ii. $\int_{\text{all } x} \mathrm{f}_X(x)\,\mathrm{d}x = 1$;
iii. For $-\infty < a < b < +\infty$, $\mathrm{pr}(a < X < b) = \int_a^b \mathrm{f}_X(x)\,\mathrm{d}x$; and
iv. $\mathrm{pr}(X = x) = 0$ for any particular value x, since $\int_x^x \mathrm{f}_X(x)\,\mathrm{d}x = 0$.

1.2.2 Cumulative Distribution Functions

In general, the cumulative distribution function (CDF) for a univariate random variable X is the function $\mathrm{F}_X(x) = \mathrm{pr}(X \leq x), -\infty < x < +\infty$, which possesses the following properties:

i. $0 \leq \mathrm{F}_X(x) \leq 1, -\infty < x < +\infty$;
ii. $\mathrm{F}_X(x)$ is a monotonically nondecreasing function of x; and
iii. $\lim_{x \to -\infty} \mathrm{F}_X(x) = 0$ and $\lim_{x \to +\infty} \mathrm{F}_X(x) = 1$.

For an integer-valued discrete random variable X, it follows that

i. $\mathrm{F}_X(x) = \sum_{\text{all } x^* \leq x} \mathrm{p}_X(x^*)$;
ii. $\mathrm{p}_X(x) = \mathrm{pr}(X = x) = \mathrm{F}_X(x) - \mathrm{F}_X(x-1)$; and
iii. $[\mathrm{dF}_X(x)]/\mathrm{d}x \neq \mathrm{p}_X(x)$ since $\mathrm{F}_X(x)$ is a discontinuous function of x.

For a continuous random variable X, it follows that

i. $\mathrm{F}_X(x) = \int_{\text{all} x^* \leq x} \mathrm{f}_X(x^*)\,\mathrm{d}x^*$;
ii. For $-\infty < a < b < +\infty, \mathrm{pr}(a < X < b) = \mathrm{F}_X(b) - \mathrm{F}_X(a)$; and
iii. $[\mathrm{dF}_X(x)]/\mathrm{d}x = \mathrm{f}_X(x)$ since $\mathrm{F}_X(x)$ is an absolutely continuous function of x.

1.2.3 Median and Mode

For any discrete distribution $\mathrm{p}_X(x)$ or density function $\mathrm{f}_X(x)$, the population *median* ξ satisfies the two inequalities

$$\mathrm{pr}(X \leq \xi) \geq \tfrac{1}{2} \quad \text{and} \quad \mathrm{pr}(X \geq \xi) \geq \tfrac{1}{2}.$$

For a density function $\mathrm{f}_X(x), \xi$ is that value of X such that

$$\int_{-\infty}^{\xi} \mathrm{f}_X(x)\,\mathrm{d}x = \frac{1}{2}.$$

The population *mode* for either a discrete probability distribution $\mathrm{p}_X(x)$ or a density function $\mathrm{f}_X(x)$ is a value of x that maximizes $\mathrm{p}_X(x)$ or $\mathrm{f}_X(x)$. The population mode is not necessarily unique since $\mathrm{p}_X(x)$ or $\mathrm{f}_X(x)$ may achieve its maximum for several different values of x; in this situation, all these local maxima are called modes.

1.2.4 Expectation Theory

Let $\mathrm{g}(X)$ be any scalar function of a univariate random variable X. Then, the *expected value* $\mathrm{E}[\mathrm{g}(X)]$ of $\mathrm{g}(X)$ is defined to be

$$\mathrm{E}[\mathrm{g}(X)] = \sum_{\text{all } x} \mathrm{g}(x)\mathrm{p}_X(x) \text{ when } X \text{ is a discrete random variable,}$$

and is defined to be

$$\mathrm{E}[\mathrm{g}(X)] = \int_{\text{all } x} \mathrm{g}(x)\mathrm{f}_X(x)\,\mathrm{d}x \text{ when } X \text{ is a continuous random variable.}$$

Note that $\mathrm{E}[\mathrm{g}(X)]$ is said to exist if $|\mathrm{E}[\mathrm{g}(X)]| < +\infty$; otherwise, $\mathrm{E}[\mathrm{g}(X)]$ is said not to exist.

Some general rules for computing expectations are:

i. If C is a constant independent of X, then $\mathrm{E}(\mathrm{C}) = \mathrm{C}$;

ii. $\mathrm{E}[\mathrm{Cg}(X)] = \mathrm{CE}[\mathrm{g}(X)]$;

iii. If $\mathrm{C}_1, \mathrm{C}_2, \ldots, \mathrm{C}_k$ are k constants all independent of X, and if $\mathrm{g}_1(X), \mathrm{g}_2(X), \ldots, \mathrm{g}_k(X)$ are k scalar functions of X, then

$$\mathrm{E}\left[\sum_{i=1}^{k} \mathrm{C}_i \mathrm{g}_i(X)\right] = \sum_{i=1}^{k} \mathrm{C}_i \mathrm{E}[\mathrm{g}_i(X)];$$

iv. If $k \to \infty$, then

$$\mathrm{E}\left[\sum_{i=1}^{\infty} \mathrm{C}_i \mathrm{g}_i(X)\right] = \sum_{i=1}^{\infty} \mathrm{C}_i \mathrm{E}[\mathrm{g}_i(X)]$$

when $|\sum_{i=1}^{\infty} \mathrm{C}_i \mathrm{E}[\mathrm{g}_i(X)]| < +\infty$.

1.2.5 Some Important Expectations

1.2.5.1 Mean

$\mu = \mathrm{E}(X)$ is the *mean* of X.

1.2.5.2 Variance

$\sigma^2 = \mathrm{V}(X) = \mathrm{E}\{[X - \mathrm{E}(X)]^2\}$ is the *variance* of X, and $\sigma = +\sqrt{\sigma^2}$ is the *standard deviation* of X.

1.2.5.3 Moments

More generally, if r is a positive integer, a binomial expansion of $[X - \mathrm{E}(X)]^r$ gives

$$\mathrm{E}\{[X-\mathrm{E}(X)]^r\} = \mathrm{E}\left\{\sum_{j=0}^{r} \mathrm{C}_j^r X^j [-\mathrm{E}(X)]^{r-j}\right\} = \sum_{j=0}^{r} \mathrm{C}_j^r (-1)^{r-j} \mathrm{E}(X^j)[\mathrm{E}(X)]^{r-j},$$

where $\mathrm{E}\{[X - \mathrm{E}(X)]^r\}$ is the *rth moment about the mean.*

For example, for $r = 2$, we obtain

$$\mathrm{E}\{[X - \mathrm{E}(X)]^2\} = \mathrm{V}(X) = \mathrm{E}(X^2) - [\mathrm{E}(X)]^2;$$

and, for $r = 3$, we obtain

$$\mathrm{E}\{[X - \mathrm{E}(X)]^3\} = \mathrm{E}(X^3) - 3\mathrm{E}(X^2)\mathrm{E}(X) + 2[\mathrm{E}(X)]^3,$$

which is a measure of the *skewness* of the distribution of X.

1.2.5.4 Moment Generating Function

$\mathrm{M}_X(t) = \mathrm{E}(\mathrm{e}^{tX})$ is called the *moment generating function* for the random variable X, provided that $\mathrm{M}_X(t) < +\infty$ for t in some neighborhood of 0 [i.e., for all $t \in (-\epsilon, \epsilon), \epsilon > 0$]. For r a positive integer, and with $\mathrm{E}(X^r)$ defined as the *rth moment about the origin* (i.e., about 0) for the random variable X, then $\mathrm{M}_X(t)$ can be used to generate moments about the origin via the algorithm

$$\frac{\mathrm{d}^r \mathrm{M}_X(t)}{\mathrm{d}t^r}_{|t=0} = \mathrm{E}(X^r).$$

More generally, for r a positive integer, the function

$$\mathrm{M}_X^*(t) = \mathrm{E}\left\{\mathrm{e}^{t[X-\mathrm{E}(X)]}\right\} = \mathrm{e}^{-t\mathrm{E}(X)}\mathrm{M}_X(t)$$

can be used to generate moments about the mean via the algorithm

$$\frac{\mathrm{d}^r \mathrm{M}_X^*(t)}{\mathrm{d}t^r}_{|t=0} = \mathrm{E}\{[X - \mathrm{E}(X)]^r\}.$$

1.2.5.5 Probability Generating Function

If we let e^t equal s in $\mathrm{M}_X(t) = \mathrm{E}(\mathrm{e}^{tX})$, we obtain the *probability generating function* $\mathrm{P}_X(s) = \mathrm{E}(s^X)$. Then, for r a positive integer, and with

$$\mathrm{E}\left[\frac{X!}{(X-r)!}\right] = \mathrm{E}[X(X-1)(X-2)\cdots(X-r+1)]$$

defined as the *rth factorial moment* for the random variable X, then $\mathrm{P}_X(s)$ can be used to generate factorial moments via the algorithm

$$\frac{\mathrm{d}^r \mathrm{P}_X(s)}{\mathrm{d}s^r}_{|s=1} = \mathrm{E}\left[\frac{X!}{(X-r)!}\right].$$

As an example, the probability generating function $\mathrm{P}_X(s)$ can be used to find the variance of X when $\mathrm{V}(X)$ is written in the form

$$\mathrm{V}(X) = \mathrm{E}[X(X-1)] + \mathrm{E}(X) - [\mathrm{E}(X)]^2.$$

1.2.6 *Inequalities Involving Expectations*

1.2.6.1 *Markov's Inequality*

If X is a nonnegative random variable [i.e., $\text{pr}(X \geq 0) = 1$], then $\text{pr}(X > k) \leq \text{E}(X)/k$ for any constant $k > 0$. As a special case, for $r > 0$, if $X = |Y - \text{E}(Y)|^r$ when Y is any random variable, then, with $\nu_r = \text{E}\left[|Y - \text{E}(Y)|^r\right]$, we have

$$\text{pr}\left[|Y - \text{E}(Y)|^r > k\right] \leq \frac{\nu_r}{k},$$

or equivalently with $k = t^r \nu_r$,

$$\text{pr}\left[|Y - \text{E}(Y)| > t\nu_r^{1/r}\right] \leq t^{-r}, \quad t > 0.$$

For $r = 2$, we obtain *Tchebyshev's Inequality*, namely,

$$\text{pr}\left[|Y - \text{E}(Y)| > t\sqrt{\text{V}(Y)}\right] \leq t^{-2}, \quad t > 0.$$

1.2.6.2 *Jensen's Inequality*

Let X be a random variable with $|\text{E}(X)| < \infty$. If $\text{g}(X)$ is a *convex* function of X, then $\text{E}[\text{g}(X)] \geq \text{g}[\text{E}(X)]$, provided that $|\text{E}[\text{g}(X)]| < \infty$. If $\text{g}(X)$ is a *concave* function of X, then the inequality is reversed, namely, $\text{E}[\text{g}(X)] \leq \text{g}[\text{E}(X)]$.

1.2.6.3 *Hölder's Inequality*

Let X and Y be random variables, and let $p, 1 < p < \infty$, and $q, 1 < q < \infty$, satisfy the restriction $1/p + 1/q = 1$. Then,

$$\text{E}(|XY|) \leq \left[\text{E}(|X|^p)\right]^{1/p} \left[\text{E}(|Y|^q)\right]^{1/q}.$$

As a special case, when $p = q = 2$, we obtain the *Cauchy–Schwartz Inequality*, namely,

$$\text{E}(|XY|) \leq \sqrt{\text{E}(X^2)\text{E}(Y^2)}.$$

1.2.7 Some Important Probability Distributions for Discrete Random Variables

1.2.7.1 Binomial Distribution

If X is the number of successes in n trials, where the trials are conducted independently with the probability π of success remaining the same from trial to trial, then

$$\mathrm{p}_X(x) = \mathrm{C}_x^n \pi^x (1-\pi)^{n-x}, \quad x = 0, 1, \ldots, n \quad \text{and} \quad 0 < \pi < 1.$$

When $X \sim \mathrm{BIN}(n, \pi)$, then $\mathrm{E}(X) = n\pi, V(X) = n\pi(1-\pi)$, and $\mathrm{M}_X(t) = [\pi \mathrm{e}^t + (1-\pi)]^n$.

When $n = 1, X$ has the *Bernoulli* distribution.

1.2.7.2 Negative Binomial Distribution

If Y is the number of trials required to obtain exactly k successes, where k is a specified positive integer, and where the trials are conducted independently with the probability π of success remaining the same from trial to trial, then

$$\mathrm{p}_Y(y) = \mathrm{C}_{k-1}^{y-1} \pi^k (1-\pi)^{y-k}, \quad y = k, k+1, \ldots, \infty \quad \text{and} \quad 0 < \pi < 1.$$

When $Y \sim \mathrm{NEGBIN}(k, \pi)$, then $\mathrm{E}(Y) = k/\pi, \mathrm{V}(Y) = k(1-\pi)/\pi^2$, and

$$\mathrm{M}_Y(t) = \left[\frac{\pi \mathrm{e}^t}{1 - (1-\pi)\mathrm{e}^t} \right]^k.$$

In the special case when $k = 1$, then Y has a *geometric* distribution, namely,

$$\mathrm{p}_Y(y) = \pi(1-\pi)^{y-1}, \quad y = 1, 2, \ldots, \infty \quad \text{and} \quad 0 < \pi < 1.$$

When $Y \sim \mathrm{GEOM}(\pi)$, then $\mathrm{E}(Y) = 1/\pi, \mathrm{V}(Y) = (1-\pi)/\pi^2$, and $\mathrm{M}_Y(t) = \pi \mathrm{e}^t / [1 - (1-\pi)\mathrm{e}^t]$.

When $X \sim \mathrm{BIN}(n, \pi)$ and when $Y \sim \mathrm{NEGBIN}(k, \pi)$, then $\mathrm{pr}(X < k) = \mathrm{pr}(Y > n)$.

1.2.7.3 Poisson Distribution

As a model for rare events, the Poisson distribution can be derived as a limiting case of the binomial distribution as $n \to \infty$ and $\pi \to 0$ with $\lambda = n\pi$ held constant; this limit is

$$\mathrm{p}_X(x) = \frac{\lambda^x \mathrm{e}^{-\lambda}}{x!}, \quad x = 0, 1, \ldots, \infty \quad \text{and} \quad \lambda > 0.$$

When $X \sim \mathrm{POI}(\lambda)$, then $\mathrm{E}(X) = \mathrm{V}(X) = \lambda$ and $\mathrm{M}_X(t) = \mathrm{e}^{\lambda(\mathrm{e}^t - 1)}$.

1.2.7.4 Hypergeometric Distribution

Suppose that a finite-sized population of size $N(< +\infty)$ contains a items of Type A and b items of Type B, with $(a+b) = N$. If a sample of $n(< N)$ items is randomly selected *without replacement* from this population of N items, then the number X of items of Type A contained in this sample of n items has the hypergeometric distribution, namely,

$$\mathrm{p}_X(x) = \frac{\mathrm{C}_x^a \mathrm{C}_{n-x}^b}{\mathrm{C}_n^{a+b}} = \frac{\mathrm{C}_x^a \mathrm{C}_{n-x}^{N-a}}{\mathrm{C}_n^N}, \quad \max(0, n-b) \leq X \leq \min(n, a).$$

When $X \sim \mathrm{HG}(a, N-a, n)$, then

$$\mathrm{E}(X) = n\left(\frac{a}{N}\right) \quad \text{and} \quad \mathrm{V}(X) = n\left(\frac{a}{N}\right)\left(\frac{N-a}{N}\right)\left(\frac{N-n}{N-1}\right).$$

1.2.8 Some Important Distributions (i.e., Density Functions) for Continuous Random Variables

1.2.8.1 Normal Distribution

The normal distribution density function is

$$\mathrm{f}_X(x) = \frac{1}{\sqrt{2\pi}\sigma}\mathrm{e}^{-(x-\mu)^2/2\sigma^2}, \qquad -\infty < x < \infty, \quad -\infty < \mu < \infty, \quad 0 < \sigma^2 < \infty.$$

When $X \sim \mathrm{N}(\mu, \sigma^2)$, then $\mathrm{E}(X) = \mu, \mathrm{V}(X) = \sigma^2$, and $\mathrm{M}_X(t) = \mathrm{e}^{\mu t + \sigma^2 t^2/2}$. Also, when $X \sim \mathrm{N}(\mu, \sigma^2)$, then the standardized variable $Z = (X - \mu)/\sigma \sim \mathrm{N}(0, 1)$, with density function

$$\mathrm{f}_Z(z) = \frac{1}{\sqrt{2\pi}}\mathrm{e}^{-z^2/2}, \quad -\infty < z < \infty.$$

1.2.8.2 Lognormal Distribution

When $X \sim \mathrm{N}(\mu, \sigma^2)$, then the random variable $Y = \mathrm{e}^X$ has a *lognormal distribution*, with density function

$$\mathrm{f}_Y(y) = \frac{1}{\sqrt{2\pi}\sigma y}\mathrm{e}^{-[\ln(y)-\mu]^2/2\sigma^2}, \quad 0 < y < \infty, \quad -\infty < \mu < \infty, \quad 0 < \sigma^2 < \infty.$$

When $Y \sim \mathrm{LN}(\mu, \sigma^2)$, then $\mathrm{E}(Y) = \mathrm{e}^{\mu + (\sigma^2/2)}$ and $\mathrm{V}(Y) = [\mathrm{E}(Y)]^2(\mathrm{e}^{\sigma^2} - 1)$.

1.2.8.3 Gamma Distribution

The gamma distribution density function is

$$\mathrm{f}_X(x) = \frac{x^{\beta-1}\mathrm{e}^{-x/\alpha}}{\Gamma(\beta)\alpha^\beta}, \qquad 0 < x < \infty, \quad 0 < \alpha < \infty, \quad 0 < \beta < \infty.$$

When $X \sim \mathrm{GAMMA}(\alpha, \beta)$, then $\mathrm{E}(X) = \alpha\beta, \mathrm{V}(X) = \alpha^2\beta$, and $\mathrm{M}_X(t) = (1 - \alpha t)^{-\beta}$. The Gamma distribution has two important special cases:

i. When $\alpha = 2$ and $\beta = \nu/2$, then $X \sim \chi^2_\nu$ (i.e., X has a *chi-squared distribution* with ν degrees of freedom). When $X \sim \chi^2_\nu$, then

$$\mathrm{f}_X(x) = \frac{x^{\frac{\nu}{2}-1}\mathrm{e}^{-x/2}}{\Gamma\left(\frac{\nu}{2}\right)2^{\nu/2}}, \qquad 0 < x < \infty \quad \text{and} \quad \nu \text{ a positive integer;}$$

also, $\mathrm{E}(X) = \nu, \mathrm{V}(X) = 2\nu$, and $\mathrm{M}_X(t) = (1-2t)^{-\nu/2}$. And, if $Z \sim \mathrm{N}(0,1)$, then $Z^2 \sim \chi^2_1$.

ii. When $\beta = 1$, then X has a *negative exponential* distribution with density function

$$\mathrm{f}_X(x) = \frac{1}{\alpha}\mathrm{e}^{-x/\alpha}, \quad 0 < x < \infty, \quad 0 < \alpha < \infty.$$

When $X \sim \mathrm{NEGEXP}(\alpha)$, then $\mathrm{E}(X) = \alpha, \mathrm{V}(X) = \alpha^2$, and $\mathrm{M}_X(t) = (1 - \alpha t)^{-1}$.

1.2.8.4 Beta Distribution

The Beta distribution density function is

$$\mathrm{f}_X(x) = \frac{\Gamma(\alpha+\beta)}{\Gamma(\alpha)\Gamma(\beta)}x^{\alpha-1}(1-x)^{\beta-1}, \qquad 0 < x < 1, \quad 0 < \alpha < \infty, \quad 0 < \beta < \infty.$$

When $X \sim \mathrm{BETA}(\alpha, \beta)$, then $\mathrm{E}(X) = \frac{\alpha}{\alpha+\beta}$ and $\mathrm{V}(X) = \frac{\alpha\beta}{(\alpha+\beta)^2(\alpha+\beta+1)}$.

1.2.8.5 Uniform Distribution

The uniform distribution density function is

$$\mathrm{f}_X(x) = \frac{1}{(\theta_2 - \theta_1)}, \quad -\infty < \theta_1 < x < \theta_2 < \infty.$$

When $X \sim \mathrm{UNIF}(\theta_1, \theta_2)$, then $\mathrm{E}(X) = \frac{(\theta_1+\theta_2)}{2}$, $\mathrm{V}(X) = \frac{(\theta_2-\theta_1)^2}{12}$ and $\mathrm{M}_X(t) = \frac{(\mathrm{e}^{t\theta_2} - \mathrm{e}^{t\theta_1})}{t(\theta_2-\theta_1)}$.

1.3 Multivariate Distribution Theory

1.3.1 Discrete and Continuous Multivariate Distributions

A *discrete multivariate* probability distribution for k discrete random variables $X_1, X_2, \ldots, X_k$ is denoted

$$\mathrm{p}_{X_1,X_2,\ldots,X_k}(x_1, x_2, \ldots, x_k) = \mathrm{pr}\left[\cap_{i=1}^k (X_i = x_i)\right] \equiv \mathrm{p}_{\boldsymbol{X}}(\boldsymbol{x}) = \mathrm{pr}(\boldsymbol{X} = \boldsymbol{x}), \quad \boldsymbol{x} \in \mathcal{D},$$

where the row vector $\boldsymbol{X} = (X_1, X_2, \ldots, X_k)$, the row vector $\boldsymbol{x} = (x_1, x_2, \ldots, x_k)$, and $\mathcal{D}$ is the *domain* (i.e., the set of all permissible values) of the discrete random vector $\boldsymbol{X}$. A valid multivariate discrete probability distribution has the following properties:

(i) $0 \leq \mathrm{p}_{\boldsymbol{X}}(\boldsymbol{x}) \leq 1$ for all $\mathbf{x} \in \mathcal{D}$;

(ii) $\sum\sum_{\mathcal{D}}\cdots\sum \mathrm{p}_{\boldsymbol{X}}(\boldsymbol{x}) = 1$;

(iii) If $\mathcal{D}_1$ is a subset of $\mathcal{D}$, then

$$\mathrm{pr}[\boldsymbol{X} \in \mathcal{D}_1] = \sum\sum_{\mathcal{D}_1}\cdots\sum \mathrm{p}_{\boldsymbol{X}}(\boldsymbol{x}).$$

A *continuous multivariate* probability distribution (i.e., a multivariate density function) for k continuous random variables $X_1, X_2, \ldots, X_k$ is denoted

$$\mathrm{f}_{X_1,X_2,\ldots,X_k}(x_1, x_2, \ldots, x_k) \equiv \mathrm{f}_{\boldsymbol{X}}(\boldsymbol{x}), \quad \boldsymbol{x} \in \mathcal{D},$$

where $\mathcal{D}$ is the domain of the continuous random vector $\boldsymbol{X}$. A valid multivariate density function has the following properties:

(i) $0 \leq \mathrm{f}_{\boldsymbol{X}}(\boldsymbol{x}) < +\infty$ for all $\boldsymbol{x} \in \mathcal{D}$;

(ii) $\int\int_{\mathcal{D}}\cdots\int \mathrm{f}_{\boldsymbol{X}}(\boldsymbol{x})\,\mathbf{d}\boldsymbol{x} = 1$, where $\mathbf{d}\boldsymbol{x} = \mathrm{d}x_1\mathrm{d}x_2\ldots\mathrm{d}x_k$;

(iii) If $\mathcal{D}_1$ is a subset of $\mathcal{D}$, then

$$\text{pr}[\boldsymbol{X} \in \mathcal{D}_1] = \int\int_{\mathcal{D}_1}\cdots\int \text{f}_{\boldsymbol{X}}(\boldsymbol{x})\, \mathbf{d}\boldsymbol{x}.$$

1.3.2 *Multivariate Cumulative Distribution Functions*

In general, the multivariate CDF for a random vector $\boldsymbol{X}$ is the scalar function

$$\text{F}_{\boldsymbol{X}}(\boldsymbol{x}) = \text{pr}(\boldsymbol{X} \leq \boldsymbol{x}) = \text{pr}\left[\cap_{i=1}^{k}(X_i \leq x_i)\right].$$

For a discrete random vector, $\text{F}_{\boldsymbol{X}}(\boldsymbol{x})$ is a discontinuous function of $\boldsymbol{x}$. For a continuous random vector, $\text{F}_{\boldsymbol{X}}(\boldsymbol{x})$ is an absolutely continuous function of $\boldsymbol{x}$, so that

$$\frac{\partial^k \text{F}_{\boldsymbol{X}}(\boldsymbol{x})}{\partial x_1 \partial x_2 \cdots \partial x_k} = \text{f}_{\boldsymbol{X}}(\boldsymbol{x}).$$

1.3.3 *Expectation Theory*

Let $\text{g}(\boldsymbol{X})$ be a scalar function of $\boldsymbol{X}$. If $\boldsymbol{X}$ is a discrete random vector with probability distribution $\text{p}_{\boldsymbol{X}}(\boldsymbol{x})$, then

$$\text{E}[\text{g}(\boldsymbol{X})] = \sum\sum_{\mathcal{D}}\cdots\sum \text{g}(\boldsymbol{x})\text{p}_{\boldsymbol{X}}(\boldsymbol{x}).$$

And, if $\boldsymbol{X}$ is a continuous random vector with density function $\text{f}_{\boldsymbol{X}}(\mathbf{x})$, then

$$\text{E}[\text{g}(\boldsymbol{X})] = \int\int_{\mathcal{D}}\cdots\int \text{g}(\boldsymbol{x})\text{f}_{\boldsymbol{X}}(\boldsymbol{x})\, \mathbf{d}\boldsymbol{x}.$$

Some important expectations of interest in the multivariate setting are as follows:

1.3.3.1 *Covariance*

For $i \neq j$, the *covariance* between the two random variables X_i and X_j is defined as

$$\begin{aligned}\text{cov}(X_i, X_j) &= \text{E}\{[X_i - \text{E}(X_i)][X_j - \text{E}(X_j)]\} \\ &= \text{E}(X_i X_j) - \text{E}(X_i)\text{E}(X_j), \quad -\infty < \text{cov}(X_i, X_j) < +\infty.\end{aligned}$$

1.3.3.2 Correlation

For $i \neq j$, the *correlation* between the two random variables X_i and X_j is defined as

$$\operatorname{corr}(X_i, X_j) = \frac{\operatorname{cov}(X_i, X_j)}{\sqrt{\mathrm{V}(X_i)\mathrm{V}(X_j)}}, \quad -1 \leq \operatorname{corr}(X_i, X_j) \leq +1.$$

1.3.3.3 Moment Generating Function

With the row vector $\boldsymbol{\theta} = (t_1, t_2, \ldots, t_k)$,

$$\mathrm{M}_{\boldsymbol{X}}(\boldsymbol{\theta}) = \mathrm{E}\left(\mathrm{e}^{\boldsymbol{\theta}\boldsymbol{X}'}\right) = \mathrm{E}\left(\mathrm{e}^{\sum_{i=1}^{k} t_i X_i}\right)$$

is called the *multivariate moment generating function* for the random vector $\boldsymbol{X}$. In particular, with $r_1, r_2, \ldots, r_k$ being nonnegative integers satisfying the restriction $\sum_{i=1}^{k} r_i = r$, we have

$$\mathrm{E}[X_1^{r_1} X_2^{r_2} \cdots X_k^{r_k}] = \frac{\partial^r \mathrm{M}_{\boldsymbol{X}}(\boldsymbol{\theta})}{\partial t_1^{r_1} \partial t_2^{r_2} \cdots \partial t_k^{r_k}}_{|\boldsymbol{\theta}=\mathbf{0}},$$

where the notation $\boldsymbol{\theta} = \mathbf{0}$ means that $t_i = 0$, $i = 1, 2, \ldots, k$.

1.3.4 Marginal Distributions

When $\boldsymbol{X}$ is a discrete random vector, the *marginal distribution* of any proper subset of the k random variables $X_1, X_2, \ldots, X_k$ can be found by summing over all the random variables *not* in the subset of interest. In particular, for $1 \leq j < k$, the marginal distribution of the random variables $X_1, X_2, \ldots, X_j$ is equal to

$$\mathrm{p}_{X_1, X_2, \ldots, X_j}(x_1, x_2, \ldots, x_j) = \sum_{\text{all } x_{j+1}} \sum_{\text{all } x_{j+2}} \cdots \sum_{\text{all } x_{k-1}} \sum_{\text{all } x_k} \mathrm{p}_{\boldsymbol{X}}(\boldsymbol{x}).$$

When $\boldsymbol{X}$ is a continuous random vector, the *marginal distribution* of any proper subset of the k random variables $X_1, X_2, \ldots, X_k$ can be found by integrating over all the random variables *not* in the subset of interest. In particular, for $1 \leq j < k$, the marginal distribution of the random variables $X_1, X_2, \ldots, X_j$ is equal to

$$\begin{aligned}&\mathrm{f}_{X_1, X_2, \ldots, X_j}(x_1, x_2, \ldots, x_j) \\ &= \int\limits_{\text{all } x_{j+1}} \int\limits_{\text{all } x_{j+2}} \cdots \int\limits_{\text{all } x_{k-1}} \int\limits_{\text{all } x_k} \mathrm{f}_{\boldsymbol{X}}(\boldsymbol{x}) \mathrm{d}x_k \, \mathrm{d}x_{k-1} \cdots \mathrm{d}x_{j+2} \, \mathrm{d}x_{j+1}.\end{aligned}$$

1.3.5 *Conditional Distributions and Expectations*

For $\boldsymbol{X}$ a discrete random vector, let $\boldsymbol{X}_1$ denote a proper subset of the k discrete random variables $X_1, X_2, \ldots, X_k$, let $\boldsymbol{X}_2$ denote another proper subset of $X_1, X_2, \ldots, X_k$, and assume that the subsets $\boldsymbol{X}_1$ and $\boldsymbol{X}_2$ have no elements in common. Then, the *conditional distribution* of $\boldsymbol{X}_2$ given that $\boldsymbol{X}_1 = \boldsymbol{x}_1$ is defined as the joint distribution of $\boldsymbol{X}_1$ and $\boldsymbol{X}_2$ divided by the marginal distribution of $\boldsymbol{X}_1$, namely,

$$\begin{aligned} \mathrm{p}_{\boldsymbol{X}_2}(\boldsymbol{x}_2|\boldsymbol{X}_1 = \boldsymbol{x}_1) &= \frac{\mathrm{p}_{\boldsymbol{X}_1,\boldsymbol{X}_2}(\boldsymbol{x}_1, \boldsymbol{x}_2)}{\mathrm{p}_{\boldsymbol{X}_1}(\mathbf{x}_1)} \\ &= \frac{\mathrm{pr}[(\boldsymbol{X}_1 = \boldsymbol{x}_1) \cap (\boldsymbol{X}_2 = \boldsymbol{x}_2)]}{\mathrm{pr}(\boldsymbol{X}_1 = \boldsymbol{x}_1)}, \quad \mathrm{pr}(\boldsymbol{X}_1 = \boldsymbol{x}_1) > 0. \end{aligned}$$

Then, if $\mathrm{g}(\boldsymbol{X}_2)$ is a scalar function of $\boldsymbol{X}_2$, it follows that

$$\mathrm{E}[\mathrm{g}(\boldsymbol{X}_2)|\boldsymbol{X}_1 = \boldsymbol{x}_1] = \underset{\text{all } \boldsymbol{x}_2}{\sum\sum\cdots\sum} \mathrm{g}(\boldsymbol{x}_2)\mathrm{p}_{\boldsymbol{X}_2}(\boldsymbol{x}_2|\boldsymbol{X}_1 = \boldsymbol{x}_1).$$

For $\boldsymbol{X}$ a continuous random vector, let $\boldsymbol{X}_1$ denote a proper subset of the k continuous random variables $X_1, X_2, \ldots, X_k$, let $\boldsymbol{X}_2$ denote another proper subset of $X_1, X_2, \ldots, X_k$, and assume that the subsets $\boldsymbol{X}_1$ and $\boldsymbol{X}_2$ have no elements in common. Then, the *conditional density function* of $\boldsymbol{X}_2$ given that $\boldsymbol{X}_1 = \boldsymbol{x}_1$ is defined as the joint density function of $\boldsymbol{X}_1$ and $\boldsymbol{X}_2$ divided by the marginal density function of $\boldsymbol{X}_1$, namely,

$$\mathrm{f}_{\boldsymbol{X}_2}(\boldsymbol{x}_2|\boldsymbol{X}_1 = \boldsymbol{x}_1) \quad = \quad \frac{\mathrm{f}_{\boldsymbol{X}_1,\boldsymbol{X}_2}(\boldsymbol{x}_1, \boldsymbol{x}_2)}{\mathrm{f}_{\boldsymbol{X}_1}(\boldsymbol{x}_1)}, \quad \mathrm{f}_{\boldsymbol{X}_1}(\boldsymbol{x}_1) > 0.$$

Then, if $\mathrm{g}(\boldsymbol{X}_2)$ is a scalar function of $\boldsymbol{X}_2$, it follows that

$$\mathrm{E}[\mathrm{g}(\boldsymbol{X}_2)|\boldsymbol{X}_1 = \boldsymbol{x}_1] = \underset{\text{all } \boldsymbol{x}_2}{\int\int\cdots\int} \mathrm{g}(\boldsymbol{x}_2)\mathrm{f}_{\boldsymbol{X}_2}(\boldsymbol{x}_2|\boldsymbol{X}_1 = \boldsymbol{x}_1)\,\mathbf{d}\boldsymbol{x}_2.$$

More generally, if $\mathrm{g}(\boldsymbol{X}_1, \boldsymbol{X}_2)$ is a scalar function of $\boldsymbol{X}_1$ and $\boldsymbol{X}_2$, then useful *iterated expectation* formulas are:

$$\mathrm{E}[\mathrm{g}(\boldsymbol{X}_1, \boldsymbol{X}_2)] = \mathrm{E}_{\boldsymbol{x}_1}\{\mathrm{E}[\mathrm{g}(\boldsymbol{X}_1, \boldsymbol{X}_2)|\boldsymbol{X}_1 = \boldsymbol{x}_1]\} = \mathrm{E}_{\boldsymbol{x}_2}\{\mathrm{E}[\mathrm{g}(\boldsymbol{X}_1, \boldsymbol{X}_2)|\boldsymbol{X}_2 = \boldsymbol{x}_2]\}$$

and

$$\begin{aligned} \mathrm{V}[\mathrm{g}(\boldsymbol{X}_1, \boldsymbol{X}_2)] &= \mathrm{E}_{\boldsymbol{x}_1}\{\mathrm{V}[\mathrm{g}(\boldsymbol{X}_1, \boldsymbol{X}_2)|\boldsymbol{X}_1 = \boldsymbol{x}_1]\} + \mathrm{V}_{\boldsymbol{x}_1}\{\mathrm{E}[\mathrm{g}(\boldsymbol{X}_1, \boldsymbol{X}_2)|\boldsymbol{X}_1 = \boldsymbol{x}_1]\} \\ &= \mathrm{E}_{\boldsymbol{x}_2}\{\mathrm{V}[\mathrm{g}(\boldsymbol{X}_1, \boldsymbol{X}_2)|\boldsymbol{X}_2 = \boldsymbol{x}_2]\} + \mathrm{V}_{\boldsymbol{x}_2}\{\mathrm{E}[\mathrm{g}(\boldsymbol{X}_1, \boldsymbol{X}_2)|\boldsymbol{X}_2 = \boldsymbol{x}_2]\}. \end{aligned}$$

Also,

$$\mathrm{p}_{\boldsymbol{X}}(\boldsymbol{x}) \equiv \mathrm{p}_{X_1,X_2,\ldots,X_k}(x_1,x_2,\ldots,x_k) = \mathrm{p}_{X_1}(x_1)\prod_{i=2}^{k}\mathrm{p}_{X_i}\left[x_i\middle|\cap_{j=1}^{i-1}(X_j=x_j)\right]$$

and

$$\mathrm{f}_{\boldsymbol{X}}(\boldsymbol{x}) \equiv \mathrm{f}_{X_1,X_2,\ldots,X_k}(x_1,x_2,\ldots,x_k) = \mathrm{f}_{X_1}(x_1)\prod_{i=2}^{k}\mathrm{f}_{X_i}\left[x_i\middle|\cap_{j=1}^{i-1}(X_j=x_j)\right].$$

Note that there are $k!$ ways of writing each of the above two expressions.

1.3.6 Mutual Independence among a Set of Random Variables

The random vector $\boldsymbol{X}$ is said to consist of a set of k mutually independent random variables if and only if

$$\mathrm{F}_{\boldsymbol{X}}(\boldsymbol{x}) = \prod_{i=1}^{k}\mathrm{F}_{X_i}(x_i) = \prod_{i=1}^{k}\mathrm{pr}(X_i \leq x_i)$$

for all possible choices of $x_1, x_2, \ldots, x_k$.

Given mutual independence, then

$$\mathrm{p}_{\boldsymbol{X}}(\boldsymbol{x}) \equiv \mathrm{p}_{X_1,X_2,\ldots,X_k}(x_1,x_2,\ldots,x_k) = \prod_{i=1}^{k}\mathrm{p}_{X_i}(x_i)$$

when $\boldsymbol{X}$ is a discrete random vector, and

$$\mathrm{f}_{\boldsymbol{X}}(\boldsymbol{x}) \equiv \mathrm{f}_{X_1,X_2,\ldots,X_k}(x_1,x_2,\ldots,x_k) = \prod_{i=1}^{k}\mathrm{f}_{X_i}(x_i)$$

when $\boldsymbol{X}$ is a continuous random vector.

Also, for $i = 1, 2, \ldots, k$, let $\mathrm{g}_i(X_i)$ be a scalar function of X_i. Then, if $X_1, X_2, \ldots, X_k$ constitute a set of k mutually independent random variables, it follows that

$$\mathrm{E}\left[\prod_{i=1}^{k}\mathrm{g}_i(X_i)\right] = \prod_{i=1}^{k}\mathrm{E}[\mathrm{g}_i(X_i)].$$

And, if $X_1, X_2, \ldots, X_k$ are mutually independent random variables, then any subset of these k random variables also constitutes a group of mutually independent random variables. Also, for $i \neq j$, if X_i and X_j are independent random variables, then $\mathrm{corr}(X_i, X_j) = 0$; however, if $\mathrm{corr}(X_i, X_j) = 0$, it does *not* necessarily follow that X_i and X_j are independent random variables.

1.3.7 *Random Sample*

Using the notation $\boldsymbol{X}_i = (X_{i1}, X_{i2}, \ldots, X_{ik})$, the random vectors $\boldsymbol{X}_1, \boldsymbol{X}_2, \ldots, \boldsymbol{X}_n$ are said to constitute a random sample of size n from the discrete parent population $\mathrm{p}_{\boldsymbol{X}}(\boldsymbol{x})$ if the following two conditions hold:

(i) $\boldsymbol{X}_1, \boldsymbol{X}_2, \ldots, \boldsymbol{X}_n$ constitute a set of mutually independent random vectors;

(ii) For $i = 1, 2, \ldots, n, \mathrm{p}_{\boldsymbol{X}_i}(\boldsymbol{x}_i) = \mathrm{p}_{\boldsymbol{X}}(\boldsymbol{x}_i)$; in other words, $\boldsymbol{X}_i$ follows the discrete parent population distribution $\mathrm{p}_{\boldsymbol{X}}(\boldsymbol{x})$.

A completely analogous definition holds for a random sample from a continuous parent population $\mathrm{f}_{\boldsymbol{X}}(\boldsymbol{x})$.

Standard statistical terminology describes a random sample $\boldsymbol{X}_1, \boldsymbol{X}_2, \ldots, \boldsymbol{X}_n$ of size n as consisting of a set of *independent and identically distributed* (i.i.d.) random vectors. In this regard, it is important to note that the mutual independence property pertains to the relationship among the *random vectors*, *not* to the relationship among the k (*possibly mutually dependent*) scalar random variables within a random vector.

1.3.8 *Some Important Multivariate Discrete and Continuous Probability Distributions*

1.3.8.1 *Multinomial*

The *multinomial* distribution is often used as a statistical model for the analysis of categorical data. In particular, for $i = 1, 2, \ldots, k$, suppose that π_i is the probability that an observation falls into the ith of k distinct categories, where $0 < \pi_i < 1$ and where $\sum_{i=1}^{k} \pi_i = 1$. If the discrete random variable X_i is the number of observations out of n that fall into the ith category, then the k random variables $X_1, X_2, \ldots, X_k$ jointly follow a k-variate multinomial distribution, namely,

$$\mathrm{p}_{\boldsymbol{X}}(\boldsymbol{x}) \equiv \mathrm{p}_{X_1, X_2, \ldots, X_k}(x_1, x_2, \ldots, x_k) = \frac{n!}{x_1! x_2! \cdots x_k!} \pi_1^{x_1} \pi_2^{x_2} \cdots \pi_k^{x_k}, \quad \mathbf{x} \in \mathcal{D},$$

where $\mathcal{D} = \{\boldsymbol{x} : 0 \leq x_i \leq n, i = 1, 2, \ldots, k, \text{ and } \sum_{i=1}^{k} x_i = n\}$.

When $(X_1, X_2, \ldots, X_k) \sim \text{MULT}(n; \pi_1, \pi_2, \ldots, \pi_k)$, then $X_i \sim \text{BIN}(n, \pi_i)$ for $i = 1, 2, \ldots, k$, and $\text{cov}(X_i, X_j) = -n\pi_i\pi_j$ for $i \neq j$.

1.3.8.2 Multivariate Normal

The *multivariate normal* distribution is often used to model the joint behavior of k possibly mutually correlated continuous random variables. The multivariate normal density function for k continuous random variables $X_1, X_2, \ldots, X_k$ is defined as

$$\mathrm{f}_{\boldsymbol{X}}(\boldsymbol{x}) \equiv \mathrm{f}_{X_1,X_2,\ldots,X_k}(x_1, x_2, \ldots, x_k) = \frac{1}{(2\pi)^{k/2}|\boldsymbol{\Sigma}|^{1/2}} \mathrm{e}^{-(1/2)(\boldsymbol{x}-\boldsymbol{\mu})\boldsymbol{\Sigma}^{-1}(\boldsymbol{x}-\boldsymbol{\mu})'},$$

where $-\infty < x_i < \infty$ for $i = 1, 2, \ldots, k$, where $\boldsymbol{\mu} = (\mu_1, \mu_2, \ldots, \mu_k) = [\mathrm{E}(X_1), \mathrm{E}(X_2), \ldots, \mathrm{E}(X_k)]$, and where $\boldsymbol{\Sigma}$ is the $(k \times k)$ covariance matrix of $\boldsymbol{X}$ with ith diagonal element equal to $\sigma_i^2 = \mathrm{V}(X_i)$ and with (i, j)th element σ_{ij} equal to $\mathrm{cov}(X_i, X_j)$ for $i \neq j$.

Also, when $\boldsymbol{X} \sim \mathrm{MVN}_k(\boldsymbol{\mu}, \boldsymbol{\Sigma})$, then the moment generating function for $\boldsymbol{X}$ is

$$\mathrm{M}_{\boldsymbol{X}}(\boldsymbol{\theta}) = \mathrm{e}^{\boldsymbol{\theta}\boldsymbol{\mu}' + (1/2)\boldsymbol{\theta}\boldsymbol{\Sigma}\boldsymbol{\theta}'}.$$

And, for $i = 1, 2, \ldots, k$, the marginal distribution of X_i is normal with mean μ_i and variance σ_i^2.

As an important special case, when $k = 2$, we obtain the *bivariate normal* distribution, namely,

$$\mathrm{f}_{X_1,X_2}(x_1, x_2) = \frac{1}{2\pi\sigma_1\sigma_2\sqrt{(1-\rho^2)}} \mathrm{e}^{-\frac{1}{2(1-\rho^2)}\left[\left(\frac{x_1-\mu_1}{\sigma_1}\right)^2 - 2\rho\left(\frac{x_1-\mu_1}{\sigma_1}\right)\left(\frac{x_2-\mu_2}{\sigma_2}\right) + \left(\frac{x_2-\mu_2}{\sigma_2}\right)^2\right]},$$

where $-\infty < x_1 < \infty$ and $-\infty < x_2 < \infty$, and where $\rho = \mathrm{corr}(X_1, X_2)$.

When $(X_1, X_2) \sim \mathrm{BVN}(\mu_1, \mu_2; \sigma_1^2, \sigma_2^2; \rho)$, then the moment generating function for X_1 and X_2 is

$$\mathrm{M}_{X_1,X_2}(t_1, t_2) = \mathrm{e}^{t_1\mu_1 + t_2\mu_2 + (1/2)(t_1^2\sigma_1^2 + 2t_1t_2\rho\sigma_1\sigma_2 + t_2^2\sigma_2^2)}.$$

The conditional distribution of X_2 given $X_1 = x_1$ is normal with

$$\mathrm{E}(X_2|X_1 = x_1) = \mu_2 + \rho\frac{\sigma_2}{\sigma_1}(x_1 - \mu_1) \quad \text{and} \quad \mathrm{V}(X_2|X_1 = x_1) = \sigma_2^2(1 - \rho^2).$$

And, the conditional distribution of X_1 given $X_2 = x_2$ is normal with

$$\mathrm{E}(X_1|X_2 = x_2) = \mu_1 + \rho\frac{\sigma_1}{\sigma_2}(x_2 - \mu_2) \quad \text{and} \quad \mathrm{V}(X_1|X_2 = x_2) = \sigma_1^2(1 - \rho^2).$$

These conditional expectation expressions for the bivariate normal distribution are special cases of a more general result. More generally, for a pair of either discrete or continuous random variables X_1 and X_2, if the conditional expectation of X_2 given $X_1 = x_1$ is a *linear* (or straight line) function of x_1, namely $\mathrm{E}(X_2|X_1 = x_1) = \alpha_1 + \beta_1 x_1, -\infty < \alpha_1 < +\infty, -\infty < \beta_1 < +\infty$, then $\mathrm{corr}(X_1, X_2) = \rho = \beta_1\sqrt{[\mathrm{V}(X_1)]/[\mathrm{V}(X_2)]}$. Analogously, if $\mathrm{E}(X_1|X_2 = x_2) = \alpha_2 + \beta_2 x_2, -\infty < \alpha_2 < +\infty, -\infty < \beta_2 < +\infty$, then $\rho = \beta_2\sqrt{[\mathrm{V}(X_2)]/[\mathrm{V}(X_1)]}$.

1.3.9 *Special Topics of Interest*

1.3.9.1 *Mean and Variance of a Linear Function of Random Variables*

For $i = 1, 2, \ldots, k$, let $\mathrm{g}_i(X_i)$ be a scalar function of the random variable X_i. Then, if $a_1, a_2, \ldots, a_k$ are known constants, and if $L = \sum_{i=1}^{k} a_i \mathrm{g}_i(X_i)$, we have

$$\mathrm{E}(L) = \sum_{i=1}^{k} a_i \mathrm{E}[\mathrm{g}_i(X_i)],$$

and

$$\mathrm{V}(L) = \sum_{i=1}^{k} a_i^2 \mathrm{V}[\mathrm{g}_i(X_i)] + 2\sum_{i=1}^{k-1} \sum_{j=i+1}^{k} a_i a_j \mathrm{cov}[\mathrm{g}_i(X_i), \mathrm{g}_j(X_j)].$$

In the special case when the random variables X_i and X_j are uncorrelated for all $i \neq j$, then

$$\mathrm{V}(L) = \sum_{i=1}^{k} a_i^2 \mathrm{V}[\mathrm{g}_i(X_i)].$$

1.3.9.2 *Convergence in Distribution*

A sequence of random variables $U_1, U_2, \ldots, U_n, \ldots$ *converges in distribution* to a random variable U if

$$\lim_{n \to \infty} \mathrm{F}_{U_n}(u) = \mathrm{F}_U(u)$$

for all values of u where $\mathrm{F}_U(u)$ is continuous. Notationally, we write $U_n \xrightarrow{\mathrm{D}} U$.

As an important example, suppose that $X_1, X_2, \ldots, X_n$ constitute a random sample of size n from either a univariate discrete probability distribution $\mathrm{p}_X(x)$ or a univariate density function $\mathrm{f}_X(x)$, where $\mathrm{E}(X) = \mu(-\infty < \mu < +\infty)$ and $\mathrm{V}(X) = \sigma^2(0 < \sigma^2 < +\infty)$. With $\bar{X} = n^{-1}\sum_{i=1}^{n} X_i$, consider the standardized random variable

$$U_n = \frac{\bar{X} - \mu}{\sigma/\sqrt{n}} = \frac{\sum_{i=1}^{n} X_i - n\mu}{\sqrt{n}\sigma}.$$

Then, it can be shown that $\lim_{n\to\infty} \mathrm{M}_{U_n}(t) = \mathrm{e}^{t^2/2}$, leading to the conclusion that $U_n \xrightarrow{\mathrm{D}} Z$, where $Z \sim$N(0,1). This is the well-known *Central Limit Theorem.*

1.3.9.3 Order Statistics

Let $X_1, X_2, \ldots, X_n$ constitute a random sample of size n from a univariate density function $\mathrm{f}_X(x), -\infty < x < +\infty$, with corresponding cumulative distribution function $\mathrm{F}_X(x) = \int_{-\infty}^{x} \mathrm{f}_X(t)\,dt$. Then, the n *order statistics* $X_{(1)}, X_{(2)}, \ldots, X_{(n)}$ satisfy the relationship

$$-\infty < X_{(1)} < X_{(2)} < \cdots < X_{(n-1)} < X_{(n)} < +\infty.$$

For $r = 1, 2, \ldots, n$, the random variable $X_{(r)}$ is called the rth order statistic. In particular, $X_{(1)} = \min\{X_1, X_2, \ldots, X_n\}$, $X_{(n)} = \max\{X_1, X_2, \ldots, X_n\}$, and $X_{((n+1)/2)} = \text{median}\{X_1, X_2, \ldots, X_n\}$ when n is an odd positive integer.

For $r = 1, 2, \ldots, n$, the distribution of $X_{(r)}$ is

$$\mathrm{f}_{X_{(r)}}(x_{(r)}) = n\mathrm{C}_{r-1}^{n-1}[\mathrm{F}_X(x_{(r)})]^{r-1}[1-\mathrm{F}_X(x_{(r)})]^{n-r}\mathrm{f}_X(x_{(r)}), -\infty < x_{(r)} < +\infty.$$

For $1 \leq r < s \leq n$, the joint distribution of $X_{(r)}$ and $X_{(s)}$ is equal to

$$\begin{aligned}\mathrm{f}_{X_{(r)},X_{(s)}}(x_{(r)}, x_{(s)}) = {} & \frac{n!}{(r-1)!(s-r-1)!(n-s)!}[\mathrm{F}_X(x_{(r)})]^{r-1} \\ & \times [\mathrm{F}_X(x_{(s)}) - \mathrm{F}_X(x_{(r)})]^{s-r-1} \\ & \times [1 - \mathrm{F}_X(x_{(s)})]^{n-s}\mathrm{f}_X(x_{(r)})\mathrm{f}_X(x_{(s)}), \\ -\infty < x_{(r)} < x_{(s)} < +\infty. &\end{aligned}$$

And, the joint distribution of $X_{(1)}, X_{(2)}, \ldots, X_{(n)}$ is

$$\mathrm{f}_{X_{(1)},X_{(2)},\ldots,X_{(n)}}(x_{(1)}, x_{(2)}, \ldots, x_{(n)}) = n!\prod_{i=1}^{n}\mathrm{f}_X(x_{(i)}),$$
$$-\infty < x_{(1)} < x_{(2)} < \cdots < x_{(n-1)} < x_{(n)} < +\infty.$$

1.3.9.4 Method of Transformations

With $k = 2$, let X_1 and X_2 be two continuous random variables with joint density function $\mathrm{f}_{X_1,X_2}(x_1, x_2), (x_1, x_2) \in \mathcal{D}$. Let $Y_1 = \mathrm{g}_1(X_1, X_2)$ and $Y_2 = \mathrm{g}_2(X_1, X_2)$ be random variables, where the functions $y_1 = \mathrm{g}_1(x_1, x_2)$ and $y_2 = \mathrm{g}_2(x_1, x_2)$ define a one-to-one transformation from the domain $\mathcal{D}$ in the (x_1, x_2)-plane to the domain $\mathcal{D}^*$ in the (y_1, y_2)-plane. Further, let $x_1 = \mathrm{h}_1(y_1, y_2)$ and $x_2 = \mathrm{h}_2(y_1, y_2)$ be the inverse functions expressing x_1 and x_2 as functions of y_1 and y_2. Then, the joint density function of the random variables Y_1 and Y_2 is

$$\mathrm{f}_{Y_1,Y_2}(y_1, y_2) = \mathrm{f}_{X_1,X_2}[\mathrm{h}_1(y_1, y_2), \mathrm{h}_2(y_1, y_2)]|J|, \quad (y_1, y_2) \in \mathcal{D}^*,$$

where the *Jacobian* $J, J \neq 0$, of the transformation is the second-order determinant

$$J = \begin{vmatrix} \dfrac{\partial h_1(y_1, y_2)}{\partial y_1} & \dfrac{\partial h_1(y_1, y_2)}{\partial y_2} \\ \dfrac{\partial h_2(y_1, y_2)}{\partial y_1} & \dfrac{\partial h_2(y_1, y_2)}{\partial y_2} \end{vmatrix}.$$

For the special case $k = 1$ when $Y_1 = \mathrm{g}_1(X_1)$ and $X_1 = \mathrm{h}_1(Y_1)$, it follows that

$$\mathrm{f}_{Y_1}(y_1) = \mathrm{f}_{X_1}[\mathrm{h}_1(y_1)] \left| \frac{\mathrm{dh}_1(y_1)}{\mathrm{d}y_1} \right|, \quad y_1 \in \mathcal{D}^*.$$

It is a direct generalization to the situation when $Y_i = \mathrm{g}_i(X_1, X_2, \ldots, X_k)$, $i = 1, 2, \ldots, k$, with the Jacobian J being the determinant of a $(k \times k)$ matrix.

1.4 Estimation Theory

1.4.1 Point Estimation of Population Parameters

Let the random variables $X_1, X_2, \ldots, X_n$ constitute a sample of size n from some population with properties depending on a row vector $\boldsymbol{\theta} = (\theta_1, \theta_2, \ldots, \theta_p)$ of p unknown parameters, where the *parameter space* is the set Ω of all possible values of $\boldsymbol{\theta}$. In the most general situation, the n random variables $X_1, X_2, \ldots, X_n$ are allowed to be mutually dependent and to have different distributions (e.g., different means and different variances).

A *point estimator* or a *statistic* is any scalar function $U(X_1, X_2, \ldots, X_n) \equiv U(\boldsymbol{X})$ of the random variables $X_1, X_2, \ldots, X_n$, but *not* of $\boldsymbol{\theta}$. A point estimator or statistic is itself a random variable since it is a function of the random vector $\boldsymbol{X} = (X_1, X_2, \ldots, X_n)$. In contrast, the corresponding *point estimate* or *observed statistic* $U(x_1, x_2, \ldots, x_n) \equiv U(\boldsymbol{x})$ is the realized (or observed) numerical value of the point estimator or statistic that is computed using the realized (or observed) numerical values $x_1, x_2, \ldots, x_n$ of $X_1, X_2, \ldots, X_n$ for the particular sample obtained.

Some popular methods for obtaining a row vector $\hat{\boldsymbol{\theta}} = (\hat{\theta}_1, \hat{\theta}_2, \ldots, \hat{\theta}_p)$ of point estimators of the elements of the row vector $\boldsymbol{\theta} = (\theta_1, \theta_2, \ldots, \theta_p)$, where $\hat{\theta}_j \equiv \hat{\theta}_j(\boldsymbol{X})$ for $j = 1, 2, \ldots, p$, are the following.

1.4.1.1 *Method of Moments (MM)*

For $j = 1, 2, \ldots, p$, let

$$M_j = \frac{1}{n}\sum_{i=1}^{n} X_i^j \quad \text{and} \quad \mathrm{E}(M_j) = \frac{1}{n}\sum_{i=1}^{n} \mathrm{E}(X_i^j),$$

where $\mathrm{E}(M_j)$, $j = 1, 2, \ldots, p$, is a function of the elements of $\boldsymbol{\theta}$.

Then, $\hat{\boldsymbol{\theta}}_{\mathrm{mm}}$, the MM estimator of $\boldsymbol{\theta}$, is obtained as the solution of the p equations

$$M_j = \mathrm{E}(M_j), j = 1, 2, \ldots, p.$$

1.4.1.2 *Unweighted Least Squares (ULS)*

Let $Q_{\mathrm{u}} = \sum_{i=1}^{n}[X_i - \mathrm{E}(X_i)]^2$. Then, $\hat{\boldsymbol{\theta}}_{\mathrm{uls}}$, the ULS estimator of $\boldsymbol{\theta}$, is chosen to *minimize* Q_{u} and is defined as the solution of the p equations

$$\frac{\partial Q_{\mathrm{u}}}{\partial \theta_j} = 0, \quad j = 1, 2, \ldots, p.$$

1.4.1.3 *Weighted Least Squares (WLS)*

Let $Q_{\mathrm{w}} = \sum_{i=1}^{n} w_i[X_i - \mathrm{E}(X_i)]^2$, where $w_1, w_2, \ldots, w_n$ are weights. Then, $\hat{\boldsymbol{\theta}}_{\mathrm{wls}}$, the WLS estimator of $\boldsymbol{\theta}$, is chosen to minimize Q_{w} and is defined as the solution of the p equations

$$\frac{\partial Q_{\mathrm{w}}}{\partial \theta_j} = 0, \quad j = 1, 2, \ldots, p.$$

1.4.1.4 *Maximum Likelihood (ML)*

Let $\mathcal{L}(\boldsymbol{x}; \boldsymbol{\theta})$ denote the *likelihood function*, which is often simply the joint distribution of the random variables $X_1, X_2, \ldots, X_n$. Then, $\hat{\boldsymbol{\theta}}_{\mathrm{ml}}$, the ML estimator (MLE) of $\boldsymbol{\theta}$, is chosen to *maximize* $\mathcal{L}(\boldsymbol{x}; \boldsymbol{\theta})$ and is defined as the solution of the p equations

$$\frac{\partial \ln \mathcal{L}(\boldsymbol{x}; \boldsymbol{\theta})}{\partial \theta_j} = 0, \quad j = 1, 2, \ldots, p.$$

If $\tau(\boldsymbol{\theta})$ is a scalar function of $\boldsymbol{\theta}$, then $\tau(\hat{\boldsymbol{\theta}}_{\mathrm{ml}})$ is the MLE of $\tau(\boldsymbol{\theta})$; this is known as the *invariance property* of MLEs.

1.4.2 Data Reduction and Joint Sufficiency

The goal of any statistical analysis is to quantify the information contained in a sample of size n by making valid and precise statistical inferences using the smallest possible number of point estimators or statistics. This data reduction goal leads to the concept of joint sufficiency.

1.4.2.1 Joint Sufficiency

The statistics $U_1(\boldsymbol{X}),\ U_2(\boldsymbol{X}), \ldots, U_k(\boldsymbol{X}),\ k \geq p$, are jointly sufficient for the parameter vector $\boldsymbol{\theta}$ if and only if the conditional distribution of $\boldsymbol{X}$ given $U_1(\boldsymbol{X}) = U_1(\boldsymbol{x}),\ U_2(\boldsymbol{X}) = U_2(\boldsymbol{x}), \ldots, U_k(\boldsymbol{X}) = U_k(\boldsymbol{x})$ does not *in any way* depend on $\boldsymbol{\theta}$. More specifically, the phrase "in any way" means that the conditional distribution of $\boldsymbol{X}$, including the domain of $\boldsymbol{X}$, given the k sufficient statistics is not a function of $\boldsymbol{\theta}$. In other words, the jointly sufficient statistics $U_1(\boldsymbol{X}), U_2(\boldsymbol{X}), \ldots, U_k(\boldsymbol{X})$ utilize all the information about $\boldsymbol{\theta}$ that is contained in the sample $\boldsymbol{X}$.

1.4.2.2 Factorization Theorem

To demonstrate joint sufficiency, the *Factorization Theorem* (Halmos and Savage, 1949) is quite useful: Let $\boldsymbol{X}$ be a discrete or continuous random vector with distribution $\mathcal{L}(\boldsymbol{x}; \boldsymbol{\theta})$. Then, $U_1(\boldsymbol{X}), U_2(\boldsymbol{X}), \ldots, U_k(\boldsymbol{X})$ are jointly sufficient for $\boldsymbol{\theta}$ if and only if there are nonnegative functions $g[U_1(\boldsymbol{x}), U_2(\boldsymbol{x}), \ldots, U_k(\boldsymbol{x}); \boldsymbol{\theta}]$ and $h(\boldsymbol{x})$ such that

$$\mathcal{L}(\boldsymbol{x}; \boldsymbol{\theta}) = g[U_1(\boldsymbol{x}), U_2(\boldsymbol{x}), \ldots, U_k(\boldsymbol{x}); \boldsymbol{\theta}]h(\boldsymbol{x}),$$

where, given $U_1(\boldsymbol{X}) = U_1(\boldsymbol{x}), U_2(\boldsymbol{X}) = U_2(\boldsymbol{x}), \ldots, U_k(\boldsymbol{X}) = U_k(\boldsymbol{x})$, the function $h(\mathbf{x})$ in no way depends on $\boldsymbol{\theta}$. Also, any one-to-one function of a sufficient statistic is also a sufficient statistic.

As an important example, a *family* $\mathcal{F}_d = \{\mathrm{p}_X(x; \boldsymbol{\theta}), \boldsymbol{\theta} \in \Omega\}$ of discrete probability distributions is a member of the *exponential* family of distributions if $\mathrm{p}_X(x; \boldsymbol{\theta})$ can be written in the general form

$$\mathrm{p}_X(x; \boldsymbol{\theta}) = h(x)b(\boldsymbol{\theta})\mathrm{e}^{\sum_{j=1}^{k} w_j(\boldsymbol{\theta})v_j(x)},$$

where $h(x) \geq 0$ does not in any way depend on $\boldsymbol{\theta}$, $b(\boldsymbol{\theta}) \geq 0$ does not depend on x, $w_1(\boldsymbol{\theta}), w_2(\boldsymbol{\theta}), \ldots, w_k(\boldsymbol{\theta})$ are real-valued functions of $\boldsymbol{\theta}$ but not of x, and $v_1(x), v_2(x), \ldots, v_k(x)$ are real-valued functions of x but not of $\boldsymbol{\theta}$. Then, if $X_1, X_2, \ldots, X_n$ constitute a random sample of size n from $\mathrm{p}_X(x; \boldsymbol{\theta})$, so that

$\mathrm{p}_{\boldsymbol{X}}(\boldsymbol{x};\boldsymbol{\theta}) = \prod_{i=1}^{n} \mathrm{p}_{X}(x_i;\boldsymbol{\theta})$, it follows that

$$\mathrm{p}_{\boldsymbol{X}}(\boldsymbol{x};\boldsymbol{\theta}) = \left\{ [b(\boldsymbol{\theta})]^n \mathrm{e}^{\sum_{j=1}^{k} w_j(\boldsymbol{\theta})\left[\sum_{i=1}^{n} v_j(x_i)\right]} \right\} \left\{ \prod_{i=1}^{n} h(x_i) \right\};$$

so, by the Factorization Theorem, the p statistics $U_j(\boldsymbol{X}) = \sum_{i=1}^{n} v_j(X_i)$, $j = 1, 2, \ldots, k$, are jointly sufficient for $\boldsymbol{\theta}$. The above results also hold when considering a family $\mathcal{F}_c = \{\mathrm{f}_X(x;\boldsymbol{\theta}), \boldsymbol{\theta} \in \Omega\}$ of continuous probability distributions. Many important families of distributions are members of the exponential family; these include the binomial, Poisson, and negative binomial families in the discrete case, and the normal, gamma, and beta families in the continuous case.

1.4.3 Methods for Evaluating the Properties of a Point Estimator

For now, consider the special case of one unknown parameter θ.

1.4.3.1 Mean-Squared Error (MSE)

The *mean-squared error* of $\hat{\theta}$ as an estimator of the parameter θ is defined as

$$\mathrm{MSE}(\hat{\theta}, \theta) = \mathrm{E}[(\hat{\theta} - \theta)^2] = \mathrm{V}(\hat{\theta}) + [\mathrm{E}(\hat{\theta}) - \theta]^2,$$

where $\mathrm{V}(\hat{\theta})$ is the variance of $\hat{\theta}$ and $[\mathrm{E}(\hat{\theta}) - \theta]^2$ is the squared-bias of $\hat{\theta}$ as an estimator of the parameter θ. An estimator with small MSE has both a small variance and a small squared-bias.

Using MSE as the criterion for choosing among a class of possible estimators of θ is problematic because this class is too large. Hence, it is common practice to limit the class of possible estimators of θ to those estimators that are *unbiased* estimators of θ. More formally, $\hat{\theta}$ is an *unbiased* estimator of the parameter θ if $\mathrm{E}(\hat{\theta}) = \theta$ for all $\theta \in \Omega$. Then, if $\hat{\theta}$ is an unbiased estimator of θ, we have $\mathrm{MSE}(\hat{\theta}, \theta) = \mathrm{V}(\hat{\theta})$, so that the criterion for choosing among competing unbiased estimators of θ is based solely on variance considerations.

1.4.3.2 Cramér–Rao Lower Bound (CRLB)

Let $\mathcal{L}(\boldsymbol{x};\theta)$ denote the distribution of the random vector $\boldsymbol{X}$, and let $\hat{\theta}$ be *any* unbiased estimator of the parameter θ. Then, under certain mathematical regularity conditions, it can be shown (Rao, 1945; Cramér, 1946) that

$$\mathrm{V}(\hat{\theta}) \geq \frac{1}{\mathrm{E}_{\mathbf{x}}\left[(\partial \ln \mathcal{L}(\boldsymbol{x};\theta)/\partial\theta)^2\right]} = \frac{1}{-\mathrm{E}_{\mathbf{x}}\left[\partial^2 \ln \mathcal{L}(\boldsymbol{x};\theta)/\partial\theta^2\right]}.$$

In the important special case when $X_1, X_2, \ldots, X_n$ constitute a random sample of size n from the discrete probability distribution $\mathrm{p}_X(x;\theta)$, so that $\mathcal{L}(\boldsymbol{x};\theta) = \prod_{i=1}^{n} \mathrm{p}_X(x_i;\theta)$, then we obtain

$$\mathrm{V}(\hat{\theta}) \geq \frac{1}{n\mathrm{E}_x\left\{(\partial \ln[\mathrm{p}_X(x;\theta)]/\partial\theta)^2\right\}} = \frac{1}{-n\mathrm{E}_x\left\{\partial^2 \ln[\mathrm{p}_X(x;\theta)]/\partial\theta^2\right\}}.$$

A completely analogous result holds when $X_1, X_2, \ldots, X_n$ constitute a random sample of size n from the density function $\mathrm{f}_X(x;\theta)$. For further discussion, see Lehmann (1983).

1.4.3.3 Efficiency

The *efficiency* of any unbiased estimator $\hat{\theta}$ of θ relative to the CRLB is defined as

$$\mathrm{EFF}(\hat{\theta},\theta) = \frac{\mathrm{CRLB}}{\mathrm{V}(\hat{\theta})}, \quad 0 \leq \mathrm{EFF}(\hat{\theta},\theta) \leq 1,$$

and the corresponding *asymptotic efficiency* is $\lim_{n\to\infty} \mathrm{EFF}(\hat{\theta},\theta)$.

There are situations when no unbiased estimator of θ achieves the CRLB. In such a situation, we can utilize the *Rao–Blackwell Theorem* (Rao, 1945; Blackwell, 1947) to aid in the search for that unbiased estimator with the smallest variance (i.e., the minimum variance unbiased estimator or MVUE).

First, we need to introduce the concept of a *complete* sufficient statistic.

1.4.3.4 Completeness

The family $\mathcal{F}_u = \{\mathrm{p}_U(u;\theta), \theta \in \Omega\}$, or $\mathcal{F}_u = \{\mathrm{f}_U(u;\theta), \theta \in \Omega\}$, for the sufficient statistic U is called *complete* (or, equivalently, U is a complete sufficient statistic) if the condition $\mathrm{E}[g(U)] = 0$ for all $\theta \in \Omega$ implies that $\mathrm{pr}[g(U) = 0] = 1$ for all $\theta \in \Omega$.

As an important special case, for an exponential family with $U_j(\boldsymbol{X}) = \sum_{i=1}^{n} v_j(X_i)$ for $j = 1, 2, \ldots, k$, the vector of sufficient statistics

$$\boldsymbol{U}(\boldsymbol{X}) = [U_1(\boldsymbol{X}), U_2(\boldsymbol{X}), \ldots, U_k(\boldsymbol{X})]$$

is complete if $\{w_1(\boldsymbol{\theta}), w_2(\boldsymbol{\theta}), \ldots, w_k(\boldsymbol{\theta}) : \boldsymbol{\theta} \in \Omega\}$ contains an open set in $\Re^k$.

1.4.3.5 Rao–Blackwell Theorem

Let $U^* \equiv U^*(\boldsymbol{X})$ be *any* unbiased point estimator of θ, and let $U \equiv U(\boldsymbol{X})$ be a sufficient statistic for θ. Then, $\hat{\theta} = \mathrm{E}(U^*|U = u)$ is an unbiased point estimator of θ, and $\mathrm{V}(\hat{\theta}) \leq \mathrm{V}(U^*)$. If U is a *complete* sufficient statistic for θ, then $\hat{\theta}$ is the *unique* (with probability one) MVUE of θ.

It is important to emphasize that the variance of the MVUE of θ may not achieve the CRLB.

1.4.4 Interval Estimation of Population Parameters

1.4.4.1 Exact Confidence Intervals

An *exact* $100(1-\alpha)\%$ confidence interval (CI) for a parameter θ involves two random variables, L (called the *lower limit*) and U (called the *upper limit*), defined so that

$$\mathrm{pr}(L < \theta < U) = (1-\alpha),$$

where typically $0 < \alpha \leq 0.10$.

The construction of exact CIs often involves the properties of statistics based on random samples from normal populations. Some illustrations are as follows.

1.4.4.2 Exact CI for the Mean of a Normal Distribution

Let $X_1, X_2, \ldots, X_n$ constitute a random sample from a $\mathrm{N}(\mu, \sigma^2)$ parent population. The sample mean is $\bar{X} = n^{-1}\sum_{i=1}^{n} X_i$ and the sample variance is $S^2 = (n-1)^{-1}\sum_{i=1}^{n}(X_i - \bar{X})^2$.

Then,

$$\bar{X} \sim \mathrm{N}\left(\mu, \frac{\sigma^2}{n}\right),$$

$$\frac{(n-1)S^2}{\sigma^2} = \frac{\sum_{i=1}^{n}(X_i - \bar{X})^2}{\sigma^2} \sim \chi^2_{n-1},$$

and $\bar{X}$ and S^2 are independent random variables.

In general, if $Z \sim \mathrm{N}(0,1), U \sim \chi^2_\nu$, and Z and U are independent random variables, then the random variable $T_\nu = Z/\sqrt{U/\nu} \sim t_\nu$; that is, T_ν has a

t-distribution with ν degrees of freedom (df). Thus, the random variable

$$T_{n-1} = \frac{(\bar{X} - \mu)/(\sigma/\sqrt{n})}{\sqrt{[(n-1)S^2/\sigma^2]/(n-1)}} = \frac{\bar{X} - \mu}{S/\sqrt{n}} \sim t_{n-1}.$$

With $t_{n-1,1-\alpha/2}$ defined so that $\text{pr}(T_{n-1} < t_{n-1,1-\alpha/2}) = 1 - \alpha/2$, we then have

$$\begin{aligned}
(1-\alpha) &= \text{pr}(-t_{n-1,1-\alpha/2} < T_{n-1} < t_{n-1,1-\alpha/2}) \\
&= \text{pr}\left[-t_{n-1,1-\alpha/2} < \frac{\bar{X} - \mu}{S/\sqrt{n}} < t_{n-1,1-\alpha/2}\right] \\
&= \text{pr}\left[\bar{X} - t_{n-1,1-\alpha/2}\frac{S}{\sqrt{n}} < \mu < \bar{X} + t_{n-1,1-\alpha/2}\frac{S}{\sqrt{n}}\right].
\end{aligned}$$

Thus,

$$L = \bar{X} - t_{n-1,1-\alpha/2}\frac{S}{\sqrt{n}} \quad \text{and} \quad U = \bar{X} + t_{n-1,1-\alpha/2}\frac{S}{\sqrt{n}},$$

giving

$$\bar{X} \pm t_{n-1,1-\alpha/2}\frac{S}{\sqrt{n}}$$

as the *exact* $100(1-\alpha)\%$ CI for μ based on a random sample $X_1, X_2, \ldots, X_n$ of size n from a $\text{N}(\mu, \sigma^2)$ parent population.

1.4.4.3 Exact CI for a Linear Combination of Means of Normal Distributions

More generally, for $i = 1, 2, \ldots, k$, let $X_{i1}, X_{i2}, \ldots, X_{in_i}$ constitute a random sample of size n_i from a $\text{N}(\mu_i, \sigma_i^2)$ parent population. Then,

i. For $i = 1, 2, \ldots, k$, $\bar{X}_i = n_i^{-1}\sum_{j=1}^{n_i} X_{ij} \sim \text{N}\left(\mu_i, \frac{\sigma_i^2}{n_i}\right)$;

ii. For $i = 1, 2, \ldots, k$, $\frac{(n_i-1)S_i^2}{\sigma_i^2} = \frac{\sum_{j=1}^{n_i}(X_{ij}-\bar{X}_i)^2}{\sigma_i^2} \sim \chi^2_{n_i-1}$;

iii. The $2k$ random variables $\{\bar{X}_i, S_i^2\}_{i=1}^k$ are mutually independent.

Now, assuming $\sigma_i^2 = \sigma^2$ for all i (i.e., assuming *variance homogeneity*), if $c_1, c_2, \ldots, c_k$ are known constants, then the random variable

$$\sum_{i=1}^k c_i\bar{X}_i \sim \text{N}\left[\sum_{i=1}^k c_i\mu_i, \sigma^2\left(\sum_{i=1}^k \frac{c_i^2}{n_i}\right)\right];$$

and, with $N = \sum_{i=1}^{k} n_i$, the random variable

$$\frac{\sum_{i=1}^{k}(n_i - 1)S_i^2}{\sigma^2} = \frac{\sum_{i=1}^{k}\sum_{j=1}^{n_i}(X_{ij} - \bar{X}_i)^2}{\sigma^2} \sim \chi^2_{N-k};$$

Thus, the random variable

$$T_{N-k} = \frac{\sum_{i=1}^{k} c_i\bar{X}_i - \sum_{i=1}^{k} c_i\mu_i}{S_p\sqrt{\sum_{i=1}^{k}\frac{c_i^2}{n_i}}} \sim t_{N-k},$$

where the pooled sample variance is $S_p^2 = \sum_{i=1}^{k}(n_i - 1)S_i^2/(N-k)$.

This gives

$$\sum_{i=1}^{k} c_i\bar{X}_i \pm t_{N-k,1-\frac{\alpha}{2}} S_p\sqrt{\sum_{i=1}^{k}\frac{c_i^2}{n_i}}$$

as the *exact* $100(1-\alpha)\%$ CI for the parameter $\sum_{i=1}^{k} c_i\mu_i$.

In the special case when $k = 2, c_1 = +1$, and $c_2 = -1$, we obtain the well-known two-sample CI for $(\mu_1 - \mu_2)$, namely,

$$(\bar{X}_1 - \bar{X}_2) \pm t_{n_1+n_2-2,1-\alpha/2} S_p\sqrt{\frac{1}{n_1} + \frac{1}{n_2}}.$$

1.4.4.4 Exact CI for the Variance of a Normal Distribution

For $i = 1, 2, \ldots, k$, since $(n_i - 1)S_i^2/\sigma_i^2 \sim \chi^2_{n_i-1}$, we have

$$(1-\alpha) = \text{pr}\left[\chi^2_{n_i-1,\alpha/2} < \frac{(n_i-1)S_i^2}{\sigma_i^2} < \chi^2_{n_i-1,1-\alpha/2}\right] = \text{pr}(L < \sigma_i^2 < U),$$

where

$$L = \frac{(n_i-1)S_i^2}{\chi^2_{n_i-1,1-\alpha/2}} \quad \text{and} \quad U = \frac{(n_i-1)S_i^2}{\chi^2_{n_i-1,\alpha/2}},$$

and where $\chi^2_{n_i-1,\alpha/2}$ and $\chi^2_{n_i-1,1-\alpha/2}$ are, respectively, the $100\,(\alpha/2)$ and $100\,(1-\alpha/2)$ percentiles of the $\chi^2_{n_i-1}$ distribution.

1.4.4.5 Exact CI for the Ratio of Variances of Two Normal Distributions

In general, if $U_1 \sim \chi^2_{\nu_1}, U_2 \sim \chi^2_{\nu_2}$, and U_1 and U_2 are independent random variables, then the random variable

$$F_{\nu_1,\nu_2} = \frac{U_1/\nu_1}{U_2/\nu_2} \sim f_{\nu_1,\nu_2};$$

that is, F_{ν_1,ν_2} follows an f-distribution with ν_1 numerator df and ν_2 denominator df. As an example, when $k = 2$, the random variable

$$F_{n_1-1,n_2-1} = \frac{\left[[(n_1-1)S_1^2]/\sigma_1^2\right]/(n_1-1)}{\left[[(n_2-1)S_2^2]/\sigma_2^2\right]/(n_2-1)} = \left(\frac{S_1^2}{S_2^2}\right)\left(\frac{\sigma_2^2}{\sigma_1^2}\right) \sim f_{n_1-1,n_2-1}.$$

So, since $f_{n_1-1,n_2-1,\alpha/2} = f^{-1}_{n_2-1,n_1-1,1-\alpha/2}$, we have

$$\begin{aligned}(1-\alpha) &= \text{pr}\left[f^{-1}_{n_2-1,n_1-1,1-\alpha/2} < \left(\frac{S_1^2}{S_2^2}\right)\left(\frac{\sigma_2^2}{\sigma_1^2}\right) < f_{n_1-1,n_2-1,1-\alpha/2}\right] \\ &= \text{pr}\left[L < \left(\frac{\sigma_2^2}{\sigma_1^2}\right) < U\right],\end{aligned}$$

where

$$L = f^{-1}_{n_2-1,n_1-1,1-\alpha/2}\left(\frac{S_2^2}{S_1^2}\right) \quad \text{and} \quad U = f_{n_1-1,n_2-1,1-\alpha/2}\left(\frac{S_2^2}{S_1^2}\right),$$

and where $f_{n_1-1,n_2-1,1-\alpha/2}$ is the $100\,(1-\alpha/2)$ percentile of the f-distribution with $(n_1 - 1)$ numerator df and $(n_2 - 1)$ denominator df.

1.4.4.6 Large-Sample Approximate CIs

By an *approximate* CI for a parameter θ, we mean that the random variables L and U satisfy

$$\text{pr}(L < \theta < U) \approx (1-\alpha),$$

where typically $0 < \alpha \leq 0.10$.

The concepts of *convergence in distribution* (discussed in Section 1.3: Multivariate Distribution Theory) and *consistency*, coupled with the use of *Slutsky's Theorem* (see Serfling, 2002), are typically used for the development of ML-based approximate CIs.

1.4.4.7 Consistency

A point estimator $\hat{\theta}$ is a *consistent* estimator of a parameter θ if, for every $\epsilon > 0$,

$$\lim_{n\to\infty} \text{pr}(|\hat{\theta} - \theta| > \epsilon) = 0.$$

In this case, we say that $\hat{\theta}$ *converges in probability* to θ, and we write $\hat{\theta} \xrightarrow{\text{P}} \theta$.

Two *sufficient* conditions so that $\hat{\theta} \xrightarrow{\text{P}} \theta$ are

$$\lim_{n\to\infty} \text{E}(\hat{\theta}) = \theta \text{ and } \lim_{n\to\infty} \text{V}(\hat{\theta}) = 0.$$

1.4.4.8 Slutsky's Theorem

If $V_n \xrightarrow{\text{P}} c$, where c is a constant, and if $W_n \xrightarrow{\text{D}} W$, then

$$V_n W_n \xrightarrow{\text{D}} cW \quad \text{and} \quad (V_n + W_n) \xrightarrow{\text{D}} (c + W).$$

To develop ML-based large-sample approximate CIs, we make use of the following properties of the MLE $\hat{\boldsymbol{\theta}}_{\text{ml}} \equiv \hat{\boldsymbol{\theta}}$ of $\boldsymbol{\theta}$, assuming $\mathcal{L}(\boldsymbol{x};\boldsymbol{\theta})$ is the correct likelihood function and assuming that certain regularity conditions hold:

i. For $j = 1, 2, \ldots, p$, $\hat{\theta}_j$ is a consistent estimator of θ_j. More generally, if the scalar function $\tau(\boldsymbol{\theta})$ is a continuous function of $\boldsymbol{\theta}$, then $\tau(\hat{\boldsymbol{\theta}})$ is a consistent estimator of $\tau(\boldsymbol{\theta})$.

ii.
$$\sqrt{n}(\hat{\boldsymbol{\theta}} - \boldsymbol{\theta}) \xrightarrow{\text{D}} \text{MVN}_p[\mathbf{0}, n\boldsymbol{\mathcal{I}}^{-1}(\boldsymbol{\theta})],$$
where $\boldsymbol{\mathcal{I}}(\boldsymbol{\theta})$ is the $(p \times p)$ *expected information matrix*, with (j, j') element equal to
$$-\text{E}_{\boldsymbol{x}}\left[\frac{\partial^2 \ln\mathcal{L}(\boldsymbol{x};\boldsymbol{\theta})}{\partial\theta_j \partial\theta_{j'}}\right],$$
and where $\boldsymbol{\mathcal{I}}^{-1}(\boldsymbol{\theta})$ is the large-sample covariance matrix of $\hat{\boldsymbol{\theta}}$ based on expected information. In particular, the (j, j') element of $\boldsymbol{\mathcal{I}}^{-1}(\boldsymbol{\theta})$ is denoted $v_{jj'}(\boldsymbol{\theta}) = \text{cov}(\hat{\theta}_j, \hat{\theta}_{j'}), j = 1, 2, \ldots, p$ and $j' = 1, 2, \ldots, p$.

1.4.4.9 Construction of ML-Based CIs

As an illustration, properties (i) and (ii) will now be used to construct a large-sample ML-based approximate $100(1-\alpha)\%$ CI for the parameter θ_j.

First, with the (j, j) diagonal element $v_{jj}(\boldsymbol{\theta})$ of $\boldsymbol{\mathcal{I}}^{-1}(\boldsymbol{\theta})$ being the large-sample variance of $\hat{\theta}_j$ based on expected information, it follows that

$$\frac{\hat{\theta}_j - \theta_j}{\sqrt{v_{jj}(\boldsymbol{\theta})}} \xrightarrow{\text{D}} \text{N}(0, 1) \quad \text{as } n \longrightarrow \infty.$$

Then, with $\boldsymbol{\mathcal{I}}^{-1}(\hat{\boldsymbol{\theta}})$ denoting the *estimated* large-sample covariance matrix of $\hat{\boldsymbol{\theta}}$ based on expected information, and with the (j, j) diagonal element $v_{jj}(\hat{\boldsymbol{\theta}})$ of $\boldsymbol{\mathcal{I}}^{-1}(\hat{\boldsymbol{\theta}})$ being the estimated large-sample variance of $\hat{\theta}_j$ based on expected information, it follows by Sluksky's Theorem that

$$\frac{\hat{\theta}_j - \theta_j}{\sqrt{v_{jj}(\hat{\boldsymbol{\theta}})}} = \sqrt{\frac{v_{jj}(\boldsymbol{\theta})}{v_{jj}(\hat{\boldsymbol{\theta}})}}\left[\frac{\hat{\theta}_j - \theta_j}{\sqrt{v_{jj}(\boldsymbol{\theta})}}\right] \xrightarrow{\text{D}} \text{N}(0, 1) \quad \text{as } n \longrightarrow \infty$$

since $v_{jj}(\hat{\boldsymbol{\theta}})$ is a consistent estimator of $v_{jj}(\boldsymbol{\theta})$.

Thus, it follows from the above results that

$$\frac{\hat{\theta}_j - \theta_j}{\sqrt{v_{jj}(\hat{\boldsymbol{\theta}})}} \dot{\sim} \mathrm{N}(0,1) \quad \text{for large } n.$$

Finally, with $Z_{1-\alpha/2}$ defined so that $\mathrm{pr}(Z < Z_{1-\alpha/2}) = (1-\alpha/2)$ when $Z \sim \mathrm{N}(0,1)$, we have

$$\begin{aligned}(1-\alpha) &= \mathrm{pr}(-Z_{1-\alpha/2} < Z < Z_{1-\alpha/2}) \\ &\approx \mathrm{pr}\left[-Z_{1-\alpha/2} < \frac{\hat{\theta}_j - \theta_j}{\sqrt{v_{jj}(\hat{\boldsymbol{\theta}})}} < Z_{1-\alpha/2}\right] \\ &= \mathrm{pr}\left[\hat{\theta}_j - Z_{1-\alpha/2}\sqrt{v_{jj}(\hat{\boldsymbol{\theta}})} < \theta_j < \hat{\theta}_j + Z_{1-\alpha/2}\sqrt{v_{jj}(\hat{\boldsymbol{\theta}})}\right].\end{aligned}$$

Thus,

$$\hat{\theta}_j \pm Z_{1-\alpha/2}\sqrt{v_{jj}(\hat{\boldsymbol{\theta}})}$$

is the large-sample ML-based approximate $100(1-\alpha)\%$ CI for the parameter θ_j based on expected information.

In practice, instead of the estimated expected information matrix, the estimated *observed* information matrix $\boldsymbol{I}(\boldsymbol{x};\hat{\boldsymbol{\theta}})$ is used, with its (j,j') element equal to

$$-\left[\frac{\partial^2 \ln \mathcal{L}(\boldsymbol{x};\boldsymbol{\theta})}{\partial\theta_j \partial\theta_{j'}}\right]_{|\boldsymbol{\theta}=\hat{\boldsymbol{\theta}}}.$$

Then, with $\boldsymbol{I}^{-1}(\boldsymbol{x};\hat{\boldsymbol{\theta}})$ denoting the estimated large-sample covariance matrix of $\hat{\boldsymbol{\theta}}$ based on observed information, and with the (j,j) diagonal element $v_{jj}(\boldsymbol{x};\hat{\boldsymbol{\theta}})$ of $\boldsymbol{I}^{-1}(\boldsymbol{x};\hat{\boldsymbol{\theta}})$ being the estimated large-sample variance of $\hat{\theta}_j$ based on observed information, it follows that

$$\hat{\theta}_j \pm Z_{1-\alpha/2}\sqrt{v_{jj}(\boldsymbol{x};\hat{\boldsymbol{\theta}})}$$

is the large-sample ML-based approximate $100(1-\alpha)\%$ CI for the parameter θ_j based on observed information.

1.4.4.10 ML-Based CI for a Bernoulli Distribution Probability

As a simple one-parameter ($p = 1$) example, let $X_1, X_2, \ldots, X_n$ constitute a random sample of size n from the Bernoulli parent population

$$\mathrm{p}_X(x;\theta) = \theta^x (1-\theta)^{1-x}, \quad x = 0, 1 \quad \text{and} \quad 0 < \theta < 1,$$

and suppose that it is desired to develop a large-sample ML-based approximate $100(1-\alpha)\%$ CI for the parameter θ. First, the appropriate likelihood function is

$$\mathcal{L}(\boldsymbol{x};\theta) = \prod_{i=1}^{n} \left[\theta^{x_i}(1-\theta)^{1-x_i}\right] = \theta^s (1-\theta)^{n-s},$$

where $s = \sum_{i=1}^{n} x_i$ is a sufficient statistic for θ.

Now,

$$\ln\mathcal{L}(\boldsymbol{x};\theta) = s\ln\theta + (n-s)\ln(1-\theta),$$

so that the equation

$$\frac{\partial \ln\mathcal{L}(\boldsymbol{x};\theta)}{\partial\theta} = \frac{s}{\theta} - \frac{(n-s)}{(1-\theta)} = 0$$

gives $\hat{\theta} = \bar{X} = n^{-1}\sum_{i=1}^{n} X_i$ as the MLE of θ.

And,

$$\frac{\partial^2 \ln\mathcal{L}(\boldsymbol{x};\theta)}{\partial\theta^2} = \frac{-s}{\theta^2} - \frac{(n-s)}{(1-\theta)^2},$$

so that

$$-\mathrm{E}\left[\frac{\partial^2 \ln\mathcal{L}(\boldsymbol{x};\theta)}{\partial\theta^2}\right] = \frac{n\theta}{\theta^2} + \frac{(n-n\theta)}{(1-\theta)^2} = \frac{n}{\theta(1-\theta)}.$$

Hence,

$$\begin{aligned} v_{11}(\hat{\theta}) &= \left\{-\mathrm{E}\left[\frac{\partial^2 \ln\mathcal{L}(\boldsymbol{x};\theta)}{\partial\theta^2}\right]\right\}^{-1}_{|\theta=\hat{\theta}} \\ &= v_{11}(\mathbf{x};\hat{\theta}) = \left\{-\frac{\partial^2 \ln\mathcal{L}(\boldsymbol{x};\theta)}{\partial\theta^2}\right\}^{-1}_{|\theta=\hat{\theta}} \\ &= \frac{\bar{X}(1-\bar{X})}{n}, \end{aligned}$$

so that the large-sample ML-based approximate $100(1-\alpha)\%$ CI for θ is equal to

$$\bar{X} \pm Z_{1-\alpha/2}\sqrt{\frac{\bar{X}(1-\bar{X})}{n}}.$$

In this simple example, the same CI is obtained using either expected information or observed information. In more complicated situations, this will typically *not* happen.

1.4.4.11 Delta Method

Let $Y = g(\boldsymbol{X})$, where $\boldsymbol{X} = (X_1, X_2, \ldots, X_k)$, $\boldsymbol{\mu} = (\mu_1, \mu_2, \ldots, \mu_k)$, $\mathrm{E}(X_i) = \mu_i, \mathrm{V}(X_i) = \sigma_i^2$, and $\mathrm{cov}(X_i, X_j) = \sigma_{ij}$ for $i \neq j, i = 1, 2, \ldots, k$ and $j = 1, 2, \ldots, k$. Then, a *first-order* (or *linear*) multivariate Taylor series approximation to Y around $\boldsymbol{\mu}$ is

$$Y \approx g(\boldsymbol{\mu}) + \sum_{i=1}^{k} \frac{\partial g(\boldsymbol{\mu})}{\partial X_i}(X_i - \mu_i),$$

where

$$\frac{\partial g(\boldsymbol{\mu})}{\partial X_i} = \frac{\partial g(\boldsymbol{X})}{\partial X_i}\bigg|_{\boldsymbol{X}=\boldsymbol{\mu}}.$$

Thus, using the above linear approximation for Y, it follows that $\mathrm{E}(Y) \approx g(\boldsymbol{\mu})$ and that

$$\mathrm{V}(Y) \approx \sum_{i=1}^{k} \left[\frac{\partial g(\boldsymbol{\mu})}{\partial X_i}\right]^2 \sigma_i^2 + 2\sum_{i=1}^{k-1}\sum_{j=i+1}^{k} \left[\frac{\partial g(\boldsymbol{\mu})}{\partial X_i}\right]\left[\frac{\partial g(\boldsymbol{\mu})}{\partial X_j}\right]\sigma_{ij}.$$

The delta method for MLEs is as follows. For $q \leq p$, suppose that the $(1 \times q)$ row vector

$$\boldsymbol{\Phi}(\boldsymbol{\theta}) = [\tau_1(\boldsymbol{\theta}), \tau_2(\boldsymbol{\theta}), \ldots, \tau_q(\boldsymbol{\theta})]$$

involves q scalar parametric functions of the parameter vector $\boldsymbol{\theta}$. Then,

$$\boldsymbol{\Phi}(\hat{\boldsymbol{\theta}}) = [\tau_1(\hat{\boldsymbol{\theta}}), \tau_2(\hat{\boldsymbol{\theta}}), \ldots, \tau_q(\hat{\boldsymbol{\theta}})]$$

is the MLE of $\boldsymbol{\Phi}(\boldsymbol{\theta})$.

Then, the $(q \times q)$ large-sample covariance matrix of $\boldsymbol{\Phi}(\hat{\boldsymbol{\theta}})$ based on expected information is

$$[\boldsymbol{\Delta}(\boldsymbol{\theta})]\boldsymbol{\mathcal{I}}^{-1}(\boldsymbol{\theta})[\boldsymbol{\Delta}(\boldsymbol{\theta})]',$$

where the (i, j) element of the $(q \times p)$ matrix $\boldsymbol{\Delta}(\boldsymbol{\theta})$ is equal to $\partial\tau_i(\boldsymbol{\theta})/\partial\theta_j, i = 1, 2, \ldots, q$ and $j = 1, 2, \ldots, p$.

Hence, the corresponding estimated large-sample covariance matrix of $\boldsymbol{\Phi}(\hat{\boldsymbol{\theta}})$ based on expected information is

$$[\boldsymbol{\Delta}(\hat{\boldsymbol{\theta}})]\boldsymbol{\mathcal{I}}^{-1}(\hat{\boldsymbol{\theta}})[\boldsymbol{\Delta}(\hat{\boldsymbol{\theta}})]'.$$

Analogous expressions based on observed information are obtained by substituting $\boldsymbol{I}^{-1}(\boldsymbol{x};\boldsymbol{\theta})$ for $\boldsymbol{\mathcal{I}}^{-1}(\boldsymbol{\theta})$ and by substituting $\boldsymbol{I}^{-1}(\boldsymbol{x};\hat{\boldsymbol{\theta}})$ for $\boldsymbol{\mathcal{I}}^{-1}(\hat{\boldsymbol{\theta}})$ in the above two expressions.

The special case $q = p = 1$ gives

$$\mathrm{V}[\tau_1(\hat{\theta}_1)] \approx \left[\frac{\partial \tau_1(\theta_1)}{\partial \theta_1}\right]^2 \mathrm{V}(\hat{\theta}_1).$$

The corresponding large-sample ML-based approximate $100(1-\alpha)\%$ CI for $\tau_1(\theta_1)$ based on expected information is equal to

$$\tau_1(\hat{\theta}_1) \pm Z_{1-\alpha/2}\sqrt{\left[\frac{\partial \tau_1(\theta_1)}{\partial \theta_1}\right]^2_{|\theta_1=\hat{\theta}_1} v_{11}(\hat{\theta}_1)}.$$

The corresponding CI based on observed information is obtained by substituting $v_{11}(\boldsymbol{x};\hat{\theta}_1)$ for $v_{11}(\hat{\theta}_1)$ in the above expression.

1.4.4.12 *Delta Method CI for a Function of a Bernoulli Distribution Probability*

As a simple illustration, for the Bernoulli population example considered earlier, suppose that it is now desired to use the delta method to obtain a large-sample ML-based approximate $100(1-\alpha)\%$ CI for the "odds"

$$\tau(\theta) = \frac{\theta}{(1-\theta)} = \frac{\mathrm{pr}(X=1)}{[1-\mathrm{pr}(X=1)]}.$$

So, by the invariance property, $\tau(\hat{\theta}) = \bar{X}/(1-\bar{X})$ is the MLE of $\tau(\theta)$ since $\hat{\theta} = \bar{X}$ is the MLE of θ. And, via the delta method, the large-sample estimated variance of $\tau(\hat{\theta})$ is equal to

$$\begin{aligned}
\hat{\mathrm{V}}\left[\tau(\hat{\theta})\right] &\approx \left[\frac{\partial \tau(\theta)}{\partial \theta}\right]^2_{|\theta=\hat{\theta}} \hat{\mathrm{V}}(\hat{\theta}) \\
&= \left[\frac{1}{(1-\bar{X})^2}\right]^2 \left[\frac{\bar{X}(1-\bar{X})}{n}\right] \\
&= \frac{\bar{X}}{n(1-\bar{X})^3}.
\end{aligned}$$

Finally, the large-sample ML-based approximate $100(1-\alpha)\%$ CI for $\tau(\theta) =$

$\theta/(1-\theta)$ using the delta method is equal to

$$\frac{\bar{X}}{(1-\bar{X})} \pm Z_{1-\alpha/2}\sqrt{\frac{\bar{X}}{n(1-\bar{X})^3}}.$$

1.5 Hypothesis Testing Theory

1.5.1 Basic Principles

1.5.1.1 Simple and Composite Hypotheses

A *statistical hypothesis* is an assertion about the distribution of one or more random variables. If the statistical hypothesis completely specifies the distribution (i.e., the hypothesis assigns numerical values to all unknown population parameters), then it is called a *simple* hypothesis; otherwise, it is called a *composite* hypothesis.

1.5.1.2 Null and Alternative Hypotheses

In the typical statistical hypothesis testing situation, there are two hypotheses of interest: the *null* hypothesis (denoted H_0) and the *alternative* hypothesis (denoted H_1). The statistical objective is to use the information in a sample from the distribution under study to make a decision about whether H_0 or H_1 is more likely to be true (i.e., is more likely to represent the true "state of nature").

1.5.1.3 Statistical Tests

A statistical test of H_0 versus H_1 consists of a rule which, when operationalized using the available information in a sample, leads to a decision either to *reject*, or *not to reject*, H_0 in favor of H_1. It is important to point out that a decision *not to reject* H_0 does *not* imply that H_0 is, in fact, true; in particular, the decision not to reject H_0 is often due to data inadequacies (e.g., too small a sample size, erroneous or missing information, etc.)

1.5.1.4 Type I and Type II Errors

For any statistical test, there are two possible decision errors that can be made. A "Type I" error occurs when the decision is made to reject H_0 in favor of H_1 when, in fact, H_0 is true; the probability of a Type I error is denoted as $\alpha = \text{pr}(\text{test rejects } H_0|H_0 \text{ true})$. A "Type II" error occurs when the decision is made not to reject H_0 when, in fact, H_0 is false and H_1 is true; the probability of a Type II error is denoted as $\beta = \text{pr}(\text{test does not reject } H_0|H_0 \text{ false})$.

1.5.1.5 Power

The *power* of a statistical test is the probability of rejecting H_0 when, in fact, H_0 is false and H_1 is true; in particular,

$$\text{POWER} = \text{pr}(\text{test rejects } H_0|H_0 \text{ false}) = (1 - \beta).$$

Type I error rate α is controllable and is typically assigned a value satisfying the inequality $0 < \alpha \leq 0.10$. For a given value of α, Type II error rate β, and hence the power $(1 - \beta)$, will generally vary as a function of the values of population parameters allowable under a composite alternative hypothesis H_1.

In general, for a specified value of α, the power of any reasonable statistical testing procedure should increase as the sample size increases. Power is typically used as a very important criterion for choosing among several statistical testing procedures in any given situation.

1.5.1.6 Test Statistics and Rejection Regions

A statistical test of H_0 versus H_1 is typically carried out by using a *test statistic.* A test statistic is a random variable with the following properties: (i) its distribution, assuming the null hypothesis H_0 is true, is known either exactly or to a close approximation (i.e., for large sample sizes); (ii) its numerical value can be computed using the information in a sample; and (iii) its computed numerical value leads to a decision either to reject or not to reject H_0 in favor of H_1. More specifically, for a given statistical test and associated test statistic, the set of all possible numerical values of the test statistic under H_0 is divided into two disjoint subsets (or "regions")—the *rejection region* $\mathcal{R}$ and the *non-rejection region* $\bar{\mathcal{R}}$. The statistical test decision rule is then defined as follows: if the computed numerical value of the test statistic is in the rejection region $\mathcal{R}$, then reject H_0 in favor of H_1; otherwise, do not reject H_0. The rejection region $\mathcal{R}$ is chosen so that, under H_0, the probability that the test statistic falls in the rejection region $\mathcal{R}$ is equal to (or approximately equal to) α (in which case the rejection region and the associated statistical test are both said to be of "size" α).

Almost all popular statistical testing procedures use test statistics that, under H_0, follow (either exactly or approximately) well-tabulated distributions such as the standard normal distribution, the t-distribution, the chi-squared distribution, and the f-distribution.

1.5.1.7 P-Values

The *P-value* for a statistical test is the probability of observing a test statistic value at least as rare as the value actually observed under the assumption that the null hypothesis H_0 is true. Thus, for a size α test, when the decision is made to reject H_0, then the P-value is less than α; and when the decision is made not to reject H_0, then the P-value is greater than α.

1.5.2 Most Powerful (MP) and Uniformly Most Powerful (UMP) Tests

Let $\boldsymbol{X} = (X_1, X_2, \ldots, X_n)$ be a random row vector with likelihood function (or joint distribution) $\mathcal{L}(\boldsymbol{x}; \boldsymbol{\theta})$ depending on a row vector $\boldsymbol{\theta} = (\theta_1, \theta_2, \ldots, \theta_p)$ of p unknown parameters. Let $\mathcal{R}$ denote some subset of all the possible realizations $\boldsymbol{x} = (x_1, x_2, \ldots, x_n)$ of the random vector $\boldsymbol{X}$. Then, $\mathcal{R}$ is the *most powerful* (or MP) rejection region of size α for testing the simple null hypothesis $\mathrm{H}_0 : \boldsymbol{\theta} = \boldsymbol{\theta}_0$ versus the simple alternative hypothesis $\mathrm{H}_1 : \boldsymbol{\theta} = \boldsymbol{\theta}_1$ if, for every subset $\mathcal{A}$ of all possible realizations $\boldsymbol{x}$ of $\boldsymbol{X}$ for which $\mathrm{pr}(\boldsymbol{X} \in \mathcal{A} | \mathrm{H}_0 : \boldsymbol{\theta} = \boldsymbol{\theta}_0) = \alpha$, we have

$$\mathrm{pr}(\boldsymbol{X} \in \mathcal{R} | \mathrm{H}_0 : \boldsymbol{\theta} = \boldsymbol{\theta}_0) = \alpha$$

and

$$\mathrm{pr}(\boldsymbol{X} \in \mathcal{R} | \mathrm{H}_1 : \boldsymbol{\theta} = \boldsymbol{\theta}_1) \geq \mathrm{pr}(\boldsymbol{X} \in \mathcal{A} | \mathrm{H}_1 : \boldsymbol{\theta} = \boldsymbol{\theta}_1).$$

Given $\mathcal{L}(\boldsymbol{x}; \boldsymbol{\theta})$, the determination of the structure of the MP rejection region $\mathcal{R}$ of size α for testing $\mathrm{H}_0 : \boldsymbol{\theta} = \boldsymbol{\theta}_0$ versus $\mathrm{H}_1 : \boldsymbol{\theta} = \boldsymbol{\theta}_1$ can be made using the *Neyman–Pearson Lemma* (Neyman and Pearson, 1933).

Lemma 1 (Neyman–Pearson Lemma) *Let $\boldsymbol{X} = (X_1, X_2, \ldots, X_n)$ be a random row vector with likelihood function (or joint distribution) of known form $\mathcal{L}(\boldsymbol{x}; \boldsymbol{\theta})$ that depends on a row vector $\boldsymbol{\theta} = (\theta_1, \theta_2, \ldots, \theta_p)$ of p unknown parameters. Let $\mathcal{R}$ be a subset of all possible realizations $\boldsymbol{x} = (x_1, x_2, \ldots, x_n)$ of $\boldsymbol{X}$. Then, $\mathcal{R}$ is the most powerful (MP) rejection region of size α (and the associated test using $\mathcal{R}$ is the most powerful test of size α) for testing the simple null hypothesis $H_0 : \boldsymbol{\theta} = \boldsymbol{\theta}_0$ versus the simple alternative hypothesis*

$H_1 : \boldsymbol{\theta} = \boldsymbol{\theta}_1$ *if, for some* $k > 0$, *the following three conditions are satisfied:*

$$\frac{\mathcal{L}(\boldsymbol{x};\boldsymbol{\theta}_0)}{\mathcal{L}(\boldsymbol{x};\boldsymbol{\theta}_1)} < k \quad \textit{for every } \boldsymbol{x} \in \mathcal{R};$$

$$\frac{\mathcal{L}(\boldsymbol{x};\boldsymbol{\theta}_0)}{\mathcal{L}(\boldsymbol{x};\boldsymbol{\theta}_1)} \geq k \quad \textit{for every } \boldsymbol{x} \in \bar{\mathcal{R}};$$

and

$$pr(\boldsymbol{X} \in \mathcal{R} | H_0 : \boldsymbol{\theta} = \boldsymbol{\theta}_0) = \alpha.$$

A rejection region $\mathcal{R}$ *is a* uniformly most powerful *(UMP) rejection region of size* α *(and the associated test using* $\mathcal{R}$ *is a* uniformly most powerful *test of size* α*) for testing a simple null hypothesis* H_0 *versus a composite alternative hypothesis* H_1 *if the region* $\mathcal{R}$ *is a most powerful region of size* α for every *simple alternative hypothesis contained in* H_1.

1.5.2.1 *Review of Notation*

In the subsections to follow, we will utilize the following quantities, which were introduced in Section 4.1.

$\hat{\boldsymbol{\theta}} = (\hat{\theta}_1, \hat{\theta}_2, \ldots, \hat{\theta}_p)$, the MLE of $\boldsymbol{\theta} = (\theta_1, \theta_2, \ldots, \theta_p)$ based on the likelihood $\mathcal{L}(\boldsymbol{x};\boldsymbol{\theta})$;

$\mathcal{I}(\hat{\boldsymbol{\theta}})$, the estimated expected information matrix based on the likelihood $\mathcal{L}(\boldsymbol{x};\boldsymbol{\theta})$;

$\mathcal{I}^{-1}(\hat{\boldsymbol{\theta}})$, the estimated large-sample covariance matrix of $\hat{\boldsymbol{\theta}}$ based on expected information for the likelihood $\mathcal{L}(\boldsymbol{x};\boldsymbol{\theta})$;

$\boldsymbol{I}(\boldsymbol{x};\hat{\boldsymbol{\theta}})$, the estimated observed information matrix based on the likelihood $\mathcal{L}(\boldsymbol{x};\boldsymbol{\theta})$;

$\boldsymbol{I}^{-1}(\boldsymbol{x};\hat{\boldsymbol{\theta}})$, the estimated large-sample covariance matrix of $\hat{\boldsymbol{\theta}}$ based on observed information for the likelihood $\mathcal{L}(\boldsymbol{x};\boldsymbol{\theta})$.

1.5.3 *Large-Sample ML-Based Methods for Testing the Simple Null Hypothesis* $H_0 : \boldsymbol{\theta} = \boldsymbol{\theta}_0$ *(i.e.,* $\boldsymbol{\theta} \in \omega$*) versus the Composite Alternative Hypothesis* $H_1 : \boldsymbol{\theta} \in \bar{\omega}$

In general, a null hypothesis places a set of restrictions on the *unrestricted parameter space* Ω, where Ω is the set of all possible values of the parameter vector $\boldsymbol{\theta}$. Let ω denote the *restricted* parameter space, where $\omega \subset \Omega$. Then, for the simple null hypothesis $\mathrm{H}_0 : \boldsymbol{\theta} = \boldsymbol{\theta}_0$, it follows that $\omega = \{\boldsymbol{\theta} : \boldsymbol{\theta} = \boldsymbol{\theta}_0\}$, and $\Omega = \omega \cup \bar{\omega}$, where $\bar{\omega}$ is the complement of ω.

1.5.3.1 Likelihood Ratio Test

The *likelihood ratio test* statistic $\hat{\lambda}, 0 < \hat{\lambda} < 1$, for testing $\mathrm{H}_0 : \boldsymbol{\theta} = \boldsymbol{\theta}_0$ (i.e., $\boldsymbol{\theta} \in \omega$) versus $\mathrm{H}_1 : \boldsymbol{\theta} \in \bar{\omega}$ is defined as

$$\hat{\lambda} = \frac{\max\limits_{\boldsymbol{\theta} \in \omega} \mathcal{L}(\boldsymbol{x}; \boldsymbol{\theta})}{\max\limits_{\boldsymbol{\theta} \in \Omega} \mathcal{L}(\boldsymbol{x}; \boldsymbol{\theta})} = \frac{\mathcal{L}(\boldsymbol{x}; \boldsymbol{\theta}_0)}{\mathcal{L}(\boldsymbol{x}; \hat{\boldsymbol{\theta}})} \equiv \frac{\mathcal{L}_\omega}{\hat{\mathcal{L}}_\Omega}.$$

Clearly, small values of $\hat{\lambda}$ favor H_1, and a size α likelihood ratio test of H_0 versus H_1 using $\hat{\lambda}$ rejects H_0 in favor of H_1 when $\hat{\lambda} < k_\alpha$, where $\mathrm{pr}(\hat{\lambda} < k_\alpha | \mathrm{H}_0) = \alpha$.

Since the exact distribution of $\hat{\lambda}$ is often difficult to determine (either under H_0 or under H_1), the following large-sample approximation is typically used (Neyman and Pearson, 1928).

Under certain regularity conditions, for large n and under $\mathrm{H}_0 : \boldsymbol{\theta} = \boldsymbol{\theta}_0$,

$$-2\ln\hat{\lambda} = 2\left[\ln\mathcal{L}(\boldsymbol{x}; \hat{\boldsymbol{\theta}}) - \ln\mathcal{L}(\boldsymbol{x}; \boldsymbol{\theta}_0)\right] \dot{\sim} \chi^2_p.$$

Thus, for a likelihood ratio test of approximate size α, one would reject $\mathrm{H}_0 : \boldsymbol{\theta} = \boldsymbol{\theta}_0$ in favor of $\mathrm{H}_1 : \boldsymbol{\theta} \neq \boldsymbol{\theta}_0$ when $-2\ln\hat{\lambda} > \chi^2_{p,1-\alpha}$.

1.5.3.2 Wald Test

The Wald test statistic $\hat{W}, 0 < \hat{W} < +\infty$, for testing $\mathrm{H}_0 : \boldsymbol{\theta} = \boldsymbol{\theta}_0$ versus $\mathrm{H}_1 : \boldsymbol{\theta} \in \bar{\omega}$ is defined as

$$\hat{W} = (\hat{\boldsymbol{\theta}} - \boldsymbol{\theta}_0)\boldsymbol{\mathcal{I}}(\hat{\boldsymbol{\theta}})(\hat{\boldsymbol{\theta}} - \boldsymbol{\theta}_0)'$$

when using expected information, and is defined as

$$\hat{W} = (\hat{\boldsymbol{\theta}} - \boldsymbol{\theta}_0)\boldsymbol{I}(\mathbf{x}; \hat{\boldsymbol{\theta}})(\hat{\boldsymbol{\theta}} - \boldsymbol{\theta}_0)'$$

when using observed information.

Under certain regularity conditions (e.g., see Wald, 1943) for large n and under $\mathrm{H}_0 : \boldsymbol{\theta} = \boldsymbol{\theta}_0, \hat{W} \dot{\sim} \chi^2_p$. Thus, for a Wald test of approximate size α, one would reject $\mathrm{H}_0 : \boldsymbol{\theta} = \boldsymbol{\theta}_0$ in favor of $\mathrm{H}_1 : \boldsymbol{\theta} \in \bar{\omega}$ when $\hat{W} > \chi^2_{p,1-\alpha}$.

1.5.3.3 Score Test

With the row vector $\mathbf{S}(\boldsymbol{\theta})$ defined as

$$\mathbf{S}(\boldsymbol{\theta}) = \left[\frac{\partial\ln\mathcal{L}(\boldsymbol{x}; \boldsymbol{\theta})}{\partial\theta_1}, \frac{\partial\ln\mathcal{L}(\boldsymbol{x}; \boldsymbol{\theta})}{\partial\theta_2}, \ldots, \frac{\partial\ln\mathcal{L}(\boldsymbol{x}; \boldsymbol{\theta})}{\partial\theta_p}\right],$$

the score test statistic $\hat{S}, 0 < \hat{S} < +\infty$, for testing $\mathrm{H}_0 : \boldsymbol{\theta} = \boldsymbol{\theta}_0$ versus $\mathrm{H}_1 : \boldsymbol{\theta} \in \bar{\omega}$ is defined as

$$\hat{S} = \mathbf{S}(\boldsymbol{\theta}_0)\boldsymbol{\mathcal{I}}^{-1}(\boldsymbol{\theta}_0)\mathbf{S}'(\boldsymbol{\theta}_0)$$

when using expected information, and is defined as

$$\hat{S} = \mathbf{S}(\boldsymbol{\theta}_0)\boldsymbol{I}^{-1}(\boldsymbol{x}; \boldsymbol{\theta}_0)\mathbf{S}'(\boldsymbol{\theta}_0)$$

when using observed information. For the simple null hypothesis $\mathrm{H}_0 : \boldsymbol{\theta} = \boldsymbol{\theta}_0$, note that the computation of the value of $\hat{S}$ involves no parameter estimation.

Under certain regularity conditions (e.g., see Rao, 1947) for large n and under $\mathrm{H}_0 : \boldsymbol{\theta} = \boldsymbol{\theta}_0$, $\hat{S} \dot{\sim} \chi^2_p$. Thus, for a score test of approximate size α, one would reject $\mathrm{H}_0 : \boldsymbol{\theta} = \boldsymbol{\theta}_0$ in favor of $\mathrm{H}_1 : \boldsymbol{\theta} \in \bar{\omega}$ when $\hat{S} > \chi^2_{p,1-\alpha}$.

For further discussion concerning likelihood ratio, Wald, and score tests, see Rao (1973).

Example. As an example, let $X_1, X_2, \ldots, X_n$ constitute a random sample of size n from the parent population $\mathrm{p}_X(x; \theta) = \theta^x(1-\theta)^{1-x}, x = 0, 1$ and $0 < \theta < 1$. Consider testing $\mathrm{H}_0 : \theta = \theta_0$ versus $\mathrm{H}_1 : \theta \neq \theta_0$. Then, with $\hat{\theta} = \bar{X} = n^{-1}\sum_{i=1}^{n} X_i$, it can be shown that

$$-2\ln\hat{\lambda} = 2n\left[\bar{X}\ln\left(\frac{\bar{X}}{\theta_0}\right) + (1-\bar{X})\ln\left(\frac{1-\bar{X}}{1-\theta_0}\right)\right]$$

that

$$\hat{W} = \left[\frac{(\bar{X}-\theta_0)}{\sqrt{\bar{X}(1-\bar{X})/n}}\right]^2,$$

and that

$$\hat{S} = \left[\frac{(\bar{X}-\theta_0)}{\sqrt{\theta_0(1-\theta_0)/n}}\right]^2.$$

This simple example highlights an important general difference between Wald tests and score tests. Wald tests use parameter variance estimates assuming that $\boldsymbol{\theta} \in \Omega$ is true (i.e., assuming no restrictions on the parameter space Ω), and score tests use parameter variance estimates assuming that $\boldsymbol{\theta} \in \omega$ (i.e., assuming that H_0 is true).

1.5.4 Large Sample ML-Based Methods for Testing the Composite Null Hypothesis $H_0 : \boldsymbol{\theta} \in \omega$ versus the Composite Alternative Hypothesis $H_1 : \boldsymbol{\theta} \in \bar{\omega}$

Let $R_i(\boldsymbol{\theta}) = 0, i = 1, 2, \ldots, r$, represent r $(<p)$ *independent* restrictions placed on the parameter vector $\boldsymbol{\theta}$, and consider the null hypothesis $\mathrm{H}_0 : \boldsymbol{\theta} \in \omega$, where

$\omega = \{\boldsymbol{\theta} : R_i(\boldsymbol{\theta}) = 0, i = 1, 2, \ldots, r\}$. For example, with $\boldsymbol{\theta} = (\theta_1, \theta_2, \theta_3, \theta_4)$ for $p = 4$, consider the $r = 3$ linearly independent linear restrictions

$$R_1(\boldsymbol{\theta}) = (\theta_1 - \theta_2) = 0, \quad R_2(\boldsymbol{\theta}) = (\theta_1 - \theta_3) = 0$$

and

$$R_3(\boldsymbol{\theta}) = (\theta_1 - \theta_4) = 0.$$

Then, the null hypothesis $\mathrm{H}_0 : R_i(\boldsymbol{\theta}) = 0, i = 1, 2, 3$, is equivalent to the null hypothesis $\mathrm{H}_0 : \theta_1 = \theta_2 = \theta_3 = \theta_4$.

In what follows, let $\hat{\boldsymbol{\theta}}_\omega$ denote the *restricted* MLE of $\boldsymbol{\theta}$ under the null hypothesis $\mathrm{H}_0 : \boldsymbol{\theta} \in \omega$.

1.5.4.1 Likelihood Ratio Test

The likelihood ratio test statistic $\hat{\lambda}, 0 < \hat{\lambda} < 1$, for testing $\mathrm{H}_0 : \boldsymbol{\theta} \in \omega$ versus $\mathrm{H}_1 : \boldsymbol{\theta} \in \bar{\omega}$ is defined as

$$\hat{\lambda} = \frac{\max\limits_{\boldsymbol{\theta} \in \omega} \mathcal{L}(\boldsymbol{x}; \boldsymbol{\theta})}{\max\limits_{\boldsymbol{\theta} \in \Omega} \mathcal{L}(\boldsymbol{x}; \boldsymbol{\theta})} = \frac{\mathcal{L}(\boldsymbol{x}; \hat{\boldsymbol{\theta}}_\omega)}{\mathcal{L}(\boldsymbol{x}; \hat{\boldsymbol{\theta}})} \equiv \frac{\hat{\mathcal{L}}_\omega}{\hat{\mathcal{L}}_\Omega}.$$

Under certain regularity conditions, for large n and under $\mathrm{H}_0 : \boldsymbol{\theta} \in \omega$,

$$-2\ln\hat{\lambda} = 2\left[\ln\mathcal{L}(\boldsymbol{x}; \hat{\boldsymbol{\theta}}) - \ln\mathcal{L}(\boldsymbol{x}; \hat{\boldsymbol{\theta}}_\omega)\right] \dot{\sim} \chi^2_r.$$

Thus, for a likelihood ratio test of approximate size α, one would reject $\mathrm{H}_0 : \boldsymbol{\theta} \in \omega$ in favor of $\mathrm{H}_1 : \boldsymbol{\theta} \in \bar{\omega}$ when $-2\ln\hat{\lambda} > \chi^2_{r,1-\alpha}$.

1.5.4.2 Wald Test

Let the $(1 \times r)$ row vector $\boldsymbol{R}(\boldsymbol{\theta})$ be defined as

$$\boldsymbol{R}(\boldsymbol{\theta}) = [R_1(\boldsymbol{\theta}), R_2(\boldsymbol{\theta}), \ldots, R_r(\boldsymbol{\theta})].$$

Also, let the $(r \times p)$ matrix $\boldsymbol{T}(\boldsymbol{\theta})$ have (i, j) element equal to $[\partial R_i(\boldsymbol{\theta})]/\partial\theta_j, i = 1, 2, \ldots, r$ and $j = 1, 2, \ldots, p$.

And, let the $(r \times r)$ matrix $\boldsymbol{\Lambda}(\boldsymbol{\theta})$ have the structure

$$\boldsymbol{\Lambda}(\boldsymbol{\theta}) = \boldsymbol{T}(\boldsymbol{\theta})\boldsymbol{\mathcal{I}}^{-1}(\boldsymbol{\theta})\boldsymbol{T}'(\boldsymbol{\theta})$$

when using expected information, and have the structure

$$\boldsymbol{\Lambda}(\boldsymbol{x};\boldsymbol{\theta}) = \boldsymbol{T}(\boldsymbol{\theta})\boldsymbol{I}^{-1}(\boldsymbol{x};\boldsymbol{\theta})\boldsymbol{T}'(\boldsymbol{\theta})$$

when using observed information.

Then, the Wald test statistic $\hat{W}, 0 < \hat{W} < +\infty$, for testing $\mathrm{H}_0 : \boldsymbol{\theta} \in \omega$ versus $\mathrm{H}_1 : \boldsymbol{\theta} \in \bar{\omega}$ is defined as

$$\hat{W} = \boldsymbol{R}(\hat{\boldsymbol{\theta}})\boldsymbol{\Lambda}^{-1}(\hat{\boldsymbol{\theta}})\boldsymbol{R}'(\hat{\boldsymbol{\theta}})$$

when using expected information, and is defined as

$$\hat{W} = \boldsymbol{R}(\hat{\boldsymbol{\theta}})\boldsymbol{\Lambda}^{-1}(\boldsymbol{x};\hat{\boldsymbol{\theta}})\boldsymbol{R}'(\hat{\boldsymbol{\theta}})$$

when using observed information.

Under certain regularity conditions, for large n and under $\mathrm{H}_0 : \boldsymbol{\theta} \in \omega$, $\hat{W} \dot{\sim} \chi^2_r$. Thus, for a Wald test of approximate size α, one would reject $\mathrm{H}_0 : \boldsymbol{\theta} \in \omega$ in favor of $\mathrm{H}_1 : \boldsymbol{\theta} \in \bar{\omega}$ when $\hat{W} > \chi^2_{r,1-\alpha}$.

1.5.4.3 Score Test

The score test statistic $\hat{S}, 0 < \hat{S} < +\infty$, for testing $\mathrm{H}_0 : \boldsymbol{\theta} \in \omega$ versus $\mathrm{H}_1 : \boldsymbol{\theta} \in \bar{\omega}$ is defined as

$$\hat{S} = \mathbf{S}(\hat{\boldsymbol{\theta}}_\omega)\boldsymbol{\mathcal{I}}^{-1}(\hat{\boldsymbol{\theta}}_\omega)\mathbf{S}'(\hat{\boldsymbol{\theta}}_\omega)$$

when using expected information, and is defined as

$$\hat{S} = \mathbf{S}(\hat{\boldsymbol{\theta}}_\omega)\boldsymbol{I}^{-1}(\boldsymbol{x};\hat{\boldsymbol{\theta}}_\omega)\mathbf{S}'(\hat{\boldsymbol{\theta}}_\omega)$$

when using observed information.

Under certain regularity conditions, for large n and under $\mathrm{H}_0 : \boldsymbol{\theta} \in \omega$, $\hat{S} \dot{\sim} \chi^2_r$. Thus, for a score test of approximate size α, one would reject $\mathrm{H}_0 : \boldsymbol{\theta} \in \omega$ in favor of $\mathrm{H}_1 : \boldsymbol{\theta} \in \bar{\omega}$ when $\hat{S} > \chi^2_{r,1-\alpha}$.

Example. As an example, let $X_1, X_2, \ldots, X_n$ constitute a random sample of size n from a $\mathrm{N}(\mu, \sigma^2)$ parent population. Consider testing the composite null hypothesis $\mathrm{H}_0 : \mu = \mu_0, 0 < \sigma^2 < +\infty$, versus the composite alternative hypothesis $\mathrm{H}_1 : \mu \neq \mu_0, 0 < \sigma^2 < +\infty$. Note that this test is typically called a test of $\mathrm{H}_0 : \mu = \mu_0$ versus $\mathrm{H}_1 : \mu \neq \mu_0$.

It is straightforward to show that the vector $\hat{\boldsymbol{\theta}}$ of MLEs of μ and σ^2 for the unrestricted parameter space Ω is equal to

$$\hat{\boldsymbol{\theta}} = (\hat{\mu}, \hat{\sigma}^2) = \left[\bar{X}, \left(\frac{n-1}{n}\right) S^2\right],$$

where $\bar{X} = n^{-1}\sum_{i=1}^{n} X_i$ and $S^2 = (n-1)^{-1}\sum_{i=1}^{n}(X_i - \bar{X})^2$.

Then, it can be shown directly that

$$-2\ln\hat{\lambda} = n\ln\left[1 + \frac{T_{n-1}^2}{(n-1)}\right],$$

where

$$T_{n-1} = \frac{(\bar{X} - \mu_0)}{S/\sqrt{n}} \sim t_{n-1} \text{ under } \mathrm{H}_0 : \mu = \mu_0;$$

thus, the likelihood ratio test is a function of the usual one-sample t-test in this simple situation.

In this simple situation, the Wald test is also a function of the usual one-sample t-test since

$$\hat{W} = \left(\frac{n}{n-1}\right) T_{n-1}^2.$$

In contrast, the score test statistic has the structure

$$\hat{S} = \left[\frac{(\bar{X} - \mu_0)}{\hat{\sigma}_\omega/\sqrt{n}}\right]^2,$$

where

$$\hat{\sigma}_\omega^2 = n^{-1}\sum_{i=1}^{n}(X_i - \mu_0)^2$$

is the estimator of σ^2 under the null hypothesis $\mathrm{H}_0 : \mu = \mu_0$.

Although all three of these ML-based hypothesis-testing methods (the likelihood ratio test, the Wald test, and the score test) are *asymptotically* equivalent, their use can lead to different conclusions in some actual data-analysis scenarios.

Chapter 2

Basic Probability Theory

2.1 Exercises

Exercise 2.1. Prove that

$$\text{pr}(\text{A} \cup \text{B}|\text{C}) = \text{pr}(\text{A}|\text{C}) + \text{pr}(\text{B}|\text{C}) - \text{pr}(\text{A} \cap \text{B}|\text{C}).$$

Exercise 2.2. Consider two events, denoted event A and event B.

(a) If events A and B are independent, prove that the events $\bar{\text{A}}$ and $\bar{\text{B}}$ are independent.

(b) If events A and B are independent, prove that the events A and $\bar{\text{B}}$ are independent and that the events $\bar{\text{A}}$ and B are independent.

Exercise 2.3. Out of 20 female patients with cervical cancer, 12 have Stage I cervical cancer and 8 have Stage II cervical cancer. Ten of these 20 cervical cancer patients are to be randomly chosen to receive a new chemotherapy treatment.

(a) What is the probability that at least 5 patients with Stage II cervical cancer receive the new chemotherapy treatment?

(b) What is the probability that at least 2 patients with Stage I cervical cancer receive the new chemotherapy treatment?

(c) What is the probability that only patients with Stage I cervical cancer receive the new chemotherapy treatment?

(d) What is the probability that the set of 10 patients who are chosen to receive the new chemotherapy treatment includes at least 4 patients with Stage I cervical cancer and at least 4 patients with Stage II cervical cancer?

Exercise 2.4. Suppose that a certain hospital is interested in purchasing a large lot of 100 kidney dialysis machines from a manufacturer of such machines. Suppose that the hospital agrees to buy the entire lot of machines if *none* of the machines in a random sample of $n = 10$ machines selected from the lot of 100 is found to be defective.

(a) If, actually, exactly 5 of the 100 machines are defective, what is the probability that the hospital will purchase the entire lot of 100 kidney dialysis machines?

(b) Given the sampling plan described in part (a), what is the *smallest* number of defective machines that can be in the lot of 100 machines such that the probability is no more than 0.20 that the hospital will purchase the entire lot of 100 kidney dialysis machines?

Exercise 2.5. A *binary classifier* is an algorithm for assigning an individual to one of two populations (classes) based on observed characteristics of the individual, when it is *not* known to which population the individual actually belongs. As an example, suppose that a certain population of 1,000,000 people (designated Population I) contains 100,000 people with a certain genetic trait. And, suppose that another population of 1,000,000 people (designated Population II) contains 200,000 people with this genetic trait.

(a) Suppose that an individual is observed to have this genetic trait. Assuming that it is *a priori* equally likely for this person to be a member of either Population I or Population II, find the numerical value of the *a posteriori* probability that this individual is a member of Population I.

(b) Now, suppose that this individual has an *a priori* probability of 2/3 of being a member of Population I and so has an *a priori* probability of 1/3 of being a member of Population II. Then, find the exact numerical value of the *a posteriori* probability that this individual is a member of Population I.

For numerous examples of statistical classification, see Hastie, Tibshirani, and Friedman (2009).

Exercise 2.6. In certain industrial settings, a so-called *torture test* is used to assess the durability of products (e.g., car doors, electrical switches, etc.). Suppose that a car door is opened and closed repeatedly by a mechanical arm until it breaks. Assume that the "first open-then close" repetitive trials operate independently, so that the probability of the car door breaking on any particular trial does not vary from trial to trial; further, assume that the probability of the car door breaking on any particular trial has the value 0.0005.

(a) What is the probability that the car door breaks during the 1000-th trial?

(b) What is the probability that the car door breaks before the 1001-th trial starts?

(c) Comment on the reasonableness of the assumptions being made.

Exercise 2.7. A random number N of balanced dice are tossed, where

$$\text{pr}(N = n) = \pi(1 - \pi)^{n-1}, n = 1, 2, \ldots, \infty, \text{ and } 0 < \pi < 1.$$

Find an explicit expression for the probability θ_k that the largest number shown by any of the dice does not exceed $k, k = 1, 2, \ldots, 6$.

Exercise 2.8. Consider three events, denoted A, B, and C. Suppose that $\text{pr}(\text{A}|\text{C}) = 0.90, \text{pr}(\text{A}|\bar{\text{C}}) = 0.06, \text{pr}(\text{B}|\text{C}) = 0.95, \text{pr}(\text{B}|\bar{\text{C}}) = 0.08$, and $\text{pr}(\text{C}) = 0.01$. Further, suppose that events A and B are conditionally independent given event C, and that events A and B are also conditionally independent given event $\bar{\text{C}}$. Are A and B unconditionally independent events? Provide a more general interpretation for these numerical findings.

Exercise 2.9. In the United States (U.S.), it is known that the probability of a human birth resulting in twins is about 0.012. Given that a human birth results in twins, the probability is 1/3 that they are identical (one-egg) twins, and the probability is 2/3 that they are fraternal (two-egg) twins. Identical twins are necessarily of the same sex, with male and female pairs of identical twins being equally likely to occur. Also, given that a pair of fraternal twins is born, the probability is 1/4 that they are both females, the probability is 1/4 that they are both males, and the probability is 1/2 that there is one male and one female.

Now, consider the following events:

Event T: "a U.S. birth results in twins"

Event I: "a U.S. birth results in identical twins"

Event F: "a U.S. birth results in fraternal twins"

Event M: "a U.S. birth results in twin males"

Find the numerical values of the following probabilities:

(a) pr(I)

(b) pr(F)

(c) pr(M)

(d) $\text{pr}(\text{F}|\overline{\text{M}})$

(e) pr(I∩M|T)

(f) pr(I ∪ M|T)

Exercise 2.10. Suppose that a balanced coin is tossed n times, $n \geq 2$. Let A be the event that "at least one head and at least one tail are obtained among these n tosses," and let B be the event that "there is at most one tail obtained among these n tosses." Find the value of n such that A and B are independent events.

Exercise 2.11. For families in the United States (U.S.) with at least one child, assume that any such U.S. family has exactly k children with probability $(0.50)^k, k = 1, 2, \ldots, \infty$. Also, assume that the probability that any child is male is equal to 0.50.

(a) For U.S. families with at least one child, find the probability of the event "no male children."

(b) For U.S. families with at least one child, find the probability of the event "at least one male child and at least one female child."

(c) It is known that a randomly chosen U.S. family has at least one male child and at least one female child. Find the probability that this family has at most three children.

Exercise 2.12. In a certain large area of the United States, the race distribution is 45% Caucasian (C), 25% Hispanic (H), 20% African-American (A), and 10% Native-American (N). Four *unrelated* individuals are selected *independently* from this area of the United States.

Find numerical values for the probabilities of the following events:

(a) All four individuals are of the same race.

(b) Exactly two (and only two) of the four individuals are of the same race.

(c) At least two of the four individuals are not Caucasian.

(d) Exactly two of the four individuals are Caucasian given that all four individuals are each known to be either Caucasian or Hispanic.

Exercise 2.13. Suppose that a balanced die is tossed four times.

(a) Find the numerical value of the probability that exactly one of the four numbers obtained is either a 5 or a 6.

(b) Given that all four numbers obtained are different from each other, find the probability that exactly one of the four numbers obtained is either a 5 or a 6.

Exercise 2.14. A group of 300 males is cross-classified based on the presence or absence of the following three factors:

1. carrier (C) or not ($\bar{\text{C}}$) of the AIDS virus;
2. homosexual (H) or not ($\bar{\text{H}}$);
3. IV drug user (D) or not ($\overline{\text{D}}$).

The data appear in tabular form below:

H

	D	$\overline{\text{D}}$	
C	100	10	
$\bar{\text{C}}$	50	40	
			200

$\bar{\text{H}}$

	D	$\overline{\text{D}}$	
C	40	10	
$\bar{\text{C}}$	20	30	
			100

Using these tabulated data, find numerical values for the following probabilities:

(a) $\text{pr}(\text{C}|\text{H} \cap \text{D})$

(b) $\text{pr}(\text{C} \cup \text{D}|\bar{\text{H}})$

(c) $\text{pr}(\text{H}|\bar{\text{C}})$

(d) $\text{pr}(\overline{\text{C} \cap \text{H}}|\text{D})$

(e) $\text{pr}(\text{C} \cup \text{D} \cup \text{H})$

(f) $\text{pr}[\text{C} \cup (\text{H} \cap \text{D})]$

Exercise 2.15. Suppose that $n(\geq 3)$ fair (or unbiased) coins are tossed simultaneously. Given that *at least* $(n-1)$ coins of the n coins show either all heads or all tails, find an explicit expression (as a function of n) for the probability that all n coins show either all heads or all tails.

Exercise 2.16. Suppose that n individual items coming one-by-one off a certain production line are sequentially examined, and further suppose that each item is found upon examination to be either defective (with probability π) or non-defective [with probability $(1-\pi)$], $0 < \pi < 1$. Assume that items are produced independently of one another, so that whether or not any particular item is defective has no effect on whether or not any other particular item is defective. Find expressions for the following probabilities:

(a) the n items are found to be either all defective or all non-defective;

(b) for $0 \leq r \leq n$, only the first r items are found to be defective among the n items that are sequentially examined;

(c) for $0 \leq r \leq n$, there are no more than r defective items found among the n items that are sequentially examined;

(d) the first s items are found to be defective, and there are a total of $r (0 \leq s \leq r \leq n)$ defective items found among the n items that are sequentially examined;

(e) the first s items are found to be defective, and there are at least $r (0 \leq s \leq r \leq n)$ defective items found among the n items that are sequentially examined.

Exercise 2.17. Consider the following three events, denoted A, B, and C. Find a necessary and sufficient condition for which the equality $\text{pr}(\text{A}|\text{B} \cap \text{C}) = \text{pr}(\text{A}|\text{C})$ is equivalent to the equality $\text{pr}(\text{A}|\text{B}) = \text{pr}(\text{A})$. In other words, find a necessary and sufficient condition for which the conditional independence of events A and B given event C is equivalent to the unconditional independence of events A and B. Then, use this result to find a sufficient, but not necessary, condition for such an equivalence.

Exercise 2.18. Consider the following three events, denoted A, B, and C.

(a) The events A and B are said to be *conditionally* independent given that event C has occurred if any one of the following three equalities holds: $\text{pr}(\text{A} \cap \text{B}|\text{C}) = \text{pr}(\text{A}|\text{C})\text{pr}(\text{B}|\text{C})$; $\text{pr}(\text{A}|\text{B} \cap \text{C}) = \text{pr}(\text{A}|\text{C})$; and $\text{pr}(\text{B}|\text{A} \cap \text{C}) = \text{pr}(\text{B}|\text{C})$. Show that these equalities are equivalent (i.e., if any one of the three equalities holds, show that the other two equalities must also hold).

(b) Suppose that a balanced die is rolled exactly one time. Let A be the event that "an even number is rolled"; let B be the event that "a number greater than 3 is rolled"; and let C be the event that "a number greater than 4 is rolled". Are events A and B conditionally independent given that event C has occurred? Are events A and B conditionally independent given that event $\bar{\text{C}}$ has occurred?

Exercise 2.19. In a certain chemical industry, it is known that 5% of all workers are exposed to a high (H) daily concentration level of a certain potential carcinogen (i.e., are members of Group H), that 15% of all workers are exposed to an intermediate (I) daily concentration level (i.e., are members of Group I), that 20% of all workers are exposed to a low (L) daily concentration level (i.e., are members of Group L), and that the remaining 60% of all workers are unexposed (U) to this potential carcinogen (i.e., are members of Group U). Suppose that four workers are randomly chosen from a very large population of such chemical industry workers.

(a) What is the probability that all four randomly chosen workers are members of the same group?

(b) Given that all four randomly chosen workers are exposed to non-zero levels of this potential carcinogen, what is the probability that exactly two of these four workers are members of Group H?

(c) Suppose that C is the event that a worker in this chemical industry develops cancer. Let π_{H} =pr(C|H)= 0.002 be the conditional probability that a worker in Group H develops cancer. Similarly, let π_{I} =pr(C|I)= 0.001, π_{L} =pr(C|L)= 0.0001, and π_{U} =pr(C|U)= 0.00001. If a worker in this chemical industry develops cancer, what is the probability that this worker is a member of either Group H or Group I?

Exercise 2.20. One of the greatest athletic feats ever is the 56-game hitting streak by Joe DiMaggio of the New York Yankees major league baseball team in 1941. During that hitting streak, "Joltin' Joe" had *at least* one hit (either a single, a double, a triple, or a homerun) in each game for 56 consecutive games, a record which still stands and which will likely never be broken. During that amazing streak, Joe DiMaggio had 223 official at bats and produced 91 hits, for a batting average of 91/223=0.408. To appreciate the rarity of this phenomenal performance, calculate the probability π of the occurrence of this hitting streak under the following (only approximately valid) set of assumptions:

1) Joe's performance during any one official at bat (i.e., opportunity to be credited with a hit) is independent of his performance during any other official at bat during this 56 game hitting streak.
2) Joe had 3 official at bats in 13 of the 56 games, 4 official at bats in 31 of the 56 games, and 5 official at bats in 12 of the 56 games.
3) Joe's probability of getting a hit for any official at bat during this streak is 0.408.

Exercise 2.21. An urn contains a total of N balls, of which B are black in color, R are red in color, and W are white in color. One ball (call it Ball #1) is randomly selected from this urn and its color is noted. If Ball #1 is black, then this black ball is returned to the urn, along with K additional black balls; if Ball #1 is red, then this red ball is returned to the urn, along with K additional red balls; and if Ball #1 is white, then this white ball is returned to the urn, along with K additional white balls. Then, a second ball (call it Ball #2) is randomly selected from this urn.

(a) Develop an explicit expression for the probability α that one of the two balls randomly selected from the urn is black and that the other ball randomly selected is red.

(b) Develop an explicit expression for the probability β that Ball #2 is black.

(c) Develop an explicit expression for the probability γ that Ball #1 is black given that Ball #2 is black.

Exercise 2.22. A box contains two coins of identical size. One coin has heads on both its sides, while the other coin has heads on one side and tails on the other. A coin is selected randomly from the box, and the upturned face of this coin shows heads. Find the numerical value of the probability that the other side of this coin is also heads.

Exercise 2.23. A common design for epidemiologic studies is the case-control design in which investigators sample subjects with disease (cases) and without disease (controls), and then determine the proportion of subjects in each group previously exposed to a certain agent. A primary aim of such studies is to determine if there is an association between exposure status and disease status. A popular parametric measure of such an association is the *exposure odds ratio* ($\mathrm{OR_e}$), a parameter that compares the odds of exposure for cases to the odds of exposure for controls; specifically,

$$\mathrm{OR_e} = \frac{\mathrm{pr}(\mathrm{E}|\mathrm{D})/\mathrm{pr}(\overline{\mathrm{E}}|\mathrm{D})}{\mathrm{pr}(\mathrm{E}|\overline{\mathrm{D}})/\mathrm{pr}(\overline{\mathrm{E}}|\overline{\mathrm{D}})},$$

where the event E denotes "exposed," $\overline{\mathrm{E}}$ denotes "not exposed," D denotes "disease," and $\overline{\mathrm{D}}$ denotes "no disease." Another measure of association between exposure status and disease status is the *risk ratio*

$$\mathrm{RR} = \frac{\mathrm{pr}(\mathrm{D}|\mathrm{E})}{\mathrm{pr}(\mathrm{D}|\overline{\mathrm{E}})},$$

which compares the probability (or risk) of disease among exposed subjects to the risk of disease among non-exposed subjects. Because case-control studies involve sampling conditional on disease status rather than sampling conditional on exposure status, the risk ratio cannot be directly estimated using a case-control study design. However, under the so-called *rare disease assumption*, which states that $\mathrm{pr}(\mathrm{D}|\mathrm{E}) \approx \mathrm{pr}(\mathrm{D}|\overline{\mathrm{E}}) \approx 0$, the exposure odds ratio mathematically approximates the risk ratio. The rare disease assumption is appropriate in many practical situations, such as when evaluating risk factors for rare diseases such as cancer and certain genetic disorders.

(a) Show that $\text{OR}_\text{e} \approx \text{RR}$ under the rare disease assumption.

(b) Suppose that

$$\text{pr}(\text{E}|X = x) = \left[1 + \text{e}^{-(\alpha+\beta x)}\right]^{-1},$$

where $X = 1$ for a diseased person and $X = 0$ for a non-diseased person. How are OR_e and β related?

For detailed information about issues regarding the use of case-control study data to make statistical inferences about risk ratios, see the books by Breslow and Day (1980) and Kleinbaum, Kupper, and Morgenstern (1982).

Exercise 2.24. In a certain small rural area of the United States, there are three golf courses (designated course #1, course #2, and course #3). A survey of adult residents of this rural area indicates that 18% of these adult residents play course #1, that 15% play #2, that 12% play #3, that 9% play both #1 and #2, that 6% play both #1 and #3, that 5% play both #2 and #3, and that 2% play all three courses. If an adult resident of this small rural area is randomly chosen, find the probability that this adult resident:

(a) plays none of these three courses;

(b) plays exactly one of these three courses;

(c) plays only #1 and #2 given that this adult resident plays at least one of these three courses.

Exercise 2.25. There are three cabinets, each containing four drawers. For one of the cabinets, one drawer contains a gold coin and the other three drawers each contain a silver coin. For another cabinet, one drawer contains a silver coin and the other three drawers each contain a gold coin. And for the remaining cabinet, two of the drawers each contain a gold coin and two of the drawers each contain a silver coin.

Suppose that a cabinet is randomly chosen and then a randomly chosen drawer is opened and found to contain a silver coin. Find the numerical value of the probability θ that the next randomly chosen drawer opened for this particular cabinet contains a gold coin.

Exercise 2.26. Suppose that $k (2 \leq k \leq 6)$ balanced dice are tossed simultaneously. Given that no two of these k dice show the same number, find an explicit expression for the probability θ_k that one of the k dice shows the number 6.

Exercise 2.27. Consider a finite-sized population containing $N(1 < N < \infty)$ members. A sample of size $n(1 \leq n < N)$ is said to be *randomly selected* from this population if all samples of size n are *equally likely* to be selected.

(a) What are the probabilities of obtaining a particular sample of size n when population members are randomly selected with replacement (WR) and without replacement (WOR)?

(b) What are the probabilities of obtaining a particular member of this population when population members are randomly selected WR and WOR?

Exercise 2.28. Suppose that $k(1 \leq k \leq n)$ balls are sequentially randomly tossed at n urns, and assume that each ball must fall into exactly one of the n urns.

(a) Find an explicit expression for the probability $\theta(n, k)$ that no urn contains more than one ball.

(b) Use the result in part (a) to find the numerical value of the probability γ that at least two people in a group of five have a birthday during the same month.

Exercise 2.29. Suppose that an urn contains four white balls and two black balls. If balls are selected from this urn sequentially one at a time without replacement, provide an explicit expression (as a function of n) for the probability (say, P_n) that the n-th ball selected is the last black ball remaining in the urn. What is the minimum number (say, n^*) of balls that must be selected from the urn in this manner so that $\text{P}_n \geq \frac{1}{3}$?

Exercise 2.30*. When a balanced die is tossed repeatedly, show that the probability that the number 1 first appears on the i-th toss *and* that the number 2 first appears on the j-th toss can be written as an explicit function of $\min\{i, j\}$ and $\max\{i, j\}$ for all $i \neq j$, where i and j are non-negative integers.

Exercise 2.31*. A marine biologist is interested in studying, over a short time period of two weeks, a very large population of aquatic turtles inhabiting a particular chemically polluted pond. During this short time period, it is reasonable to assume that the size of this large population of aquatic turtles may change negligibly due to death, but *not* due to birth, immigration, or emigration. At the start of the two-week period, suppose that this marine biologist randomly selects a small set of n aquatic turtles, marks them with distinct identifying numbers, and then returns them to the pond. Then, on each of two subsequent occasions (at the end of the first week and also at the end of the second week), this marine biologist takes a small random sample of n aquatic turtles from this pond, records which of the marked turtles are contained in that particular random sample, and then returns to the pond all the turtles in that particular random sample.

Assume that each turtle functions completely independently of every other turtle in the pond, assume that each turtle in the pond has probability $\gamma(0 < \gamma < 1)$ of surviving for a week in this polluted pond, and assume that each *marked* turtle has probability $\delta(0 < \delta < 1)$ of being contained in any random sample of size n given that it is alive at the time of that sampling occasion.

As a function of $\alpha = \gamma\delta$ and $\beta = \gamma(1 - \delta)$, develop explicit expressions for the probability that a marked turtle is a member only of the random sample taken at the end of the first week, for the probability that a marked turtle is a member only of the random sample taken at the end of the second week, for the probability that a marked turtle is a member of both of these random samples, and for the probability that a marked turtle is a member of neither of these random samples.

Exercise 2.32*. An experiment consists of tossing three balanced dice simultaneously. Let A be the event that the same number appears on exactly two of the three dice when the experiment is conducted. Find the smallest number, say n^*, of mutually independent repetitions of this experiment that would be required so that the probability is at least 0.90 that event A will occur at least twice during these n^* repetitions of the experiment.

Exercise 2.33*. Suppose that a particular clinical trial is designed to compare a new chemotherapy treatment to a standard chemotherapy treatment for treating Hodgkin's disease. At the beginning of this clinical trial, suppose that patients are assigned to the new treatment with probability $\pi, 0 < \pi < 1$, and patients are assigned to the standard treatment with probability $(1 - \pi)$. If a patient receives the new treatment, then that patient has probability θ_1 of going into remission; if a patient receives the standard treatment, then that patient has probability θ_0 of going into remission.

(a) What is the probability that a patient participating in this clinical trial actually goes into remission?

(b) If a patient participating in this clinical trial actually goes into remission, what is the probability that this patient actually received the new treatment?

(c) If two patients *independently* participating in this clinical trial both go into remission, what is the probability that one patient received the new treatment and the other patient received the standard treatment?

Exercise 2.34*. Suppose that each repetition of an experiment can result in only one of two possible outcomes, say, outcome A and outcome B. If any repetition of the experiment results in outcome A, then the probability is $\alpha(0.50 < \alpha < 1)$ that the immediately following repetition of the experiment will also result in outcome A. If any repetition of the experiment results in outcome B, then the probability is $\beta(0.50 < \beta < 1)$ that the immediately following repetition of the experiment will also result in outcome B. In particular, the outcome for any repetition of the experiment is only affected by the outcome on the immediately preceeding repetition of the experiment.

Find an explicit expression for the probability θ_n that the n-th repetition of the experiment results in outcome A given that the first repetition of the experiment results in outcome A, and also find an explicit expression for θ_n given that the first repetition of the experiment results in outcome B. Find the limiting values of these two expressions for θ_n as $n \to \infty$, and comment on your findings.

Exercise 2.35*. The *attributable risk* parameter α is a widely used epidemiologic measure designed to quantify the public health consequences of an established association between a particular exposure and a particular disease. More specifically, for all subjects with this particular disease in the population under study, α is defined as the proportion of all those subjects whose disease can be directly attributed to the particular exposure in question, namely,

$$\alpha = \frac{\text{pr(D)} - \text{pr(D}|\bar{\text{E}})}{\text{pr(D)}},$$

where pr(D) is the *prevalence* of the disease in the population under study [which has a proportion pr(E) of subjects who are exposed and a proportion $\text{pr}(\bar{\text{E}}) = 1 - \text{pr(E)}$ of subjects who are not exposed] and where $\text{pr}(\text{D}|\bar{\text{E}})$ is the proportion of unexposed subjects who have the disease.

Let the *risk ratio* parameter be defined as $\theta = \text{pr(D|E)}/\text{pr}(\text{D}|\bar{\text{E}})$, where $\theta \geq 1$. Then, show that α can be equivalently written in the form

$$\alpha = \frac{\text{pr(E)}(\theta - 1)}{1 + \text{pr(E)}(\theta - 1)},$$

or in the form

$$\alpha = \frac{\text{pr(E|D)}(\theta - 1)}{\theta}.$$

For more information about the use of attributable risk (also known as the *etiologic fraction*) in public health research, see Kleinbaum, Kupper, and Morgenstern (1982), Chapter 9.

Exercise 2.36*. Suppose that two players, Player A and Player B, play a series of games, with each player betting the same dollar amount on each game. For each game, the probability that Player A wins the game is $\pi, 0 < \pi < 1$, and so the probability that Player B wins the game is $(1 - \pi)$. Further, assume that Player A has a total of a dollars to bet, and that Player B has a total of b dollars to bet. The goal is to develop explicit expressions for the probability that Player A is ruined (i.e., Player A loses all a dollars) and for the probability that Player B is ruined (i.e., Player B loses all b dollars). This is the classic *Gambler's Ruin* problem.

(a) With θ_x denoting the probability that Player A is ruined when Player A has exactly x dollars, show that

$$\theta_x = \pi\theta_{x+1} + (1-\pi)\theta_{x-1}, x = 1, 2, \ldots, (a+b-1),$$

where $\theta_0 = 1$ and $\theta_{a+b} = 0$.

(b) Show that $\theta_x = \alpha + \beta\left(\frac{1-\pi}{\pi}\right)^x$ is a general solution to the difference equation given in part (a).

(c) Find an explicit expression for θ_x as a function of x, π, a, and b.

(d) Use the result in part (c) to find explicit expressions for the probability that Player A is ruined and for the probability that Player B is ruined. Show that these two probabilities add to 1, and then comment on this finding. What are the values of these two probabilities when $\pi = 1/2$?

(e) Suppose that Player B represents a gambling casino (typically called *the house*), so that Player B has an unlimited amount of money and can play indefinitely. Find the limit of θ_a as $b \to \infty$ if $\pi \leq 0.50$ and if $\pi > 0.50$, and then comment on your findings.

Exercise 2.37*. Let D be the event that a person develops a particular disease, let E be the event that this person is exposed to an *observable* risk factor suspected of being a cause of this disease, and let U be the event that this person is exposed to an *unobservable* risk factor also suspected of being a cause of this disease. As an example, the disease of interest could be lung cancer, the observable risk factor could be cigarette smoke, and the unobservable risk factor could be a certain genetic trait. Let

$$\theta = \frac{\text{pr(D|E)}}{\text{pr(D}|\overline{\text{E}})}, \theta \geq 1,$$

be the *risk ratio* quantifying the strength of the association between the observable risk factor and the disease, and let

$$\theta^* = \frac{\text{pr(D|U)}}{\text{pr(D}|\overline{\text{U}})}, \theta^* \geq 1,$$

be the *risk ratio* quantifying the strength of the association between the unobservable risk factor and the disease.

In what follows, assume that the events D and E are independent given U; in particular, assume that

$$\pi_0 = \text{pr}(\text{D}|\text{E} \cap \overline{\text{U}}) = \text{pr}(\text{D}|\overline{\text{E}} \cap \overline{\text{U}}) = \text{pr}(\text{D}|\overline{\text{U}})$$

and

$$\pi_1 = \pi_0\theta^* = \text{pr}(\text{D}|\text{E} \cap \text{U}) = \text{pr}(\text{D}|\overline{\text{E}} \cap \text{U}) = \text{pr}(\text{D}|\text{U}).$$

(a) If $\gamma_1 = \text{pr}(\text{U}|\text{E})$ and $\gamma_0 = \text{pr}(\text{U}|\bar{\text{E}})$, show that

$$\theta = \frac{\gamma_1(\theta^* - 1) + 1}{\gamma_0(\theta^* - 1) + 1}.$$

(b) Show that $\theta \leq \theta^*$, that θ is a monotonically increasing function of θ^*, and that $\theta \leq \gamma_1/\gamma_0$.

(c) Discuss how the results in part (b) can be used to counteract arguments that the strong association between smoking and lung cancer (namely, $\theta \approx 9$) can be explained away by some unobservable risk factor. For more details, see Cornfield (1959).

Exercise 2.38*. Mega Millions is a large multi-state lottery in the United States. To play, a person chooses five numbers *without replacement* from the set (call it Set #1) of the first 56 positive integers $\{1, 2, \ldots, 56\}$, and also chooses one number (the so-called *megaball number*) from the set (call it Set #2) of the first 46 positive integers $\{1, 2, \ldots, 46\}$.

The five winning numbers in Set #1 and the winning megaball number in Set #2 are drawn at random at the Mega Millions national lottery headquarters. Let x be the number of the five winning numbers in Set #1 that the person matches with his or her choices from Set #1, so that the possible values of x are 0, 1, 2, 3, 4, and 5; also, let $y = 1$ if the person matches the winning megaball number and let $y = 0$ otherwise.

The following (x, y) pairs are winning Mega Millions pairs: (3,0), (4,0), (5,0), (0,1), (1,1), (2,1), (3,1), (4,1), and (5,1). In addition, the lower the probability of a particular pair occurring, the larger is the amount of money to be won by matching that particular pair.

(a) For each of these winning pairs, find the numerical value of the probability that a person matches that particular pair when playing the Mega Millions lottery game one time.

(b) Find the numerical value of the overall probability of winning (i.e., of matching a winning pair) if a person plays this Mega Millions lottery game one time.

(c) What is the minimum number n^* of different sets of six numbers that a person has to choose to have a probability of at least 0.90 of winning?

Exercise 2.39*. A certain small clinical trial is designed to compare a new prescription-required appetite suppressant drug (Drug 1) to an available over-the-counter appetite suppressant drug (Drug 2). In this small clinical trial, n pairs of overweight adult males are formed; members of each pair are matched on weight, height, age, dietary and exercise habits, and other relevant variables. Then, one member of each pair is randomly assigned to receive Drug 1, and the other member of the pair receives Drug 2. Each of the $2n$ adult males then takes his assigned drug (one pill each morning) for exactly 60 days. At the end of 60 days, each of these $2n$ adult males is weighed, and the weight loss (in pounds) is recorded for each adult male.

To determine whether there is statistical evidence that Drug 1 is more effective than Drug 2, it is decided to count the number x of pairs for which the weight loss using Drug 1 is more than the weight loss using Drug 2.

(a) Assuming that the two drugs are equally effective, develop an explicit expression for

$$\theta(x^*|n) = \text{pr}(x \geq x^*|n),$$

where $x^*(0 \leq x^* \leq n)$ is a fixed value of x.

(b) Suppose that 15 pairs of adult males participate in this small clinical trial, and that 9 of these 15 pairs result in more weight loss using Drug 1 than using Drug 2. Use the result in part (a) to determine whether these data provide statistical evidence that Drug 1 is more effective than Drug 2.

(c) What is the smallest value of x^* for which $\theta(x^*|15) < 0.05$?

Exercise 2.40*. Suppose that $k(\geq 2)$ people participate in the following two-person game, with each person playing each of the other $(k-1)$ persons exactly one time.

For each game, each of the two persons playing flips a coin, with one person choosing "evens" and the other person choosing "odds". If the two coins match (i.e., either two heads or two tails are obtained), then the person who chose "evens" wins the game; if the two coins do not match, then the person who chose "odds" wins the game.

(a) Develop an explicit expression for the probability θ_k that one person wins exactly $(k-1)$ games, that one person wins exactly $(k-2)$ games, that one person wins exactly $(k-3)$ games, ..., that one person wins exactly one game, and that one person wins zero games. For example, if $k = 3$, then θ_3 is the probability that one person wins two games, that one person wins one game, and that one person wins zero games. Also, find the numerical values of θ_2 and θ_6.

(b) Find $\lim_{k\to\infty}\theta_k$.

Exercise 2.41*. In the dice game known as "craps," a player competes against the casino (called the "house") according to the following rules. If the player (called the "shooter" when rolling the pair of dice) rolls either a 7 or an 11 on the first roll of the pair of dice, the player wins the game; if the player rolls either a 2, 3, or 12 on this first roll, the player loses the game. If the player rolls a 4, 5, 6, 8, 9, or 10 on the first roll (such a number is called the "point"), the player then keeps rolling the pair of dice until either the point is rolled again (in which case the player wins the game) or until a 7 is rolled (in which case the player loses the game).

Suppose that a game of craps is played with a possibly *biased* pair of dice. In particular, if X denotes the number rolled using this possibly biased pair of dice, suppose that $\text{pr}(X = 7) = \pi, 0 \leq \pi \leq 1$, and that

$$\text{pr}(X = x) = \frac{(1-\pi)\min\{x-1, 13-x\}}{30}, x = 2, 3, 4, 5, 6, 8, 9, 10, 11, 12.$$

For this probability model [considered in detail by Bryson (1973)], the pair of dice are *unbiased* when $\pi = 1/6$.

(a) For this probability model, develop an explicit expression (as a function of π) for the probability $\theta(\pi)$ that the player wins the game.

(b) What is the numerical value of $\theta(1/6)$, the probability that the player wins the game when using an unbiased pair of dice? Find the numerical values of $\theta(0)$ and $\theta(1)$, and justify why these numerical values make sense.

Exercise 2.42*. Suppose that a gambler has a dollars to bet and bets one dollar on each play of a certain game. Let $\pi (0 < \pi < 1)$ be the probability of winning any one play of the game. Further, suppose that this gambler plays this game with the goal of accumulating b dollars, where $0 \leq a \leq b$.

(a) Let

$$\theta_a = \text{pr}\left(b \text{ dollars are accumulated} | a \text{ dollars to bet}\right).$$

Clearly, $\theta_0 = 0$ and $\theta_b = 1$. Show that

$$\theta_a = \pi\theta_{a+1} + (1-\pi)\theta_{a-1}, a = 1, 2, \ldots, (b-1).$$

(b) If $\pi = 1/2$, show that $\theta_a = a/b$ is the solution to the difference equations given in part (a). And, if $\pi \neq 1/2$, show that

$$\theta_a = \frac{\left(\frac{1-\pi}{\pi}\right)^a - 1}{\left(\frac{1-\pi}{\pi}\right)^b - 1}$$

is the solution to the difference equations given in part (a).

(c) Consider the following two scenarios:

Scenario I: $\pi = 0.50, a = \$100$, and $b = \$10,000$.

Scenario II: $\pi = 0.48, a = \$100$, and $b = \$200$.

Which of these two scenarios provides the better opportunity for the gambler to accumulate b dollars? Comment on your finding.

For more details related to this problem, see Coyle and Wang (1993).

2.2 Solutions to Odd-Numbered Exercises

Solution 2.1.

$$\begin{aligned}\text{pr}(\text{A} \cup \text{B}|\text{C}) &= \frac{\text{pr}[(\text{A} \cup \text{B}) \cap \text{C}]}{\text{pr}(\text{C})} = \frac{\text{pr}[(\text{A} \cap \text{C}) \cup (\text{B} \cap \text{C})]}{\text{pr}(\text{C})} \\ &= \frac{\text{pr}(\text{A} \cap \text{C}) + \text{pr}(\text{B} \cap \text{C}) - \text{pr}(\text{A} \cap \text{B} \cap \text{C})}{\text{pr}(\text{C})} \\ &= \frac{\text{pr}(\text{A}|\text{C})\text{pr}(\text{C}) + \text{pr}(\text{B}|\text{C})\text{pr}(\text{C}) - \text{pr}(\text{A} \cap \text{B}|\text{C})\text{pr}(\text{C})}{\text{pr}(\text{C})} \\ &= \text{pr}(\text{A}|\text{C}) + \text{pr}(\text{B}|\text{C}) - \text{pr}(\text{A} \cap \text{B}|\text{C}).\end{aligned}$$

Solution 2.3.

(a)

$$\sum_{j=5}^{8} \frac{\text{C}_j^8 \text{C}_{10-j}^{12}}{\text{C}_{10}^{20}}.$$

(b) This probability is equal to 1, since the set of 10 randomly chosen cervical cancer patients must always include at least 2 patients with Stage I cervical cancer (i.e., there are only 8 patients in total with Stage II cervical cancer).

(c)

$$\frac{\text{C}_{10}^{12} \text{C}_0^8}{\text{C}_{10}^{20}}.$$

(d)

$$\sum_{j=4}^{6} \frac{C_j^{12} C_{10-j}^{8}}{C_{10}^{20}}.$$

Solution 2.5.

(a) Let P_1 be the event that an individual is a member of Population I, let P_2 be the event that an individual is a member of Population II, and let G be the event that an individual has the genetic trait. Then,

$$\begin{aligned}\text{pr}(P_1|G) &= \frac{\text{pr}(P_1 \cap G)}{\text{pr}(G)} = \frac{\text{pr}(G|P_1)\text{pr}(P_1)}{\text{pr}(G|P_1)\text{pr}(P_1) + \text{pr}(G|P_2)\text{pr}(P_2)} \\ &= \frac{(0.10)(1/2)}{(0.10)(1/2) + (0.20)(1/2)} = \frac{1}{3},\end{aligned}$$

so that $\text{pr}(P_2|G) = 2/3$.

(b) When $P_1 = 2/3$ and $P_2 = 1/3$, then

$$\text{pr}(P_1|G) = \frac{(0.10)(2/3)}{(0.10)(2/3) + (0.20)(1/3)} = \frac{1}{2},$$

so that $\text{pr}(P_2|G) = 1/2$.

Solution 2.7. We have

$$\begin{aligned}\theta_k &= \text{pr}(\text{all } n \text{ dice show a number} \leq k|N = n)\text{pr}(N = n) \\ &= \sum_{n=1}^{\infty} \left(\frac{k}{6}\right)^n \pi(1-\pi)^{n-1} \\ &= \frac{k\pi}{6} \sum_{n=1}^{\infty} \left[\frac{k(1-\pi)}{6}\right]^{n-1} \\ &= \frac{k\pi/6}{\left[1 - \frac{k(1-\pi)}{6}\right]} \\ &= \frac{k\pi}{6 - k(1-\pi)}, k = 1, 2, \ldots, 6.\end{aligned}$$

As expected, $\theta_6 = 1$.

Solution 2.9.

(a) $\text{pr}(I) = \text{pr}[I \cap (T \cup \overline{T})] = \text{pr}(I \cap T) = \text{pr}(I|T)\text{pr}(T) = (1/3)(0.012) = 0.004.$

(b) Similarly, $\text{pr}(\text{F}) = \text{pr}(\text{F} \cap \text{T}) = \text{pr}(\text{F}|\text{T})\text{pr}(\text{T}) = (2/3)(0.012) = 0.008$. Or, since $\text{T} = \text{I} \cup \text{F}$, and events I and F are mutually exclusive, $\text{pr}(\text{F}) = \text{pr}(\text{T}) - \text{pr}(\text{I}) = 0.012 - 0.004 = 0.008$.

(c)

$$\begin{aligned}\text{pr}(\text{M}) &= \text{pr}[\text{M} \cap (\text{T} \cup \overline{\text{T}})] = \text{pr}(\text{M} \cap \text{T}) = \text{pr}[\text{M} \cap (\text{I} \cup \text{F})] \\ &= \text{pr}(\text{M} \cap \text{I}) + \text{pr}(\text{M} \cap \text{F}) = \text{pr}(\text{M}|\text{I})\text{pr}(\text{I}) + \text{pr}(\text{M}|\text{F})\text{pr}(\text{F}) \\ &= (1/2)(0.004) + (1/4)(0.008) = 0.004.\end{aligned}$$

(d)

$$\begin{aligned}\text{pr}(\text{F}|\overline{\text{M}}) &= \frac{\text{pr}(\text{F} \cap \overline{\text{M}})}{\text{pr}(\overline{\text{M}})} = \frac{\text{pr}(\overline{\text{M}}|\text{F})\text{pr}(\text{F})}{\text{pr}(\overline{\text{M}})} \\ &= \frac{\left(\frac{3}{4}\right)(0.008)}{0.996} = 0.0060.\end{aligned}$$

(e)

$$\begin{aligned}\text{pr}(\text{I} \cap \text{M}|\text{T}) &= \frac{\text{pr}(\text{I} \cap \text{M} \cap \text{T})}{\text{pr}(\text{T})} = \frac{\text{pr}(\text{M}|\text{I} \cap \text{T})\text{pr}(\text{I} \cap \text{T})}{\text{pr}(\text{T})} \\ &= \frac{\text{pr}(\text{M}|\text{I})\text{pr}(\text{I})}{\text{pr}(\text{T})} = \frac{\left(\frac{1}{2}\right)(0.004)}{0.012} = 1/6.\end{aligned}$$

(f)

$$\begin{aligned}\text{pr}(\text{I} \cup \text{M}|\text{T}) &= \text{pr}(\text{I}|\text{T}) + \text{pr}(\text{M}|\text{T}) - \text{pr}(\text{I} \cap \text{M}|\text{T}) \\ &= 1/3 + \text{pr}(\text{M}|\text{T}) - 1/6 = 1/6 + \text{pr}(\text{M}|\text{T}).\end{aligned}$$

Since

$$\text{pr}(\text{M}|\text{T}) = \frac{\text{pr}(\text{M} \cap \text{T})}{\text{pr}(\text{T})} = \frac{\text{pr}(\text{T}|\text{M})\text{pr}(\text{M})}{\text{pr}(\text{T})} = \frac{(1)(0.004)}{0.012} = 1/3,$$

it follows that $\text{pr}(\text{I} \cup \text{M}|\text{T}) = 1/6 + 1/3 = 1/2$.

Solution 2.11.

(a) pr(no male children)

$$\begin{aligned}&= \sum_{k=1}^{\infty} \text{pr}(\text{no males}|k \text{ children})\text{pr}(k \text{ children}) \\ &= \sum_{k=1}^{\infty} (0.50)^k (0.50)^k \\ &= \sum_{k=1}^{\infty} (0.25)^k = \frac{1}{3}.\end{aligned}$$

(b) pr(at least one male child and at least one female child)

$$= 1 - \text{pr(no male children)} - \text{pr(no female children)} = 1 - \frac{1}{3} - \frac{1}{3} = \frac{1}{3}.$$

(c) Let C_k be the event "exactly k children" and let A be the event "at least one male child and at least one female child." For families with at least one child,

$$\begin{aligned} \text{pr}(C_2 \cup C_3|A) &= \frac{\text{pr}[(C_2 \cup C_3) \cap A]}{\text{pr}(A)} = \frac{\text{pr}[(A \cap C_2) \cup (A \cap C_3)]}{\text{pr}(A)} \\ &= \frac{\text{pr}(A|C_2)\text{pr}(C_2) + \text{pr}(A|C_3)\text{pr}(C_3)}{\text{pr}(A)} \\ &= \frac{\text{pr}(A|C_2)\left[\left(\frac{1}{2}\right)^2\right] + \text{pr}(A|C_3)\left[\left(\frac{1}{2}\right)^3\right]}{(1/3)}. \end{aligned}$$

Now, $\text{pr}(A|C_2) = 1 - \text{pr(two boys}|C_2) - \text{pr(two girls}|C_2) = 1 - 2\left(\frac{1}{2}\right)^2 = \frac{1}{2}$.

And, $\text{pr}(A|C_3) = 1 - \text{pr(three boys}|C_3) - \text{pr(three girls}|C_3) = 1 - 2\left(\frac{1}{2}\right)^3 = \frac{3}{4}$.

Finally,

$$\text{pr}(C_2 \cup C_3) = \frac{(1/2)(1/4) + (3/4)(1/8)}{(1/3)} = 21/32.$$

Solution 2.13.

(a) Let A be the event that exactly one of the four numbers obtained is either a five or a six. Then,

$$\text{pr}(A) = 4\left(\frac{2}{6}\right)\left(\frac{4}{6}\right)^3 = \frac{32}{81} = 0.395.$$

(b) Let B be the event that all four numbers obtained are different from one another. Then,

$$\text{pr}(B) = (1)\left(\frac{5}{6}\right)\left(\frac{4}{6}\right)\left(\frac{3}{6}\right) = \frac{5}{18}.$$

And,

$$\text{pr}(A \cap B) = 4\left(\frac{2}{6}\right)\left(\frac{4}{6}\right)\left(\frac{3}{6}\right)\left(\frac{2}{6}\right) = \frac{4}{27}.$$

Finally,

$$\text{pr}(A|B) = \frac{\text{pr}(A \cap B)}{\text{pr}(B)} = \frac{4/27}{5/18} = \frac{8}{15} = 0.533.$$

Solution 2.15. Let A be the event that at least $(n-1)$ coins of the n coins show either all heads or all tails, and let B be the event that all n coins show either all heads or all tails. Then, we wish to find the numerical value of

$$\mathrm{pr}(\mathrm{B}|\mathrm{A}) = \frac{\mathrm{pr}(\mathrm{A} \cap \mathrm{B})}{\mathrm{pr}(\mathrm{A})}.$$

Now,

$$\mathrm{pr}(\mathrm{A} \cap \mathrm{B}) = \mathrm{pr}(\mathrm{B}) = \left(\frac{1}{2}\right)^n + \left(\frac{1}{2}\right)^n = \left(\frac{1}{2}\right)^{n-1}.$$

Moreover, event A will occur if either exactly $(n-1)$ heads are obtained, or if exactly $(n-1)$ tails are obtained, or if exactly n heads are obtained, or if exactly n tails are obtained, so that

$$\mathrm{pr}(\mathrm{A}) = n\left(\frac{1}{2}\right)^n + n\left(\frac{1}{2}\right)^n + \left(\frac{1}{2}\right)^n + \left(\frac{1}{2}\right)^n = (1+n)\left(\frac{1}{2}\right)^{n-1}.$$

Thus,

$$\mathrm{pr}(\mathrm{B}|\mathrm{A}) = \frac{(1/2)^{n-1}}{(1+n)(1/2)^{n-1}} = \frac{1}{(1+n)}, n = 3, 4, \ldots, \infty.$$

Solution 2.17. Now,

$$\begin{aligned}
\mathrm{pr}(\mathrm{A}|\mathrm{B} \cap \mathrm{C}) &= \mathrm{pr}(\mathrm{A}|\mathrm{C}) \Leftrightarrow \frac{\mathrm{pr}(\mathrm{A} \cap \mathrm{B} \cap \mathrm{C})}{\mathrm{pr}(\mathrm{B} \cap \mathrm{C})} = \frac{\mathrm{pr}(\mathrm{A} \cap \mathrm{C})}{\mathrm{pr}(\mathrm{C})} \\
&\Leftrightarrow \frac{\mathrm{pr}(\mathrm{C}|\mathrm{A} \cap \mathrm{B})\mathrm{pr}(\mathrm{A}|\mathrm{B})\mathrm{pr}(\mathrm{B})}{\mathrm{pr}(\mathrm{C}|\mathrm{B})\mathrm{pr}(\mathrm{B})} = \frac{\mathrm{pr}(\mathrm{C}|\mathrm{A})\mathrm{pr}(\mathrm{A})}{\mathrm{pr}(\mathrm{C})} \\
&\Leftrightarrow \mathrm{pr}(\mathrm{A}|\mathrm{B}) = \mathrm{pr}(\mathrm{A}) \left[\frac{\mathrm{pr}(\mathrm{C}|\mathrm{A})\mathrm{pr}(\mathrm{C}|\mathrm{B})}{\mathrm{pr}(\mathrm{C})\mathrm{pr}(\mathrm{C}|\mathrm{A} \cap \mathrm{B})}\right].
\end{aligned}$$

Thus, the necessary and sufficient condition is

$$\left[\frac{\mathrm{pr}(\mathrm{C}|\mathrm{A})\mathrm{pr}(\mathrm{C}|\mathrm{B})}{\mathrm{pr}(\mathrm{C})\mathrm{pr}(\mathrm{C}|\mathrm{A} \cap \mathrm{B})}\right] = 1,$$

or equivalently,

$$\frac{\mathrm{pr}(\mathrm{C}|\mathrm{A})}{\mathrm{pr}(\mathrm{C})} = \frac{\mathrm{pr}(\mathrm{C}|\mathrm{A} \cap \mathrm{B})}{\mathrm{pr}(\mathrm{C}|\mathrm{B})}.$$

A sufficient (but not necessary) condition for the above equality to hold is that events A and C are both unconditionally independent [i.e., $\mathrm{pr}(\mathrm{C}|\mathrm{A}) = \mathrm{pr}(\mathrm{C})$]

and conditionally independent given event B [i.e., $\text{pr}(\text{C}|\text{A} \cap \text{B}) = \text{pr}(\text{C}|\text{B})$.

Solution 2.19.

(a) pr(all four randomly chosen workers are members of the same group)=pr(all four workers in Group H)+pr(all four workers in Group I)+pr(all four workers in Group L)+pr(all four workers in Group U)= $(0.05)^4 + (0.15)^4 + (0.20)^4 + (0.60)^4 = 0.132$.

(b) Let A be the event that "two of the four workers are members of Group H", and let B be the event that "all four workers are exposed to non-zero levels of the potential carcinogen." Then,

$$\text{pr}(\text{A}|\text{B}) = \frac{\text{pr}(\text{A} \cap \text{B})}{\text{pr}(\text{B})} = \frac{C_2^4(0.05)^2(0.35)^2}{(0.40)^4} = 0.0718.$$

(c) Let B be the event that a worker is a member of either Group H or Group I. So,

$$\begin{aligned} \text{pr}(\text{B}|\text{C}) &= \frac{\text{pr}(\text{B} \cap \text{C})}{\text{pr}(\text{C})} = \frac{\text{pr}[(\text{H} \cup \text{I}) \cap \text{C}]}{\text{pr}(\text{C})} = \frac{\text{pr}(\text{H} \cap \text{C}) + \text{pr}(\text{I} \cap \text{C})}{\text{pr}(\text{C})} \\ &= \frac{\text{pr}(\text{C}|\text{H})\text{pr}(\text{H}) + \text{pr}(\text{C}|\text{I})\text{pr}(\text{I})}{\text{pr}(\text{C})} = \frac{(0.002)(0.05) + (0.001)(0.15)}{\text{pr}(\text{C})} \\ &= \frac{0.000250}{\text{pr}(\text{C})}. \end{aligned}$$

Since pr(C)=pr(C|H)pr(H)+pr(C|I)pr(I)+pr(C|L)pr(L)+pr(C|U)pr(U) =(0.002)(0.05)+(0.001)(0.15)+(0.0001)(0.20)+(0.00001)(0.60)=0.000276, it follows that pr(B|C)=0.000250/0.000276=0.906.

Note that the solution to part (c) of this problem involves a direct application of Bayes' Theorem.

Solution 2.21.

(a) Let B_1 be the event that Ball #1 is black, and let B_2 be the event that Ball #2 is black. Also, define the events $\text{R}_1, \text{R}_2, \text{W}_1$, and W_2 analogously. Then, with $N = (B + R + W)$, it follows that

$$\begin{aligned} \alpha &= \text{pr}(\text{B}_1 \cap \text{R}_2) + \text{pr}(\text{R}_1 \cap \text{B}_2) \\ &= \text{pr}(\text{B}_1)\text{pr}(\text{R}_2|\text{B}_1) + \text{pr}(\text{R}_1)\text{pr}(\text{B}_2|\text{R}_1) \\ &= \left(\frac{\text{B}}{N}\right)\left(\frac{R}{N+K}\right) + \left(\frac{R}{N}\right)\left(\frac{B}{N+K}\right) \\ &= \frac{2BR}{N(N+K)}. \end{aligned}$$

(b) We have

$$\begin{aligned}
\beta &= \text{pr}(\text{B}_2) = \text{pr}(\text{B}_1 \cap \text{B}_2) + \text{pr}(\text{R}_1 \cap \text{B}_2) + \text{pr}(\text{W}_1 \cap \text{B}_2) \\
&= \text{pr}(\text{B}_1)\text{pr}(\text{B}_2|\text{B}_1) + \text{pr}(\text{R}_1)\text{pr}(\text{B}_2|\text{R}_1) + \text{pr}(\text{W}_1)\text{pr}(\text{B}_2|\text{W}_1) \\
&= \left(\frac{B}{N}\right)\left(\frac{B+K}{N+K}\right) + \left(\frac{R}{N}\right)\left(\frac{B}{N+K}\right) + \left(\frac{W}{N}\right)\left(\frac{B}{N+K}\right) \\
&= \frac{B(B+K+R+W)}{N(N+K)} = \frac{B}{N}.
\end{aligned}$$

(c) We have

$$\begin{aligned}
\gamma &= \text{pr}(\text{B}_1|\text{B}_2) = \frac{\text{pr}(\text{B}_1 \cap \text{B}_2)}{\text{pr}(\text{B}_2)} = \frac{\text{pr}(\text{B}_1)\text{pr}(\text{B}_2|\text{B}_1)}{\text{pr}(\text{B}_2)} \\
&= \frac{\left(\frac{B}{N}\right)\left(\frac{B+K}{N+K}\right)}{(B/N)} = \frac{(B+K)}{(N+K)}.
\end{aligned}$$

Solution 2.23.

(a)

$$\begin{aligned}
\text{OR}_\text{e} &= \frac{\text{pr}(\text{E}|\text{D})/\text{pr}(\overline{\text{E}}|\text{D})}{\text{pr}(\text{E}|\overline{\text{D}})/\text{pr}(\overline{\text{E}}|\overline{\text{D}})} \\
&= \frac{\frac{\text{pr}(\text{D}|\text{E})\text{pr}(\text{E})}{\text{pr}(\text{D})} \Big/ \frac{\text{pr}(\text{D}|\overline{\text{E}})\text{pr}(\overline{\text{E}})}{\text{pr}(D)}}{\frac{\text{pr}(\overline{\text{D}}|\text{E})\text{pr}(\text{E})}{\text{pr}(\overline{\text{D}})} \Big/ \frac{\text{pr}(\overline{\text{D}}|\overline{\text{E}})\text{pr}(\overline{\text{E}})}{\text{pr}(\overline{\text{D}})}} \\
&= \left[\frac{\text{pr}(\text{D}|\text{E})}{\text{pr}(\text{D}|\overline{\text{E}})}\right]\left[\frac{\text{pr}(\overline{\text{D}}|\overline{\text{E}})}{\text{pr}(\overline{\text{D}}|\text{E})}\right] \\
&= (\text{RR})\left[\frac{\text{pr}(\overline{\text{D}}|\overline{\text{E}})}{\text{pr}(\overline{\text{D}}|\text{E})}\right].
\end{aligned}$$

Under the rare-disease assumption, $\text{pr}(\overline{\text{D}}|\overline{\text{E}}) = 1 - \text{pr}(\text{D}|\overline{\text{E}}) \approx 1$ and $\text{pr}(\overline{\text{D}}|\text{E}) = 1 - \text{pr}(\text{D}|\text{E}) \approx 1$; thus, the exposure odds ratio OR_e mathematically approximates the risk ratio RR.

(b) Since

$$\text{pr}(\text{E}|X = 1) = \text{pr}(\text{E}|\text{D}) = \frac{\text{e}^{(\alpha+\beta)}}{1 + \text{e}^{(\alpha+\beta)}}$$

and

$$\text{pr}(\text{E}|X = 0) = \text{pr}(\text{E}|\overline{\text{D}}) = \frac{\text{e}^{\alpha}}{1 + \text{e}^{\alpha}},$$

it follows from simple algebra that $\text{OR}_\text{e} = \text{e}^{\beta}$.

Solution 2.25. Let C_{ij} be the event that the cabinet containing i gold coins and j silver coins is selected, so that the events C_{13}, C_{31}, and C_{22} are of interest. Further, let S_1 be the event that the first drawer opened for a randomly selected cabinet contains a silver coin, and let G_2 be the event that the second drawer opened for that selected cabinet contains a gold coin. Then,

$$\theta = \text{pr}(G_2|S_1) = \frac{\text{pr}(G_2 \cap S_1)}{\text{pr}(S_1)}.$$

Now,

$$\begin{aligned}
\text{pr}(G_2 \cap S_1) &= \text{pr}(G_2 \cap S_1 \cap C_{13}) \\
&\quad + \text{pr}(G_2 \cap S_1 \cap C_{31}) \\
&\quad + \text{pr}(G_2 \cap S_1 \cap C_{22}) \\
&= \text{pr}(G_2|S_1 \cap C_{13})\text{pr}(S_1|C_{13})\text{pr}(C_{13}) \\
&\quad + \text{pr}(G_2|S_1 \cap C_{31})\text{pr}(S_1|C_{31})\text{pr}(C_{31}) \\
&\quad + \text{pr}(G_2|S_1 \cap C_{22})\text{pr}(S_1|C_{22})\text{pr}(C_{22}) \\
&= \left(\frac{1}{3}\right)\left(\frac{3}{4}\right)\left(\frac{1}{3}\right) + (1)\left(\frac{1}{4}\right)\left(\frac{1}{3}\right) + \left(\frac{2}{3}\right)\left(\frac{2}{4}\right)\left(\frac{1}{3}\right) \\
&= \frac{5}{18}.
\end{aligned}$$

And,

$$\begin{aligned}
\text{pr}(S_1) &= \text{pr}(S_1 \cap C_{13}) + \text{pr}(S_1 \cap C_{31}) + \text{pr}(S_1 \cap C_{22}) \\
&= \text{pr}(S_1|C_{13})\text{pr}(C_{13}) + \text{pr}(S_1|C_{31})\text{pr}(C_{31}) + \text{pr}(S_1|C_{22})\text{pr}(C_{22}) \\
&= \left(\frac{3}{4}\right)\left(\frac{1}{3}\right) + \left(\frac{1}{4}\right)\left(\frac{1}{3}\right) + \left(\frac{2}{4}\right)\left(\frac{1}{3}\right) \\
&= \frac{1}{2}.
\end{aligned}$$

Thus,

$$\theta = \text{pr}(G_2|S_1) = \frac{(5/18)}{(1/2)} = \frac{5}{9}.$$

Solution 2.27.

(a) Under WR random selection, there are N^n possible samples of size n, all of which are equally likely to be obtained; so, the probability of obtaining a particular sample of size n is equal to $1/N^n$.

Under WOR random selection, there are $N(N-1)\cdots(N-n+1)$ possible samples of size n, all of which are equally likely to be obtained; so, the probability of obtaining a particular sample of size n is equal to $1/N(N-1)\cdots(N-n+1) = (N-n)!/N!$.

(b) Under WR random selection, the probability of obtaining a particular member of the population in a sample of size n is equal to

$$1-\left(\frac{N-1}{N}\right)^n = 1-\left(1-\frac{1}{N}\right)^n.$$

Under WOR random selection, the probability of obtaining a particular member of the population in a sample of size n is equal to

$$1-\left(\frac{N-1}{N}\right)\left(\frac{N-2}{N-1}\right)\cdots\left(\frac{N-n}{N-n+1}\right)$$

$$=1-\frac{(N-1)!/(N-n-1)!}{N!/(N-n)!}=1-\frac{(N-n)}{N}=\frac{n}{N}.$$

Solution 2.29. Let A be the event that "the n-th ball selected is the last black ball remaining in the urn," let B be the event that "exactly one black ball is among the first $(n-1)$ balls selected," and let C be the event that "the n-th ball selected is a black ball." Then, it follows that

$$\begin{aligned}
\mathrm{P}_n &= \mathrm{pr}(\mathrm{A}) = \mathrm{pr}(\mathrm{B}\cap\mathrm{C}) = \mathrm{pr}(\mathrm{B})\mathrm{pr}(\mathrm{C}|\mathrm{B}) \\
&= \frac{\mathrm{C}_1^2\mathrm{C}_{n-2}^4}{\mathrm{C}_{n-1}^6}\cdot\left(\frac{1}{6-(n-1)}\right) \\
&= \frac{2\,[4!/(n-2)!(6-n)!]}{[6!/(n-1)!(7-n)!]}\left(\frac{1}{7-n}\right) \\
&= \frac{(n-1)}{15},\ n=2,3,4,5,6.
\end{aligned}$$

Thus, $n^*=6$, since $\mathrm{P}_6 = 5/15 = 1/3$.

Note that the above probability is *not* the same as

pr(both black balls are obtained in a sample of size n selected without replacement)

$$=1-\sum_{x=0}^{1}\frac{\mathrm{C}_x^2\mathrm{C}_{n-x}^4}{\mathrm{C}_n^6}=\frac{n(n-1)}{30},$$

where the permissible values of the sample size n are 2, 3, 4, 5, and 6.

Solution 2.31*. Consider the following four events: M_1: "A marked turtle is a member of the random sample selected at the end of the first week"; M_2: "A marked turtle is a member of the random sample selected at the end of the second week"; and S: a marked turtle survives during any one-week time period.

So, the goal is to develop explicit expressions, as functions of α and β, for $\text{pr}(\text{M}_1 \cap \bar{\text{M}}_2), \text{pr}(\bar{\text{M}}_1 \cap \text{M}_2), \text{pr}(\text{M}_1 \cap \text{M}_2)$, and $\text{pr}(\bar{\text{M}}_1 \cap \bar{\text{M}}_2)$.

First,

$$\begin{aligned}
\text{pr}(\text{M}_1) &= \text{pr}(\text{M}_1 \cap \text{S}) + \text{pr}(\text{M}_1 \cap \bar{\text{S}}) \\
&= \text{pr}(\text{M}_1|\text{S})\text{pr}(\text{S}) + \text{pr}(\text{M}_1|\bar{\text{S}})\text{pr}(\bar{\text{S}}) \\
&= \delta\gamma + (0)(1-\gamma) = \delta\gamma = \alpha.
\end{aligned}$$

Clearly, then, $\text{pr}(\text{M}_2|\text{M}_1) = \alpha$.

So,

$$\text{pr}(\text{M}_1 \cap \text{M}_2) = \text{pr}(\text{M}_1)\text{pr}(\text{M}_2|\text{M}_1) = \alpha^2.$$

And,

$$\text{pr}(\text{M}_1 \cap \bar{\text{M}}_2) = \text{pr}(\text{M}_1)\text{pr}(\bar{\text{M}}_2|\text{M}_1) = \alpha(1-\alpha).$$

Also,

$$\begin{aligned}
\text{pr}(\bar{\text{M}}_1 \cap \text{M}_2) &= \text{pr}(\bar{\text{M}}_1 \cap \text{M}_2 \cap \text{S}) + \text{pr}(\bar{\text{M}}_1 \cap \text{M}_2 \cap \bar{\text{S}}) \\
&= \text{pr}(\text{M}_2|\bar{\text{M}}_1 \cap \text{S})\text{pr}(\bar{\text{M}}_1 \cap \text{S}) + 0 \\
&= \text{pr}(\text{M}_2|\bar{\text{M}}_1 \cap \text{S})\text{pr}(\bar{\text{M}}_1|\text{S})\text{pr}(\text{S}) \\
&= \alpha(1-\delta)\gamma = \alpha\beta.
\end{aligned}$$

Finally,

$$\begin{aligned}
\text{pr}(\bar{\text{M}}_1 \cap \bar{\text{M}}_2) &= 1 - \text{pr}(\text{M}_1 \cap \text{M}_2) - \text{pr}(\text{M}_1 \cap \bar{\text{M}}_2) - \text{pr}(\bar{\text{M}}_1 \cap \text{M}_2) \\
&= 1 - \alpha^2 - \alpha(1-\alpha) - \alpha\beta \\
&= (1 - \alpha - \alpha\beta).
\end{aligned}$$

Solution 2.33*.

(a) Let N be the event that a patient receives the new treatment, let S be the event that a patient receives the standard treatment, and let R be the event that a patient goes into remission. Then,

$$\begin{aligned}
\text{pr}(\text{R}) &= \text{pr}(\text{R} \cap \text{N}) + \text{pr}(\text{R} \cap \text{S}) = \text{pr}(\text{R}|\text{N})\text{pr}(\text{N}) + \text{pr}(\text{R}|\text{S})\text{pr}(\text{S}) \\
&= \theta_1\pi + \theta_0(1-\pi) = \theta_0 + (\theta_1 - \theta_0)\pi.
\end{aligned}$$

(b) Now,

$$\begin{aligned}
\mathrm{pr(N|R)} &= \frac{\mathrm{pr(N \cap R)}}{\mathrm{pr(R)}} = \frac{\mathrm{pr(R|N)pr(N)}}{\mathrm{pr(R)}} \\
&= \frac{\theta_1 \pi}{\theta_0 + (\theta_1 - \theta_0)\pi}.
\end{aligned}$$

(c) For $i = 1, 2$, let $\mathrm{N}_i, \mathrm{R}_i$, and S_i be the events N, R, and S specific to the ith patient; for example, N_1 is the event that patient #1 received the new treatment, etc. Then, letting α denote the probability of interest, we have

$$\begin{aligned}
\alpha &= \mathrm{pr}\left[(\mathrm{N}_1 \cap \mathrm{S}_2) \cup (\mathrm{S}_1 \cap \mathrm{N}_2)|\mathrm{R}_1 \cap \mathrm{R}_2\right] \\
&= \frac{\mathrm{pr}\left[(\mathrm{N}_1 \cap \mathrm{S}_2 \cap \mathrm{R}_1 \cap \mathrm{R}_2) \cup (\mathrm{S}_1 \cap \mathrm{N}_2 \cap \mathrm{R}_1 \cap \mathrm{R}_2)\right]}{\mathrm{pr}(\mathrm{R}_1 \cap \mathrm{R}_2)} \\
&= \frac{\mathrm{pr}(\mathrm{R}_1 \cap \mathrm{R}_2|\mathrm{N}_1 \cap \mathrm{S}_2)\mathrm{pr}(\mathrm{N}_1 \cap \mathrm{S}_2) + \mathrm{pr}(\mathrm{R}_1 \cap \mathrm{R}_2|\mathrm{S}_1 \cap \mathrm{N}_2)\mathrm{pr}(\mathrm{S}_1 \cap \mathrm{N}_2)}{\mathrm{pr}(\mathrm{R}_1)\mathrm{pr}(\mathrm{R}_2)} \\
&= \frac{\mathrm{pr}(\mathrm{R}_1|\mathrm{N}_1)\mathrm{pr}(\mathrm{R}_2|\mathrm{S}_2)\mathrm{pr}(\mathrm{N}_1)\mathrm{pr}(\mathrm{S}_2) + \mathrm{pr}(\mathrm{R}_1|\mathrm{S}_1)\mathrm{pr}(\mathrm{R}_2|\mathrm{N}_2)\mathrm{pr}(\mathrm{S}_1)\mathrm{pr}(\mathrm{N}_2)}{\left[\theta_0 + (\theta_1 - \theta_0)\pi\right]^2} \\
&= \frac{\theta_1\theta_0\pi(1-\pi) + \theta_0\theta_1(1-\pi)\pi}{\left[\theta_0 + (\theta_1 - \theta_0)\pi\right]^2} \\
&= \frac{2\pi(1-\pi)\theta_0\theta_1}{\left[\theta_0 + (\theta_1 - \theta_0)\pi\right]^2}.
\end{aligned}$$

Solution 2.35*. First, note that

$$\begin{aligned}
\mathrm{pr(D)} &= \mathrm{pr(D|E)pr(E)} + \mathrm{pr(D|\bar{E})pr(\bar{E})} \\
&= \mathrm{pr(D|E)pr(E)} + \mathrm{pr(D|\bar{E})[1 - pr(E)]} \\
&= \mathrm{pr(E)[pr(D|E) - pr(D|\bar{E})] + pr(D|\bar{E})}.
\end{aligned}$$

Then, using this expression for pr(D), it follows directly that

$$\begin{aligned}
\alpha &= \frac{\mathrm{pr(D) - pr(D|\bar{E})}}{\mathrm{pr(D)}} \\
&= \frac{\mathrm{pr(E)[pr(D|E) - pr(D|\bar{E})]}}{\mathrm{pr(E)[pr(D|E) - pr(D|\bar{E})] + pr(D|\bar{E})}}.
\end{aligned}$$

Finally, dividing the numerator and denominator of this expression by $\mathrm{pr(D|\bar{E})}$ gives

$$\alpha = \frac{\mathrm{pr(E)}(\theta - 1)}{1 + \mathrm{pr(E)}(\theta - 1)}.$$

Now,

$$\begin{aligned}
\mathrm{pr}(\mathrm{E}|\mathrm{D}) &= \frac{\mathrm{pr}(\mathrm{D}\cap\mathrm{E})}{\mathrm{pr}(\mathrm{D})} = \frac{\mathrm{pr}(\mathrm{D}|\mathrm{E})\mathrm{pr}(\mathrm{E})}{\mathrm{pr}(\mathrm{D}|\mathrm{E})\mathrm{pr}(\mathrm{E}) + \mathrm{pr}(\mathrm{D}|\bar{\mathrm{E}})\mathrm{pr}(\bar{\mathrm{E}})} \\
&= \frac{\theta\mathrm{pr}(\mathrm{E})}{\theta\mathrm{pr}(\mathrm{E}) + \mathrm{pr}(\bar{\mathrm{E}})} = \frac{\theta\mathrm{pr}(\mathrm{E})}{(\theta-1)\mathrm{pr}(\mathrm{E})+1},
\end{aligned}$$

so that

$$\mathrm{pr}(\mathrm{E}) = \frac{\mathrm{pr}(\mathrm{E}|\mathrm{D})}{\mathrm{pr}(\mathrm{E}|\mathrm{D}) + [1-\mathrm{pr}(\mathrm{E}|\mathrm{D})]\theta}.$$

Finally, using the above expression, we obtain

$$\begin{aligned}
\alpha &= \frac{\mathrm{pr}(\mathrm{E})(\theta-1)}{1+\mathrm{pr}(\mathrm{E})(\theta-1)} \\
&= \frac{\left\{\frac{\mathrm{pr}(\mathrm{E}|\mathrm{D})}{\mathrm{pr}(\mathrm{E}|\mathrm{D})+[1-\mathrm{pr}(\mathrm{E}|\mathrm{D})]\theta}\right\}(\theta-1)}{1+\left\{\frac{\mathrm{pr}(\mathrm{E}|\mathrm{D})}{\mathrm{pr}(\mathrm{E}|\mathrm{D})+[1-\mathrm{pr}(\mathrm{E}|\mathrm{D})]\theta}\right\}(\theta-1)} \\
&= \frac{\mathrm{pr}(\mathrm{E}|\mathrm{D})(\theta-1)}{\mathrm{pr}(\mathrm{E}|\mathrm{D}) + [1-\mathrm{pr}(\mathrm{E}|\mathrm{D})]\theta + \mathrm{pr}(\mathrm{E}|\mathrm{D})(\theta-1)} \\
&= \frac{\mathrm{pr}(\mathrm{E}|\mathrm{D})(\theta-1)}{\theta}.
\end{aligned}$$

Note that $0 \leq \alpha \leq 1$.

Solution 2.37*.

(a) Now,

$$\begin{aligned}
\mathrm{pr}(\mathrm{D}|\mathrm{E}) &= \mathrm{pr}(\mathrm{D}\cap\mathrm{U}|\mathrm{E}) + \mathrm{pr}(\mathrm{D}\cap\overline{\mathrm{U}}|\mathrm{E}) \\
&= \mathrm{pr}(\mathrm{D}|\mathrm{E}\cap\mathrm{U})\mathrm{pr}(\mathrm{U}|\mathrm{E}) + \mathrm{pr}(\mathrm{D}|\mathrm{E}\cap\overline{\mathrm{U}})\mathrm{pr}(\overline{\mathrm{U}}|\mathrm{E}) \\
&= \pi_1\gamma_1 + \pi_0(1-\gamma_1) \\
&= \gamma_1(\pi_1-\pi_0)+\pi_0.
\end{aligned}$$

And, using completely analogous arguments, it follows that

$$\mathrm{pr}(\mathrm{D}|\bar{\mathrm{E}}) = \gamma_0(\pi_1-\pi_0)+\pi_0.$$

Thus,

$$\begin{aligned}
\theta &= \frac{\gamma_1(\pi_1-\pi_0)+\pi_0}{\gamma_0(\pi_1-\pi_0)+\pi_0} \\
&= \frac{\gamma_1(\theta^*-1)+1}{\gamma_0(\theta^*-1)+1}.
\end{aligned}$$

(b) Since $\theta^* \geq 1$, the expression for θ in part (a) is maximized (and takes the maximum value θ^*) when $\gamma_1 = 1$ and $\gamma_0 = 0$, so that $\theta \leq \theta^*$.

And,

$$\begin{aligned}(\theta - 1) &= \frac{[\gamma_1(\theta^* - 1) + 1] - [\gamma_0(\theta^* - 1) + 1]}{\gamma_0(\theta^* - 1) + 1} \\ &= \frac{(\gamma_1 - \gamma_0)}{\gamma_0 + (\theta^* - 1)^{-1}}.\end{aligned}$$

Clearly, θ is a monotonically increasing function of θ^* with $\lim_{\theta^* \to \infty} \theta = \gamma_1/\gamma_0$, so that $\theta \leq \frac{\gamma_1}{\gamma_0}$.

(c) From the results in part (b), in order for some unobservable risk factor to completely explain away the established strong association ($\theta \approx 9$) between smoking and lung cancer, we must have

$$\theta^* \geq 9 \text{ and } \frac{\gamma_1}{\gamma_0} = \frac{\text{pr}(\text{U}|\text{E})}{\text{pr}(\text{U}|\bar{\text{E}})} \geq 9.$$

These two inequalities imply that:

(i) Such an unobservable risk factor would have to be at least as strong a risk factor for lung cancer as is cigarette smoke

(ii) Such an unobservable risk factor would have to be at least nine times more prevalent among smokers than among non-smokers.

The existence of such an unobservable risk factor is extremely unlikely; if it did exist, it would almost certainly have been identified and carefully studied.

Solution 2.39*.

(a) Under the assumption that the two drugs are equally effective, there are $2^n = \sum_{x=0}^{n} \text{C}_x^n$ equally likely outcomes for the n pairs. Among these 2^n equally likely outcomes, there are $\sum_{x=x^*}^{n} \text{C}_x^n$ outcomes for which $x \geq x^*$. So, it follows that

$$\theta(x^*|n) = \frac{\sum_{x=x^*}^{n} \text{C}_x^n}{2^n}, 0 \leq x^* \leq n$$

(b) When $n = 15$ and $x^* = 9$, we have

$$\theta(9|15) = \frac{\sum_{x=9}^{15} \text{C}_x^n}{2^{15}} = \frac{9,949}{32,768} = 0.3036.$$

The value 0.3036 is fairly large, indicating that observing a finding at least as extreme as the one found is reasonably likely when the two drugs are actually equally effective. Therefore, these data do not provide any statistical evidence that Drug 1 is more effective than Drug 2. In statistical terminology, we have computed what is known as a P-value, and standard statistical practice considers only small P-values (say, less than 0.05 in value) to be providing reasonably strong statistical evidence.

(c) When $x^* = 11, \theta(11|15) = 0.0592$, and when $x^* = 12, \theta(12|15) = 0.0176$. So, $x^* = 12$.

Solution 2.41*.

(a) Let W be the event that the player wins the game, and let A_x be the event that the number x is obtained on the first roll of the pair of dice. Then, noting that the two numbers 4 and 10 each have the same probability of occurring (as do the two numbers 5 and 9, and the two numbers 6 and 8), we have

$$\begin{aligned}
\theta(\pi) &= \text{pr(W)} = \sum_{x=2}^{12} \text{pr(W|A}_x\text{)pr(A}_x\text{)} \\
&= (0)\left[\frac{(1-\pi)}{30}\right] + (0)\left[\frac{2(1-\pi)}{30}\right] + 2\text{pr(W|A}_4\text{)}\left[\frac{3(1-\pi)}{30}\right] \\
&+ 2\text{pr(W|A}_5\text{)}\left[\frac{4(1-\pi)}{30}\right] + 2\text{pr(W|A}_6\text{)}\left[\frac{5(1-\pi)}{30}\right] \\
&+ (1)(\pi) + (1)\left[\frac{2(1-\pi)}{30}\right] + (0)\left[\frac{(1-\pi)}{30}\right] \\
&= \left[\pi + \frac{2(1-\pi)}{30}\right] + 2\sum_{x=4}^{6} \text{pr(W|A}_x\text{)}\left[\frac{(x-1)(1-\pi)}{30}\right] \\
&= \frac{1}{15}\left[1 + 14\pi + (1-\pi)\sum_{x=4}^{6}(x-1)\text{pr(W|A}_x\text{)}\right].
\end{aligned}$$

Now, for $x = 4, 5, 6$,

$$
\begin{aligned}
\text{pr}(\text{W}|\text{A}_x) &= \sum_{j=1}^{\infty} [\text{pr}(\text{any number but } x \text{ or } 7 \text{ is rolled})]^{j-1} \\
&\quad \times \text{pr}(\text{the number } x \text{ is rolled}) \\
&= \sum_{j=1}^{\infty} \left[1 - \frac{(x-1)(1-\pi)}{30} - \pi\right]^{j-1} \left[\frac{(x-1)(1-\pi)}{30}\right] \\
&= \frac{(x-1)(1-\pi)}{30} \sum_{j=1}^{\infty} \left[(1-\pi)\left(\frac{30-x+1}{30}\right)\right]^{j-1} \\
&= \frac{(x-1)(1-\pi)}{30-(31-x)(1-\pi)}.
\end{aligned}
$$

So,

$$
\begin{aligned}
\text{pr}(\text{W}) &= \frac{1}{15}\left\{1 + 14\pi + (1-\pi)\sum_{x=4}^{6}(x-1)\left[\frac{(x-1)(1-\pi)}{30-(31-x)(1-\pi)}\right]\right\} \\
&= \frac{1}{15}\left\{1+14\pi+(1-\pi)^2\left[\left(\frac{3}{1+9\pi}\right)+\left(\frac{8}{2+13\pi}\right)+\left(\frac{5}{1+5\pi}\right)\right]\right\}.
\end{aligned}
$$

It is easy to verify that $\theta(1/6) = 0.493$. And, $\theta(1) = 1$ is correct because a 7 will always come up on the first roll. Finally, $\theta(0) = 13/15$; this answer is correct because since a 7 can never be obtained, the only way for the player to *lose* the game is to roll a 2, 3, or 12 on the first roll. This probability is equal to $\left(\frac{1}{30} + \frac{2}{30} + \frac{1}{30}\right) = \frac{2}{15}$.

Chapter 3

Univariate Distribution Theory

3.1 Exercises

Exercise 3.1. Consider two urns (denoted Urn 1 and Urn 2). Urn 1 contains 2 white balls and 1 black ball; Urn 2 contains 1 white ball and 2 black balls.

Suppose that one ball is randomly drawn from Urn 1 and is put into Urn 2; then, balls are selected one-at-a-time *without replacement* from Urn 2 until a white ball is obtained. Let Y denote the number of balls selected from Urn 2 until a white ball is obtained (e.g., if the first ball selected from Urn 2 is black and the second one is white, then $Y = 2$). Provide a *formula*, not a table, for the probability distribution $\mathrm{p}_Y(y)$ of the random variable Y, and then use this formula to find numerical values for $\mathrm{E}(Y)$ and $\mathrm{V}(Y)$.

Exercise 3.2. After extensive atmospheric sampling and data analysis, an environmental scientist decides that the distribution of measurements of the sulfur dioxide concentration X (in parts per million) in the air near a certain oil refinery can be closely approximated by the density function

$$\mathrm{f}_X(x) = \begin{cases} \frac{4}{5}x^3, & 0 \leq x < 1; \\ \frac{4}{5}\mathrm{e}^{1-x}, & 1 \leq x < +\infty \end{cases}.$$

(a) Find $\mathrm{F}_X(x)$, the cumulative distribution function (CDF) of X.

(b) Find the numerical value of $\mathrm{E}(X)$.

(c) Find the numerical value of

$$\mathrm{pr}\left(\frac{1}{2} < X < 2 | X \geq \frac{1}{3}\right).$$

Exercise 3.3. Consider the double exponential distribution

$$\mathrm{f}_X(x) = \frac{1}{2\alpha}\mathrm{e}^{-|x-\beta|/\alpha}, -\infty < x < \infty, 0 < \alpha < \infty, -\infty < \beta < \infty.$$

Derive an explicit expression for $\nu_1 = \mathrm{E}[|X - \mathrm{E}(X)|]$, the first absolute moment about the mean.

Exercise 3.4. Among 100 kidney dialysis machines, suppose that exactly 5 of these 100 machines are defective. If machines are randomly sampled one at a time without replacement and tested, what is the exact probability distribution $\mathrm{p}_X(x) = \mathrm{pr}(X = x)$ of the discrete random variable X, the number of machines that have to be examined until the first defective machine is found?

Exercise 3.5. A circular-shaped archery target has three concentric circles painted on it. The innermost circle has a radius of $1/\sqrt{3}$ feet (measured from the center of the circular target), the middle circle has a radius of 1 foot, and the outermost circle has a radius of $\sqrt{3}$ feet. An arrow hitting within the innermost circle counts 4 points, an arrow hitting in the area between the innermost circle and the middle circle counts 3 points, an arrow hitting in the area between the middle circle and the outermost circle counts 2 points, and an arrow not hitting within the outermost circle counts 0 points. Suppose that the distance R (in feet) from the exact center of the target that any arrow shot by a certain archer hits the target follows the distribution

$$\mathrm{f}_R(r) = \frac{2}{\pi}(1 + r^2)^{-1}, 0 < r < \infty.$$

Let the random variable S denote the score received by this archer based on any one shot at the target.

Find numerical values for $\mathrm{E}(S)$ and $\mathrm{V}(S)$.

Exercise 3.6. A hospital is interested in purchasing a lot of 25 kidney dialysis machines from a certain manufacturing company. Suppose that this lot contains some defective machines and some non-defective machines. To determine the number of defective machines in the lot of 25 machines, two machines are selected at random and tested. If the probability that these two machines are either both defective or both non-defective is equal to the probability that one of the machines is defective and the other machine is non-defective, provide a reasonable numerical value (or values) for the number of defective machines in the lot of 25 machines.

Exercise 3.7. Suppose that $X \sim \mathrm{N}(\theta, \theta), \theta > 0$, and let $U = |X|$. Develop an explicit expression for $\mathrm{F}_U(u)$, the cumulative distribution function (CDF) of the random variable U, and then use this result to develop an explicit expression for $\mathrm{f}_U(u)$, the density function of the random variable U.

Exercise 3.8. Suppose that a point moves along the x-axis in jumps of one unit each, starting at the origin. Each jump may be to the right or left, with respective probabilities θ and $(1-\theta), 0 < \theta < 1$. Furthermore, each jump is assumed to be independent of all other jumps. Let X be the coordinate of the point on the x-axis after n jumps, where $n(> 0)$ is an *odd* positive integer.

(a) Derive an *explicit general formula* for the probability distribution $\mathrm{p}_X(x)$ of the random variable X for each of the special cases $n = 1$ and $n = 3$.

(b) Based on the findings in part (a), provide an *explicit general formula* for $\mathrm{p}_X(x)$ when n is allowed to be any *odd* positive integer, and also find explicit expressions for $\mathrm{E}(X)$ and $\mathrm{V}(X)$.

Exercise 3.9. Suppose that the continuous random variable X has the distribution $\mathrm{f}_X(x), -\infty < x < \infty$, which is symmetric about the value $x = 0$. Evaluate the integral

$$\int_{-k}^{k} \mathrm{F}_X(x)\mathrm{d}x,$$

where $\mathrm{F}_X(x)$ is the CDF for X and where k is a non-negative real number.

Exercise 3.10. Let

$$\mathrm{f}_P(p) = 6p - 6p^2,\ 0 < p < 1.$$

(a) Find an explicit expression for $\mathrm{F}_P(p)$, the CDF of the random variable P, and then use this result to find the numerical value of $\mathrm{pr}(0.60 < P < 0.80)$.

(b) Find the numerical value of $\mathrm{pr}(0.70 < P < 0.80|0.60 < P < 0.80)$.

(c) For $k \geq 0$, find an explicit expression for $\mathrm{E}(P^k)$, and then use this expression to find explicit expressions for $\mathrm{E}(P)$ and $\mathrm{V}(P)$.

Exercise 3.11. After examining relevant air pollution data for a certain city in the United States, an environmental scientist postulates that the distribution of the carbon monoxide concentration level X (measured in parts per million, or ppm) above k ppm (where k is a known positive constant) can be accurately modeled by the one-parameter Pareto density function

$$\mathrm{f}_X(x) = \theta k^{\theta}/x^{(\theta+1)}, \quad 0 < k < x < +\infty,\ \theta > 3.$$

(a) Find an *explicit expression* for $\mathrm{F}_X(x)$, the CDF of X, and then use this CDF to find the numerical value of $\mathrm{pr}\left[(k+1) < X < (k+3)|X > (k+1)\right]$ when $k = 1$ and $\theta = 4$.

(b) Develop an *explicit expression* for $\mu_3 = \mathrm{E}\{[X - \mathrm{E}(X)]^3\}$. Find the limiting value of μ_3 as $\theta \to +\infty$, and then provide analytical justification for why this limiting value of μ_3 makes sense.

(c) After careful thought, this environmental scientist suggests that the distribution of the random variable $Y = \ln(X)$ has more scientific relevance than the distribution of X itself. Develop an *explicit expression* for the moment generating function $\mathrm{M}_Y(t)$ of Y, and then use $\mathrm{M}_Y(t)$ directly to find an *explicit expression* for $\mathrm{E}(Y)$.

Exercise 3.12. A very large research study was conducted to investigate the possible relationship between adolescent diabetes and body mass index (BMI) among teenage children in the United States. Each teenager in the study was classified as being diabetic (D) or not being diabetic ($\overline{\mathrm{D}}$) based on a thorough clinical diagnosis. For teenagers participating in this study, 50% had normal (N) BMI values, 25% had mildly (M) elevated BMI values, 20% had severely (S) elevated BMI values, and only 5% had lower (L) than normal BMI values. *Conditional on BMI status* (i.e., conditional on being a member of one of the four groups N, M, S, or L), the percentages of teenagers in each group having diabetes were as follows: 1% of the N group had diabetes; 2% of the M group had diabetes; 5% of the S group had diabetes; and 0.30% of the L group had diabetes.

In answering the questions that follow, you may assume that the number of teenagers included in the study (i.e., the study population) is very large, so that the probabilistic attributes of "sampling with replacement" are operable.

(a) What is the numerical value of the probability that a teenager selected randomly from the study population actually has diabetes?

(b) Given that a randomly chosen teenager selected from the study population does *not* have diabetes, what is the numerical value of the probability that this particular teenager is *not* in either the L group or the N group?

(c) If 10 teenagers are randomly chosen from the study population, what is the numerical value of the probability π that *at least* two of these 10 teenagers have *both* of the following two characteristics: (1) they are *not* members of either the L group or the N group; and, (2) they do not have diabetes.

(d) Let the random variable X be the number of teenagers randomly chosen one at a time from the study population until at least one member of the M group *and* at least one member of the S group are obtained. Derive an *explicit expression* for the probability distribution $\mathrm{p}_X(x)$, and show directly that $\mathrm{p}_X(x)$ is a valid discrete probability distribution.

Exercise 3.13. Suppose that X is a positive random variable with density function $\mathrm{f}_X(x)$, $x > 0$, and moment generating function $\mathrm{M}_X(t)$.

(a) Using the fact that $x^{-1} = \int_{-\infty}^{0} e^{ux}\, du$, $x > 0$, prove rigorously that

$$\mathrm{E}(X^{-1}) = \int_{0}^{\infty} \mathrm{M}_X(-t)\, dt.$$

(b) Use the result in part (a) to find $\mathrm{E}(X^{-1})$ if $X \sim \mathrm{GAMMA}(\alpha, \beta)$. For which values of α and β does $\mathrm{E}(X^{-1})$ exist?

Exercise 3.14. Let X be a univariate random variable. Use Jensen's Inequality to establish inequalities between components of each of the following pairs of expected value functions: (a) $\mathrm{E}(X^2)$ and $[\mathrm{E}(X)]^2$; (b) $\mathrm{E}(e^X)$ and $e^{\mathrm{E}(X)}$; (c) $\mathrm{E}[\ln(X)]$ and $\ln[\mathrm{E}(X)]$; (d) $\mathrm{E}(1/X)$ and $1/\mathrm{E}(X)$.

Exercise 3.15. Suppose that the continuous random variable X has the distribution

$$\mathrm{f}_X(x) = k e^{-(x-\theta)^{2m}}, -\infty < x < \infty, -\infty < \theta < \infty, k > 0,$$

where m is a known positive integer.

(a) For r a non-negative integer, develop an explicit expression for $\mathrm{E}\left[(X-\theta)^{2r}\right]$.

(b) When $m = 1$ and $r = 1$, determine the numerical value of $\mathrm{E}\left[(X-\theta)^{2r}\right]$, and provide a rationale for why this numerical answer makes sense.

Exercise 3.16. It is a standard result in statistical theory that the Poisson distribution

$$\mathrm{p}_X(x; \lambda) = \frac{\lambda^x e^{-\lambda}}{x!}, x = 0, 1, 2, \ldots, \infty \text{ and } \lambda > 0,$$

can be derived as a limiting case of the binomial distribution as $n \to \infty$ $\pi \to 0$, with $\lambda = n\pi$ held constant.

The above derivation suggests that the Poisson distribution can serve as a useful model for the occurrences over time of rare events (e.g., occurrences of certain chronic diseases like cancer, occurrences of catastrophic events like plane crashes and floods, etc.). In fact, the Poisson distribution can be alternatively derived from a few basic assumptions about how rare events occur randomly over time. In particular, suppose that $\mathrm{h}(x, t)$ is the probability of observing x rare events during the time interval $(0, t)$ of length t, and suppose that the following assumptions are valid:

i) the probability of a rare event occurring during the time interval $(t, t+\Delta t)$ is $\theta\Delta t (0 < \theta\Delta t < 1)$, where $\theta > 0$ and where Δt is a very small positive real number;

ii) the probability of more than one rare event occurring during the time interval $(t, t+\Delta t)$ is zero; and

iii) the probability of a rare event occurring during the time interval $(t, t+\Delta t)$ does not depend on what happened prior to time t.

(a) Given the above three assumptions, show that

$$\mathrm{h}(x, t+\Delta t) = \mathrm{h}(x,t)[1-\theta\Delta t] + \mathrm{h}(x-1,t)\theta\Delta t,$$

and hence that

$$\frac{\mathrm{d}[\mathrm{h}(x,t)]}{\mathrm{d}t} = \theta[\mathrm{h}(x-1,t) - \mathrm{h}(x,t)].$$

(b) Show that the Poisson distribution

$$\mathrm{p}_X(x;\theta t) = \mathrm{h}(x,t) = \frac{(\theta t)^x \mathrm{e}^{-\theta t}}{x!}, x = 0, 1, \ldots, \infty \text{ and } \theta t > 0,$$

satisfies the above differential equation.

Exercise 3.17. A history professor at the University of North Carolina believes that two students (say, Student 1 and Student 2) cheated on a multiple-choice examination, and she wants to know if there is statistical evidence to support her belief. Suppose that this multiple-choice examination involved 50 questions; each question listed 5 possible answers, only one of which was correct. Suppose that Student 1 answered 37 questions correctly and 13 questions incorrectly, and that Student 2 answered 35 questions correctly and 15 questions incorrectly. Further, suppose that there were 32 questions that both students answered correctly, and that there were 10 questions that both students answered incorrectly. Among these 10 incorrectly answered questions, there were 5 questions for which these two students gave exactly the same wrong answer. Do these data provide statistical evidence that these two students cheated on this multiple-choice examination?

Exercise 3.18. Suppose that $X \sim \mathrm{BIN}(n, \pi)$. For r a positive integer, derive an explicit expression for

$$\mu_{(r)} = \mathrm{E}\left[X(X-1)(X-2)\cdots(X-r+1)\right],$$

and then use this result to find $\mathrm{E}(X)$ and $\mathrm{V}(X)$. Then, find the limiting value of $\mu_{(r)}$ as $n \to +\infty$ and $\pi \to 0$ subject to the restriction $n\pi = \lambda$, and comment on your finding.

Exercise 3.19. In a particular dice game, a person bets B dollars and rolls 5 balanced die simultaneously. If the 5 die all show the same number, the person wins A (> B) dollars; if the 5 die do not all show the same number, the person loses B dollars. Suppose that the person plays this game $k(\geq 1)$ consecutive times.

(a) Let the random variable G be the person's gain (in dollars) based on k plays of this game. Develop an explicit expression for $\mathrm{E}(G)$, the expected gain. Then, find a sufficient condition so that $\mathrm{E}(G) \geq k\mathrm{B}$, where $k\mathrm{B}$ is the maximum possible dollar loss for k plays of this game.

(b) Find the smallest value of k, say k^*, such that the probability of winning the game at least once in k^* plays is at least 0.80. For this value of k^*, and for $\mathrm{A} = 3000$ and $\mathrm{B} = 1$, what is the numerical value of $\mathrm{E}(G)$? How does the value of $\mathrm{E}(G)$ change if the person plays the game exactly one time (i.e., $k = 1$)?

Exercise 3.20. Suppose that $X \sim \mathrm{N}(\mu, \sigma^2)$. Develop an explicit expression for

$$\mathrm{E}\left\{[X - \mathrm{E}(X)][X^2 - \mathrm{E}(X^2)]\right\}.$$

Exercise 3.21. Suppose that $Y \sim \mathrm{N}(\mu, \sigma^2)$. For $r = 0, 1, \ldots, \infty$, find an explicit expression for $\mu_{2r} = \mathrm{E}\left[(Y - \mu)^{2r}\right]$ by first finding an explicit expression for $\mathrm{E}\left(Z^{2r}\right)$ when $Z = (Y - \mu)/\sigma$. Then, find explicit expressions for μ_2, μ_4, and μ_6.

Exercise 3.22. Consider the same probability model as used in Exercise 2.41 to model the game of craps when played with a possibly biased pair of dice.

Let the discrete random variable N denote the number of rolls of this biased pair of dice needed until a game of craps ends (i.e., until either the player wins or loses the game). Develop an explicit expression (as a function of π) for $\mathrm{E}(N)$; as a check, show that $\mathrm{E}(N) = 3.376$ when $\pi = 1/6$ (i.e., when using an unbiased pair of dice).

Exercise 3.23. Suppose that the discrete random variable X_n has the geometric distribution

$$\mathrm{p}_{X_n}(x_n) = \pi_n(1 - \pi_n)^{x_n}, x_n = 0, 1, \ldots, \infty,$$

where $\pi_n = \lambda/n$ and $0 < \lambda < n$.

Find the limiting value of the moment generating function of $Y_n = X_n/n$ as $n \to \infty$, and then use this result to determine the asymptotic distribution of Y_n.

Exercise 3.24*. Suppose that $Y \sim \mathrm{Binomial}(n, \pi)$. Consider the standardized random variable

$$U = \frac{Y - n\pi}{\sqrt{n\pi(1 - \pi)}}.$$

(a) Show that the moment generating function $\mathrm{M}_U(t) = \mathrm{E}(\mathrm{e}^{tU})$ of the random variable U converges to $\mathrm{e}^{t^2/2}$ as $n \to \infty$, thus demonstrating the legitimacy of the so-called "normal approximation to the binomial distribution."

(b) Use the result in part (a) to find a reasonable value for $\text{pr}(148 \leq Y \leq 159)$ if $Y \sim \text{Binomial}(n = 500, \pi = 0.30)$.

Exercise 3.25*. Suppose that X is a normally distributed random variable with mean $\mu(> 0)$ and with variance $\sigma^2 = 1$. If

$$\text{g}(X) = \text{e}^{X^2/2} \int_X^{\infty} \text{e}^{-t^2/2} dt,$$

derive an explicit expression for $\text{E}[\text{g}(X)]$.

Exercise 3.26*. Suppose that the discrete random variable X has probability distribution $\text{p}_X(x), x = 0, 1, \ldots, \infty$.

(a) Prove that $\text{E}(X) = \sum_{u=0}^{\infty} [1 - \text{F}_X(u)]$, where $\text{F}_X(x) = \text{pr}(X \leq x)$ is the CDF for the random variable X.

(b) Use the result in part (a) to find $\text{E}(X)$ when $\text{p}_X(x) = (1 - \pi)\pi^x, x = 0, 1, \ldots, \infty$ and $0 < \pi < 1$.

Exercise 3.27*. Let Y be a random variable for which $\text{E}(Y) = 0$ and for which $\mu'_r = \text{E}(Y^r)$ satisfies the inequality

$$|\mu'_r| \leq \frac{r!}{2} K^{r-2} \text{V}(Y),$$

where $K > 0, r \geq 0$, and $\mu'_2 = \text{E}(Y^2) = \text{V}(Y)$.

For any constant $A > 1$, show that

$$\text{E}(A^Y) \leq 1 + \frac{1}{2}(\ln A)^2 (1 - K \ln A)^{-1} \text{V}(Y),$$

provided that $0 < K \ln A < 1$.

Exercise 3.28*. Suppose that $X \sim \text{POI}(\lambda)$. Show that

$$\nu_1 = \text{E}\left(|X - \text{E}(X)|\right) = (2\lambda)\text{pr}\left(X = [\lambda]\right),$$

where $[\lambda]$ is the greatest integer less than or equal to λ.

Exercise 3.29*. An urn contains $N(\geq 3)$ balls, of which $B(\geq 1)$ are black in color, R (≥ 1) are red in color, and W (≥ 1) are white in color. One ball (call it Ball #1) is randomly selected from this urn and its color is noted. If Ball #1 is black, then this black ball is returned to the urn, along with K (≥ 1)

additional black balls; if Ball #1 is red, then this red ball is returned to the urn, along with K (≥ 1) additional red balls; and if Ball #1 is white, then this white ball is returned to the urn, along with K (≥ 1) additional white balls. Then, a second ball (call it Ball #2) is randomly selected from this urn.

(a) Let the discrete random variable X be the total number of black balls selected from this urn. Find explicit expressions for $\mathrm{E}(X)$ and $\mathrm{V}(X)$.

(b) Find the limiting values of $\mathrm{E}(X)$ and $\mathrm{V}(X)$ as $K \to \infty$, and then provide a logical argument for why these limiting values make sense.

Exercise 3.30*. Suppose that $Y \sim \mathrm{N}(\mu, \sigma^2)$. Show that

$$\mathrm{E}\left(|Y - c|\right) = 2\sigma\left[\phi(\beta) + \beta\Phi(\beta)\right] - \sigma\beta, -\infty < c < \infty,$$

where

$$\phi(\beta) = \frac{1}{\sqrt{2\pi}}\mathrm{e}^{-\beta^2/2}, \ \Phi(\beta) = \int_{-\infty}^{\beta} \frac{1}{\sqrt{2\pi}}\mathrm{e}^{-z^2/2}\mathrm{d}z,$$

and $\beta = (c - \mu)/\sigma$.

Also, find the value of c that *minimizes* $\mathrm{E}\left(|Y - c|\right)$.

Exercise 3.31*. Suppose that the probability distribution $\mathrm{p}_X(x)$ for a discrete random variable X is of the form

$$\mathrm{p}_X(x) = k(x+1)^{-1}\mathrm{C}_x^n, x = 0, 1, \ldots, n,$$

where k is an appropriately chosen constant.

(a) Develop an explicit expression for $\mathrm{M}_X(t) = \mathrm{E}\left(\mathrm{e}^{tX}\right)$, the moment generating function for the random variable X, and then use this result to develop an explicit expression for the constant k.

(b) Use the results from part (a) to develop an explicit expression for $\mathrm{E}(X)$.

Exercise 3.32*. Suppose that an automobile insurance company insures a very large number of drivers, with each insured driver being in one of three classes, designated L, M, and H. Class L drivers are drivers at *low* risk for automobile accidents; Class M drivers are drivers at *moderate* risk for automobile accidents; and, Class H drivers are drivers at *high* risk for automobile accidents. Thirty percent of all drivers insured by this company are in Class L, 50% are in Class M, and the remaining 20% are in Class L.

The number of automobile accidents per year for any driver in Class L is assumed to have a Poisson distribution with mean 0.02. The number of automobile accidents per year for any driver in Class M is assumed to have a

Poisson distribution with mean 0.10. And, the number of automobile accidents per year for any driver in Class H is assumed to have a Poisson distribution with mean 0.20. Further, assume that all drivers insured by this company act completely independently of one another with regard to their involvement in automobile accidents.

(a) Find the numerical value of the probability that a randomly selected driver insured by this company will be involved in at least two accidents during any 12-month period of time.

(b) Suppose that two drivers insured by this company are randomly selected, and it is determined that neither driver has been involved in an automobile accident during a particular 12-month period of time. Find the numerical value of the probability that one of these two drivers belongs to Class L and that the other driver belongs to Class M.

(c) For any randomly chosen insured driver belonging to Class H, let the continuous random variable W_1 be the waiting time (in years) from the start of insurance coverage until such a driver is involved in his or her first automobile accident. By expressing $F_{W_1}(w_1)$, the CDF of W_1, in terms of a probability statement about an appropriately chosen discrete random variable, derive an explicit expression for the density function of W_1. Thus, use this density function to find that particular value w_1^* of W_1 such that the probability of waiting at least as long as w_1^* for the first automobile accident by such an insured driver is no greater than 0.50.

Exercise 3.33*. Suppose that the random variable X has a *logistic* distribution defined by the CDF

$$F_X(x) = \frac{1}{1 + e^{-\pi(x-\mu)/\sigma\sqrt{3}}}, -\infty < x < \infty, -\infty < \mu < \infty, 0 < \sigma < \infty.$$

Notationally, we write $X \sim \text{LOGISTIC}(\mu, \sigma)$.

(a) Find explicit expressions for $E(X)$ and $V(X)$. You may have use for the following result:

$$\int_0^1 \frac{\ln x}{(1-x)} dx = \int_0^1 \frac{\ln(1-x)}{x} dx = -\frac{\pi^2}{6}.$$

(b) Suppose that $X_1 \sim \text{LOGISTIC}(\mu_1, \sigma)$ and that $X_2 \sim \text{LOGISTIC}(\mu_2, \sigma)$. Let $C(-\infty < C < \infty)$ be a known constant. With $\pi_1 = \text{pr}(X_1 \leq C)$ and $\pi_2 = \text{pr}(X_2 \leq C)$, show that

$$\ln\psi = -\frac{\pi}{\sqrt{3}}\left(\frac{\mu_1 - \mu_2}{\sigma}\right),$$

where the odds ratio parameter $\psi = \frac{\pi_1/(1-\pi_1)}{\pi_2/(1-\pi_2)}$. For meta-analysis applications using this result, see Hasselblad and Hedges (1995).

Exercise 3.34*. A certain private medical insurance company models the amount X of individual claims per year (in thousands of dollars) using the distribution

$$\mathrm{f}_X(x) = \frac{k}{\sqrt{2\pi}\sigma x}\mathrm{e}^{-(\ln x-\mu)^2/2\sigma^2},$$
$$0 < x < M < \infty, -\infty < \mu < \infty, 0 < \sigma^2 < \infty,$$

where M denotes an upper bound on the allowed amount of any individual claim and where k is chosen so that $\mathrm{f}_X(x)$ is a valid density function.

Find a general expression for $\mathrm{E}(X)$, and then find the numerical value of $\mathrm{E}(X)$ when $M = 25$, $\mu = 1.90$, and $\sigma^2 = 0.34$.

Exercise 3.35*. In the field of *actuarial science*, the strategy of *proportional reinsurance* has been extensively studied. Under proportional reinsurance, an insurance company (called the "insurer") partners with a reinsurance company (called the "reinsurer"). The reinsurer agrees to pay the excess of any claim over an agreed upon amount A.

More specifically, suppose that the random variable X denotes the amount (in dollars) of a claim, and assume that X has the distribution $\mathrm{f}_X(x), 0 < x < \infty$. Then, $U = \min(X, A)$ is the amount paid by the insurer, and $V = \max(0, X - A)$ is the amount paid by the reinsurer.

(a) Show that

$$\mathrm{E}(U) = \mathrm{E}(X) - \int_0^\infty y\mathrm{f}_X(y + A)\mathrm{d}y,$$

so that the expected amount paid per claim by the insurer is reduced under a proportional reinsurance strategy.

(b) If

$$\mathrm{f}_X(x) = \frac{1}{\alpha}\mathrm{e}^{-x/\alpha}, 0 < x < \infty, 0 < \alpha < \infty,$$

find the value of A, say A^*, which will be a function of α, such that the average claim amount paid by the insurer is reduced by exactly 20% using a proportional reinsurance strategy.

Exercise 3.36*. Suppose that a continuous random variable X, $0 < X < 1$, has the distribution

$$\mathrm{f}_X(x) = \begin{cases} \alpha\left(\frac{x}{\beta}\right)^{\alpha-1} & \text{for } 0 < x \le \beta \\ \alpha\left(\frac{1-x}{1-\beta}\right)^{\alpha-1} & \text{for } \beta \le x < 1 \end{cases},$$

where $\alpha \geq 1$ and $0 < \beta < 1$.

(a) Develop an explicit expression for the CDF of the random variable X.

(b) Show how the median ξ of $f_X(x)$ varies as a function of the parameter β.

(c) For r a non-negative integer, show that

$$\mathrm{E}(X^r) = \frac{\alpha\beta^{r+1}}{(\alpha+r)} + \alpha(1-\beta)\sum_{j=0}^{r} \mathrm{C}_j^r \frac{(\beta-1)^{r-j}}{(\alpha+r-j)}.$$

Then, use this result to find explicit expressions for $\mathrm{E}(X)$ and $\mathrm{V}(X)$, and show that these expressions for $\mathrm{E}(X)$ and $\mathrm{V}(X)$ give the proper answers when $\alpha = 1$.

3.2 Solutions to Odd-Numbered Exercises

Solution 3.1. Define the following events: W="white ball is put into Urn 2"; B="black ball is put into Urn 2". Then,

$$\begin{aligned}
\mathrm{pr}(Y=1) &= \mathrm{pr}(Y=1|\mathrm{W})\mathrm{pr}(\mathrm{W}) + \mathrm{pr}(Y=1|\mathrm{B})\mathrm{pr}(\mathrm{B}) \\
&= \left(\frac{2}{4}\right)\left(\frac{2}{3}\right) + \left(\frac{1}{4}\right)\left(\frac{1}{3}\right) = 5/12; \\
\mathrm{pr}(Y=2) &= \mathrm{pr}(Y=2|\mathrm{W})\mathrm{pr}(\mathrm{W}) + \mathrm{pr}(Y=2|\mathrm{B})\mathrm{pr}(\mathrm{B}) \\
&= \left(\frac{2}{4}\right)\left(\frac{2}{3}\right)\left(\frac{2}{3}\right) + \left(\frac{3}{4}\right)\left(\frac{1}{3}\right)\left(\frac{1}{3}\right) = 11/36; \\
\text{similarly, } \mathrm{pr}(Y=3) &= \left(\frac{2}{4}\right)\left(\frac{1}{3}\right)\left(\frac{2}{2}\right)\left(\frac{2}{3}\right) + \left(\frac{3}{4}\right)\left(\frac{2}{3}\right)\left(\frac{1}{2}\right)\left(\frac{1}{3}\right) = 7/36; \\
\text{and } \mathrm{pr}(Y=4) &= (0)\left(\frac{2}{3}\right) + \left(\frac{3}{4}\right)\left(\frac{2}{3}\right)\left(\frac{1}{2}\right)(1)\left(\frac{1}{3}\right) = 1/12 \\
&= 1 - \mathrm{pr}(Y=1) - \mathrm{pr}(Y=2) - \mathrm{pr}(Y=3).
\end{aligned}$$

Thus, the probability distribution of Y is

$$\mathrm{p}_Y(y) = \frac{(19-4y)}{36} = \frac{19}{36} - \frac{y}{9}, \ y = 1, 2, 3, 4.$$

Thus,

$$\begin{aligned}
\mathrm{E}(Y) &= \sum_{y=1}^{4} y\left[\frac{19}{36} - \frac{y}{9}\right] = \frac{19}{36}\sum_{y=1}^{4} y - \frac{1}{9}\sum_{y=1}^{4} y^2 \\
&= \frac{19}{36}\left[\frac{4(5)}{2}\right] - \frac{1}{9}\left[\frac{4(5)(9)}{6}\right] = 1.944.
\end{aligned}$$

And, since

$$\begin{aligned}\mathrm{E}(Y^2) &= \sum_{y=1}^{4} y^2\left[\frac{19}{36}-\frac{y}{9}\right] = \frac{19}{36}\sum_{y=1}^{4} y^2 - \frac{1}{9}\sum_{y=1}^{4} y^3 \\ &= \frac{19}{36}\left[\frac{4(5)(9)}{6}\right] - \frac{1}{9}\left[\frac{4(5)}{2}\right]^2 = 4.722,\end{aligned}$$

we have

$$\mathrm{V}(Y) = \mathrm{E}(Y^2) - [\mathrm{E}(Y)]^2 = 4.722 - (1.944)^2 = 4.722 - 3.779 = 0.943.$$

Solution 3.3. Clearly, this double exponential density is symmetric around $\mathrm{E}(X) = \beta$. So,

$$\begin{aligned}\nu_1 &= \mathrm{E}[|X-\beta|] = \int_{-\infty}^{\infty} |x-\beta|\frac{1}{2\alpha}\mathrm{e}^{-|x-\beta|/\alpha}\mathrm{d}x \\ &= \int_{-\infty}^{\beta}(\beta-x)\frac{1}{2\alpha}\mathrm{e}^{-(\beta-x)/\alpha}\mathrm{d}x + \int_{\beta}^{\infty}(x-\beta)\frac{1}{2\alpha}\mathrm{e}^{-(x-\beta)/\alpha}\mathrm{d}x \\ &= \frac{1}{2}\int_0^{\infty} u\frac{1}{\alpha}\mathrm{e}^{-u/\alpha}\mathrm{d}u + \frac{1}{2}\int_0^{\infty} v\frac{1}{\alpha}\mathrm{e}^{-v/\alpha}\mathrm{d}v \\ &= \frac{1}{2}(\alpha) + \frac{1}{2}(\alpha) = \alpha.\end{aligned}$$

Solution 3.5. In general,

$$\begin{aligned}\mathrm{pr}(0 \le a < R < b < \infty) &= \int_a^b \frac{2}{\pi}(1+r^2)^{-1}\mathrm{d}r \\ &= \frac{2}{\pi}\left[\tan^{-1}(b) - \tan^{-1}(a)\right].\end{aligned}$$

So,

$$\begin{aligned}\mathrm{pr}(S=4) &= \mathrm{pr}\left(0 < R \le \frac{1}{\sqrt{3}}\right) \\ &= \frac{2}{\pi}\left[\tan^{-1}\left(\frac{1}{\sqrt{3}}\right) - \tan^{-1}(0)\right] \\ &= \frac{2}{\pi}\left(\frac{\pi}{6} - 0\right) = \frac{1}{3};\end{aligned}$$

$$\begin{aligned}\mathrm{pr}(S=3) &= \mathrm{pr}\left(\frac{1}{\sqrt{3}} < R \le 1\right) \\ &= \frac{2}{\pi}\left[\tan^{-1}(1) - \tan^{-1}\left(\frac{1}{\sqrt{3}}\right)\right] \\ &= \frac{2}{\pi}\left(\frac{\pi}{4} - \frac{\pi}{6}\right) = \frac{1}{6};\end{aligned}$$

$$\begin{aligned}
\mathrm{pr}(S=2) &= \mathrm{pr}(1 < R \leq \sqrt{3}) \\
&= \frac{2}{\pi}\left[\tan^{-1}(\sqrt{3}) - \tan^{-1}(1)\right] \\
&= \frac{2}{\pi}\left(\frac{\pi}{3} - \frac{\pi}{4}\right) = \frac{1}{6};
\end{aligned}$$

and $\mathrm{pr}(S=0) = \mathrm{pr}(R > \sqrt{3}) = 1 - \mathrm{pr}(S=4) - \mathrm{pr}(S=3) - \mathrm{pr}(S=2) = 1/3$.

So,

$$\mathrm{E}(S) = 4\left(\frac{1}{3}\right) + 3\left(\frac{1}{6}\right) + 2\left(\frac{1}{6}\right) + (0)\left(\frac{1}{3}\right) = \frac{13}{6} = 2.167.$$

And

$$\mathrm{E}(S^2) = (4)^2\left(\frac{1}{3}\right) + (3)^2\left(\frac{1}{6}\right) + (2)^2\left(\frac{1}{6}\right) + (0)^2\left(\frac{1}{3}\right) = \frac{45}{6} = 7.500,$$

so that $\mathrm{V}(S) = \mathrm{E}(S^2) - [\mathrm{E}(S)]^2 = 7.500 - (2.167)^2 = 2.804$.

Solution 3.7. For $0 < u < \infty$, we have

$$\begin{aligned}
\mathrm{F}_U(u) &= \mathrm{pr}(U \leq u) = \mathrm{pr}(|X| \leq u) = \mathrm{pr}(-u \leq X \leq u) \\
&= \mathrm{pr}\left[\frac{-u-\theta}{\sqrt{\theta}} \leq \frac{X-\theta}{\sqrt{\theta}} \leq \frac{u-\theta}{\sqrt{\theta}}\right] \\
&= \mathrm{F}_Z\left(\frac{u-\theta}{\sqrt{\theta}}\right) - \mathrm{F}_Z\left(\frac{-u-\theta}{\sqrt{\theta}}\right), Z \sim \mathrm{N}(0,1).
\end{aligned}$$

So,

$$\begin{aligned}
\mathrm{f}_U(u) &= \frac{\mathrm{dF}_U(u)}{\mathrm{d}u} \\
&= (2\pi\theta)^{-1/2}\mathrm{e}^{-(u-\theta)^2/2\theta} + (2\pi\theta)^{-1/2}\mathrm{e}^{-(u+\theta)^2/2\theta} \\
&= (2\pi\theta)^{-1/2}\mathrm{e}^{-\theta/2}e^{-u^2/2\theta}\left(\mathrm{e}^{u} + \mathrm{e}^{-u}\right), 0 < u < \infty.
\end{aligned}$$

Solution 3.9. Using integration by parts, we let $u = \mathrm{F}_X(x)$, so that $\mathrm{d}u = \mathrm{f}_X(x)\mathrm{d}x$; and we let $\mathrm{d}v = \mathrm{d}x$, so that $v = x$. Then, using the fact that $\mathrm{f}_X(x)$ is symmetric about $x = 0$, we have

$$\begin{aligned}
\int_{-k}^{k} \mathrm{F}_X(x)\mathrm{d}x &= [x\mathrm{F}_X(x)]_{-k}^{k} - \int_{-k}^{k} x\mathrm{f}_X(x)\mathrm{d}x \\
&= [k\mathrm{F}_X(k) - (-k)\mathrm{F}_X(-k)] - 0 \\
&= k\mathrm{F}_X(k) + k\left[1 - \mathrm{F}_X(k)\right] = k.
\end{aligned}$$

Solution 3.11.

(a)

$$\begin{aligned}
F_X(x) &= \int_k^x \theta k^{\theta} t^{-(\theta+1)} dt \\
&= \theta k^{\theta} \left[\frac{-t^{-\theta}}{\theta} \right]_k^x \\
&= k^{\theta} \left[k^{-\theta} - x^{-\theta} \right] \\
&= 1 - \left(\frac{k}{x} \right)^{\theta}, \ 0 < k < x < +\infty.
\end{aligned}$$

And

$$\begin{aligned}
&\text{pr}\left[(k+1) < X < (k+3) | X > (k+1)\right] \\
&= \frac{\text{pr}[(k+1) < X < (k+3)]}{\text{pr}[X > (k+1)]} \\
&= \frac{F_X(k+3) - F_X(k+1)}{1 - F_X(k+1)} \\
&= \frac{\left[1 - \left(\frac{k}{k+3}\right)^{\theta}\right] - \left[1 - \left(\frac{k}{k+1}\right)^{\theta}\right]}{1 - \left[1 - \left(\frac{k}{k+1}\right)^{\theta}\right]} \\
&= \frac{\left(\frac{k}{k+1}\right)^{\theta} - \left(\frac{k}{k+3}\right)^{\theta}}{\left(\frac{k}{k+1}\right)^{\theta}} \\
&= 1 - \left(\frac{k+1}{k+3}\right)^{\theta}.
\end{aligned}$$

When $k = 1$ and $\theta = 4$,

$$1 - \left(\frac{k+1}{k+3}\right)^{\theta} = 1 - \left(\frac{1}{2}\right)^4 = \frac{15}{16}.$$

(b) Now, for $0 \leq r < \theta$,

$$\begin{aligned}
\mathrm{E}(X^r) &= \int_k^{\infty} x^r \theta k^{\theta} x^{-(\theta+1)} \mathrm{d}x \\
&= \theta k^{\theta} \int_k^{\infty} x^{(r-\theta)-1} \mathrm{d}x \\
&= \theta k^{\theta} \left[\frac{x^{(r-\theta)}}{(r-\theta)} \right]_k^{\infty} \\
&= \theta k^{\theta} \left[0 - \frac{k^{(r-\theta)}}{(r-\theta)} \right] \\
&= \frac{\theta k^r}{(\theta - r)}, \quad 0 \leq r < \theta.
\end{aligned}$$

Thus,

$$\begin{aligned}
\mu_3 &= \mathrm{E}(X^3) - 3\mathrm{E}(X^2)\mathrm{E}(X) + 2[\mathrm{E}(X)]^3 \\
&= \frac{\theta k^3}{(\theta-3)} - 3\left[\frac{\theta k^2}{(\theta-2)}\right]\left[\frac{\theta k}{(\theta-1)}\right] + 2\left[\frac{\theta k}{(\theta-1)}\right]^3 \\
&= k^3 \left[\frac{\theta}{(\theta-3)} - \frac{3\theta^2}{(\theta-1)(\theta-2)} + \frac{2\theta^3}{(\theta-1)^3} \right].
\end{aligned}$$

Clearly, as $\theta \to +\infty$, the limiting of μ_3 is $k^3(1-3+2) = 0$. Now, note that

$$\mathrm{E}(X) = \frac{k\theta}{(\theta-1)} \text{ and}$$

$$\mathrm{V}(X) = \frac{\theta k^2}{(\theta-2)} - \left[\frac{\theta k}{(\theta-1)}\right]^2 = k^2 \left[\frac{\theta}{(\theta-2)} - \frac{\theta^2}{(\theta-1)^2} \right].$$

Thus, as $\theta \to +\infty$, $\mathrm{E}(X) \to k$ and $\mathrm{V}(X) \to 0$. Hence, as $\theta \to +\infty$, the limiting distribution of X becomes "degenerate," namely, X takes the value k with probability 1. Note also that $\mathrm{F}_X(x) = \mathrm{pr}(X \leq x) \to 1$ as $\theta \to +\infty$, which is as expected since $0 < k < x < +\infty$.

(c) $\mathrm{E}(\mathrm{e}^{tY}) = \mathrm{E}(\mathrm{e}^{t\ln X}) = \mathrm{E}(\mathrm{e}^{\ln X^t}) = \mathrm{E}(X^t) = \frac{\theta k^t}{(\theta-t)} = \mathrm{M}_Y(t),\ t < \theta$.
So, $\mathrm{E}(Y) = \mathrm{E}[\ln(X)] = \left\{\frac{\mathrm{dM}_Y(t)}{\mathrm{d}t}\right\}_{|t=0}$. Now,

$$\begin{aligned}
\frac{\mathrm{d}\left[\frac{\theta k^t}{(\theta-t)}\right]}{\mathrm{d}t} &= \theta \frac{\mathrm{d}}{\mathrm{d}t}\left[\frac{\mathrm{e}^{t\ln k}}{(\theta-t)}\right] \\
&= \theta \left\{ \frac{(\ln k)\mathrm{e}^{t\ln k}(\theta-t) - (-1)\mathrm{e}^{t\ln k}}{(\theta-t)^2} \right\}.
\end{aligned}$$

Finally,

$$\begin{aligned} \mathrm{E}(Y) &= \theta\left[\frac{(\ln k)\mathrm{e}^0(\theta)+\mathrm{e}^0}{\theta^2}\right] \\ &= \ln k + \frac{1}{\theta}. \end{aligned}$$

To illustrate an alternative and more time-consuming approach for finding $\mathrm{M}_Y(t)$, note that

$$\begin{aligned} \mathrm{F}_Y(y) &= \mathrm{pr}(Y \leq y) = \mathrm{pr}[\ln(X) \leq y] = \mathrm{pr}[X \leq \mathrm{e}^y] \\ &= 1 - \left(\frac{k}{\mathrm{e}^y}\right)^{\theta} = 1 - k^{\theta}\mathrm{e}^{-\theta y}, \end{aligned}$$

so that $\mathrm{f}_Y(y) = \theta k^{\theta}\mathrm{e}^{-\theta y}, -\infty < \ln(k) < y < +\infty$. Then, it follows directly that

$$\mathrm{M}_Y(t) = \int_{\ln(k)}^{\infty} \mathrm{e}^{ty}\theta k^{\theta}\mathrm{e}^{-\theta y}\mathrm{d}y = \frac{\theta k^t}{(\theta - t)}, t < \theta.$$

Solution 3.13.

(a) With $u = -t$, so that $\mathrm{d}u = -\mathrm{d}t$, and noting that switching the order of integration is legitimate here, we have

$$\begin{aligned} \mathrm{E}(X^{-1}) &= \int_0^{\infty} x^{-1}\mathrm{f}_X(x)\,\mathrm{d}x = \int_0^{\infty}\left(\int_{-\infty}^{0} \mathrm{e}^{ux}\,\mathrm{d}u\right)\mathrm{f}_X(x)\,\mathrm{d}x \\ &= \int_{-\infty}^{0}\left(\int_0^{\infty} \mathrm{e}^{ux}\mathrm{f}_X(x)\,\mathrm{d}x\right)\mathrm{d}u \\ &= \int_{-\infty}^{0} \mathrm{M}_X(u)\,\mathrm{d}u \\ &= \int_{\infty}^{0} \mathrm{M}_X(-t)(-\mathrm{d}t) \\ &= \int_0^{\infty} \mathrm{M}_X(-t)\,\mathrm{d}t. \end{aligned}$$

(b) If $X \sim$GAMMA(α, β), then $\mathrm{M}_X(t) = (1 - \alpha t)^{-\beta}$, so that $\mathrm{M}_X(-t) = (1 + \alpha t)^{-\beta}$.

So,

$$\begin{aligned} \mathrm{E}(X^{-1}) &= \int_0^{\infty} (1+\alpha t)^{-\beta}\,\mathrm{d}t \\ &= \left[\frac{-(1+\alpha t)^{-(\beta-1)}}{\alpha(\beta-1)}\right]_0^{\infty} = \frac{1}{\alpha(\beta-1)},\ \alpha > 0,\ \beta > 1. \end{aligned}$$

Solution 3.15.

(a) With $u = (x-\theta)^{2m}$, so that $(x-\theta) = u^{1/2m}$ and $dx = \frac{1}{2m}u^{\frac{1}{2m}-1}$, and appealing to properties of the gamma distribution, we have

$$\begin{aligned}
\mathrm{E}\left[(X-\theta)^{2r}\right] &= \int_{-\infty}^{\infty}(x-\theta)^{2r}k\mathrm{e}^{-(x-\theta)^{2m}}dx \\
&= 2k\int_{0}^{\infty}(x-\theta)^{2r}\mathrm{e}^{-(x-\theta)^{2m}}dx \\
&= 2k\int_{0}^{\infty}u^{r/m}\mathrm{e}^{-u}\frac{1}{2m}u^{\frac{1}{2m}-1}du \\
&= \frac{k}{m}\int_{0}^{\infty}u^{\left(\frac{2r+1}{2m}\right)-1}\mathrm{e}^{-u}du \\
&= \left(\frac{k}{m}\right)\Gamma\left(\frac{2r+1}{2m}\right).
\end{aligned}$$

When $r = 0$, it follows that

$$\left(\frac{k}{m}\right)\Gamma\left(\frac{1}{2m}\right) = 1,$$

so that

$$k = \frac{m}{\Gamma\left(\frac{1}{2m}\right)}.$$

Finally, we have

$$\mathrm{E}\left[(X-\theta)^{r}\right] = \frac{\Gamma\left(\frac{2r+1}{2m}\right)}{\Gamma\left(\frac{1}{2m}\right)}, r = 0, 1, 2, \ldots$$

(b) When $m = 1$ and $r = 1$, it follows that

$$\mathrm{E}\left[(X-\theta)^{2}\right] = \frac{\Gamma\left(\frac{3}{2}\right)}{\Gamma\left(\frac{1}{2}\right)} = \frac{\sqrt{\pi}/2}{\sqrt{\pi}} = \frac{1}{2} = \mathrm{V}(X),$$

since $X \sim \mathrm{N}(\theta, 1/2)$.

Solution 3.17. Since these two students each may have reasonable knowledge of the course material on which the examination is based, it is problematic to use the pattern in correct answers to assess the possibility of cheating. However, given some reasonable assumptions, the pattern in wrong answers can be used to make such an assessment.

First, for each question answered incorrectly, assume that the two students were not cheating and were simply guessing at the right answer. Then, since there were four possible wrong answers to any question, the probability that these two students randomly chose the same wrong answer to a particular question is equal to $4\left(\frac{1}{4}\right)^2 = \frac{1}{4}$. Then, assuming mutual independence among the students' responses to all questions,

$$X \sim \text{BIN}\left(n = 10, \pi = \frac{1}{4}\right),$$

where X is the number of questions out of 10 for which the two students gave the same wrong answer.

Now,

$$\begin{aligned}
\text{pr}(X \geq 5|\text{no cheating}) &= \sum_{x=5}^{10} \text{C}_x^{10}\left(\frac{1}{4}\right)^x\left(\frac{3}{4}\right)^{10-x} \\
&= 1 - \sum_{x=0}^{4} \text{C}_x^{10}\left(\frac{1}{4}\right)^x\left(\frac{3}{4}\right)^{10-x} = 0.0781.
\end{aligned}$$

The probability value of 0.0781 is suggestive of the possibility of cheating, but it is probably not a sufficiently small enough value to warrant a confrontation with the two students.

Solution 3.19.

(a) Since the probability of the 5 die all showing a particular number is $(1/6)^5$, and since there are six possible numbers, the probability of winning the game is equal to $6(1/6)^5 = 1/1296$.

Now, for k plays of the game, and using the binomial distribution, we have

$$\begin{aligned}
&\text{pr}\left[G = j(\text{A} - \text{B}) - (k - j)\text{B}\right] \\
&\qquad = \text{C}_j^k\left(\frac{1}{1296}\right)^j\left(\frac{1295}{1296}\right)^{k-j}, \quad j = 0, 1, \ldots, k.
\end{aligned}$$

So, since $G = (j\text{A} - k\text{B})$, it follows that

$$\begin{aligned}
\text{E}(G) &= \text{A}\sum_{j=0}^{k} j\text{C}_j^k\left(\frac{1}{1296}\right)^j\left(\frac{1295}{1296}\right)^{k-j} - k\text{B} \\
&= \text{A}\left(\frac{k}{1296}\right) - k\text{B} \\
&= k\left(\frac{\text{A}}{1296} - \text{B}\right).
\end{aligned}$$

We require

$$k\left(\frac{\text{A}}{1296} - \text{B}\right) \geq k\text{B}, \text{ or A} \geq (2592)\text{B}.$$

(b) We require the smallest value of k such that

$$1 - \left(\frac{1295}{1296}\right)^k \geq 0.80,$$

which gives $k^* = 2,085$.
When $k^* = 2085, \text{A} = 3000$, and $\text{B} = 1$, we have

$$\text{E}(G) = (2,012)\left(\frac{3000}{1296} - 1\right) = 2741.39.$$

If $k = 1$, then $\text{E}(G) = (1)\left(\frac{3000}{1296} - 1\right) = 1.31$.

Solution 3.21. Since $Z = (Y - \mu)/\sigma \sim \text{N}(0,1)$, the moment generating function of Z is $\text{M}_Z(t) = \text{E}\left(\text{e}^{tZ}\right) = \text{e}^{t^2/2}$. Thus, we have

$$\begin{aligned} \text{M}_Z(t) &= \text{e}^{t^2/2} = \sum_{r=0}^{\infty} \frac{(t^2/2)^r}{r!} \\ &= \sum_{r=0}^{\infty} \left[\frac{(2r)!}{r!2^r}\right] \frac{t^{2r}}{(2r)!}, \end{aligned}$$

so that

$$\text{E}\left(Z^{2r}\right) = \frac{(2r)!}{r!2^r}, r = 0, 1, \ldots, \infty.$$

From this expansion of $\text{M}_Z(t)$, it also follows directly that $\text{E}(Z^k) = 0$ if k is an odd positive integer.

Now, since

$$\text{E}\left(Z^{2r}\right) = \text{E}\left[\left(\frac{Y - \mu}{\sigma}\right)^{2r}\right] = \sigma^{-2r}\text{E}\left[(Y - \mu)^{2r}\right],$$

we have

$$\mu_{2r} = \text{E}\left[(Y - \mu)^{2r}\right] = \frac{(2r)!\sigma^{2r}}{r!2^r}, r = 0, 1, \ldots, \infty.$$

It then follows directly that $\mu_2 = \text{V}(Y) = \sigma^2, \mu_4 = 3\sigma^4$, and $\mu_6 = 15\sigma^6$.

Solution 3.23. We have

$$\begin{aligned}
\mathrm{M}_{X_n}(t) &= \mathrm{E}\left(\mathrm{e}^{tX_n}\right) = \sum_{x_n=0}^{\infty} \mathrm{e}^{tx_n}\pi_n(1-\pi_n)^{x_n} \\
&= \pi_n \sum_{x_n=0}^{\infty} \left[\mathrm{e}^t(1-\pi_n)\right]^{x_n} \\
&= \frac{\pi_n}{1-\mathrm{e}^t(1-\pi_n)}, 0 < \mathrm{e}^t(1-\pi_n) < 1.
\end{aligned}$$

So,

$$\begin{aligned}
\mathrm{M}_{Y_n}(t) &= \mathrm{E}\left(\mathrm{e}^{tY_n}\right) = \mathrm{E}\left(\mathrm{e}^{\frac{t}{n}X_n}\right) \\
&= \frac{\pi_n}{1-\mathrm{e}^{t/n}(1-\pi_n)} \\
&= \frac{\lambda/n}{1-\mathrm{e}^{t/n}\left(1-\frac{\lambda}{n}\right)}.
\end{aligned}$$

Since $\lim_{n\to\infty}\mathrm{M}_{Y_n}(t) = \frac{0}{0}$, we can employ L'Hôpital's Rule. In particular, we have

$$\begin{aligned}
&\frac{\mathrm{d}(\lambda/n)/\mathrm{d}n}{\mathrm{d}\left[1-\mathrm{e}^{t/n}\left(1-\frac{\lambda}{n}\right)\right]/\mathrm{d}n} \\
&= \frac{-\lambda/n^2}{-\mathrm{e}^{t/n}\left(-\frac{t}{n^2}\right)\left(1-\frac{\lambda}{n}\right) - \mathrm{e}^{t/n}\left(\frac{\lambda}{n^2}\right)} \\
&= \frac{-\lambda}{t\mathrm{e}^{\lambda/n}\left(1-\frac{\lambda}{n}\right) - \lambda\mathrm{e}^{t/n}}.
\end{aligned}$$

As $n \to \infty$, this quantity converges to

$$\frac{\lambda}{(\lambda - t)} = \left(1 - \frac{t}{\lambda}\right)^{-1},$$

which is the moment generating function of a NEGEXP($\alpha = \lambda^{-1}$) random variable.

Solution 3.25*. For $Z \sim \mathrm{N}(0,1)$,

$$\mathrm{F}_Z(x) = \frac{1}{\sqrt{2\pi}}\int_{-\infty}^{x} \mathrm{e}^{-t^2/2}\mathrm{d}t, \text{ so that } \mathrm{g}(X) = \sqrt{2\pi}\mathrm{e}^{X^2/2}[1-\mathrm{F}_Z(X)].$$

Thus,

$$\begin{aligned}
\mathrm{E}[\mathrm{g}(X)] &= \int_{-\infty}^{\infty} \mathrm{g}(x)\frac{1}{\sqrt{2\pi}}\mathrm{e}^{-(x-\mu)^2/2}\mathrm{d}x \\
&= \int_{-\infty}^{\infty} \mathrm{e}^{x^2/2}[1-\mathrm{F}_Z(x)]\mathrm{e}^{-(x-\mu)^2/2}\mathrm{d}x \\
&= \mathrm{e}^{-\mu^2/2}\int_{-\infty}^{\infty}[1-\mathrm{F}_Z(x)]\mathrm{e}^{\mu x}\mathrm{d}x.
\end{aligned}$$

To evaluate this integral, we use integration by parts with

$$u = [1-\mathrm{F}_Z(x)], \mathrm{d}u = -\mathrm{f}_Z(x)\mathrm{d}x = \frac{-1}{\sqrt{2\pi}}\mathrm{e}^{-x^2/2}\mathrm{d}x,$$

$$\mathrm{d}v = \mathrm{e}^{\mu x}\mathrm{d}x, \text{ and } v = \mu^{-1}\mathrm{e}^{\mu x}.$$

So, $\mathrm{E}[\mathrm{g}(X)] = \mathrm{e}^{-\mu^2/2}\int_{-\infty}^{\infty} u\mathrm{d}v = \mathrm{e}^{-\mu^2/2}\left\{[uv]_{-\infty}^{\infty} - \int_{-\infty}^{\infty} v\mathrm{d}u\right\}$

$$\begin{aligned}
&= \mathrm{e}^{-\mu^2/2}\left\{\left[\{1-\mathrm{F}_Z(x)\}\mu^{-1}\mathrm{e}^{\mu x}\right]_{-\infty}^{\infty} + \int_{-\infty}^{\infty}\mu^{-1}\mathrm{e}^{\mu x}\frac{1}{\sqrt{2\pi}}\mathrm{e}^{-x^2/2}\mathrm{d}x\right\} \\
&= \frac{\mathrm{e}^{-\mu^2/2}}{\mu}\left\{[\{1-\mathrm{F}_Z(x)\}\mathrm{e}^{\mu x}]_{-\infty}^{\infty} + \int_{-\infty}^{\infty}\mathrm{e}^{\mu x}\frac{1}{\sqrt{2\pi}}\mathrm{e}^{-x^2/2}\mathrm{d}x\right\} \\
&= \frac{\mathrm{e}^{-\mu^2/2}}{\mu}\left\{[\{1-\mathrm{F}_Z(x)\}\mathrm{e}^{\mu x}]_{-\infty}^{\infty} + \mathrm{e}^{\mu^2/2}\right\} \\
&= \mu^{-1} + \frac{\mathrm{e}^{-\mu^2/2}}{\mu}\left\{[1-\mathrm{F}_Z(x)]\,\mathrm{e}^{\mu x}\right\}_{-\infty}^{\infty}.
\end{aligned}$$

Now, since $\mu > 0$, it follows directly that $\lim_{x\to-\infty}[1-\mathrm{F}_Z(x)]\mathrm{e}^{\mu x} = 0$ since $\mathrm{F}_Z(-\infty) = 0$.

And, since $\mathrm{F}_Z(\infty) = 1$, it follows that $lim_{x\to\infty}\left[\frac{1-\mathrm{F}_Z(x)}{\mathrm{e}^{-\mu x}}\right] = \frac{0}{0}$ for $\mu > 0$, so that we can employ L'Hospital's Rule. Thus,

$$\begin{aligned}
lim_{x\to\infty}\left[\frac{1-\mathrm{F}_Z(x)}{\mathrm{e}^{-\mu x}}\right] &= lim_{x\to\infty}\left[\frac{-\mathrm{f}_Z(x)}{-\mu\mathrm{e}^{-\mu x}}\right] \\
&= \mu^{-1}lim_{x\to\infty}\left[\frac{\frac{1}{\sqrt{2\pi}}\mathrm{e}^{-x^2/2}}{\mathrm{e}^{-\mu x}}\right] \\
&= \frac{1}{\mu\sqrt{2\pi}}lim_{x\to\infty}\left[\mathrm{e}^{-\frac{1}{2}(x-\mu)^2+\frac{\mu^2}{2}}\right] = 0.
\end{aligned}$$

Thus, $\mathrm{E}[\mathrm{g}(X)] = \mu^{-1}$.

Solution 3.27*.

$$\begin{aligned}
\mathrm{E}(A^Y) &= \mathrm{E}(\mathrm{e}^{Y\ln A}) = \mathrm{E}\left[\sum_{r=0}^{\infty} \frac{(Y\ln A)^r}{r!}\right] \\
&= 1 + \mathrm{E}(Y\ln A) + \sum_{r=2}^{\infty} \frac{(\ln A)^r}{r!}\mathrm{E}(Y^r) \\
&= 1 + \sum_{r=2}^{\infty} \frac{(\ln A)^r}{r!}\mu_r',
\end{aligned}$$

since $\mathrm{E}(Y) = 0$.

Now, with $A > 1$ so that $\ln A > 0$, and with $\mu_r' \leq |\mu_r'|$, we have

$$\begin{aligned}
\mathrm{E}(A^Y) &\leq 1 + \sum_{r=2}^{\infty} \frac{(\ln A)^r}{r!}|\mu_r'| \\
&\leq 1 + \sum_{r=2}^{\infty} \frac{(\ln A)^r}{r!}\left[\frac{r!}{2}K^{r-2}\mathrm{V}(Y)\right] \\
&= 1 + \frac{1}{2}\mathrm{V}(Y)\sum_{r=2}^{\infty}(\ln A)^r K^{r-2}.
\end{aligned}$$

Now, the infinite series $\sum_{r=2}^{\infty}(\ln A)^r K^{r-2}$ will be a convergent geometric series with sum equal to $(\ln A)^2(1 - K\ln A)^{-1}$ if $0 < K\ln A < 1$.

Thus,

$$\mathrm{E}(A^Y) \leq 1 + \frac{1}{2}(\ln A)^2(1 - K\ln A)^{-1}\mathrm{V}(Y) \text{ if } 0 < K\ln A < 1,$$

which is the desired result.

Solution 3.29*.

(a) Clearly, the possible values of X are 0, 1, and 2. Also, let B_1 be the event that Ball #1 is black, and let B_2 be the event that Ball #2 is black. Then, with $N = (B + R + W)$, it follows that

$$\begin{aligned}
\mathrm{pr}(X = 1) &= \mathrm{pr}(\mathrm{B}_1 \cap \overline{\mathrm{B}_2}) + \mathrm{pr}(\overline{\mathrm{B}_1} \cap \mathrm{B}_2) \\
&= \mathrm{pr}(\mathrm{B}_1)\mathrm{pr}(\overline{\mathrm{B}_2}|\mathrm{B}_1) + \mathrm{pr}(\overline{\mathrm{B}_1})\mathrm{pr}(\mathrm{B}_2|\overline{\mathrm{B}_1}) \\
&= \left(\frac{B}{N}\right)\left(\frac{R+W}{N+K}\right) + \left(\frac{R+W}{N}\right)\left(\frac{B}{N+K}\right) \\
&= \frac{2B(R+W)}{N(N+K)},
\end{aligned}$$

and that

$$\begin{aligned}\text{pr}(X=2) &= \text{pr}(\text{B}_1 \cap \text{B}_2) = \text{pr}(\text{B}_1)\text{pr}(\text{B}_2|\text{B}_1)\\ &= \left(\frac{B}{N}\right)\left(\frac{B+K}{N+K}\right)\\ &= \frac{B(B+K)}{N(N+K)}.\end{aligned}$$

Thus, we have

$$\begin{aligned}\text{E}(X) &= \sum_{x=0}^{2}(x)\text{pr}(X=x) = (1)\text{pr}(X=1) + (2)\text{pr}(X=2)\\ &= (1)\left[\frac{2B(R+W)}{N(N+K)}\right] + (2)\left[\frac{B(B+K)}{N(N+K)}\right]\\ &= \frac{2B(R+W+B+K)}{N(N+K)} = \frac{2B(N+K)}{N(N+K)}\\ &= \frac{2B}{N}.\end{aligned}$$

And, since

$$\begin{aligned}\text{E}(X^2) &= \sum_{x=0}^{2}(x^2)\text{pr}(X=x) = (1)^2\text{pr}(X=1) + (2)^2\text{pr}(X=2)\\ &= (1)^2\left[\frac{2B(R+W)}{N(N+K)}\right] + (4)\left[\frac{B(B+K)}{N(N+K)}\right]\\ &= \frac{2B(R+W+2B+2K)}{N(N+K)} = \frac{2B(N+B+2K)}{N(N+K)},\end{aligned}$$

we have

$$\begin{aligned}\text{V}(X) &= \text{E}\left(X^2\right) - [\text{E}(X)]^2\\ &= \frac{2B(N+B+2K)}{N(N+K)} - \left[\frac{2B}{N}\right]^2\\ &= \frac{2B(N-B)(N+2K)}{N^2(N+K)}.\end{aligned}$$

(b) First, clearly $\lim_{K\to\infty}\text{E}(X) = \text{E}(X) = 2B/N$, and

$$\begin{aligned}\lim_{K\to\infty}\text{V}(X) &= \lim_{K\to\infty}\left[\frac{2B(N-B)\left(\frac{N}{K}+2\right)}{N^2\left(\frac{N}{K}+1\right)}\right]\\ &= \frac{4B(N-B)}{N^2} = 4\left(\frac{B}{N}\right)\left(1-\frac{B}{N}\right).\end{aligned}$$

Note that these results follow directly since

$$\lim_{K\to\infty}\text{pr}(X=1) = \lim_{K\to\infty}\left[\frac{2B(R+W)}{N(N+K)}\right] = 0,$$

and

$$\begin{aligned}\lim_{K\to\infty}\text{pr}(X=2) &= \lim_{K\to\infty}\left[\frac{B(B+K)}{N(N+K)}\right]\\ &= \lim_{K\to\infty}\left[\frac{B\left(\frac{B}{K}+1\right)}{N\left(\frac{N}{K}+1\right)}\right]\\ &= \frac{B}{N}.\end{aligned}$$

Solution 3.31*.

(a) We have

$$\begin{aligned}\text{M}_X(t) &= \text{E}\left(\text{e}^{tX}\right) = k\sum_{x=0}^{n}\text{e}^{tx}(x+1)^{-1}\text{C}_x^n\\ &= k\sum_{x=0}^{n}\text{e}^{tx}(x+1)^{-1}\left[\frac{n!}{x!(n-x)!}\right]\\ &= \frac{k}{(n+1)}\sum_{x=0}^{n}\text{e}^{tx}\text{C}_{x+1}^{n+1}\\ &= \frac{k}{(n+1)}\sum_{u=1}^{n+1}\text{e}^{t(u-1)}\text{C}_u^{n+1}\\ &= \frac{k\text{e}^{-t}}{(n+1)}\left[\sum_{u=0}^{n+1}\text{e}^{tu}\text{C}_u^{n+1}-1\right]\\ &= \frac{k\text{e}^{-t}}{(n+1)}\left[\sum_{u=0}^{n+1}\text{C}_u^{n+1}(\text{e}^t)^u(1)^{(n+1)-u}-1\right]\\ &= \frac{k\text{e}^{-t}}{(n+1)}\left[(\text{e}^t+1)^{n+1}-1\right].\end{aligned}$$

When $t=0$, we obtain

$$\text{M}_X(0) = 1 = \frac{k}{(n+1)}\left(2^{n+1}-1\right),$$

so that

$$k = \frac{(n+1)}{(2^{n+1}-1)}.$$

(b) Now,

$$\frac{\mathrm{dM}_X(t)}{\mathrm{d}t} = \frac{k}{(n+1)}\left\{-\mathrm{e}^{-t}\left[(\mathrm{e}^t+1)^{n+1}-1\right]+\mathrm{e}^{-t}(n+1)(\mathrm{e}^t+1)^n\mathrm{e}^t\right\},$$

so that

$$\begin{aligned}\mathrm{E}(X) &= \left.\frac{\mathrm{dM}_X(t)}{\mathrm{d}t}\right|_{t=0} = \frac{k}{(n+1)}\left[-(2^{n+1}-1)+(n+1)2^n\right]\\ &= \frac{1+(n-1)2^n}{(2^{n+1}-1)}.\end{aligned}$$

Solution 3.33*.

(a) First, the density function $\mathrm{f}_X(x)$ has the structure

$$\mathrm{f}_X(x) = \frac{\mathrm{dF}_X(x)}{\mathrm{d}x} = \frac{\left(\frac{\pi}{\sigma\sqrt{3}}\right)\mathrm{e}^{\pi(x-\mu)/\sigma\sqrt{3}}}{\left[1+\mathrm{e}^{\pi(x-\mu)/\sigma\sqrt{3}}\right]^2}, -\infty < x < \infty.$$

Clearly, $\mathrm{f}_X(x)$ is symmetric about μ, so that $\mathrm{E}(X) = \mu$. Now, let $Y = \pi(X-\mu)/\sigma\sqrt{3}$; then,

$$\mathrm{E}(Y) = 0 \text{ and } \mathrm{E}(Y^2) = \left(\frac{\pi^2}{3\sigma^2}\right)\mathrm{E}[(X-\mu)^2] = \left(\frac{\pi^2}{3\sigma^2}\right)\mathrm{V}(X),$$

so that we can find $\mathrm{V}(X)$ indirectly by first finding $\mathrm{E}(Y^2)$. So, since $\mathrm{d}Y = \left(\frac{\pi}{\sigma\sqrt{3}}\right)\mathrm{d}X$, it follows that

$$\mathrm{f}_Y(y) = \frac{\mathrm{e}^y}{(1+\mathrm{e}^y)^2}, -\infty < y < \infty, \text{ and } \mathrm{E}(Y^2) = \int_{-\infty}^{\infty}(y^2)\frac{\mathrm{e}^y}{(1+\mathrm{e}^y)^2}\mathrm{d}y.$$

Now, with

$$w = \frac{\mathrm{e}^y}{(1+\mathrm{e}^y)}, 0 < w < 1, \text{ so that}$$

$$y = [\ln w - \ln(1-w)] \text{ and } \mathrm{d}w = \left[\frac{\mathrm{e}^y}{(1+\mathrm{e}^y)^2}\right]\mathrm{d}y,$$

it follows that

$$\mathrm{E}(Y^2) = \int_0^1 [\ln w - \ln(1-w)]^2\,\mathrm{d}w.$$

Now, using integration by parts with

$$u = [\ln w - \ln(1-w)] \text{ and } \mathrm{d}v = [\ln w - \ln(1-w)]\,\mathrm{d}w,$$

so that

$$du = dw/w(1-w) \text{ and } v = w\ln w + (1-w)\ln(1-w),$$

and applying L'Hôpital's Rule, we have

$$\begin{aligned}
E(Y^2) &= [uv]_0^1 - \int_0^1 v du \\
&= \{[\ln w - \ln(1-w)][w\ln w + (1-w)\ln(1-w)]\}_0^1 \\
&- \int_0^1 [w\ln w + (1-w)\ln(1-w)] \frac{dw}{w(1-w)} \\
&= 0 - \int_0^1 \frac{\ln w}{(1-w)} dw - \int_0^1 \frac{\ln(1-w)}{w} dw \\
&= 0 - \left(-\frac{\pi^2}{6}\right) - \left(-\frac{\pi^2}{6}\right) = \frac{\pi^2}{3}.
\end{aligned}$$

Finally, we have

$$V(X) = \left(\frac{3\sigma^2}{\pi^2}\right) E(Y^2) = \sigma^2.$$

(b) For $i = 1, 2$, since

$$F_{X_i}(C) = \frac{1}{1 + e^{-\pi(C-\mu_i)/\sigma\sqrt{3}}} = \frac{e^{\pi(C-\mu_i)/\sigma\sqrt{3}}}{1 + e^{\pi(C-\mu_i)/\sigma\sqrt{3}}},$$

it follows that

$$\ln\left(\frac{\pi_i}{1-\pi_i}\right) = \ln\left[\frac{F_{X_i}(C)}{1 - F_{X_i}(C)}\right] = \frac{\pi(C-\mu_i)}{\sigma\sqrt{3}}.$$

Thus,

$$\begin{aligned}
\ln\psi &= \ln\left(\frac{\pi_1}{1-\pi_1}\right) - \ln\left(\frac{\pi_2}{1-\pi_2}\right) \\
&= \frac{\pi(C-\mu_1)}{\sigma\sqrt{3}} - \frac{\pi(C-\mu_2)}{\sigma\sqrt{3}} = -\frac{\pi}{\sqrt{3}}\left(\frac{\mu_1-\mu_2}{\sigma}\right).
\end{aligned}$$

Solution 3.35*.

(a) With $\mathrm{F}_X(x) = \int_0^x \mathrm{f}_X(t)\mathrm{d}t$, we have

$$\begin{aligned}
\mathrm{E}(U) &= \mathrm{E}(U|X \leq A)\mathrm{pr}(X \leq A) + \mathrm{E}(U|X > A)\mathrm{pr}(X > A) \\
&= \mathrm{E}(X|X \leq A)\mathrm{pr}(X \leq A) + A[1 - \mathrm{F}_X(A)] \\
&= \left[\int_0^A x\frac{\mathrm{f}_X(x)}{\mathrm{F}_X(A)}\mathrm{d}x\right][\mathrm{F}_X(A)] + A[1 - \mathrm{F}_X(A)] \\
&= \int_0^A x\mathrm{f}_X(x)\mathrm{d}x + A[1 - \mathrm{F}_X(A)] \\
&= \int_0^\infty x\mathrm{f}_X(x)\mathrm{d}x - \int_A^\infty x\mathrm{f}_X(x)\mathrm{d}x + A\int_A^\infty \mathrm{f}_X(x)\mathrm{d}x \\
&= \mathrm{E}(X) - \int_A^\infty (x - A)\mathrm{f}_X(x)\mathrm{d}x \\
&= \mathrm{E}(X) - \int_0^\infty y\mathrm{f}_X(y + A)\mathrm{d}y, \text{ where } y = (x - A).
\end{aligned}$$

(b) Since $X \sim \mathrm{NEGEXP}(\alpha), \mathrm{E}(X) = \alpha$, and so

$$\begin{aligned}
\mathrm{E}(U) &= \alpha - \int_0^\infty y\frac{1}{\alpha}\mathrm{e}^{-(y+A)/\alpha}\mathrm{d}y \\
&= \alpha - \mathrm{e}^{-A/\alpha}\int_0^\infty y\frac{1}{\alpha}\mathrm{e}^{-y/\alpha}\mathrm{d}y \\
&= \alpha\left(1 - \mathrm{e}^{-A/\alpha}\right).
\end{aligned}$$

We want

$$\mathrm{E}(U) = \alpha\left(1 - \mathrm{e}^{A/\alpha}\right) = 0.80\mathrm{E}(X) = 0.80\alpha,$$

so that

$$\left(1 - \mathrm{e}^{A/\alpha}\right) = 0.80,$$

giving $A^* = (1.6094)\alpha = (1.6094)\mathrm{E}(X)$.

For this choice for A, we obtain

$$\mathrm{E}(U) = \alpha\left(1 - \mathrm{e}^{-1.6094}\right) = \alpha(1 - 0.20) = 0.80\alpha = 0.80\mathrm{E}(X).$$

Chapter 4

Multivariate Distribution Theory

4.1 Exercises

Exercise 4.1. To assess whether there is genetic predisposition to becoming a cigarette smoker, epidemiologic studies have been done in which sets of monozygotic twins separated at birth and raised to adulthood in totally different environments are located and personally interviewed regarding their current smoking habits. For the i-th adult member of such a set of monozygotic twins $(i = 1, 2)$, let $X_i = 1$ if that adult member is currently a smoker, and let $X_i = 0$ if not. Because of the distinct possibility that responses from monozygotic twins may tend to be correlated, a certain biostatistician is *not* willing to assume that X_1 and X_2 are independent random variables. So, she suggests using the following two-parameter bivariate discrete probability distribution $\text{p}_{X_1,X_2}(x_1, x_2)$ for X_1 and X_2, where $\theta > 0$ and where $0 < \pi < 1$:

$$\begin{aligned} \text{pr}[X_1 = X_2 = 0] &= K(1-\pi)^2; \\ \text{pr}[(X_1 = 1) \cap (X_2 = 0)] &= \text{pr}[(X_1 = 0) \cap (X_2 = 1)] = K\pi(1-\pi)\theta; \\ \text{pr}[X_1 = X_2 = 1] &= K\pi^2. \end{aligned}$$

(a) Find the value of K which makes $\text{p}_{X_1,X_2}(x_1, x_2)$ a valid bivariate discrete probability distribution.

(b) Find the marginal distributions of X_1 and X_2.

(c) Find explicit expressions for $\text{E}(X_1|X_2 = 1)$ and $\text{V}(X_1|X_2 = 1)$.

(d) Find an explicit expression for $\text{corr}(X_1, X_2)$. For what specific sets of values of θ is this correlation positive, negative, or zero?

(e) If $L = (3X_1 - 4X_2)$, find explicit expressions for $\text{E}(L)$ and $\text{V}(L)$.

Exercise 4.2. Suppose that a random variable X has a distribution symmetric about zero, and let $Y = \beta_0 + \beta_1 X + \beta_2 X^2$. Develop an explicit expression for $\text{corr}(X, Y)$ as a function of $\text{V}(X), \text{V}(X^2), \beta_1$, and β_2, and then interpret the value of $\text{corr}(X, Y)$ when $\beta_1 = 0$ and when $\beta_2 = 0$.

Exercise 4.3. Consider the bivariate density

$$\text{f}_{X,Y}(x, y) = \frac{y}{(1+x)^4}\text{e}^{-y/(1+x)}, \quad x > 0,\ y > 0.$$

(a) Find the exact value of $\text{pr}\{(X < 1) \cap (Y > 0)|(X < 2)\}$.

(b) For r a positive integer, show that

$$\text{E}(Y^r|X = x) = (r+1)!\,(1+x)^r.$$

(c) Given $X = x_i$, let Y_i be a randomly selected observation from the conditional density $\text{f}_Y(y|X = x_i)$, $i = 1, 2, \ldots, n$. In other words, consider the set of n pairs (x_i, Y_i), $i = 1, 2, \ldots, n$, where the x_is are fixed constants and the Y_is are mutually independent random variables. Under the above conditions, find $\text{E}(L)$ and $\text{V}(L)$, where $L = \sum_{i=1}^{n}(x_i - \bar{x})Y_i$ and $\bar{x} = n^{-1}\sum_{i=1}^{n} x_i$.

Exercise 4.4. Suppose that X and Y are independent random variables. Develop an expression for $\text{V}(XY)$ as a function of $\text{E}(X)$, $\text{E}(Y)$, $\text{V}(X)$, and $\text{V}(Y)$, and then comment on the relationship between this expression and the product $\text{V}(X)\text{V}(Y)$.

Exercise 4.5. Suppose that the random variable X_1 denotes the total time (in minutes) between a typical patient's arrival and departure from a certain health clinic. Further, suppose that the random variable X_2 denotes the total time (in minutes) that a typical patient spends in the clinic waiting room before seeing a health professional. Empirical research suggests that the joint distribution of X_1 and X_2 can be reasonably represented by the bivariate density function

$$\text{f}_{X_1,X_2}(x_1, x_2; \theta) = (2\theta^3)^{-1} x_1\, e^{-x_1/\theta}, 0 < x_2 < x_1 < \infty, 0 < \theta < \infty,$$

where θ is an unknown parameter. The following two random variables are of interest:

$P = X_2/X_1$ is the proportion of the total time in this health clinic that a typical patient spends in the waiting room before seeing a health professional.

$S = (X_1 - X_2)$ is the time (in minutes) that a typical patient spends with a health professional at this health clinic.

(a) Derive an explicit expression for $\text{f}_{P,S}(p, s; \theta)$, the joint density function of the random variables P and S.

(b) Derive explicit expressions for the marginal distributions of the random variables P and S, and also provide explicit expressions for $\mathrm{E}(P)$, $\mathrm{V}(P)$, $\mathrm{E}(S)$, and $\mathrm{V}(S)$.

Exercise 4.6. Suppose that the random variable X represents the time (in weeks) from an initial diagnosis of advanced stage leukemia until the first chemotherapy treatment, and that Y represents the time (in weeks) from the initial diagnosis until death. The joint distribution of the random variables X and Y is assumed to have the structure

$$\mathrm{f}_{X,Y}(x, y; \theta) = 2\theta^{-2}\mathrm{e}^{-(x+y)/\theta}, \quad 0 < x < y < \infty;\ \theta > 0.$$

Cancer researchers are interested in the random variable

$$P = \frac{X}{Y},$$

which is the proportion of the total time between initial diagnosis and death that an advanced stage leukemia patient spends prior to starting chemotherapy treatments.

Develop an explicit expression for the density function $\mathrm{f}_P(p)$ of the random variable P.

Exercise 4.7. Let $X_1, X_2, \ldots, X_n$ constitute a random sample of size $n(\geq 2)$ from a $\mathrm{N}(\mu, \sigma^2)$ population. Find the expected value of the random variable

$$U = \frac{\sqrt{\pi}}{n(n-1)} \sum_{i=1}^{n-1} \sum_{j=i+1}^{n} |X_i - X_j|.$$

Exercise 4.8. Suppose that the number Y_i of automobile accidents occurring during year i $(i = 1, 2, \ldots, n)$ at a certain dangerous intersection is assumed to have a Poisson distribution with parameter λ_i. Further assume that $Y_1, Y_2, \ldots, Y_n$ are *mutually independent* random variables.

If n is an even positive integer, provide an expression for the probability that more total accidents occur during years $1, 2, \ldots, n/2$ (namely, the first $n/2$ years) than during years $(n/2 + 1), (n/2 + 2), \ldots, n$ (namely, the last $n/2$ years).

Exercise 4.9. Let $X_1, X_2, \ldots, X_n$ constitute a random sample of size n from the discrete probability distribution $\mathrm{p}_X(x) = \pi^x(1-\pi), x = 0, 1, \ldots, \infty$ and $0 < \pi < 1$. Derive an explicit expression for the distribution of the random variable $U = \min\{X_1, X_2, \ldots, X_n\}$.

Exercise 4.10. Suppose that the time X to death of non-smoking heart transplant patients follows the distribution

$$\mathrm{f}_X(x;\alpha) = \alpha^{-1}\mathrm{e}^{-x/\alpha},\ 0 < x < +\infty,\ \alpha > 0.$$

Further, suppose that the time Y to death of smoking heart transplant patients follows the density function

$$\mathrm{f}_Y(y;\beta) = \beta^{-1}\mathrm{e}^{-y/\beta},\ 0 < y < +\infty,\ \beta > 0.$$

(a) Assuming X and Y are independent random variables, find an explicit expression for the joint distribution of the random variables

$$U = (X - Y) \text{ and } V = (X + Y).$$

(b) Find an explicit expression for the distribution of U. Hence, or otherwise, find explicit expressions for $\mathrm{E}(U)$ and $\mathrm{V}(U)$.

Exercise 4.11. Suppose that the random variables X and Y follow a bivariate normal distribution with $\mathrm{E}(X) = \mathrm{E}(Y) = 0$, with $\mathrm{V}(X) = \mathrm{V}(Y) = \sigma^2$, and with $\mathrm{corr}(X, Y) = \rho$. If $U = \mathrm{e}^X$ and $V = \mathrm{e}^Y$, derive an explicit expression for $\mathrm{corr}(U, V)$ as a function of σ^2 and ρ. Also, investigate the range of $\mathrm{corr}(U, V)$ as ρ ranges from -1 to $+1$, and then comment on your findings. For a related discussion, see Greenland (1996).

Exercise 4.12. Let X_1 and X_2 constitute a random sample from the discrete parent population

$$\mathrm{p}_X(x) = (1-\theta)^{1-x}\theta^x,\ x = 0, 1;\ 0 < \theta < 1.$$

(a) Derive the joint distribution of the random variables

$$\bar{X} = (X_1 + X_2)/2 \text{ and } S^2 = \sum_{i=1}^{2}(X_i - \bar{X})^2.$$

Are $\bar{X}$ and S^2 independent random variables?

(b) Let $L = (2\bar{X} - 3S^2)$. Find explicit expressions for $\mathrm{E}(L)$ and $\mathrm{V}(L)$.

Exercise 4.13. A random number N of balanced dice are tossed, where N has the geometric distribution

$$\mathrm{p}_N(n) = \mathrm{pr}(N = n) = \pi(1-\pi)^{n-1}, n = 1, 2, \ldots, \infty \text{ and } 0 < \pi < 1.$$

Let the random variable X be the largest number observed on any of the dice. Develop an explicit expression for $\mathrm{p}_X(x)$, the marginal distribution of X.

Exercise 4.14. Let $X_1, X_2, \ldots, X_n$ be $n(n > 1)$ random variables for which $\text{corr}(X_i, X_j) = \rho_{X_i,X_j} = \rho$ for every $i \neq j$ (i.e., all pairs of variables have the same correlation ρ). Prove that ρ must satisfy the inequality

$$-\frac{1}{(n-1)} \leq \rho \leq 1.$$

HINT: With $Z_i = \frac{X_i - \text{E}(X_i)}{\sqrt{\text{V}(X_i)}}$, consider $\text{V}(L)$, where $L = \sum_{i=1}^{n} Z_i$.

Exercise 4.15. Let $X_1 \sim \text{BIN}(n_1, \theta)$, let $X_2 \sim \text{BIN}(n_2, \theta)$, and assume that X_1 and X_2 are independent random variables. Develop an explicit expression for the conditional distribution of X_1 given that $(X_1 + X_2) = k$, where k is a fixed positive integer.

Exercise 4.16. Suppose that $Y_1, Y_2, \ldots, Y_k$ are a set of k mutually independent random variables with $\text{E}(Y_i) = \mu_i$ and $\text{V}(Y_i) = \sigma^2, i = 1, 2, \ldots, k$. Consider the two linear functions

$$L_1 = \sum_{i=1}^{k} a_i Y_i \quad \text{and } L_2 = \sum_{i=1}^{k} b_i Y_i,$$

where $a_1, a_2, \ldots, a_k$ and $b_1, b_2, \ldots, b_k$ are non-zero constants.

Find a sufficient condition involving the $\{a_i\}_{i=1}^{k}$ and the $\{b_i\}_{i=1}^{k}$ such that $\text{cor}(L_1, L_2) = 0$.

Exercise 4.17. Assume that the n random variables $X_1, X_2, \ldots, X_n$ have the following properties: (i) $\text{E}(X_i) = \mu$, $i = 1, 2, \ldots, n$; (ii) $\text{V}(X_i) = \sigma^2$, $i = 1, 2, \ldots, n$; and (iii) $\text{corr}(X_i, X_{i'}) = \rho$ for all $i \neq i'$, $i = 1, 2, \ldots, n$ and $i' = 1, 2, \ldots, n$. Develop an explicit expression for $\text{E}(S^2)$, where $S^2 = (n-1)^{-1} \sum_{i=1}^{n} (X_i - \bar{X})^2$, and then comment on your finding.

Exercise 4.18. Suppose that X, the time (in years) to first failure for a certain type of kidney dialysis machine, has a gamma distribution with mean $\alpha\beta$ and variance $\alpha^2\beta$ where $\alpha > 0$ and $\beta > 2$. This machine is repaired after its first failure, and then runs for an additional time Y (in years) before it fails again. Given that $X = x$, the *conditional* distribution of Y is gamma with mean $\alpha = \gamma/x$ and $\beta = 1$, where $\gamma > 0$. Thus, if the observed time to first failure is large, then one would expect the second failure to occur in relatively less time.

(a) Prove that

$$\text{E}(X^r) = \alpha^r \frac{\Gamma(\beta + r)}{\Gamma(\beta)}, \ (\beta + r) > 0.$$

(b) Consider the random variable $T = (X + Y)$, the total time (in years) before the second failure occurs. Using conditional expectation theory, find explicit expressions for $\mathrm{E}(T)$ and $\mathrm{V}(T)$ as a function of α, β, and γ.

(c) Find an explicit expression for $\mathrm{f}_Y(y)$, the (marginal) density function of the random variable Y.

Exercise 4.19. To protect against potentially life-threatening situations in case of electric power failures, hospital incubators for very premature infants are individually backed up by a set of n micro-batteries $\{\mathrm{B}_1, \mathrm{B}_2, \ldots, \mathrm{B}_n\}$, and only one of these n batteries is needed to supply sufficient power for an incubator to stay in operation for some limited period of time (in days). In other words, when battery B_1 fails, then battery B_2 takes over; when battery B_2 fails, then battery B_3 takes over, etc. Suppose that the times to failure (in days) of the individual batteries are independently and identically distributed negative exponential random variables, each with mean $\lambda = 1.5$ days. Use the Central Limit Theorem to provide a reasonable value for the smallest number (say n^*) of batteries needed so that the probability is at least 0.95 that an incubator will operate continuously on battery power for at least 125 days.

Exercise 4.20. Suppose that the proportion X of a certain protein in a cubic centimeter of human blood is assumed to have the distribution $\mathrm{f}_X(x;\theta) = \theta(1-x)^{\theta-1}$, $0 < x < 1$, $\theta > 0$. If $\theta = 2$, provide a reasonable value for the minimum number n^* of randomly selected human subjects needed so that the sample mean of the n^* proportions for these randomly selected subjects deviates from its expected value by no more than 0.10 with probability at least 0.95. Comment on your findings.

Exercise 4.21. Suppose that two continuous random variables X and Y have the joint density

$$\mathrm{f}_{X,Y}(x,y) = \mathrm{e}^{-(\theta x + \theta^{-1} y)}, \; x > 0, \; y > 0, \; \theta > 0.$$

(a) Develop explicit expressions for the joint distribution $\mathrm{f}_{U,V}(u,v)$ of the two random variables $U = (Y/X)^{1/2}$ and $V = (XY)^{1/2}$ and for the marginal distribution $\mathrm{f}_U(u)$ of U.

(b) Let $U_1, U_2, \ldots, U_n$ constitute a random sample from $\mathrm{f}_U(u)$. Develop an explicit expression for a function of $\bar{U} = n^{-1}\sum_{i=1}^{n} U_i$, say $\mathrm{h}(\bar{U})$, such that $\mathrm{E}[\mathrm{h}(\bar{U})] = \theta$.

Exercise 4.22. Suppose that $X_1, X_2, \ldots, X_n$ constitute a random sample of size n $(n > 1)$ from the discrete probability distribution

$$\mathrm{p}_X(x;\theta) = \left(\frac{\theta}{2}\right)^{|x|} (1-\theta)^{1-|x|}, x = -1, 0, 1, \text{ and } 0 < \theta < 1.$$

Derive the probability distribution $\mathrm{p}_{X_{(n)}}(x_{(n)})$ of the random variable $X_{(n)} = \max\{X_1, X_2, \ldots, X_n\}$.

Exercise 4.23. For a certain complex chemical process, it is of interest to model the joint behavior of the times X and Y to equilibrium (in minutes) for two competing chemical reactions. Based on an analysis of a large amount of data, a statistician postulates that the conditional density of X given $Y = y$ is

$$\mathrm{f}_X(x|Y = y) = \frac{(x + y)}{(1 + y)}\mathrm{e}^{-x},\ x > 0,$$

and that the marginal density of Y is

$$\mathrm{f}_Y(y) = \frac{1}{2}(1 + y)\mathrm{e}^{-y}, y > 0.$$

(a) Find an explicit expression for $\mathrm{E}(Y|X = x)$.

(b) Determine the density function of the random variable $S = (X + Y)$ by first finding the moment generating function of S.

Exercise 4.24. Let X_1 and X_2 constitute a random sample of size $n = 2$ from the Bernoulli population

$$\mathrm{p}_X(x) = \pi^x(1 - \pi)^{1-x}, x = 0, 1 \text{ and } 0 < \pi < 1.$$

Let $U = (X_1 + X_2)$ and let $V = |X_1 - X_2|$. Develop an explicit expression for $\mathrm{cov}(U, V)$. For what value of π are U and V uncorrelated? For this particular value of π, are U and V independent random variables? Comment on your findings.

Exercise 4.25. Suppose that $Y_1, Y_2, \ldots, Y_n$ constitute a random sample of size n from the density function

$$\mathrm{f}_Y(y; \theta) = \theta y^{\theta-1}, 0 < y < 1, \theta > 0.$$

Consider the random variable $U = nY_{(1)}{}^{\theta}$, where $Y_{(1)} = \min\{Y_1, Y_2, \ldots, Y_n\}$. By directly evaluating $\lim_{n\to\infty} \mathrm{F}_U(u; \theta) = \mathrm{pr}(U \leq u)$, determine the asymptotic distribution of the random variable U.

Exercise 4.26. In any particular year, suppose that a particular health insurance company provides health insurance coverage for 100,000 adults, and suppose that this insurance company classifies each such insured adult as being in one of three health categories, designated G, A, and P. Category G adults are classified as being in “good” health, category A adults are classified as being in “average” health, and category P adults are classified as being in "poor" health. In any particular year, 20% of all 100,000 insured adults are in category G, 50% of all 100,000 insured adults are in category A, and 30% of all 100,000 insured adults are in category P. The number of health insurance claims per year for any insured adult is assumed to have a Poisson distribution with mean 1.00 for category G adults, 2.00 for category A adults, and 4.00 for category P adults. It is reasonable to assume that all adults insured by this company act independently of one another with regard to making health insurance claims.

(a) Let the random variable T denote the total number of insurance claims that this insurance company receives in any particular year. Find numerical values for $\mathrm{E}(T)$ and $\mathrm{V}(T)$.

(b) Develop an expression that can be used to compute the probability that, in any particular year, the total number of insurance claims made by members of Category A exceeds the total number of insurance claims made by members of Category P.

Exercise 4.27. Suppose that the random variables X and Y have a bivariate normal distribution with $\mathrm{E}(X) = \mathrm{E}(Y) = 0, \mathrm{V}(X) = \mathrm{V}(Y) = 1$, and $\mathrm{corr}(X, Y) = \rho, -1 \leq \rho \leq 1$. If $-\infty < \theta < \infty$, use Tchebyshev's Inequality to find a lower bound for

$$\mathrm{pr}\left[|Y - \theta X| \leq \delta\right], \delta > 0.$$

What value θ^* of θ maximizes this lower bound?

Exercise 4.28. Suppose that $X_1, X_2, \ldots, X_n$ constitute a random sample of size n from a $\mathrm{N}(\mu, 1)$ parent population. Consider the random variable $S = \sum_{i=1}^{n} X_i^2$.

(a) Find explicit expressions for $\mathrm{E}(S)$ and $\mathrm{V}(S)$, the mean and variance of the random variable S.

(b) When $\mu = 0$, determine the exact distribution of the random variable S.

(c) When $\mu \neq 0$, the exact distribution of the random variable S is more complicated to determine. Suppose that it is desired to approximate the distribution of S when $\mu \neq 0$ with the distribution of the random variable aY, where $Y \sim \chi_b^2$ and where a is a positive constant. Find values for a and b so that the mean and variance of the random variable aY are the same as the mean and variance of the random variable S.

Exercise 4.29. A balanced coin is tossed $n(\geq 1)$ times; let the random variable X_1 be the number of heads obtained among these n tosses. Then, a balanced six-sided die is rolled X_1 times; let the random variable X_2 be the number of "ones" obtained among these X_1 rolls.

(a) Develop an explicit expression for $\text{corr}(X_1, X_2)$.

(b) With $S = (X_1 + X_2)$, develop explicit expressions for $\text{E}(S)$ and $\text{V}(S)$.

Exercise 4.30. Suppose that n randomly selected Olympic athletes are each given a drug test based on a urine sample. For $i = 1, 2, \ldots, n$, suppose that the i-th athlete has probability $\pi_i, 0 < \pi_i < 1$, of producing a positive drug test.

(a) Let the random variable X denote the total number of these n athletes who produce a positive drug test. Develop explicit expressions for $\text{E}(X)$ and $\text{V}(X)$.

(b) Find explicit expressions for $\pi_1, \pi_2, \ldots, \pi_n$ such that $\text{V}(X)$ is maximized subject to the constraint that $\text{E}(X) = k$, where $k, 0 < k < n$, is a known constant.

Exercise 4.31. Suppose that there are $n(\geq 1)$ original members of a semi-private golf course in a certain large city. In an effort to increase the size of the membership, suppose that each of these n original members attempts to recruit eligible golfers in this city to become new members of this golf course, and assume that each of these new members will then also immediately become involved in this recruiting process, and so on. Membership in this golf course is restricted to a total of $N(> n)$ individuals.

Assume that each of the original n members has probability $1/n$ of being the individual who successfully recruits the $(n + 1)$-th member. And, once this new member is recruited, each of these $(n + 1)$ members then has probability $1/(n + 1)$ of being the individual who recruits the $(n + 2)$-th member.

Given this scenario, develop an expression (which may involve summation signs) for the expected total number of members recruited by any particular one of the original n members. Find the numerical value of this expected number if $N = 200$ and $n = 190$.

Exercise 4.32. Suppose that $X \sim \text{N}(0, 1)$, that $Y \sim \text{N}(0, 1)$, and that X and Y are independent random variables. Let $U = X/Y$.

(a) Use the method of transformations to show that U has a *standard Cauchy* distribution. Is there another name for this distribution?

(b) Develop an explicit expression for $\text{F}_U(u)$, the cumulative distribution function of the random variable U. Also, find the numerical value of $\text{pr}(-\sqrt{3} < U < 1)$.

Exercise 4.33. In a table of random numbers, each row lists 60 digits, with each digit taking one of the ten integer values $0, 1, 2, \ldots, 9$. Find a reasonable value for the probability that any such row contains between 25 and 35 odd digits (i.e., at least 25, but no more than 35, of the digits 1, 3, 5, 7, and 9).

Exercise 4.34. Suppose that the two continuous random variables Y_1 and Y_2 have the joint distribution

$$\text{f}_{Y_1,Y_2}(y_1, y_2) = 1, 0 < y_1 < 2, 0 < y_2 < 1, 2y_2 < y_1.$$

(a) Derive an explicit expression for the cumulative distribution function $\text{F}_U(u) = \text{pr}(U \leq u)$ of the random variable $U = (Y_1 - Y_2)$, and then use this result to find an explicit expression for $\text{f}_U(u)$.

(b) Find $\text{pr}\left[(Y_2 - Y_1)^2 > \frac{1}{4}\right]$.

Exercise 4.35. Let $X_1, X_2, \ldots, X_i, \ldots$ be a sequence of mutually independent dichotomous random variables, where $\text{pr}(X_i = 1) = \pi, 0 < \pi < 1$, and $\text{pr}(X_i = 0) = (1 - \pi)$. Consider the following sequential sampling procedure. Let k be a positive integer, and let $S_j = (X_1 + X_2 + \cdots + X_j)$. Define the random variable N to be the number of X_is that have to be sequentially sampled until $S_N = k$.

(a) Find the distribution of the random variable N.

(b) Find the expected value of the random variable $(k - 1)/(N - 1)$.

(c) For $k = 1$, find a function $\text{g}(N)$ such that $\text{E}[\text{g}(N)] = \pi \text{e}^{(1-\pi)}$.

Exercise 4.36. The radius R of spherically shaped steel ball bearings manufactured by a certain industrial process is normally distributed with mean $\mu = 3.0$ cm and variance $\sigma^2 = 0.02$ cm^2. If the density of steel is 7.85 grams per cubic centimeter, find the expected value of the total weight W of 200 randomly chosen steel ball bearings manufactured by this industrial process.

Exercise 4.37. Suppose, for $j = 1, 2, \ldots, n$, that

$$Y_{1j} = \mu_1 + X_1 + U_j \text{ and } Y_{2j} = \mu_2 + X_2 + W_j,$$

where $\text{E}(X_1) = \text{E}(X_2) = \text{E}(U_j) = \text{E}(W_j) = 0$, $\text{V}(X_1) = \sigma_1^2$, $\text{V}(X_2) = \sigma_2^2$, $\text{corr}(X_1, X_2) = \rho$, $\text{V}(U_j) = \sigma_u^2$, $\text{V}(W_j) = \sigma_w^2$, and the $2n$ random variables $U_1, U_2, \ldots, U_n, W_1, W_2, \ldots, W_n$ are mutually independent and are also independent of the random variables X_1 and X_2.

With $\bar{Y}_1 = n^{-1}\sum_{j=1}^{n} Y_{1j}$ and $\bar{Y}_2 = n^{-1}\sum_{j=1}^{n} Y_{2j}$, show that $\text{corr}(\bar{Y}_1, \bar{Y}_2)$ may be expressed as $\theta\rho$, where $0 < \theta < 1$.

Exercise 4.38. Suppose that the random variables X_1 and X_2 have the joint bivariate discrete probability distribution

$$\text{p}_{X_1,X_2}(x_1, x_2) = \pi_{11}^{x_1 x_2}\pi_{10}^{x_1(1-x_2)}\pi_{01}^{(1-x_1)x_2}\pi_{00}^{(1-x_1)(1-x_2)},$$

for $x_1 = 0, 1$ and $x_2 = 0, 1$, where $\pi_{11} = \theta^2 + \rho\theta(1-\theta), \pi_{10} = \pi_{01} = (1-\rho)\theta(1-\theta)$, and $\pi_{00} = (1-\theta)^2 + \rho\theta(1-\theta)$, $0 < \theta < 1$, and $-1 < \rho < 1$.

(a) Develop explicit expressions for $\text{p}_{X_1}(x_1)$ and $\text{p}_{X_2}(x_2)$, the marginal distributions of X_1 and X_2.

(b) Develop an explicit expression for $\text{corr}(X_1, X_2)$.

(c) Develop explicit expressions for $\text{E}(X_1|X_2 = 1)$ and $\text{V}(X_1|X_2 = 1)$.

Exercise 4.39. Certain biological evidence has suggested that some tumors are monoclonal in origin (i.e., develop from a single cell), and that the development of a monoclonal-type tumor is a very rare event occurring among an extremely large population of cells at risk. Thus, the number of monoclonal-type tumors that develop in a laboratory animal (e.g., a mouse or a rat) during some specified time period after exposure to a potential carcinogen might be expected to follow the Poisson distribution

$$\text{p}_X(x|\lambda) = \frac{\lambda^x \text{e}^{-\lambda}}{x!}, x = 0, 1, \ldots, \infty; \lambda > 0.$$

However, as has been shown by several groups of researchers, the observed variability in laboratory animal tumor multiplicity data generally *exceeds* the observed mean number of tumors per animal. This excess (or "extra-Poisson") variation has been attributed to the variability among the animals in their inherent susceptibilities to monoclonal-type tumor development. Furthermore, the presence of this excess variability argues against the use of the Poisson distribution for modeling such data, since the Poisson distribution characteristic $\text{E}(X|\lambda) = \text{V}(X|\lambda) = \lambda$ appears to be too restrictive.

It has been suggested that a better description of laboratory animal tumor multiplicity data would be obtained by considering a "generalized Poisson" model, in which the inherent susceptibility to monoclonal-type tumor development (as measured by the Poisson parameter λ) varies according to the gamma density

$$\text{f}(\lambda) = \frac{\lambda^{\beta-1}\text{e}^{-\lambda/\alpha}}{\Gamma(\beta)\alpha^\beta}, \lambda > 0; \alpha > 0, \beta \text{ a positive integer},$$

with $\text{E}(\lambda) = \alpha\beta$ and $\text{V}(\lambda) = \alpha^2\beta$.

(a) Given $\mathrm{p}_X(x|\lambda)$ and $\mathrm{f}(\lambda)$, develop an explicit expression for the unconditional distribution $\mathrm{p}_X(x)$ of X (i.e., the so-called "generalized Poisson" distribution mentioned previously). Does the distribution $\mathrm{p}_X(x)$ have a name?

(b) Use conditional expectation theory directly to develop explicit expressions for $\mathrm{E}(X)$ and $\mathrm{V}(X)$. Which distribution, $\mathrm{p}_X(x|\lambda)$ or $\mathrm{p}_X(x)$, would you expect to be a better statistical model for laboratory animal tumor multiplicity data?

Exercise 4.40. Let $X_1, X_2, \ldots, X_n$ constitute a random sample of size n from the discrete parent population

$$\mathrm{p}_X(x) = (1-\pi)\pi^x, x = 0, 1, \ldots, \infty \text{ and } 0 < \pi < 1.$$

(a) If $n = 3, \pi = 0.40$, and $\bar{X} = n^{-1}\sum_{i=1}^{n} X_i$, find the numerical value of $\mathrm{pr}\left(\bar{X} \leq 1\right)$.

(b) Show that

$$\theta = \mathrm{pr}\left(X_1 \leq X_2 \leq \cdots \leq X_{n-1} \leq X_n\right) = \frac{(1-\pi)^n}{\prod_{i=1}^{n}(1-\pi^i)}.$$

Exercise 4.41. The following statistical model has been proposed by highway safety researchers. For a particular heavily traveled section of highway in a major metropolitan city, let Y_k be a discrete random variable denoting the number of automobile accidents occurring during a specified time period that involve exactly k people (counting drivers and any passengers), $k = 1, 2, \ldots, \infty$. For example, $Y_3 = 2$ means that exactly two automobile accidents, with *each* such automobile accident involving exactly three people (counting drivers and any passengers), occurred within the particular heavily traveled section of highway during the specified time period. Assume that Y_k has a Poisson distribution with mean $\mathrm{E}(Y_k) = \lambda\theta^k/k!$, with $\lambda > 0, 0 < \theta < 1$, and $k = 1, 2, \ldots, \infty$. Further, assume that the $\{Y_k\}$ constitute a set of *mutually independent* random variables.

Given that $\lambda = 1.0$ and that $\theta = 0.40$, find the numerical value of the probability that, for this particular heavily traveled section of highway during the specified time period, exactly three automobile accidents, each involving only one person [i.e., single-car and one-occupant (the driver) automobile accidents], occur given that a total of exactly 10 automobile accidents occur.

Exercise 4.42. An environmental scientist postulates that the joint density function representing the concentrations X and Y of two air pollutants in a certain metropolitan city in the United States is of the form

$$\mathrm{f}_{X,Y}(x, y) = \theta^{-1}(\theta - x)^{-1}, 0 < y < (\theta - x), 0 < x < \theta,$$

where θ is an unknown parameter.

(a) Derive explicit expressions for $\mathrm{f}_X(x)$ and $\mathrm{f}_Y(y)$, the marginal distributions of the random variables X and Y.

(b) Find an explicit expression for $\rho_{X,Y}$, the correlation between the random variables X and Y.

(c) Find an explicit expression for $\mathrm{pr}(X > Y)$.

(d) Let $(X_1, Y_1), (X_2, Y_2), \ldots, (X_n, Y_n)$ constitute a random sample of size $n(> 1)$ from $\mathrm{f}_{X,Y}(x, y)$. Consider the following two estimators of the unknown parameter θ:

$$\hat{\theta}_1 = k_1(\bar{X} - \bar{Y}), \text{ where } \bar{X} = n^{-1}\sum_{i=1}^{n} X_i \text{ and } \bar{Y} = n^{-1}\sum_{i=1}^{n} Y_i;$$

and

$$\hat{\theta}_2 = k_2 U, \text{ where } U = \max\{X_1, X_2, \ldots, X_n\}.$$

Find explicit expressions for k_1 and k_2 such that $\hat{\theta}_1$ and $\hat{\theta}_2$ are both unbiased estimators of the unknown parameter θ.

(e) Derive explicit expressions for the variances of the unbiased estimators $\hat{\theta}_1$ and $\hat{\theta}_2$. Which of these two unbiased estimators of θ do you prefer and why?

Exercise 4.43. Suppose that the number N of genes hit by gamma rays from a radioactive source has the probability distribution

$$\mathrm{p}_N(n; \theta) = \theta(1 - \theta)^n, \quad n = 0, 1, \ldots, \infty; \; 0 < \theta < 1.$$

(a) Find the probability generating function for the random variable N and use it to find $\mathrm{E}[N(N + 1)]$.

(b) Given that n genes are hit, suppose that the number X of genes out of n which suffer genetic damage has the conditional probability distribution

$$\mathrm{p}_X(X = x|N = n) = \mathrm{C}_x^n \pi^x (1 - \pi)^{n-x}, \quad x = 0, 1, \ldots, n; \; 0 < \pi < 1.$$

Find an *explicit expression* for $\mathrm{p}_X(x) = \mathrm{pr}(X = x)$.

Exercise 4.44. Let $X_1, X_2, \ldots, X_n$ constitute a random sample of size n from a population with *unknown* mean μ and *unknown* variance σ^2. Consider estimating μ with the estimator

$$\hat{\mu} = \sum_{i=1}^{n} c_i X_i,$$

where the set of constants $\{c_1, c_2, \ldots, c_n\}$ satisfies the constraint $\sum_{i=1}^{n} c_i = 1$.

If $\bar{X} = n^{-1}\sum_{i=1}^{n} X_i$, derive an *explicit expression* for $\text{corr}(\bar{X}, \hat{\mu})$ that is an *explicit function* of $\text{V}(\bar{X})$ and $\text{V}(\hat{\mu})$. What is the value of this correlation when $n = 5, \sigma^2 = 3$, and $c_i = i^2, i = 1, 2, \ldots, 5$? Is the value of σ^2 actually needed?

Exercise 4.45. For a certain psychiatric clinic in a large city in the United States, suppose that the random variable X represents the *total time* (in minutes) that a typical patient spends in this clinic during a typical visit (where this *total time* is the sum of the *waiting time* and the *treatment time*), and that the random variable Y represents the *waiting time* (in minutes) that a typical patient spends in the waiting room before starting treatment with a psychiatrist. Further, suppose that X and Y can be assumed to follow the bivariate density function

$$\text{f}_{X,Y}(x, y) = \lambda^2 \text{e}^{-\lambda x}, 0 < y < x < +\infty, \lambda > 0.$$

(a) Develop an explicit expression for $\text{F}_U(u) = \text{pr}(U \leq u)$, where $U = (X - Y)$ is the random variable representing the length of time (in minutes) that a typical patient spends with a psychiatrist at this clinic (i.e., U is the *treatment time* for a typical patient at this clinic). Hence, or otherwise, find $\text{E}(U)$ and $\text{V}(U)$.

(b) Develop an explicit expression for

$$\text{M}_{X,Y}(s, t) = \text{E}\left(\text{e}^{sX + tY}\right),$$

the joint moment generating function for the random variables X and Y. Then, use this result to find the marginal distributions of X and Y and to find $\text{corr}(X, Y)$.

(c) If six patients visit this psychiatric clinic on different days (so that these six patients can be assumed to be mutually independent of one another with regard to their waiting and treatment times), provide an explicit expression for the probability that at least two of these six patients have waiting times that exceed their corresponding treatment times.

Exercise 4.46. Let $X_1, X_2, \ldots, X_n$ constitute a random sample from the uniform density $\text{f}_X(x) = 1, 0 < x < 1$. Derive an explicit expression for the density function $\text{f}_G(g)$ of the geometric mean $G = \left(\prod_{i=1}^{n} X_i\right)^{1/n}$.

Exercise 4.47. In a two-component signal transduction system for modeling transduction across biological membranes, suppose that the lifetime X (in days) of the "primary component" can be adequately modeled by the density function

$$\text{f}_X(x) = (2\beta^3)^{-1} x^2 \text{e}^{-x/\beta}, \ x > 0 \text{ and } \beta > 0;$$

and given that $X = x$, the lifetime Y (in days) of the "secondary component" can be reasonably modeled by the density function

$$\mathrm{f}_Y(y|X = x) = (\alpha x)\mathrm{e}^{-(\alpha x)y}, \ y > 0, \alpha > 0, x > 0,$$

so that $\mathrm{E}(Y|X = x) = (\alpha x)^{-1}$ and $\mathrm{V}(Y|X = x) = (\alpha x)^{-2}$.

(a) Use conditional expectation theory to find explicit expressions for $\mathrm{E}(Y)$ and $\mathrm{V}(Y)$.

(b) Find an explicit expression for $\mathrm{corr}(X, Y)$; then, use the "correlation-linear regression" connection to find an explicit expression for $\mathrm{corr}(X^{-1}, Y)$.

(c) Find an explicit expression for $\mathrm{F}_Y(y)$, the cumulative distribution function (CDF) of Y.

(d) Set up appropriate integral expressions that, if evaluated, would allow one to determine $\mathrm{pr}[(X^2 + Y^2) < 1|X > Y]$.

Exercise 4.48. In a certain politically conservative county in rural North Carolina, suppose that an election for county commissioner involves two candidates, one candidate being pro-life and one candidate being pro-choice with regard to decisions about pregnancy terminations. If this county has n residents who vote (where n is a large and even positive integer), assume that a large subset of s ($< n$) of these voting residents will always vote for the pro-life candidate. The other remaining large subset of size $r = (n - s)$ contains voting residents who each have probability π ($0 < \pi < 1$) of voting for the pro-life candidate [and hence probability $(1 - \pi)$ of voting for the pro-choice candidate]. Find an expression that can be used to determine (with reasonable accuracy for large values of r and s) the minimum value of s such that there is at least a probability of θ ($0 < \theta < 1$) for the pro-life candidate to win the election (i.e., to receive the majority of the n votes). Your answer will necessarily be a function of one or more of the quantities r, π, and θ. If $\theta = 0.841$ and $\pi = 0.50$, how does your answer simplify?

Exercise 4.49. Suppose that the continuous variable X represents the time (in months) from the initial diagnosis of leukemia until the first chemotherapy treatment, and that the continuous variable Y represents the time (in months) from the initial diagnosis of leukemia until death. The joint density function of the random variables X and Y is assumed to be of the form

$$\mathrm{f}_{X,Y}(x, y) = 2\theta^{-2}\mathrm{e}^{-(x+y)/\theta}, 0 < x < y < \infty, \theta > 0.$$

(a) For r a non-negative integer, prove that

$$\mathrm{E}(Y^r|X = x) = \sum_{j=0}^{r} C_j^r x^{r-j}\Gamma(j + 1)\theta^j.$$

(b) Make use of the formula given in part (a) to derive an explicit expression for $\rho = \text{corr}(X, Y)$, the correlation between the two random variables X and Y.

(c) Cancer researchers are interested in the random variable $P = X/Y$. By relating the CDF $\text{F}_P(p) = \text{pr}(P \leq p)$ of the random variable P to the random variables X and Y, derive an explicit expression for the density function of the random variable P, and then find $\text{E}(P)$.

Exercise 4.50. Suppose that $X_1, X_2, \ldots, X_n$ constitute a random sample from $\text{p}_X(x) = \lambda^x \text{e}^{-\lambda}/x!$, $x = 0, 1, \ldots, \infty, \lambda > 0$, and let $\bar{X} = n^{-1}\sum_{i=1}^{n} X_i$.

(a) If $n = 4$ and $\lambda = 0.20$, what is the numerical value of $\text{pr}(\bar{X} \geq 0.40)$?

(b) Let $Z = \left[\bar{X} - \text{E}(\bar{X})\right] / \left[\text{V}(\bar{X})\right]^{1/2}$. Find $\lim_{n \to \infty} \text{E}(\text{e}^{tZ})$, and then comment on your finding.

Exercise 4.51. Two machines (Machine 1 and Machine 2), installed side-by-side in the same plant, are designed to punch holes in metal sheeting. For $i = 1, 2$, assume that the probability distribution of X_i, the number of holes punched until the i-th machine breaks down, is given by the geometric distribution with parameter θ_i:

$$\text{p}_{X_i}(x) = \text{pr}(X_i = x) = \theta_i(1 - \theta_i)^{x-1}, \; x = 1, 2, \ldots, \infty; \; 0 < \theta_i < 1.$$

(a) On a given day, suppose that the two machines start punching holes at the same time. Prove rigorously that

$$\text{pr}(X_1 = X_2) = \theta_1\theta_2/(\theta_1 + \theta_2 - \theta_1\theta_2),$$

where $\text{pr}(X_1 = X_2)$ is the probability that the two machines break down at exactly the same time (i.e., that they both fail after punching exactly the same number of holes). You may assume that the two machines operate completely independently of one another, so that the events $\{X_1 = x_1\}$ and $\{X_2 = x_2\}$ are statistically independent of one another for all permissible values of x_1 and x_2.

(b) The plant supervisor decides to conduct an experiment to see if there is evidence that $\theta_1 \neq \theta_2$. On the j-th of four consecutive days ($j = 1, 2, 3, 4$), he starts the two machines at exactly the same time and records the value of the random variable T_j, where

$$T_j = \begin{cases} 0, & \text{if } X_1 = X_2 \text{ on day } j; \\ 1, & \text{if } X_1 \neq X_2 \text{ on day } j. \end{cases}$$

Assuming that $\theta_1 = \theta_2 = \theta$ (say), what is the probability distribution of T_j?

(c) Assuming that $\theta_1 = \theta_2 = \theta$ and that the outcomes from day-to-day are independent of one another, give an explicit formula for $\mathrm{p}_T(t)$, the probability distribution of the discrete random variable

$$T = \sum_{j=1}^{4} T_j.$$

(d) Suppose that the outcome of the plant supervisor's experiment is that $T_j = 1$ on each of the four days. After seeing this outcome, the plant supervisor claims that this is "very strong evidence" that $\theta_1 \neq \theta_2$. Prove that the supervisor's conclusion is unjustified by finding a range of values for θ such that

$$\mathrm{pr}\left\{\bigcap_{j=1}^{4}(T_j = 1)\right\} \geq 0.50,$$

where $\theta_1 = \theta_2 = \theta$.

Exercise 4.52. In a two-component system with the two components *not* operating independently of each other (e.g., two synchronized electronic components, two lungs or two kidneys in a human being), the lifetime of one component affects the lifetime of the other component. If X and Y are continuous random variables denoting the individual lifetimes of two such components, consider the following bivariate density function for X and Y:

$$\begin{aligned} \mathrm{f}_{X,Y}(x,y) &= \alpha_1\beta_2 \exp[-\beta_2 y - (\alpha_1+\beta_1-\beta_2)x], 0 < x < y < \infty; \\ \mathrm{f}_{X,Y}(x,y) &= \alpha_2\beta_1 \exp[-\alpha_2 x - (\alpha_1+\beta_2-\alpha_2)y], 0 < y < x < \infty. \end{aligned}$$

Here, the population parameters α_1, α_2, β_1, and β_2 are all positive, with $(\alpha_1 + \beta_1 - \beta_2) > 0$ and $(\alpha_1 + \beta_2 - \alpha_2) > 0$.

(a) Prove that $\mathrm{pr}(X < Y) = \alpha_1/(\alpha_1 + \beta_1)$.

(b) Develop an explicit expression for the conditional joint density function $\mathrm{f}_{X,Y}(x, y|X < Y)$ of X and Y.

(c) Develop an explicit expression for $\mathrm{f}_X(x|X < Y)$, the marginal density function of X given that $X < Y$.

(d) Develop an explicit expression for $\mathrm{f}_Y[y|(X = x) \cap (X < Y)]$, the conditional density function of Y given both $X = x$ and $X < Y$.

(e) Use conditional expectation theory to find explicit expressions for $\mathrm{E}(Y|X < Y)$ and $\mathrm{V}(Y|X < Y)$.

Exercise 4.53. In a certain very large human population (which can be assumed to be infinitely large for all practical purposes), suppose that each member of this population is in one (and only one) of four distinct categories of risk for HIV infection. These four categories are as follows:

Category 1: neither homosexual nor an intravenous drug user

Category 2: homosexual but not an intravenous drug user

Category 3: an intravenous drug user but not homosexual

Category 4: both homosexual and an intravenous drug user

The proportions of this population in these four mutually exclusive and exhaustive categories are, respectively, $(2+\theta)/4$ for Category 1, $(1-\theta)/4$ for Category 2, $(1-\theta)/4$ for Category 3, and $\theta/4$ for Category 4. Here, θ $(0 < \theta < 1)$ is an unknown parameter. Suppose that a random sample of n people is selected from this population. For $i = 1, 2, 3, 4$, let the random variable X_i be the number of people in this random sample who belong to Category i.

(a) Provide an explicit expression for the joint distribution of the random variables X_1, X_2, X_3, and X_4.

(b) Develop an explicit expression for the probability distribution of the number of homosexuals contained in this random sample of size n.

(c) Given that this random sample of size n contains k $(0 < k < n)$ intravenous drug users, how many homosexuals, on average, would you expect to find among these k intravenous drug users?

(d) Consider the following two linear functions of X_1, X_2, X_3, and X_4:

$$L_1 = n^{-1}(X_1 - X_2 - X_3 + X_4)$$
$$\text{and } L_2 = (2n)^{-1}(X_1 - X_2 - X_3 + 5X_4).$$

Find explicit expressions for $\mathrm{E}(L_1)$ and $\mathrm{E}(L_2)$.

(e) Develop explicit expressions for $\mathrm{V}(L_1)$ and $\mathrm{V}(L_2)$. How do $\mathrm{V}(L_1)$ and $\mathrm{V}(L_2)$ compare for different values of $\theta, 0 < \theta < 1$?

Exercise 4.54. Let X_1 and X_2 constitute a random sample of size $n = 2$ from the parent population

$$\mathrm{p}_X(x) = \theta(1-\theta)^{x-1},\ x = 1, 2, \ldots, \infty \text{ and } 0 < \theta < 1.$$

(a) Use moment generating function theory to show that the distribution of $S = (X_1 + X_2)$ is

$$\mathrm{p}_S(s) = (s-1)\theta^2(1-\theta)^{s-2},\ s = 2, 3, \ldots, \infty.$$

(b) Find an explicit expression for the conditional distribution of X_1 given $S = s$.

(c) Determine the numerical value of $\text{cor}(X_1, S)$.

Exercise 4.55. Let X_1 and X_2 constitute a random sample of size $n = 2$ from the discrete parent population

$$\mathrm{p}_X(x) = \theta(1-\theta)^{x-1}, x = 1, 2, \ldots, \infty \text{ and } 0 < \theta < 1.$$

(a) Develop an explicit expression for the probability distribution $\mathrm{p}_Y(y)$ of the discrete random variable $Y = \max\{X_1, X_2\}$. In particular, show that $\mathrm{p}_Y(y)$ can be written as a linear function of two different geometric distributions.

(b) Use the result in part (a) to develop explicit expressions for $\mathrm{E}(Y)$ and $\mathrm{V}(Y) = 0$.

Exercise 4.56. Let the random variables X and Y have the bivariate density function

$$\mathrm{h}_{X,Y}(x, y) = \pi \mathrm{f}_1(x)\mathrm{g}_1(y) + (1-\pi)\mathrm{f}_2(x)\mathrm{g}_2(y),$$

where $0 < \pi < 1$, $-\infty < x < \infty$, $-\infty < y < \infty$. Here, $\mathrm{f}_1(x)$ and $\mathrm{f}_2(x)$ are valid density functions defined over the range $-\infty < x < \infty$, and $\mathrm{g}_1(y)$ and $\mathrm{g}_2(y)$ are valid density functions defined over the range $-\infty < y < \infty$.

(a) Prove that X and Y are independent random variables if and only if

$$[\mathrm{f}_1(x) - \mathrm{f}_2(x)][\mathrm{g}_1(y) - \mathrm{g}_2(y)] = 0.$$

(b) For $i = 1, 2$, define

$$\begin{aligned} \alpha_i &= \int_{-\infty}^{\infty} x\mathrm{f}_i(x)\mathrm{d}x, \\ \beta_i &= \int_{-\infty}^{\infty} y\mathrm{g}_i(x)\mathrm{d}y. \end{aligned}$$

Derive an explicit expression for $\text{cov}(X, Y)$, the covariance between the random variables X and Y, as a function of the parameters π, α_1, α_2, β_1, and β_2.

(c) Provide a complete set of *sufficient conditions* for which the random variables X and Y are *uncorrelated* but still *dependent.*

For additional details related to this problem, see Behboodian (1990).

Exercise 4.57. In environmental health applications, it is a very common occurrence that the concentration X of an environmental contaminant of interest is unobservable when that concentration falls below a certain known detection limit L. In this situation, X is said to be "censored" (more specifically, left-censored) when $X < L$, and so only values of X for which $X \geq L$ are observable. Hence, it is often of interest to study characteristics of X given that $X \geq L$. In particular, suppose that $X \sim N(\mu, \sigma^2)$. If $\mu = 3$, $\sigma^2 = 1$, and $L = 1.60$, find the numerical value of $\mathrm{E}(X|X \geq 1.60)$.

Exercise 4.58. Suppose that two continuous random variables X and Y have the joint density function

$$\mathrm{f}_{X,Y}(x, y) = \mathrm{e}^{-(\theta x + \theta^{-1} y)}, \ x > 0, \ y > 0, \ \theta > 0.$$

(a) Derive an explicit expression for the marginal distribution $\mathrm{f}_U(u)$ of the random variable $U = (Y/X)^{1/2}$.

(b) Let $U_1, U_2, \ldots, U_n$ constitute a random sample from $\mathrm{f}_U(u)$. Develop an explicit expression for the expected value of the random variable $\bar{U} = n^{-1} \sum_{i=1}^{n} U_i$.

Exercise 4.59. The time T to failure (in years) of a certain brand of electronic component has the negative exponential density function $\mathrm{f}_T(t) = \mathrm{e}^{-t}$, $t > 0$. Provide a reasonable numerical value for the smallest number n^* of electronic components needed so that, with probability no smaller than 0.95, at least 30% of these n^* components will each have a time to failure exceeding one year.

Exercise 4.60. For $i = 1, 2, 3$, suppose that U_i has the Bernoulli distribution

$$\mathrm{p}_{U_i}(u_i) = \pi^{u_i}(1 - \pi)^{1-u_i}, u_i = 0, 1 \text{ and } 0 < \pi < 1.$$

Now, let

$$X = WU_1 + (1 - W)U_2 \text{ and } Y = WU_1 + (1 - W)U_3,$$

where the random variable W has the Bernoulli distribution

$$\mathrm{p}_W(w) = \theta^w(1 - \theta)^{1-w}, w = 0, 1 \text{ and } 0 < \theta < 1.$$

Further, assume that U_1, U_2, U_3, and W are mutually independent random variables.

(a) Develop an explicit expression for $\mathrm{corr}(X, Y)$.

(b) Show that the conditional expectation of Y given $X = x$ can be expressed as $\mathrm{E}(Y|X = x) = \alpha + \beta x$, and find explicit expressions for α and β.

Exercise 4.61. Conditional on θ fixed, suppose that X and Y are independent discrete random variables, each having a Poisson distribution with parameter

θ. Also, assume that the variation in θ is described by the density function

$$\mathrm{f}(\theta) = [\Gamma(\alpha)]^{-1}\, \theta^{\alpha-1}\mathrm{e}^{-\theta}, 0 < \theta < \infty,$$

where α is a positive integer.

(a) Develop an explicit expression for $\mathrm{p}_{X,Y}(x, y)$, the joint distribution of the random variables X and Y, and then show directly that $\mathrm{p}_{X,Y}(x, y)$ is a valid bivariate discrete probability distribution.

(b) Use the results in part (a) to find explicit expressions for $\mathrm{E}(Y|X = x)$ and for $\mathrm{corr}(X, Y)$.

Exercise 4.62. Let $X_1, X_2, \ldots, X_n$ constitute a random sample of size $n(\geq 2)$ from the parent population defined by the CDF

$$\mathrm{F}_X(x) = \left(\frac{x}{\theta} - k\right), k\theta < x < (k+1)\theta,$$

where k is a known non-negative number and where $\theta(> 0)$ is an unknown parameter.

Let $X_{(1)} = \min\{X_1, X_2, \ldots, X_n\}$, let $X_{(n)} = \max\{X_1, X_2, \ldots, X_n\}$, and consider the random variable $U = X_{(n)} - X_{(1)}$.

(a) Develop an explicit expression for the density function $\mathrm{f}_U(u)$ of the random variable U.

(b) Find a function $\mathrm{g}(U)$ such that $\mathrm{E}[\mathrm{g}(U)] = \theta$.

Exercise 4.63. Suppose that X_1 and X_2 constitute a random sample of size $n = 2$ from the uniform density function $\mathrm{f}_X(x) = 1, 0 < x < 1$. Derive an explicit expression for $\mathrm{E}\left(|X_1 - X_2|^k\right)$, where k is a fixed non-negative number.

Exercise 4.64. Suppose that the number N of automobiles passing through a certain rural intersection between the hours of 5:00 pm and 6:00 pm on any weekday has the probability distribution

$$\mathrm{p}_N(n) = (1 - \mathrm{e}^{-\lambda})^{-1}\lambda^n\mathrm{e}^{-\lambda}/n!, n = 1, 2, \ldots, \infty \text{ and } \lambda > 0.$$

Further, given that n automobiles pass through this rural intersection between the hours of 5:00 pm and 6:00 pm on a weekday, the conditional distribution of the number X of these n drivers who are wearing seatbelts is given by the expression

$$\mathrm{p}_X(x|N = n) = \mathrm{C}_x^n\pi^x(1 - \pi)^{n-x}, x = 0, 1, \ldots, n \text{ and } 0 < \pi < 1.$$

(a) Derive the probability distribution $\mathrm{p}_X(x)$ of the random variable X, and show directly that $\mathrm{p}_X(x)$ is a valid discrete probability distribution.

(b) Develop an explicit expression for $\mathrm{E}(X)$.

Exercise 4.65. Suppose that $X_1 \sim \mathrm{BIN}(n_1, \pi_1)$, that $X_2 \sim \mathrm{BIN}(n_2, \pi_2)$, and that X_1 and X_2 are independent random variables.

(a) Find the conditional distribution of X_1 given that $S = (X_1 + X_2) = s$. Show that this conditional distribution, known as the *non-central hypergeometric distribution*, can be expressed as a function of the odds ratio parameter $\theta = \pi_1(1-\pi_2)/\pi_2(1-\pi_1)$.

(b) If $n_1 = 3$ and $n_2 = 2$, find an explicit expression for $\mathrm{E}(X_1|S=4)$.

For applications involving the use of the non-central hypergeometric distribution, see Breslow and Day (1980) and Kleinbaum, Kupper, and Morgenstern (1982).

Exercise 4.66. Suppose that the 24-hour fine particulate matter concentration X (in micrograms per cubic meter, or $\mu\mathrm{g/m}^3$) near a certain industrial site follows a lognormal distribution; more specifically, the random variable $\ln X$ follows a normal distribution with mean $\mu = 3.22$ and $\sigma^2 = 0.03$. Further, suppose that the Environmental Protection Agency dictates that such an industrial site will be in violation of the Clean Air Act when any measured 24-hour fine particulate matter concentration level exceeds 35 $\mu\mathrm{g/m}^3$.

(a) If an environmental engineer makes one reading of the 24-hour fine particulate matter concentration at this industrial site, what is the probability π that she will find this industrial site to be in violation of the Clean Air Act?

(b) If this environmental engineer makes three mutually independent readings of the 24-hour fine particulate matter concentration at this site, what is the probability that at least two of these three readings will exceed 35 $\mu\mathrm{g/m}^3$?

(c) Let $X_1, X_2, \ldots, X_n$ be n mutually independent readings of the 24-hour fine particulate matter concentration at this industrial site. Then, for $i = 1, 2, \ldots, n$, let the dichotomous random variable $Y_i = 1$ if $X_i > 35$ $\mu\mathrm{g/m}^3$ and let $Y_i = 0$ if not. If Z denotes the number of the n Y_i values that are *at least* as large as Y_1, derive the probability distribution $\mathrm{p}_Z(z)$ of the discrete random variable Z, and also develop an explicit expression for $\mathrm{E}(Z)$.

Exercise 4.67. In an experiment designed to test subjects for evidence of extrasensory perception (ESP), each subject is asked to identify the number that appears on the *back* of each of 100 cards. Each card has one of the five numbers 1, 2, 3, 4, or 5 on its back, and one of these five numbers was assigned randomly to each of the cards. So, if the subject being tested is just guessing about the number on the back of a card, then this subject has probability 1/5 of guessing the number correctly. Suppose that $n = 50$ randomly selected subjects are each tested in this manner, and that the highest scoring subject correctly identifies the numbers on 30 of the 100 cards. Does this result statistically support the hypothesis that all 50 subjects participating in this ESP experiment have made purely random guesses about the numbers on the backs of the cards?

Exercise 4.68. In a small U.S. college town containing a number of homeless people, let the random variable Y be the nightly number of homeless people who have no shelter, and assume that $Y \sim \text{POI}(\lambda)$. Information concerning $\text{E}(Y) = \lambda$ would be helpful to town planners for assessing requirements for new homeless shelters.

Suppose that town employees attempt to count the number of homeless people without shelter on any particular night, and further suppose that each homeless person without nighttime shelter has probability $\pi (0 < \pi < 1)$ of being counted. Also, assume that whether or not a particular homeless person is counted is not affected by whether or not any other homeless person is counted. Let the random variable X denote the number of homeless persons without nighttime shelter who are *actually* counted on any particular night.

(a) Develop explicit expressions for $\text{p}_X(x)$, the marginal distribution of X, and for $\text{E}(Y|X = x)$.

(b) Find an explicit expression for $\rho_{X,Y}$, the correlation between the random variables X and Y.

Exercise 4.69. Suppose that the density X of a ball bearing produced by a certain manufacturing process follows a normal distribution with unknown mean μ and known variance $\sigma^2 = 10$. Crates of these ball bearings are delivered to a certain company, and this company proposes to utilize the following sampling plan for deciding whether to purchase any particular crate of ball bearings. More specifically, for each crate, a random sample of n ball bearings is selected from the crate, where n is very small compared to the total number of ball bearings in the crate. Then, the average density $\bar{X}$ of these n ball bearings is computed, where $\bar{X} = n^{-1} \sum_{i=1}^{n} X_i$ and where X_i is the density of the i-th ball bearing in the random sample of n ball bearings. With k an appropriately chosen constant, if $\bar{X} > k$, then the crate is rejected (i.e., is not purchased); and if $\bar{X} \leq k$, then the crate is accepted (i.e., is purchased).

If $\mu > 46$, then this company wants to reject any crate with probability at least equal to 0.95; and if $\mu < 42$, then this company wants to accept any crate with probability at least equal to 0.98. As a consulting statistician for this company, find the smallest value of n, say n^*, that satisfies these requirements. Then, find an appropriate value k^* for k, and verify directly that these choices for n^* and k^* meet the stated requirements.

Exercise 4.70. Let X_1 and X_2 constitute a random sample of size $n = 2$ from the geometric distribution

$$\mathrm{p}_X(x) = \pi(1-\pi)^{x-1}, x = 1, 2, \ldots, \infty \text{ and } 0 < \pi < 1.$$

Let the range $R = \max\{X_1, X_2\} - \min\{X_1, X_2\}$.

(a) Develop an explicit for $\mathrm{p}_R(r)$, the probability distribution of the random variable R.

(b) Find $\mathrm{P}_R(s)$, the probability generating function for the random variable R, and then use $\mathrm{P}_R(s)$ to find $\mathrm{E}(R)$ and $\mathrm{V}(R)$.

Exercise 4.71. Suppose that

$$(X_1, X_2, X_3) \sim \text{MULT}\left[n; (1-\pi), \pi(1-\theta), \pi\theta\right],$$

and consider the two random variables

$$U = \frac{(X_2 + X_3)}{n} \text{ and } W = \frac{X_3}{(X_2 + X_3)}.$$

Develop explicit expressions for $\mathrm{E}(U), \mathrm{V}(U), \mathrm{E}(W)$, and $\mathrm{corr}(U, W)$.

Exercise 4.72. Let X_1 and X_2 constitute a random sample of size $n = 2$ from the geometric distribution

$$\mathrm{p}_X(x) = \pi(1-\pi)^{x-1}, x = 1, 2, \ldots, \infty \text{ and } 0 < \pi < 1.$$

Further, let

$$U = \min\{X_1, X_2\} \text{ and } V = \max\{X_1, X_2\}.$$

(a) Develop an explicit expression for $\mathrm{p}_{U,V}(u, v)$, the joint distribution of the random variables U and V, and then verify directly that $\mathrm{p}_{U,V}(u, v)$ is a valid bivariate discrete probability distribution.

(b) Use the bivariate discrete probability distribution developed in part (a) to derive explicit expressions for $\mathrm{p}_U(u)$ and $\mathrm{p}_V(v)$, the marginal distributions of the random variables U and V, and then verify directly that $\mathrm{p}_U(u)$ and $\mathrm{p}_V(v)$ are valid univariate discrete probability distributions.

Exercise 4.73. Queueing theory is the statistical study of waiting lines or *queues*. Statistical queueing theory allows for the derivation and estimation of various performance indices, including average waiting times for different positions in a queue, the expected number of people waiting or receiving service at any one time, etc. For further information, see Gross et al. (2008).

As a simple example, suppose that a store has $s(\geq 1)$ service stations. When the store opens, suppose that s customers immediately fill these s service stations, and then suppose that there are at least $n(> 1)$ customers still waiting in the queue to be served. For $i = 1, 2, \ldots, s$, assume that the time T_i (in minutes) required for the i-th service station to become available follows the negative exponential distribution with mean μ. Also, further assume that the random variables $T_1, T_2, \ldots, T_s$ are mutually independent.

(a) For a multiple-server situation (i.e., $s > 1$), let the random variable W_1 be the time (in minutes) that the first customer in the queue has to wait until a service station becomes available. Find explicit expressions for the distribution of W_1, for $\mathrm{E}(W_1)$, and for $\mathrm{V}(W_1)$.

(b) For a single-server situation (i.e., $s = 1$), let the random variable W_n be the time (in minutes) that the n-th customer in the queue has to wait until the service station becomes available. Find explicit expressions for the distribution of W_n, for $\mathrm{E}(W_n)$, and for $\mathrm{V}(W_n)$.

Exercise 4.74. Consider a one-dimensional *random walk* of a particle starting at the origin (i.e., position zero) where the possible positions of the particle are $0, \pm 1, \pm 2, \pm 3, \ldots$.

For $i = 1, 2, \ldots$, let X_i be the outcome of the i-th move of the particle, where $\mathrm{pr}(X_i = +1) = \pi$ and $\mathrm{pr}(X_i = -1) = (1 - \pi), 0 < \pi < 1$. Assume that $X_1, X_2, \ldots$ are mutually independent random variables.

Let Y_n be the random variable denoting the position of the particle after n moves. Find an explicit expression for the probability generating function $\mathrm{P}_{Y_n}(s)$ of Y_n, and then use this result to find $\mathrm{p}_{Y_n}(y_n)$, the probability distribution of Y_n. Also, provide explicit expressions for $\mathrm{p}_{Y_3}(y_3)$ and $\mathrm{p}_{Y_4}(y_4)$.

Exercise 4.75. A certain manufacturing company mass-produces small hearing aid batteries. These batteries are then placed into containers by a filling machine, with each container holding between a and b batteries (where a and b are positive integers with $a < b$). The number X of batteries in each container follows the discrete probability distribution

$$\mathrm{p}_X(x) = k(1 - \pi)^{x-1}\pi, a \leq x \leq b \text{ and } 0 < \pi < 1.$$

For a container holding x batteries, the number Y of defective batteries in such a container follows the conditional binomial distribution

$$\mathrm{p}_Y(y|X=x) = \mathrm{C}_y^x \theta^y (1-\theta)^{x-y}, y = 0, 1, \ldots, x \text{ and } 0 < \theta < 1.$$

(a) Find the value of k that makes $\mathrm{p}_X(x)$ a valid discrete probability distribution.

(b) Develop an explicit expression for $\mathrm{E}(Y)$, the expected number of defective batteries in any randomly chosen container, and then find an explicit expression for $\lim_{b\to\infty}\mathrm{E}(Y)$.

Exercise 4.76. Let $X_1, X_2, \ldots, X_n$ constitute a random sample of size $n(>1)$ from a $\mathrm{N}(\mu, \sigma^2)$ population. For $r = 1, 2, \ldots$, find $\mathrm{E}(U_r)$, where

$$U_r = \sum_{i=1}^{n} (X_i - \bar{X})^r \text{ and } \bar{X} = n^{-1} \sum_{i=1}^{n} X_i.$$

Then, use this result to find an explicit expression for a constant k_j such that $\mathrm{E}(k_j U_{2j}) = \sigma^{2j}, j = 1, 2, \ldots$

Exercise 4.77. For the i-th of n_j senior high school students ($i = 1, 2, \ldots, n_j$) in the j-th of k schools ($j = 1, 2, \ldots, k$) in a certain large U.S. city, suppose that Y_{ij} is the end-of-the-year score on a certain standardized mathematics examination. Further, suppose that $x_{1ij}, x_{2ij}, \ldots, x_{pij}$ constitute a set of p non-random covariate values specific to student i in school j; these covariate values are assumed to be predictive of Y_{ij}.

More specifically, suppose that Y_{ij} is linearly related to these p covariates via the following model:

$$Y_{ij} = B_{0j} + B_{1j} x_{1ij} + \sum_{l=2}^{p} \beta_l x_{lij} + e_{ij},$$

where $B_{0j} = (\beta_0 + U_{0j}), B_{1j} = (\beta_1 + U_{1j}), \beta_0, \beta_1, \ldots, \beta_p$ are unknown parameters, $U_{0j} \sim \mathrm{N}(0, \sigma_0^2), U_{1j} \sim \mathrm{N}(0, \sigma_1^2), \mathrm{cov}(U_{0j}, U_{1j}) = \sigma_{01}, e_{ij} \sim \mathrm{N}(0, \sigma_e^2)$, the pair (U_{0j}, U_{1j}) is independent of e_{ij} for all i and j, and the $\{e_{ij}\}$ constitute a set of $N = \sum_{j=1}^{k} n_j$ mutually independent random variables.

This model for Y_{ij} is called a *multilevel* statistical model, with students (Level 1) nested (or clustered) within schools (Level 2). There are two random coefficients, B_{0j} and B_{1j}, which vary from school to school. For an excellent book on the theory and application of multilevel statistical models, see Goldstein (1995).

(a) Find the exact distribution of Y_{ij}.

(b) Develop an explicit expression for $\text{cov}(Y_{ij}, Y_{i'j}), 1 \leq i < i' \leq n_j$.

(c) When $\sigma_1^2 = 0$, develop an explicit expression for $\text{corr}(Y_{ij}, Y_{i'j})$, $1 \leq i < i' \leq n_j$. Provide an interpretation for your finding.

Exercise 4.78. Suppose that two players (designated Player 1 and Player 2) play a particular game n times, where n is large. For $i = 1, 2, \ldots, n$, let X_i be a continuous random variable representing the score obtained by Player 1 for game i, and let Y_i be a continuous random variable representing the score obtained by Player 2 for game i.

Assume that the pairs $(X_1, Y_1), (X_2, Y_2), \ldots, (X_n, Y_n)$ constitute a random sample of size n from the bivariate density function $\text{f}_{X,Y}(x, y)$, which is of unspecified structure. Further, let $\text{E}(X_i) = \mu_x, \text{V}(X_i) = \sigma_x^2, \text{E}(Y_i) = \mu_y, \text{V}(Y_i) = \sigma_y^2$, and $\text{corr}(X_i, Y_i) = \rho$.

(a) Given numerical values for $n, \mu_x, \sigma_x^2, \mu_y, \sigma_y^2$, and ρ, develop an expression that can be used to provide a reasonable value for the probability θ_n that the total score $T_{1n} = \sum_{i=1}^n X_i$ for Player 1 is larger in value than the total score $T_{2n} = \sum_{i=1}^n Y_i$ for Player 2 after the game has been played n times. If $n = 100, \mu_x = 10, \sigma_x^2 = 4, \mu_y = 9.8, \sigma_y^2 = 3$, and $\rho = 0.10$, what is the numerical value of θ_{100}?

(b) If $\mu_x = 10, \sigma_x^2 = 4, \mu_y = 9.8, \sigma_y^2 = 3$, and $\rho = 0.10$, find the smallest value of n, say n^*, such that $\text{pr}(T_{1n^*} > T_{2n^*} + 5) \geq 0.90$.

Exercise 4.79. For a certain population of size N, suppose that θ is the number of people in this population who have a difficult-to-diagnose disease. Suppose that a sample of size $n, 0 < n < N$, is randomly selected *without replacement* from this population, and then each person in the sample is tested for the presence of the disease using an imperfect diagnostic instrument. Let D be the event that a person in this population has the disease in question, and let $\overline{\text{D}}$ be the event that a person in this population does not have the disease in question.

The *sensitivity* of this imperfect diagnostic instrument is defined as

$$\pi_1 = \text{pr}(\text{a person is diagnosed as having the disease}|\text{D}),$$

and the *specificity* of this imperfect diagnostic instrument is defined as

$$\pi_0 = \text{pr}(\text{a person is diagnosed as not having the disease}|\overline{\text{D}}).$$

Here, it is assumed that $0.50 < \pi_1 < 1$ and $0.50 < \pi_0 < 1$.

Let X_1 be the number of truly diseased persons in the sample who are diagnosed as having the disease, let X_0 be the number of truly non-diseased persons in the sample who are diagnosed as having the disease, let $X = (X_1 + X_0)$ be the number of persons in the sample who are diagnosed as having the disease, and let Y be the number of persons in the sample who actually have the disease in question. Note that X is observable, but that X_0, X_1, and Y are unobservable (or latent).

Develop explicit expressions for $\mathrm{E}(X)$ and $\mathrm{V}(X)$, the mean and variance of the random variable X.

For more details about the random variable X, see Stefanski (1992).

Exercise 4.80. When conducting a *sample survey* of a defined population, it is typically the case that the population is finite in size and that population members are selected randomly. Thus, valid statistical developments must be based on the principles associated with "sampling without replacement (WOR) from a finite population." To formalize these ideas, suppose that a continuous variable y is of interest. Further, for a finite-sized population of size $N(1 < N < \infty)$, let $y_1, y_2, \ldots, y_N$ denote the N *distinct* values of y for the N members of this population.

Then, the population mean μ is defined as

$$\mu = N^{-1} \sum_{j=1}^{N} y_j,$$

and the population variance σ^2 is defined as

$$\sigma^2 = N^{-1} \sum_{j=1}^{N} (y_j - \mu)^2 = N^{-1} \sum_{j=1}^{N} y_j^2 - \mu^2.$$

A sample of size $n(1 < n < N)$ is said to be *randomly selected* WOR from this population if all such samples of size n are equally likely to be selected. Let $Y_1, Y_2, \ldots, Y_n$ constitute a randomly selected WOR sample of size n from this population.

(a) For $i = 1, 2, \ldots, n$, prove that the marginal distribution of Y_i is equal to

$$\mathrm{pr}(Y_i = y_j) = \frac{1}{N}, j = 1, 2, \ldots, N.$$

(b) Prove that

$$\mathrm{cov}\,(Y_i, Y_{i'}) = -\frac{\sigma^2}{(N-1)}, i \neq i',$$

and then interpret this finding.

(c) Develop explicit expressions for $\mathrm{E}(\bar{Y})$ and $\mathrm{V}(\bar{Y})$, where $\bar{Y} = n^{-1}\sum_{i=1}^{n} Y_i$, and then comment on your findings.

Exercise 4.81. Suppose that X and Y are continuous random variables. Further, suppose that the marginal distribution of Y is

$$\mathrm{f}_Y(y) = \frac{\left(y - \frac{1}{\alpha}\right)^{-\beta} \alpha^{-\beta}}{y\Gamma(\beta)\Gamma(1-\beta)}, \alpha^{-1} < y < \infty, 0 < \alpha < \infty, 0 < \beta < 1,$$

and that the conditional distribution of X given $Y = y$ is

$$\mathrm{f}_X(x|Y = y) = y\mathrm{e}^{-yx}, 0 < x < \infty.$$

(a) Show that $\mathrm{f}_Y(y)$ is a valid density function.

(b) Develop an explicit expression for $\mathrm{f}_X(x)$, the marginal distribution of X. Do you recognize the structure of $\mathrm{f}_X(x)$?

Exercise 4.82. Let $X_1, X_2, \ldots, X_n$ constitute a random sample from an $\mathrm{N}(\mu, \sigma^2)$ population. Develop a *proof by induction* to show that

$$U_n = \frac{\sum_{i=1}^{n}(X_i - \bar{X}_n)^2}{\sigma^2} \sim \chi^2_{n-1}, \text{ where } \bar{X}_n = n^{-1}\sum_{i=1}^{n} X_i.$$

Exercise 4.83. Let $X_1, X_2, \ldots, X_n$ constitute a random sample of size $n(\geq 1)$ from a population with mean μ and variance σ^2. For $1 \leq m \leq n$, if

$$U_n = \sum_{i=1}^{n} a_i X_i \text{ and } U_m = \sum_{i=1}^{m} a_i X_i,$$

where $a_1, a_2 \ldots, a_n$ are constants, develop an explicit expression for $\mathrm{corr}(U_n, U_m)$. If $a_i = n^{-1}, i = 1, 2, \ldots, n$, how does the expression for $\mathrm{corr}(U_n, U_m)$ simplify?

Exercise 4.84. Let $Y_1, Y_2, \ldots, Y_n$ constitute a random sample of size n from the negative exponential density function

$$\mathrm{f}_Y(y; \theta) = \theta^{-1} e^{-y/\theta}, 0 < y < +\infty, \theta > 0.$$

Find the limiting value as $n \to +\infty$ of the moment generating function $\mathrm{M}_{Z_{(1)}}(t)$ of the standardized random variable

$$Z_{(1)} = \frac{Y_{(1)} - \mathrm{E}(Y_{(1)})}{\sqrt{\mathrm{V}(Y_{(1)})}},$$

where $Y_{(1)} = \min\{Y_1, Y_2, \ldots, Y_n\}$, and then use this result to characterize the asymptotic distribution of the random variable $Z_{(1)}$. Also, provide another analytical argument as to why this asymptotic distribution for $Z_{(1)}$ makes sense.

Exercise 4.85. Patients with end-stage acute lymphocytic leukemia are given an experimental drug treatment designed to increase their survival times. After receiving this experimental drug treatment, each patient is to be followed for up to one year to measure his or her survival time. During this one-year follow-up period, each patient can experience one of three outcomes: the patient dies during the one-year follow-up period; the patient survives the entire one-year follow-up period; the patient is lost to follow-up (e.g., withdraws from the study). Patients who experience either of the latter two outcomes are said to have *censored* responses in the sense that the outcome of interest (namely, death) is unobservable. The appropriate statistical treatment of time to event data with censoring is called *survival analysis.* For an excellent treatise on applied survival analysis, see the book by Hosmer, Lemeshow, and May (2011).

Let the random variable $T, 0 < T < \infty$, be the time to death for a patient receiving this experimental drug treatment, and let $\mathrm{f}_T(t), 0 < t < \infty$, be the density function for the random variable T. Further, let the random variable $C, 0 < C < 1$, be the time until a censored response occurs, and assume that T and C are independent random variables.

Now, consider the dichotomous random variable U, defined as follows:

$$U = 1 \text{ if } C \geq T, \text{ and } U = 0 \text{ if } C < T.$$

Then, $\mathrm{pr}(U = 1)$ is the probability that a patient is known to have died during the one-year follow-up period.

a) Prove that
$$\mathrm{pr}(U = 1) = \int_0^1 \mathrm{pr}(C \geq t)\mathrm{f}_T(t)\mathrm{d}t.$$

b) Suppose that $T \sim \mathrm{NEGEXP}(\alpha)$ and that C has a uniform density on the interval $(0, 1)$. Develop an explicit expression for $\mathrm{pr}(U = 1)$, namely, the expected proportion of patients who are known to have died during the one-year follow-up period. Also, find the limit of $\mathrm{pr}(U = 1)$ as $\alpha \to 0$ and as $\alpha \to \infty$, and comment on your findings.

Exercise 4.86. Suppose that an urn contains $N(\geq 1)$ balls, numbered individually from 1 to N. Suppose that $n(\geq 1)$ balls are randomly selected *with replacement* from this urn, and let X_i be the number on the i-th ball selected, $i = 1, 2, \ldots, n$.

Find an expression (which may involve summation signs) for $\mathrm{E}(U)$ when $U = \max\{X_1, X_2, \ldots, X_n\}$.

Exercise 4.87*. Let $X(s,t)$, where $X(s,t) = 0, 1, \ldots, \infty$, be a discrete random variable denoting the number of "hits" to a website during a time interval (s,t), where $0 < s < t < \infty$. Suppose that, for all s and t, $X(s,t)$ can be characterized by the following conditions:

1. $X(0,t)$ is a non-decreasing function of t; that is, $X(0,t) \leq X(0,t')$ for all $t \leq t'$.
2. For non-overlapping (i.e., disjoint) time intervals (s,s') and (t,t'), $X(s,s')$ and $X(t,t')$ are independent random variables.
3. The probability of *no hits* during the time interval $(t, t+\Delta t)$, where $\Delta t (> 0)$ is a very small positive quantity, is

$$\text{pr}[X(t, t + \Delta t) = 0] = 1 - \lambda \Delta t, 0 < \lambda \Delta t < 1.$$

4. The probability of *exactly one hit* during the time interval $(t, t + \Delta t)$ is

$$\text{pr}[X(t, t + \Delta t) = 1] = \lambda \Delta t.$$

5. The following *boundary condition* holds: $\text{pr}[X(0,0) = 0] = 1$.

Conditions (3) and (4) together imply that

$$\text{pr}[X(t, t + \Delta) > 1] = 1 - [1 - \lambda \Delta t] - [\lambda \Delta t] = 0,$$

and hence the probability of two or more hits occurring in the time interval $(t, t + \Delta t)$ equals 0. The sequence of random variables $\{X(0,t)\}, t > 0$, is said to form a *Poisson process* with *intensity parameter* λ. For further reading about Poisson processes, and more generally about stochastic processes, see the book by Lefebvre (2006).

(a) Derive an explicit expression, as a function of k, λ, and t, for $\text{p}_k(t) = \text{pr}[X(0,t) = k]$, the probability that exactly k hits, $k = 0, 1, \ldots, \infty$, occur during the time interval $(0,t)$. [HINT: First, set up a differential equation involving $\text{p}_k(t)$ and $\text{p}_{k-1}(t)$, and use it to derive expressions for $\text{p}_0(t)$ and $\text{p}_1(t)$. Then, use mathematical induction to derive a general expression for $\text{p}_k(t)$.]

(b) As an alternative derivation method, use probability generating function theory to derive an explicit expression for $\text{p}_k(t)$ as a function of k, λ, and t. [HINT: Let $\text{P}_X(s;t) = \sum_{k=0}^{\infty} p_k(t) s^k$ denote the probability generating function for $X(0,t)$, and consider $\text{dP}_X(s;t)/\text{d}t$.]

Exercise 4.88*. Let $X_1, X_2, \ldots, X_n$ constitute a random sample of size n from the truncated Poisson distribution

$$\text{p}_X(x) = (\text{e}^{\lambda} - 1)^{-1} \frac{\lambda^x}{x!}, x = 1, 2, \ldots, \infty \text{ and } \lambda > 0.$$

For $x = 1, 2, \ldots, \infty$, let the random variable Y_x denote the number of these n observations that take the value x, in which case the set of possible values for Y_x is {0,1,...,n} subject to the restriction $\sum_{x=1}^{\infty} Y_x = n$.

For the random variable

$$U = n^{-1}(S_1 - S_2) = n^{-1}\left[\sum_{j=1}^{\infty} Y_{2j-1} - \sum_{j=1}^{\infty} Y_{2j}\right],$$

show that the mean and variance of U are, respectively,

$$\mathrm{E}(U) = \mathrm{e}^{-\lambda} \text{ and } \mathrm{V}(U) = \frac{(1 - \mathrm{e}^{-2\lambda})}{n}.$$

Exercise 4.89*. Let $X_1, X_2, \ldots, X_n$ constitute a random sample of size n from the parent population

$$\mathrm{f}_X(x;\theta) = \frac{1}{2}\mathrm{e}^{-|x-\theta|}, -\infty < x < \infty,\ -\infty < \theta < \infty.$$

Consider the random variable

$$U_n = \frac{\sum_{i=1}^{n} X_i - n\theta}{\sqrt{2n}}.$$

Derive an explicit expression for the moment generating function $M_{U_n}(t)$ of the random variable U_n, and then evaluate its limit as $n \to \infty$ to infer the form of the asymptotic distribution of U_n. Also, provide an alternative justification for the answer that you have obtained.

Exercise 4.90*. Suppose that the probability π of a tsunami occurring in any given calendar year in a particular country remains the same from year to year. Further, assume that the probability of the occurrence of more than one tsunami in any calendar year is zero, and that tsunami occurrences in different years are mutually independent. Meterologists are naturally interested in determining the probability of future tsunami occurrences in this particular country. In particular, suppose that meterologists are interested in the possible clustering of yearly occurrences of tsunamis over the next $k(> 1)$ years.

(a) Given that exactly two tsumanis occur during the next k years, what is the probability that these two tsunamis occur in consecutive years?

(b) For $j = 1, 2, \ldots, (k-1)$, define the random variable X_j as follows:

$$X_j = \begin{cases} 1 \text{ if a tsunami occurs in both years } j \text{ and } (j+1), \\ 0 \text{ if not.} \end{cases}$$

Then, consider the random variable $X = \sum_{j=1}^{k-1} X_j$. Find explicit expressions, as functions of π and k, for $\mathrm{E}(X)$ and $\mathrm{V}(X)$.

Exercise 4.91*. As a model for the effects of well water contamination by underground pollutants, suppose that X is a dichotomous random variable taking the value 1 if a well is not contaminated and taking the value 0 if a well is contaminated; also, assume that $\text{p}_X(x) = \pi^x(1-\pi)^{1-x}, x = 0, 1$ and $0 < \pi < 1$. Further, suppose that the number Y of living organisms in a cubic centimeter of well water, given $X = x$, is assumed to have the Poisson distribution $\text{p}_Y(y|X = x) = (\alpha + \beta x)^y \text{e}^{-(\alpha+\beta x)}/y!$, $y = 0, 1, \ldots, \infty$, $\alpha > 0, \beta > 0$. So, $\text{E}(Y|X = 0) = \alpha$ and $\text{E}(Y|X = 1) = (\alpha + \beta)$.

(a) Derive an explicit expression for $\text{pr}(Y \leq X)$.

(b) Let $(X_1, Y_1), (X_2, Y_2), \ldots, (X_n, Y_n)$ constitute a random sample of size n from the underlying joint distribution of X and Y. Consider the random variable $L = c\bar{X} + (1-c)\bar{Y}$, where $\bar{X} = n^{-1}\sum_{i=1}^{n} X_i$ and $\bar{Y} = n^{-1}\sum_{i=1}^{n} Y_i$, and where c is a constant. Derive an explicit expression for that value of c, say c^*, such that $\text{V}(L)$ is a minimum. Provide a reasonable interpretation for your answer when $\beta = 0$.

(c) For L as defined in part (b), derive an explicit expression for the moment generating function of the random variable L.

Exercise 4.92*. Let $X_1, X_2, \ldots, X_n$ constitute a random sample of size $n(\geq 1)$ from the uniform density function $\text{f}_X(x) = \theta^{-1}, 0 < x < \theta$, and let

$$S_n = \sum_{i=1}^{n} X_i.$$

(a) Use an inductive argument to show that

$$\text{pr}(S_n \leq t) = \frac{t^n}{n!\theta^n}, 0 \leq t \leq \theta.$$

(b) Let the discrete random variable N be defined as the *smallest* positive integer n such that $S_n > \theta$. Derive an explicit expression for the probability distribution $\text{p}_N(n)$ of the random variable N, and show directly that $\text{p}_N(n)$ is a valid discrete probability distribution.

(c) Develop explicit expressions for $\text{E}(N)$ and $\text{V}(N)$.

Exercise 4.93*. Suppose that U_1 and U_2 are two random variables such that $\text{E}(U_1) = \text{E}(U_2) = 0, \text{V}(U_1) = \text{V}(U_2) = 1$, and $\text{corr}(U_1, U_2) = \rho, -1 \leq \rho \leq 1$.

(a) For $k > 0$, use Markov's Inequality to show that

$$\text{pr}\left[(|U_1| > k) \cup (|U_2| > k)\right] \leq \frac{2}{k^2}.$$

(b) Show that

$$\mathrm{E}\left[\max(U_1^2, U_2^2)\right] \leq 1 + \sqrt{1-\rho^2}.$$

HINT: For real numbers a and b, $\max(|a|, |b|) = \frac{1}{2}\left(|a-b| + |a+b|\right)$.

(c) For $k > 0$, use the result in part (b) and Markov's Inequality to show that

$$\mathrm{pr}\left[(|U_1| > k) \cup (|U_2| > k)\right] \leq \frac{1 + \sqrt{1-\rho^2}}{k^2},$$

which is a sharper inequality than the one developed in part (a).

(d) Let Y_1 and Y_2 be two random variables such that $\mathrm{E}(Y_1) = \mu_1, \mathrm{E}(Y_2) = \mu_2, \mathrm{V}(Y_1) = \sigma_1^2, \mathrm{V}(Y_2) = \sigma_2^2$, and $\mathrm{corr}(Y_1, Y_2) = \rho, -1 \leq \rho \leq 1$. For $0 < \alpha < 1$, use the result in part (c) to develop a formula for k (as a function of ρ and α) such that

$$\mathrm{pr}\left\{\left[\left|\frac{Y_1 - \mu_1}{\sigma_1}\right| \leq k\right] \cap \left[\left|\frac{Y_2 - \mu_2}{\sigma_2}\right| \leq k\right]\right\} \geq \alpha.$$

Comment on how k varies as a function of α and ρ.

Exercise 4.94*. Suppose that a particular clinical trial is designed to compare a new chemotherapy treatment to a standard chemotherapy treatment for treating Hodgkin's disease. At the beginning of this clinical trial, suppose that each of n independently selected patients is assigned to the new treatment with probability $\pi, 0 < \pi < 1$, and to the standard treatment with probability $(1 - \pi)$. If a patient receives the new treatment, then that patient has probability θ_1 of going into remission; if a patient receives the standard treatment, then that patient has probability θ_0 of going into remission.

(a) Use conditional expectation theory to develop explcit expressions for $\mathrm{E}(Y)$ and $\mathrm{V}(Y)$, where the discrete random variable Y is the number of the n patients who go into remission.

(b) Develop an explicit expression for $\mathrm{p}_Y(y)$, the probability distribution of the random variable Y.

Exercise 4.95*. Social science researchers studying a latent (i.e., unobservable or unmeasurable) variable such as personality, job satisfaction, or customer satisfaction typically recruit a random sample of subjects to respond [often via a Likert-type scale (Likert, 1931)] to a series of items (e.g., statements, questions, etc.) designed to elicit responses that are highly correlated with this latent variable. In particular, for the ith of k items, let T_i be a true latent random variable of interest, and let Y_i be a random variable representing the observed response of a subject to the i-th item, $i = 1, 2, \ldots, k$. Then, the statistical model relating Y_i to T_i is typically assumed to be of the form

$$Y_i = T_i + U_i, i = 1, 2, \ldots, k,$$

where U_i is a random variable reflecting the error when using Y_i as a surrogate (i.e., imperfect) measure of T_i. It is assumed that $\mathrm{E}(U_i) = 0$, so that $\mathrm{E}(Y_i) = \mathrm{E}(T_i) = \mu_T$, say, and that T_i and U_i are independent random variables for all i and j, $i = 1, 2, \ldots, k$ and $j = 1, 2, \ldots, k$.

Let $\bar{Y} = k^{-1}\sum_{i=1}^{k} Y_i$ and let $\bar{T} = k^{-1}\sum_{i=1}^{k} T_i$, so that $\mathrm{E}(\bar{Y}) = \mathrm{E}(\bar{T}) = \mu_T$.

(a) The *reliability coefficient* θ associated with the use of the item responses $Y_1, Y_2, \ldots, Y_k$ as surrogates for the latent variables $T_1, T_2, \ldots, T_k$ is defined as the squared correlation between $\bar{Y}$ and $\bar{T}$ or, equivalently, between $\sum_{i=1}^{k} Y_i$ and $\sum_{i=1}^{k} T_i$ (Cronbach, 1951; Cortina, 1993). Show that

$$\theta = \frac{\mathrm{V}(\bar{T})}{\mathrm{V}(\bar{Y})} = \frac{\mathrm{V}\left(\sum_{i=1}^{k} T_i\right)}{\mathrm{V}\left(\sum_{i=1}^{k} Y_i\right)}.$$

(b) Show that

$$(k-1)\sum_{i=1}^{k} \mathrm{V}(T_i) \geq \sum_{\text{all } i \neq j} \mathrm{cov}(T_i, T_j).$$

As a hint, consider the expression $\sum_{i=1}^{k-1}\sum_{j=i+1}^{k} \mathrm{V}(T_j - T_i)$.

(c) Use the result in part (b) to show that $\theta \geq \alpha$, where

$$\alpha = \left(\frac{k}{k-1}\right)\left[\frac{\sum_{\text{all } i\neq j} \mathrm{cov}(Y_i, Y_j)}{\mathrm{V}\left(\sum_{i=1}^{k} Y_i\right)}\right] = \left(\frac{k}{k-1}\right)\left[1 - \frac{\sum_{i=1}^{k} \mathrm{V}(Y_i)}{\mathrm{V}\left(\sum_{i=1}^{k} Y_i\right)}\right]$$

is known as *Cronbach's alpha* and provides a lower bound for the reliability coefficient θ. Estimated values of α greater than or equal to 0.80 are considered to be desirable.

(d) Show that α can be equivalently written in the form

$$\alpha = \frac{k\bar{C}}{\bar{V} + (k-1)\bar{C}},$$

where

$$\bar{C} = \frac{1}{k(k-1)}\sum_{\text{all } i\neq j} \mathrm{cov}(Y_i, Y_j) \text{ and } \bar{V} = \frac{1}{k}\sum_{i=1}^{k} \mathrm{V}(Y_i).$$

(e) Assuming that $\mathrm{V}(Y_i) = \sigma_Y^2 (> 0), i = 1, 2, \ldots, k$, use the result in part (d) to find a sufficient condition such that α takes its maximum value of 1.

Exercise 4.96*. Consider a sequence of mutually independent continuous random variables $X_1, X_2, X_3, \ldots$, each having the same expected value μ and the same variance σ^2. Let $U_i = (X_i - \mu)/\sigma$ and assume that

$$\mathrm{E}\left[(U_i - U_j)^{-2}\right] = \theta \text{ for all } i \neq j, \text{ where } 0 < \theta < +\infty.$$

Now, suppose that observed values x_1 and x_2 of the random variables X_1 and X_2 are obtained. Then, a further set of N observations $X_3, X_4, \ldots, X_{N+2}$ is obtained, where N is the smallest positive integer at least as large as the quantity $K(x_1 - x_2)^2$, where K is a fixed positive number. Then, let

$$\bar{X} = N^{-1} \sum_{i=1}^{N} X_{i+2}.$$

(a) Show that $\mathrm{E}(N) \geq 2K\sigma^2$.

(b) Show that

$$\mathrm{pr}\left[|\bar{X} - \mu| < A\right] \geq 1 - \frac{\theta}{KA^2}$$

for any positive number A.

Exercise 4.97*. Suppose that the continuous random variables X and Y follow a bivariate normal distribution with parameters $\mathrm{E}(X) = \mu_x$, $\mathrm{E}(Y) = \mu_y$, $\mathrm{V}(X) = \sigma_x^2$, $\mathrm{V}(Y) = \sigma_y^2$, and $\mathrm{corr}(X, Y) = \rho$. Let $(X_1, Y_1), (X_2, Y_2), \ldots, (X_n, Y_n)$ constitute a random sample of size n from this bivariate normal distribution. Further, let $X_1^*, X_2^*, \ldots, X_m^*$ constitute a random sample of size m from the marginal distribution of X, and assume that the set of random variables $X_1^*, X_2^*, \ldots, X_m^*$ is independent of the set of n pairs $(X_1, Y_1), (X_2, Y_2), \ldots, (X_n, Y_n)$. Use rigorous arguments and known properties of the bivariate normal distribution to derive an explicit expression for the expected value of the random variable U, where

$$U = \bar{Y} + \hat{\beta}(\bar{X}' - \bar{X}),$$

where

$$\bar{X} = n^{-1} \sum_{i=1}^{n} X_i, \bar{Y} = n^{-1} \sum_{i=1}^{n} Y_i, \bar{X}' = (n+m)^{-1} \left[\sum_{i=1}^{n} X_i + \sum_{i=1}^{m} X_i^*\right],$$

and where

$$\hat{\beta} = \left[\sum_{i=1}^{n} (X_i - \bar{X})^2\right]^{-1} \sum_{i=1}^{n} (X_i - \bar{X})(Y_i - \bar{Y}).$$

Exercise 4.98*. In fields like physics, chemistry, and computer science, quadratic equations with random coefficients are sometimes encountered, and it is of interest to determine the probability that such a quadratic equation will have real roots.

To be more specific, consider the quadratic equation $Ax^2+Bx+C=0$, where A, B, and C are mutually independent random variables, each having a uniform distribution on the interval $(0, R), R > 0$. Develop an explicit expression for the probability that such a quadratic equation will have real roots.

Exercise 4.99*. Racial profiling is a controversial method that updates a subject's prior probability of criminal behavior based on his or her race, ethnicity, nationality, or religion. Racial profiling is a proposed technique for aiding in the identification of terrorists who use a country's airline transportation network. For further discussion, see Press (2009, 2010).

Suppose that n subjects utilize a country's airline transportation network over some defined period of time. Further, for $i = 1, 2, \ldots, n$, suppose that the i-th subject has a *known* probability π_i $(0 < \pi_i < 1)$ of being a terrorist, where the value of π_i is determined based on racial profiling information. At any airport security checkpoint, most of the n individuals will have very small π_i values and so will pass through the primary screening process with no difficulty. However, a key purpose of this primary screening process is to select individuals with high π_i values to undergo a rigorous secondary screening process that leads to the identification of terrorists. In particular, assume that subject i will be selected for secondary screening with probability θ_i $(0 < \theta_i < 1)$, where $\theta_i = \mathrm{g}(\pi_i)$ is a monotonically increasing function of π_i.

For $i = 1, 2, \ldots, n$, given that subject i is a terrorist, let the discrete random variable Y_i be the number of checkpoints required so that this terrorist is eventually selected for secondary screening, and assume that Y_i has the geometric distribution

$$\mathrm{p}_{Y_i}(y_i) = \theta_i^{y_i - 1}\theta_i, y_i = 1, 2, \ldots, \infty \text{ and } 0 < \theta_i < 1.$$

(a) If the random variable T is the total number of checkpoints performed on terrorists before each terrorist's first secondary screening, show that

$$\mathrm{E}(T) = \sum_{i=1}^{n} \frac{\pi_i}{\theta_i}.$$

Given known values for $\pi_1, \pi_2, \ldots, \pi_n$, it is of interest for logistical reasons (e.g., time, cost, etc.) to choose values for $\theta_1, \theta_2, \ldots, \theta_n$ that minimize $\mathrm{E}(T)$. For what obvious values of $\theta_1, \theta_2, \ldots, \theta_n$ is $\mathrm{E}(T)$ a minimum? For these choices of $\theta_1, \theta_2, \ldots, \theta_n$, what is the associated screening strategy? Do you notice any obvious disadvantages of this screening strategy?

(b) At any particular checkpoint, suppose that only $K(< n)$ subjects, on average, can be chosen for secondary screening. Since subject i has probability θ_i of being selected for secondary screening at any checkpoint, it makes sense to impose the restriction $\sum_{i=1}^{n} \theta_i = K$ for the choices of

$\theta_1, \theta_2, \ldots, \theta_n$. Use the method of Lagrange multipliers to show that, subject to the restriction $\sum_{i=1}^{n} \theta_i = K$, the choices for the $\{\theta_i\}$ that minimize $\mathrm{E}(T)$ are

$$\theta_i = \frac{K\sqrt{\pi_i}}{\sum_{l=1}^{n} \sqrt{\pi_l}}, i = 1, 2, \ldots, n,$$

where $K < \min\left\{\frac{\sum_{l=1}^{n}\sqrt{\pi_l}}{\sqrt{\pi_1}}, \frac{\sum_{l=1}^{n}\sqrt{\pi_l}}{\sqrt{\pi_2}}, \ldots, \frac{\sum_{l=1}^{n}\sqrt{\pi_l}}{\sqrt{\pi_n}}\right\}$, so that $0 < \theta_i < 1$ for all $i, i = 1, 2, \ldots, n$. Also, find an explicit expression for the minimized value of $\mathrm{E}(T)$ subject to the restriction $\sum_{i=1}^{n} \theta_i = K$, and then comment on all your findings.

Exercise 4.100*. Suppose that each of two physicians (say, Physician 1 and Physician 2) of equal proficiency *independently* perform an endoscopy on the same human subject. For $i = 1, 2$, suppose that the discrete random variable X_i represents the number of mucosal lesions found by Physician i, and further suppose that the discrete random variable X_{12} represents the number of mucosal lesions found by *both* Physician 1 and Physician 2.

Assuming that $X_i \sim \mathrm{BIN}(N, \pi)$ for $i = 1, 2$, where the parameter N is the unknown total number of mucosal lesions for this human subject, show that $\mathrm{E}(\hat{N}) = N$, where

$$\hat{N} = \frac{(X_1 + 1)(X_2 + 1)}{(X_{12} + 1)} - 1.$$

Exercise 4.101*. An electrical procedure designed to detect the presence of a tumor in a sample of human tissue involves measuring the voltage $V (0 < V < \infty)$ between two strategically placed electrodes. If the measured voltage exceeds some specified *detection value*, say $D(0 < D < \infty)$, then there is evidence for the presence of a tumor; if the measured voltage does not exceed D, then there is no evidence for the presence of a tumor. Clearly, a detection error occurs if either of the following two events occur: (i) V exceeds D and there is no tumor (a so-called *false positive*) and (ii) V does not exceed D and there is a tumor (a so-called *false negative*).

Suppose that the probability of a tumor being present in a sample of human tissue is equal to $\theta, 0 < \theta < 1$. Also, given that a tumor is present, suppose that $U = \ln V \sim \mathrm{N}(\mu_1, \sigma^2)$; and given that a tumor is not present, suppose that $U = \ln V \sim \mathrm{N}(\mu_0, \sigma^2)$, where $-\infty < \mu_0 < \mu_1 < +\infty$.

(a) Find an explicit expression for the optimal choice D^* for D that minimizes the probability of a detection error.

(b) Provide a reasonable interpretation for the limiting values of D^* as $\theta \to 1$ and as $\theta \to 0$.

Exercise 4.102*. Suppose that the continuous random variables $X_1, X_2, \ldots, X_n$ have a joint density function that can be written in the form

$$\mathrm{f}^* = \mathrm{e}^{B(\theta)T(\boldsymbol{x})+C(\theta)+D(\boldsymbol{x})}, -\infty < x_i < \infty, i = 1, 2, \ldots, n,$$

where $\theta(-\infty < \theta < \infty)$ is an unknown parameter, $\boldsymbol{x} = (x_1, x_2, \ldots, x_n)$, $B(\theta)$ and $C(\theta)$ depend only on θ and not on $\boldsymbol{x}$, $T(\boldsymbol{x})$ and $D(\boldsymbol{x})$ depend only on $\boldsymbol{x}$ and not on θ, and $\mathrm{E}[T(\boldsymbol{X})] = \theta$.

(a) Starting with the fact that

$$\int_{-\infty}^{\infty}\int_{-\infty}^{\infty}\cdots\int_{-\infty}^{\infty}\mathrm{f}^*\mathrm{d}x_1\mathrm{d}x_2\ldots\mathrm{d}x_n = \int_{-\infty}^{\infty}\int_{-\infty}^{\infty}\cdots\int_{-\infty}^{\infty}\mathrm{f}^*\mathbf{d}\boldsymbol{x} = 1,$$

and assuming sufficient regularity conditions (so that integral and differentiation operations can be switched), differentiate both sides of the above multiple integral expression with respect to θ and use the result to show that

$$\mathrm{E}[T(\boldsymbol{X})] = -\frac{C'(\theta)}{B'(\theta)}.$$

(b) Show that

$$\mathrm{V}(X) = \frac{C'(\theta)B''(\theta) - C''(\theta)B'(\theta)}{[B'(\theta)]^3}.$$

(c) Show that the above results hold for the special case when $X_1, X_2, \ldots, X_n$ constitute a random sample of size n from the density function

$$\mathrm{f}_X(x;\theta) = \frac{x^3\mathrm{e}^{-x/\theta}}{6\theta^4}, 0 < x < \infty, 0 < \theta < \infty,$$

and $T(\boldsymbol{X}) = \bar{X}/4$, where $\bar{X} = n^{-1}\sum_{i=1}^{n} X_i$.

Exercise 4.103*. A factory produces a certain type of medical instrument, and its quality of performance depends upon the *sum* $S = (X_1 + X_2)$ of the eccentricities X_1 and X_2 of two similarly shaped wheels that are used in the construction of such a medical instrument. Eccentricity is a mathematical measure of a departure from circularity (i.e., a measure of "out of roundness"). For any such wheel, the eccentricity X has a distribution that is reasonably modeled by the negative exponential density function

$$\mathrm{f}_X(x) = \mathrm{e}^{-x}, 0 < x < \infty.$$

Two alternative methods are suggested for choosing two wheels for the construction of any one of these medical instruments:

Random Assembly (denoted RA): Two wheels are chosen randomly.

Stratified Assembly (denoted SA): Wheels are first divided into two strata according to whether their eccentricities are above the median of $f_X(x)$ or below the median of $f_X(x)$, and then one wheel is chosen randomly from each of these two strata.

Which method of assembly (RA or SA) produces the smaller variability in the sum $S = (X_1 + X_2)$ of the eccentricities of two wheels chosen for the construction of any one of these medical instruments?

Exercise 4.104*. In a certain high-risk population of married couples having low socioeconomic status, suppose that the monthly number $N(\geq 1)$ of domestic violence events involving physical assault follows the geometric distribution

$$p_N(n) = \text{pr}(N = n) = \pi(1-\pi)^{n-1}, n = 1, 2, \ldots, \infty \text{ and } 0 < \pi < 1.$$

Further, suppose that any such domestic violence event leads to a 911 call with probability $\theta, 0 < \theta < 1$, and that only domestic violence events leading to 911 calls are recorded. Define the random variable X to be the monthly number of recorded domestic violence events involving physical assault (i.e., events that led to 911 calls).

It is of interest to use the *observed value* x of X in any month to predict the *unobservable number* N of domestic violence events involving physical assault that occurred during that same month in this high-risk population. Such a prediction is to be made using the expression $\text{E}(N|X = x)$.

Develop an explicit expression for $\text{E}(N|X = x)$. If $\pi = 0.10, \theta = 0.20$, and $x = 12$, what is the numerical value of $\text{E}(N|X = x)$?

Exercise 4.105*. Suppose that an individual plays a certain casino game $n(\geq 2)$ times. Let W be the event that this individual wins any particular game, let L be the event that this individual loses any particular game, and assume that the outcomes of these n games are mutually independent. Further, let $\text{pr}(\text{W}) = \pi, 0 < \pi < 1$, so that $\text{pr}(\text{L}) = (1 - \pi)$.

Suppose that the outcomes of these n games are arranged in a linear sequence, and let the discrete random variable Y_n be the number of times that the subsequence WW is observed (i.e., Y_n is the number of times that the individual wins two consecutive games). For example, if $n = 10$ and the sequence WWLLWWWLWW occurs, then the observed value of Y_{10} is equal to $y_{10} = 4$. Develop explicit expressions for $\text{E}(Y_n)$ and $\text{V}(Y_n)$ as a function of n and π.

Exercise 4.106*. A parapsychologist theorizes that there could be extrasensory perception (ESP) between monozygotic twins. To test this theory, this parapsychologist designs the following simple experiment. Each twin thinks of a particular whole number between 1 and m inclusive (i.e., each twin picks one of the numbers $1, 2, \ldots, m$, and then writes that number on a piece of paper). The two numbers written down are then compared to see whether they are *close* in value.

Let X be the number chosen by one member of a set of monozygotic twins, and let Y be the number chosen by the other member. Then, let $U = |X - Y|$ be the measure of closeness of the two chosen numbers.

(a) Under the assumption that each twin is selecting his or her number totally at random (so there is no ESP), develop an explicit expression for $\text{pr}(U \leq k)$, where k is a non-negative integer satisfying the inequality $0 \leq k \leq (m-1)$.

(b) Under the random selection assumption in part (a), develop explicit expressions for $\text{p}_U(u)$, the probability distribution of the random variable U, and for $\text{E}(U)$.

(c) If $m = 100$, use the result in part (a) to find the largest value of k, say k^*, such that $\text{pr}(U \leq k^*) \leq 0.05$.

(d) For a particular set of monozygotic twins, suppose that this experiment is *independently repeated* $n = 10$ times using $m = 100$ and the value of k^* determined in part (c). For this particular set of monozygotic twins, suppose that exactly 3 of the $n = 10$ pairs of numbers chosen do not differ by more than k^*. Do you think that these data provide statistical evidence of ESP?

Exercise 4.107*. Traffic engineers in a certain large U.S. city are interested in estimating the mean waiting time (in minutes) at a particular stop sign before a car can safely enter a heavily traveled highway. Suppose that the number N of cars on this highway that pass by this stop sign during a time interval of $t(>0)$ minutes follows a Poisson distribution with parameter $\text{E}(N) = \lambda t$, where $\lambda(>0)$ is the expected number of cars passing by this stop sign each minute.

(a) For any car that passes by this stop sign, let T be the time interval (in minutes) before the next car passes by this stop sign. Show that $T \sim \text{NEGEXP}(\alpha = \lambda^{-1})$.

(b) Given that the time interval T between two cars passing by this stop sign is greater than t minutes, show that

$$\text{pr}(T > t + s | T > t) = \text{pr}(T > s), t > 0, s > 0.$$

This probability equality is known as the *memoryless property* of the negative exponential distribution. For our particular example, this property means the following: given that it has been at least t minutes since a car passed by the stop sign, the probability that it will take at least another s minutes for the next car to pass by the stop sign does not depend on the value of t.

(c) Show that

$$\mathrm{E}(T|T \leq t^*) = \frac{1}{\lambda} - \frac{t^*}{(\mathrm{e}^{\lambda t^*} - 1)}, t^* > 0.$$

(d) Suppose that a car waiting at the stop sign needs a time interval between cars of at least $t^*(> 0)$ minutes in order to be able to safely enter this heavily traveled highway. If the random variable W is the waiting time (in minutes) until the start of safe entry, use the results in parts (b) and (c) to show that

$$\mathrm{E}(W) = \frac{(\mathrm{e}^{\lambda t^*} - \lambda t^* - 1)}{\lambda}.$$

If 10 cars per minute pass by this stop sign and if it takes a car 15 seconds to enter this highway safely, find the numerical value of $\mathrm{E}(W)$.

HINT: Let X be the number of time intervals required until the first time interval longer than t^* occurs, and write X as a sum of appropriately defined time intervals.

For further discussion on this topic, see Griffiths (2011).

Exercise 4.108*. Consider the following simple urn model for the movement (i.e., *diffusion*) of molecules of a certain compound across a membrane that separates two liquids having equal concentrations of this compound.

In particular, let Urn 1 contain n white balls and let Urn 2 contain n green balls, where $n \geq 2$. During each 1-minute time period, suppose that a ball is randomly selected from Urn 1 and simultaneously a ball is randomly selected from Urn 2; then the ball randomly selected from Urn 1 is put into Urn 2, and the ball randomly selected from Urn 2 is put into Urn 1 (i.e., the two balls switch urns). Let N_k be the number of white balls in Urn 1 after exactly $k(\geq 1)$ 1-minute time periods have elapsed. Develop an explicit for $\mathrm{E}(N_k)$. Then, find the limiting value of $\mathrm{E}(N_k)$ as $k \to \infty$, and comment on this finding.

Exercise 4.109*. For a certain county in North Carolina, let the random variable X denote the yearly number of suicides by non-pregnant women, and let the random variable Y denote the yearly number of suicides by pregnant women. Suppose that the random variable X has the geometric distribution

$$\mathrm{p}_X(x) = \alpha(1-\alpha)^x, x = 0, 1, \ldots, \infty \text{ and } 0 < \alpha < 1,$$

and suppose that the random variable Y has the geometric distribution

$$\mathrm{p}_Y(y) = \beta(1-\beta)^y, y = 0, 1, \ldots, \infty \text{ and } 0 < \beta < 1.$$

Assume that $\alpha \neq \beta$ and that X and Y are independent random variables.

Further, suppose that only the total number $Z = (X + Y)$ of suicides in this county is recorded each year. Given an observed value z of Z for a particular year, the goal is to estimate the numbers of suicides by pregnancy status for that particular year using the functions $\mathrm{E}(X|Z = z)$ and $\mathrm{E}(Y|Z = z)$.

Find explicit expressions for $\mathrm{E}(X|Z = z)$ and $\mathrm{E}(Y|Z = z)$. For this particular year, suppose that the observed value of Z is $z = 2$, that $\alpha = 0.30$, and that $\beta = 0.60$; use this information to find numerical values for $\mathrm{E}(X|Z = 2)$ and $\mathrm{E}(Y|Z = 2)$.

Exercise 4.110*. For $j = 1, 2, \ldots, n (n \geq 2)$, let Y_j be a continuous random variable measured at time t_j, where $0 \leq t_1 < t_2 < \cdots < t_n \leq 1, \mathrm{E}(Y_j) = \mu$, and $\mathrm{V}(Y_j) = \sigma^2$.

For $j < j'$, suppose that Y_j and $Y_{j'}$ are related by the *first-order autoregressive model*

$$\left(\frac{Y_{j'} - \mu}{\sigma}\right) = \rho^{(t_{j'} - t_j)} \left(\frac{Y_j - \mu}{\sigma}\right) + \epsilon_{j'},$$

where $0 < \rho < 1, \mathrm{E}(\epsilon_{j'}) = 0$, and Y_j and $\epsilon_{j'}$ are independent random variables for all j and j'.

(a) Under this first-order autoregressive model, show that

$$\mathrm{V}(\epsilon_{j'}) = 1 - \rho^{2(t_{j'} - t_j)}$$

and that

$$\mathrm{corr}(Y_j, Y_{j'}) = \rho^{(t_{j'} - t_j)}.$$

(b) Consider the situation where n equally spaced observations $\{Y_j\}_{j=1}^n$ are obtained in the time interval $[0, 1]$, so that $t_j = (j - 1)/(n - 1), j = 1, 2, \ldots, n$. With $\bar{Y} = n^{-1} \sum_{j=1}^n Y_j$, show that

$$\mathrm{V}(\bar{Y}) = \frac{\sigma^2}{n^2} \left\{ n + \frac{2\theta_n}{(1 - \theta_n)^2} \left[n(1 - \theta_n) - (1 - \theta_n^n) \right] \right\},$$

where $\theta_n = \rho^{1/(n-1)}$.

For more details about the properties of $\bar{Y}$, see Morris and Ebey (1984).

4.2 Solutions to Odd-Numbered Exercises

Solution 4.1.

(a) Since we require $K(1-\pi)^2 + 2K\pi(1-\pi)\theta + K\pi^2 = 1$, we need

$$K = [(1-\pi)^2 + 2\pi(1-\pi)\theta + \pi^2]^{-1}.$$

(b) For $i = 1, 2$, $\text{pr}(X_i = 1) = K\pi(1-\pi)\theta + K\pi^2 = K\pi[\pi + (1-\pi)\theta]$, so that X_i has the following Bernoulli distribution:

$$\text{p}_{X_i}(x_i) = \{K\pi[\pi + (1-\pi)\theta]\}^{x_i}\{1 - K\pi[\pi + (1-\pi)\theta]\}^{1-x_i}, x_i = 0, 1;$$
$$i = 1, 2;\ K = [(1-\pi)^2 + 2\pi(1-\pi)\theta + \pi^2]^{-1}.$$

(c) Now,

$$\begin{aligned}
\text{pr}(X_1 = 1|X_2 = 1) &= \frac{\text{pr}[(X_1 = 1) \cap (X_2 = 1)]}{\text{pr}(X_2 = 1)} \\
&= \frac{K\pi^2}{K\pi[\pi + (1-\pi)\theta]} \\
&= \frac{\pi}{\pi + (1-\pi)\theta}; \\
\text{and, pr}(X_1 = 0|X_2 = 1) &= 1 - \text{pr}(X_1 = 1|X_2 = 1) = \frac{(1-\pi)\theta}{\pi + (1-\pi)\theta}.
\end{aligned}$$

So, $\text{E}(X_1|X_2 = 1) = 1 \cdot \text{pr}(X_1 = 1|X_2 = 1) = \pi/[\pi + (1-\pi)\theta]$. Since $\text{E}(X_1^2|X_2 = 1) = (1)^2 \cdot \text{pr}(X_1 = 1|X_2 = 1) = \pi/[\pi + (1-\pi)\theta]$,

$$\begin{aligned}
\text{V}(X_1|X_2 = 1) &= \left[\frac{\pi}{\pi + (1-\pi)\theta}\right] - \left[\frac{\pi}{\pi + (1-\pi)\theta}\right]^2 \\
&= \frac{\pi(1-\pi)\theta}{[\pi + (1-\pi)\theta]^2}.
\end{aligned}$$

(d)

$$\begin{aligned}
\text{Cov}(X_1, X_2) &= \text{E}(X_1X_2) - \text{E}(X_1)\text{E}(X_2) \\
&= K\pi^2 - \{K\pi[\pi + (1-\pi)\theta]\}^2 \\
&= K\pi^2\{1 - K[\pi + (1-\pi)\theta]^2\}.
\end{aligned}$$

Since $\text{V}(X_1) = \text{V}(X_2) = K\pi[\pi + (1-\pi)\theta]\{1 - K\pi[\pi + (1-\pi)\theta]\}$,

$$\text{corr}(X_1, X_2) = \frac{\pi\{1 - K[\pi + (1-\pi)\theta]^2\}}{[\pi + (1-\pi)\theta]\{1 - K\pi[\pi + (1-\pi)\theta]\}}.$$

Now,

$$\begin{aligned}
\text{corr}(X_1, X_2) \;\begin{matrix} > \\ = \\ < \end{matrix}\; 0 &\Longleftrightarrow \{1 - K[\pi + (1-\pi)\theta]^2\} \;\begin{matrix} > \\ = \\ < \end{matrix}\; 0 \\
\Longleftrightarrow K^{-1} = [(1-\pi)^2 + 2\pi(1-\pi)\theta + \pi^2] &\;\begin{matrix} > \\ = \\ < \end{matrix}\; [\pi + (1-\pi)\theta]^2 \\
\Longleftrightarrow (1-\pi)^2 \;\begin{matrix} > \\ = \\ < \end{matrix}\; (1-\pi)^2\theta^2 &\Longleftrightarrow 1 \;\begin{matrix} > \\ = \\ < \end{matrix}\; \theta^2.
\end{aligned}$$

So, since $\theta > 0$, $\text{corr}(X_1, X_2) > 0$ when $0 < \theta < 1$, $\text{corr}(X_1, X_2) = 0$ when $\theta = 1$, and $\text{corr}(X_1, X_2) < 0$ when $\theta > 1$.

(e)

$$\begin{aligned}
\text{E}(L) &= 3\text{E}(X_1) - 4\text{E}(X_2) = -K\pi[\pi + (1-\pi)\theta] \\
&= \frac{-\pi[\pi + (1-\pi)\theta]}{(1-\pi)^2 + 2\pi(1-\pi)\theta + \pi^2}; \text{ and,} \\
\text{V}(L) &= (3)^2\text{V}(X_1) + (-4)^2\text{V}(X_2) + 2(3)(-4)\text{Cov}(X_1, X_2) \\
&= 25K\pi[\pi + (1-\pi)\theta]\{1 - K\pi[\pi + (1-\pi)\theta]\} \\
&\quad - 24K\pi^2\{1 - K[\pi + (1-\pi)\theta]^2\}.
\end{aligned}$$

Solution 4.3.

(a)

$$\text{pr}\{(X < 1) \cap (Y > 0) | (X < 2)\} = \frac{\text{pr}\{(X < 1) \cap (Y > 0)\}}{\text{pr}(X < 2)} = \frac{\text{pr}(X < 1)}{\text{pr}(X < 2)}.$$

Now, the marginal distribution of X is

$$\begin{aligned}
\text{f}_X(x) &= \int_0^\infty \frac{y}{(1+x)^4} \text{e}^{-y/(1+x)} \text{d}y \\
&= (1+x)^{-3} \int_0^\infty \frac{y\text{e}^{-y/(1+x)}}{(1+x)} \text{d}y \\
&= (1+x)^{-3}(1+x) \int_0^\infty \frac{y\text{e}^{-y/(1+x)}}{(1+x)^2} \text{d}y \\
&= (1+x)^{-3}(1+x) \cdot (1) = (1+x)^{-2}, \quad 0 < x < +\infty.
\end{aligned}$$

So,

$$\begin{aligned}
\text{pr}(X < k) &= \int_0^k (1+x)^{-2} \text{d}x = \left[-(1+x)^{-1}\right]_0^k = 1 - (1+k)^{-1} \\
&= \frac{k}{(1+k)}.
\end{aligned}$$

So,

$$\frac{\text{pr}(X < 1)}{\text{pr}(X < 2)} = \frac{1/(1+1)}{2/(1+2)} = \frac{3}{4}.$$

(b)

$$\begin{aligned} f_Y(y|x=x) &= \frac{f_{X,Y}(x,y)}{f_X(x)} \\ &= \frac{y(1+x)^{-4}e^{-y/(1+x)}}{(1+x)^{-2}} = \frac{y}{(1+x)^2}e^{-y/(1+x)}, \ y > 0. \end{aligned}$$

So,

$$\begin{aligned} E(Y^r|X=x) &= \int_0^{\infty} y^r f_Y(y|X=x)dy \\ &= \frac{1}{(1+x)^2}\int_0^{\infty} y^{r+1}e^{-y/(1+x)}dy \\ &= \frac{1}{(1+x)^2}\Gamma(r+2)\cdot(1+x)^{r+2}\int_0^{\infty}\frac{y^{(r+2)-1}e^{-y/(1+x)}}{\Gamma(r+2)(1+x)^{r+2}}dy \\ &= \Gamma(r+2)\cdot(1+x)^r\cdot(1) = (r+1)!\,(1+x)^r, \end{aligned}$$

since r is a positive integer.

(c) Given that the x_is are fixed constants and the Y_is are mutually independent, we have:

$$\begin{aligned} E(L) &= \sum_{i=1}^{n}(x_i-\bar{x})E(Y_i|X=x_i) = \sum_{i=1}^{n}(x_i-\bar{x})\cdot 2(1+x_i) \\ &= 2\sum_{i=1}^{n}(x_i-\bar{x})(1+x_i) = 2\sum_{i=1}^{n}x_i(x_i-\bar{x}) \\ &= 2\sum_{i=1}^{n}(x_i-\bar{x})^2; \text{ and} \\ V(L) &= \sum_{i=1}^{n}(x_i-\bar{x})^2V(Y_i|X=x_i) = \sum_{i=1}^{n}(x_i-\bar{x})^2\cdot 2(1+x_i)^2 \\ &= 2\sum_{i=1}^{n}(1+x_i)^2(x_i-\bar{x})^2. \end{aligned}$$

Solution 4.5.

(a) Since $P = X_2/X_1$ and $S = (X_1 - X_2)$, we have

$$X_1 = \frac{S}{(1-P)} \text{ and } X_2 = \frac{PS}{(1-P)}.$$

The Jacobian is

$$J = \begin{vmatrix} \frac{\partial X_1}{\partial S} & \frac{\partial X_1}{\partial P} \\ \frac{\partial X_2}{\partial S} & \frac{\partial X_2}{\partial P} \end{vmatrix} = \begin{vmatrix} \frac{1}{(1-P)} & \frac{S}{(1-P)^2} \\ \frac{P}{(1-P)} & \frac{S}{(1-P)^2} \end{vmatrix} = \frac{S}{(1-P)^2}.$$

So,

$$\begin{aligned} f_{P,S}(p,s;\theta) &= f_{X_1,X_2}\left(\frac{s}{1-p}, \frac{ps}{1-p}; \theta\right) \times J \\ &= (2\theta^3)^{-1}\left[\frac{s}{(1-p)}\right] e^{-\left(\frac{s}{1-p}\right)/\theta} \cdot [s(1-p)^{-2}] \\ &= (2\theta^3)^{-1}s^2(1-p)^{-3}e^{-s/\theta(1-p)}, 0 < s < \infty, 0 < p < 1. \end{aligned}$$

(b) Now,

$$\begin{aligned} f_P(p) &= \int_0^\infty f_{P,S}(p,s;\theta)ds \\ &= \int_0^\infty (2\theta^3)^{-1}s^2(1-p)^{-3}e^{-s/\theta(1-p)}ds \\ &= (1-p)^{-3}\int_0^\infty \frac{s^{(3-1)}}{\Gamma(3)\theta^3}e^{-s/\theta(1-p)}ds \\ &= \frac{(1-p)^3}{(1-p)^3} = 1, 0 < p < 1. \end{aligned}$$

So, $f_P(p) = 1$, $0 < p < 1$, and so $E(P) = \frac{1}{2}$ and $V(P) = \frac{1}{12}$. In addition,

$$\begin{aligned} f_S(s;\theta) &= \int_0^1 f_{P,S}(p,s;\theta)dp \\ &= \int_0^1 (2\theta^3)^{-1}\, s^2(1-p)^{-3}e^{-s/\theta(1-p)}dp \end{aligned}$$

(using the change of variable $u = 1/[1-p]$)

$$= \frac{s^2}{2\theta^3}\int_1^\infty u\, e^{-su/\theta}\, du$$

(using integration by parts)

$$\begin{aligned} &= \frac{s^2}{2\theta^3}\left[\int_0^\infty u\, e^{-su/\theta}\, du - \int_0^1 u\, e^{-su/\theta}\, du\right] \\ &= \frac{s^2}{2\theta^3}\left[\left(\frac{\theta}{s}\right)^2 + \left\{u\,\frac{\theta}{s}\, e^{-su/\theta}\right\}_0^1 - \int_0^1 \frac{\theta}{s}\, e^{-su/\theta}\, du\right] \\ &= \frac{s^2}{2\theta^3}\left\{\frac{\theta^2}{s^2} + \frac{\theta}{s}e^{-s/\theta} + \frac{\theta^2}{s^2}e^{-s/\theta} - \frac{\theta^2}{s^2}\right\} \\ &= \frac{s}{2\theta^2}e^{-s/\theta} + \frac{1}{2\theta}e^{-s/\theta}, 0 < s < \infty. \end{aligned}$$

Note that

$$\begin{aligned}\int_0^{\infty} \mathrm{f}_S(s;\theta)\mathrm{d}s &= \frac{1}{2}\int_0^{\infty} \frac{s}{\theta^2}e^{-s/\theta}\mathrm{d}s + \frac{1}{2}\int_0^{\infty} \frac{1}{\theta}e^{-s/\theta}\mathrm{d}s \\ &= \frac{1}{2} + \frac{1}{2} = 1.\end{aligned}$$

Notice also that

$$\mathrm{f}_S(s;\theta) = \frac{1}{2}\mathrm{GAMMA}(\alpha = \theta, \beta = 2) + \frac{1}{2}\mathrm{GAMMA}(\alpha = \theta, \beta = 1),$$

so that $\mathrm{f}_S(s;\theta)$ is an equally weighted mixture of two gamma distributions. So,

$$\mathrm{E}(S) = \frac{1}{2}(2\theta) + \frac{1}{2}(\theta) = \frac{3\theta}{2}.$$

Since

$$\begin{aligned}\mathrm{E}(S^2) &= \frac{1}{2}[2\theta^2 + (2\theta)^2] + \frac{1}{2}[\theta^2 + (\theta)^2] \\ &= 3\theta^2 + \theta^2 = 4\theta^2,\end{aligned}$$

we have

$$\mathrm{V}(S) = 4\theta^2 - \left(\frac{3\theta}{2}\right)^2 = \frac{7\theta^2}{4}.$$

Solution 4.7. For all $i < j$, $(X_i - X_j) \sim \mathrm{N}(0, 2\sigma^2)$. So, with $Y \sim \mathrm{N}(0, 2\sigma^2)$, we have

$$\begin{aligned}\mathrm{E}(|Y|) &= \int_{-\infty}^{\infty} |y|\frac{1}{\sqrt{2\pi}\sqrt{2\sigma^2}}\mathrm{e}^{-y^2/2(2\sigma^2)}dy \\ &= \frac{1}{2\sigma\sqrt{\pi}}\int_{-\infty}^{0}(-y)\mathrm{e}^{-y^2/4\sigma^2}\mathrm{d}y + \frac{1}{2\sigma\sqrt{\pi}}\int_0^{\infty} y\mathrm{e}^{-y^2/4\sigma^2}dy \\ &= \frac{1}{\sigma\sqrt{\pi}}\int_0^{\infty} y\mathrm{e}^{-y^2/4\sigma^2}\mathrm{d}y = \frac{1}{\sigma\sqrt{\pi}}\left[-2\sigma^2\mathrm{e}^{-y^2/4\sigma^2}\right]_0^{\infty} \\ &= \frac{2\sigma}{\sqrt{\pi}}.\end{aligned}$$

So,

$$\begin{aligned}\mathrm{E}(U) &= \frac{\sqrt{\pi}}{n(n-1)}\sum_{i=1}^{n-1}\sum_{j=i+1}^{n}\mathrm{E}(|X_i - X_j|) \\ &= \frac{\sqrt{\pi}}{n(n-1)}\sum_{i=1}^{n-1}\sum_{j=i+1}^{n}\left(\frac{2\sigma}{\sqrt{\pi}}\right) \\ &= \frac{\sqrt{\pi}}{n(n-1)}\left[\frac{n(n-1)}{2}\right]\left(\frac{2\sigma}{\sqrt{\pi}}\right) = \sigma.\end{aligned}$$

Solution 4.9. For $u = 0, 1, \ldots, \infty$, $\text{p}_U(u) = \text{pr}(U = u) = \text{pr}(U \geq u) - \text{pr}(U \geq u+1)$.

So,

$$\begin{aligned}\text{pr}(U \geq u) &= \text{pr}\left[\bigcap_{i=1}^{n}(X_i \geq u)\right] = \prod_{i=1}^{n}\text{pr}(X_i \geq u)\\ &= \prod_{i=1}^{n}\left[\sum_{x_i=u}^{\infty}\pi^{x_i}(1-\pi)\right]\\ &= \prod_{i=1}^{n}\left[(1-\pi)\frac{\pi^u}{(1-\pi)}\right] = \pi^{nu}.\end{aligned}$$

Finally,

$$\text{p}_U(u) = \pi^{nu} - \pi^{n(u+1)} = \pi^{nu}(1-\pi^n), u = 0, 1, \ldots, \infty.$$

Solution 4.11. Since $X \sim \text{N}(0, \sigma^2)$, then $\text{E}(\text{e}^{tX}) = \text{e}^{\sigma^2 t^2/2}$, so that $\text{E}(\text{e}^X) = \text{E}(U) = \text{e}^{\sigma^2/2}$; and, $\text{E}(U^2) = \text{E}(\text{e}^{2X}) = \text{e}^{2\sigma^2}$, so that $\text{Var}(U) = \text{e}^{2\sigma^2} - (\text{e}^{\sigma^2/2})^2 = \text{e}^{\sigma^2}(\text{e}^{\sigma^2}-1)$. Since $Y \sim \text{N}(0, \sigma^2)$ as well, then $\text{E}(V) = \text{e}^{\sigma^2/2}$ and $\text{Var}(V) = \text{e}^{\sigma^2}(\text{e}^{\sigma^2}-1)$.

To find $\text{E}(UV) = \text{E}\left(\text{e}^{X+Y}\right)$, we note that $X+Y$ is normal with $\text{E}(X+Y) = 0$ and $\text{Var}(X+Y) = 2\sigma^2(1+\rho)$. Hence, $\text{E}(UV) = \text{e}^{2\sigma^2(1+\rho)/2} = \text{e}^{\sigma^2(1+\rho)}$. Finally,

$$\text{corr}(U, V) = \frac{\text{e}^{\sigma^2(1+\rho)} - (\text{e}^{\sigma^2/2})(\text{e}^{\sigma^2/2})}{\sqrt{\left[\text{e}^{\sigma^2}(\text{e}^{\sigma^2}-1)\right]\left[\text{e}^{\sigma^2}(\text{e}^{\sigma^2}-1)\right]}} = \frac{\text{e}^{\rho\sigma^2}-1}{\text{e}^{\sigma^2}-1}.$$

When $\rho = +1$, $\text{corr}(U, V) = 1$; when $\rho = -1$,

$$\text{corr}(U, V) = \frac{\text{e}^{-\sigma^2}-1}{\text{e}^{\sigma^2}-1} = \frac{(1-\text{e}^{\sigma^2})/\text{e}^{\sigma^2}}{(\text{e}^{\sigma^2}-1)} = -\text{e}^{-\sigma^2},$$

which is always greater than -1 when $\sigma^2 > 0$. For example, when $\sigma^2 = 1/2$, $\text{corr}(U, V) \geq -\text{e}^{-0.50} = -0.6065$.

Solution 4.13. Given $N = n$, the probability that $X = x$ can be written as the difference between the probability that $X \leq x$ and the probability that $X \leq (x-1), x = 1, 2, \ldots, 6$.

More formally,

$$\text{pr}(X = x|N = n) = \left(\frac{x}{6}\right)^n - \left(\frac{x-1}{6}\right)^n, x = 1, 2, \ldots, 6.$$

Then, we have

$$\begin{aligned}
\mathrm{p}_X(x) &= \mathrm{pr}(X=x) = \sum_{n=1}^{\infty} \mathrm{pr}(X=x|N=n)\mathrm{pr}(N=n) \\
&= \sum_{n=1}^{\infty} \left[\left(\frac{x}{6}\right)^n - \left(\frac{x-1}{6}\right)^n \right] \pi(1-\pi)^{n-1} \\
&= \pi \left\{ \frac{x}{6} \sum_{n=1}^{\infty} \left[\frac{x(1-\pi)}{6}\right]^{n-1} - \frac{(x-1)}{6} \sum_{n=1}^{\infty} \left[\frac{(x-1)(1-\pi)}{6}\right]^{n-1} \right\} \\
&= \pi \left[\frac{x/6}{1-\frac{x(1-\pi)}{6}} - \frac{(x-1)/6}{1-\frac{(x-1)(1-\pi)}{6}} \right] \\
&= \pi \left[\frac{x}{6-x(1-\pi)} - \frac{(x-1)}{6-(x-1)(1-\pi)} \right] \\
&= \frac{6\pi}{[6-x(1-\pi)][6-(x-1)(1-\pi)]}, x=1,2,\ldots,6.
\end{aligned}$$

Solution 4.15. Note that

$$(X_1+X_2) \sim \mathrm{BIN}(n_1+n_2, \theta).$$

Now,

$$\begin{aligned}
&\mathrm{p}_{X_1}(x_1|X_1+X_2=k) \\
&\quad= \mathrm{pr}(X_1=x_1|X_1+X_2=k) \\
&\quad= \frac{\mathrm{pr}\{(X_1=x_1)\cap(X_1+X_2=k)\}}{\mathrm{pr}(X_1+X_2=k)} \\
&\quad= \frac{\mathrm{pr}\{(X_1=x_1)\cap(X_2=k-x_1)\}}{\mathrm{pr}(X_1+X_2=k)} \\
&\quad= \frac{\mathrm{pr}(X_1=x_1)\mathrm{pr}(X_2=k-x_1)}{\mathrm{pr}(X_1+X_2=k)} \\
&\quad= \frac{\left[\mathrm{C}_{x_1}^{n_1}\theta^{x_1}(1-\theta)^{n_1-x_1}\right]\left[\mathrm{C}_{k-x_1}^{n_2}\theta^{k-x_1}(1-\theta)^{n_2-(k-x_1)}\right]}{\left[\mathrm{C}_k^{n_1+n_2}\theta^k(1-\theta)^{n_1+n_2-k}\right]} \\
&\quad= \frac{\mathrm{C}_{x_1}^{n_1}\mathrm{C}_{k-x_1}^{n_2}}{\mathrm{C}_k^{n_1+n_2}}, \quad \max(0,k-n_2) \leq x_1 \leq \min(k,n_1).
\end{aligned}$$

In other words, the conditional distribution of X_1, given that $(X_1+X_2)=k$, is hypergeometric.

Solution 4.17. Since

$$\begin{aligned} S^2 &= (n-1)^{-1}\sum_{i=1}^{n}[(X_i-\mu)-(\bar{X}-\mu)]^2 \\ &= (n-1)^{-1}\left[\sum_{i=1}^{n}(X_i-\mu)^2 - 2(\bar{X}-\mu)\sum_{i=1}^{n}(X_i-\mu) + n(\bar{X}-\mu)^2\right] \\ &= (n-1)^{-1}\left[\sum_{i=1}^{n}(X_i-\mu)^2 - n(\bar{X}-\mu)^2\right], \end{aligned}$$

it follows that

$$\mathrm{E}(S^2) = (n-1)^{-1}[n\sigma^2 - n\mathrm{V}(\bar{X})].$$

Now,

$$\begin{aligned} \mathrm{V}(\bar{X}) &= \frac{1}{n^2}\mathrm{V}\left(\sum_{i=1}^{n}X_i\right) = \frac{1}{n^2}\left[\sum_{i=1}^{n}\mathrm{V}(X_i) + 2\sum_{\text{all } i<i'}\mathrm{cov}(X_i, X_{i'})\right] \\ &= \frac{1}{n^2}[n\sigma^2 + n(n-1)\rho\sigma^2] = \frac{\sigma^2}{n}[1+(n-1)\rho]. \end{aligned}$$

So,

$$\mathrm{E}(S^2) = (n-1)^{-1}\{n\sigma^2 - \sigma^2[1+(n-1)\rho]\} = \sigma^2(1-\rho).$$

So, S^2 is a *biased* estimator of σ^2 if $\rho \neq 0$. In particular, if $\rho = 0$, $\mathrm{E}(S^2) = \sigma^2$; if $\rho > 0$, $\mathrm{E}(S^2) < \sigma^2$; and, if $\rho < 0$, $\mathrm{E}(S^2) > \sigma^2$.

Solution 4.19. Let T_i be the time to failure for battery B_i, $i = 1, 2, \ldots, n$, so that $\mathrm{E}(T_i) = \lambda$ and $\mathrm{V}(T_i) = \lambda^2$, with $\lambda = 1.5$. If $T = \sum_{i=1}^{n} T_i$, where T is an incubator's total time of continuous operation on battery power, we want to find n^* such that $\mathrm{pr}(T \geq 125) \geq 0.95$.

Now, by the Central Limit Theorem, $\frac{T-\mathrm{E}(T)}{\sqrt{\mathrm{V}(T)}} \dot{\sim} \mathrm{N}(0,1)$ for large n. So

$$\begin{aligned} \mathrm{pr}\left\{\frac{T-\mathrm{E}(T)}{\sqrt{\mathrm{V}(T)}} \geq \frac{125-\mathrm{E}(T)}{\sqrt{\mathrm{V}(T)}}\right\} &= \mathrm{pr}\left\{\frac{T-n\lambda}{\sqrt{n\lambda^2}} \geq \frac{125-n\lambda}{\sqrt{n\lambda^2}}\right\} \geq 0.95 \\ \Rightarrow \frac{125-n\lambda}{\lambda\sqrt{n}} &\leq -1.645 \\ \Rightarrow n(1.5) - 1.645(1.5)\sqrt{n} &\geq 125 \\ \Rightarrow \sqrt{n}[1.5\sqrt{n} - 2.4675] &\geq 125 \\ \Rightarrow n^* &= 100. \end{aligned}$$

Solution 4.21.

(a) $U = \sqrt{Y/X}$ and $V = \sqrt{XY} \Rightarrow X = V/U$ and $Y = UV$. So, $0 < U < +\infty$ and $0 < V < +\infty$. Also, the Jacobian J is equal to

$$J = \begin{vmatrix} \frac{\partial X}{\partial U} & \frac{\partial X}{\partial V} \\ \frac{\partial Y}{\partial U} & \frac{\partial Y}{\partial V} \end{vmatrix} = \begin{vmatrix} -V/U^2 & 1/U \\ V & U \end{vmatrix} = \frac{-2V}{U},$$

so that $|J| = 2V/U$.
So,

$$\begin{aligned} \mathrm{f}_{U,V}(u,v) &= \mathrm{e}^{-\theta\left(\frac{v}{u}\right)}\mathrm{e}^{-\theta^{-1}(uv)}\left(\frac{2v}{u}\right) \\ &= 2vu^{-1}\mathrm{e}^{-\left(\frac{\theta}{u}+\frac{u}{\theta}\right)v},\ 0 < U < \infty,\ 0 < V < \infty. \end{aligned}$$

And,

$$\begin{aligned} \mathrm{f}_U(u) &= \int_0^\infty \mathrm{f}_{U,V}(u,v)\mathrm{d}v = 2u^{-1}\int_0^\infty v\mathrm{e}^{-\left(\frac{\theta}{u}+\frac{u}{\theta}\right)v}\mathrm{d}v \\ &= 2u^{-1}\left(\frac{\theta}{u}+\frac{u}{\theta}\right)^{-2},\ 0 < u < \infty. \end{aligned}$$

(b) Note that

$$\begin{aligned} \mathrm{f}_{X,Y}(x,y) &= \left(\theta\mathrm{e}^{-\theta x}\right)\left(\theta^{-1}\mathrm{e}^{-\theta^{-1}y}\right) \\ &= \mathrm{f}_X(x)\mathrm{f}_Y(y),\ 0 < x < \infty,\ 0 < y < \infty. \end{aligned}$$

In other words, X and Y are *independent* random variables with $X \sim$ GAMMA$[\alpha = \theta^{-1}, \beta = 1]$ and $Y \sim$ GAMMA$[\alpha = \theta, \beta = 1]$. So, since $U = \sqrt{Y/X}$,

$$\begin{aligned} \mathrm{E}(U) &= \mathrm{E}\left(\frac{\sqrt{Y}}{\sqrt{X}}\right) = \mathrm{E}\left(Y^{1/2}\right)\mathrm{E}\left(X^{-1/2}\right) \\ &= \left[\frac{\Gamma\left(1+\frac{1}{2}\right)}{\Gamma(1)}\theta^{1/2}\right]\left[\frac{\Gamma\left(1-\frac{1}{2}\right)}{\Gamma(1)}\left(\theta^{-1}\right)^{-1/2}\right] \\ &= \Gamma\left(\frac{3}{2}\right)\Gamma\left(\frac{1}{2}\right)\theta = \frac{\pi\theta}{2}. \end{aligned}$$

So, since

$$\mathrm{E}(\bar{U}) = n^{-1}\sum_{i=1}^n \mathrm{E}(U_i) = n^{-1}\sum_{i=1}^n\left(\frac{\pi\theta}{2}\right) = \frac{\pi\theta}{2},$$

it follows that $\mathrm{h}(\bar{U}) = 2\bar{U}/\pi$ has an expected value equal to θ.

Solution 4.23.

(a)

$$\begin{aligned}
\mathrm{f}_X(x) &= \int_0^\infty \mathrm{f}_{X,Y}(x,y)\mathrm{d}y \\
&= \int_0^\infty \frac{(x+y)}{2}\mathrm{e}^{-(x+y)}\mathrm{d}y \\
&= \frac{1}{2}\int_0^\infty x\mathrm{e}^{-(x+y)}\mathrm{d}y + \frac{1}{2}\int_0^\infty y\mathrm{e}^{-(x+y)}\mathrm{d}y \\
&= \frac{x\mathrm{e}^{-x}}{2}\int_0^\infty \mathrm{e}^{-y}\mathrm{d}y + \frac{\mathrm{e}^{-x}}{2}\int_0^\infty y\mathrm{e}^{-y}\mathrm{d}y \\
&= \frac{x\mathrm{e}^{-x}}{2}(1) + \frac{\mathrm{e}^{-x}}{2}\Gamma(2) \\
&= \frac{1}{2}(1+x)\mathrm{e}^{-x}, x > 0.
\end{aligned}$$

So,

$$\begin{aligned}
\mathrm{f}_Y(y|X=x) &= \frac{\mathrm{f}_{X,Y}(x,y)}{\mathrm{f}_X(x)} \\
&= \frac{\frac{1}{2}(x+y)\mathrm{e}^{-(x+y)}}{\frac{1}{2}(1+x)\mathrm{e}^{-x}} \\
&= \frac{(x+y)}{(1+x)}\mathrm{e}^{-y}, y > 0.
\end{aligned}$$

Finally,

$$\begin{aligned}
\mathrm{E}(Y|X=x) &= \int_0^\infty y\frac{(x+y)}{(1+x)}\mathrm{e}^{-y}\mathrm{d}y \\
&= \frac{x}{(1+x)}\int_0^\infty y\mathrm{e}^{-y}\mathrm{d}y + \frac{1}{(1+x)}\int_0^\infty y^2\mathrm{e}^{-y}\mathrm{d}y \\
&= \frac{x}{(1+x)}\Gamma(2) + \frac{1}{(1+x)}\Gamma(3) \\
&= \frac{(x+2)}{(1+x)}.
\end{aligned}$$

(b)

$$
\begin{aligned}
&\mathrm{E}(\mathrm{e}^{tS}) \\
&= \mathrm{E}\left[\mathrm{e}^{t(X+Y)}\right] \\
&= \int_0^\infty \int_0^\infty \mathrm{e}^{t(x+y)} \frac{1}{2}(x+y)\mathrm{e}^{-(x+y)} \mathrm{d}x\mathrm{d}y \\
&= \frac{1}{2}\int_0^\infty \int_0^\infty (x+y)\mathrm{e}^{-(1-t)(x+y)} \mathrm{d}x\mathrm{d}y \\
&= \frac{1}{2}\int_0^\infty \mathrm{e}^{-(1-t)y}\left[\int_0^\infty x\mathrm{e}^{-(1-t)x}\mathrm{d}x + \int_0^\infty y\mathrm{e}^{-(1-t)x}\mathrm{d}x\right]\mathrm{d}y \\
&= \frac{1}{2}\int_0^\infty \mathrm{e}^{-(1-t)y}\left[(1-t)^{-2} + y(1-t)^{-1}\right]\mathrm{d}y \\
&= \frac{1}{2}\Big\{(1-t)^{-3}\int_0^\infty (1-t)\mathrm{e}^{-(1-t)y}\mathrm{d}y \\
&\qquad + (1-t)^{-2}\int_0^\infty y(1-t)\mathrm{e}^{-(1-t)}\mathrm{d}y\Big\} \\
&= \frac{1}{2}\left[(1-t)^{-3} + (1-t)^{-2}(1-t)^{-1}\right] \\
&= (1-t)^{-3},
\end{aligned}
$$

so that $S \sim \mathrm{GAMMA}(\alpha = 1, \beta = 3)$.

Solution 4.25. Since $\mathrm{F}_Y(y;\theta) = y^\theta, 0 < y < 1$,

$$
\begin{aligned}
\mathrm{F}_{Y_{(1)}}(y_{(1)};\theta) &= \mathrm{pr}[Y_{(1)} \le y_{(1)}] = 1 - \mathrm{pr}\left[\cap_{i=1}^n (Y_i > y_{(1)})\right] \\
&= 1 - [1 - \mathrm{F}_Y(y_{(1)};\theta)]^n = 1 - \left[1 - {y_{(1)}}^\theta\right]^n, 0 < y_{(1)} < 1.
\end{aligned}
$$

Thus,

$$
\begin{aligned}
\mathrm{F}_U(u;\theta) &= \mathrm{pr}(U \le u) = \mathrm{pr}\left[n{Y_{(1)}}^\theta \le u\right] \\
&= \mathrm{pr}\left[Y_{(1)} \le \left(\frac{u}{n}\right)^{1/\theta}\right] = 1 - \left[1 - \left(\frac{u}{n}\right)\right]^n, 0 < u < n.
\end{aligned}
$$

Finally,

$$
\lim_{n\to\infty} \mathrm{F}_U(u;\theta) = 1 - \lim_{n\to\infty}\left[1 - \left(\frac{u}{n}\right)\right]^n = 1 - \mathrm{e}^{-u}, 0 < u < +\infty.
$$

Hence, the asymptotic density function of U is $\mathrm{f}_U(u) = \mathrm{e}^{-u}, 0 < u < +\infty$, which, interestingly, does not depend on the parameter θ.

Solution 4.27. Clearly, $\text{E}(Y-\theta X)=0$, and

$$\begin{aligned}\text{V}(Y-\theta X) &= \text{V}(Y)+\theta^2\text{V}(X)-2\theta\text{cov}(X,Y)\\ &= (1)+\theta^2(1)-2\theta(\rho)\\ &= 1+\theta^2-2\theta\rho.\end{aligned}$$

So, by Tchebyshev's Theorem, we have

$$\begin{aligned}\text{pr}\left[|Y-\theta X|\leq t\sqrt{\text{V}(Y-\theta X)}\right] &= \text{pr}\left[|Y-\theta X|\leq t\sqrt{(1+\theta^2-2\theta\rho)}\right]\\ &\geq 1-t^{-2}.\end{aligned}$$

If we set $t=\delta(1+\theta^2-2\theta\rho)^{-1/2}$, we obtain

$$\text{pr}\left[|Y-\theta X|\leq\delta\right]\geq 1-\frac{(1+\theta^2-2\theta\rho)}{\delta^2}.$$

To maximize this lower bound, we need to choose θ^* to minimize the expression $(1+\theta^2-2\theta\rho)$. It is easy to show that $\theta^*=\rho$, so that

$$\text{pr}\left[|Y-\rho X|\leq\delta\right]\geq 1-\frac{(1-\rho^2)}{\delta^2}.$$

Solution 4.29.

(a) Clearly, $X_1\sim\text{BIN}(n,1/2)$, so that $\text{E}(X_1)=n/2$ and $\text{V}(X_1)=n/4$. Also, given that $X_1=x_1$, it follows that $X_2\sim\text{BIN}(x_1,1/6)$, so that $\text{E}(X_2|X_1=x_1)=x_1/6$ and $\text{V}(X_2|X_1=x_1)=5x_1/36$. Now, using this information, we have

$$\text{E}(X_2)=\text{E}_{x_1}\left[\text{E}(X_2|X_1=x_1)\right]=\text{E}_{x_1}\left(\frac{x_1}{6}\right)=\frac{(n/2)}{6}=\frac{n}{12},$$

$$\begin{aligned}\text{V}(X_2) &= \text{E}_{x_1}\left[\text{V}(X_2|X_1=x_1)\right]+\text{V}_{x_1}\left[\text{E}(X_2|X_1=x_1)\right]\\ &= \text{E}_{x_1}\left(\frac{5x_1}{36}\right)+\text{V}_{x_1}\left(\frac{x_1}{6}\right)\\ &= \frac{5(n/2)}{36}+\frac{(n/4)}{36}=\frac{11n}{144},\end{aligned}$$

and

$$\begin{aligned}\text{E}(X_1X_2) &= \text{E}_{x_1}\left[\text{E}(X_1X_2|X_1=x_1)\right]\\ &= \text{E}_{x_1}\left[x_1\text{E}(X_2|X_1=x_1)\right]=\text{E}_{x_1}\left[x_1\left(\frac{x_1}{6}\right)\right]\\ &= \left(\frac{1}{6}\right)\text{E}(X_1^2)=\left(\frac{1}{6}\right)\left\{\text{V}(X_1)+\left[\text{E}(X_1)\right]^2\right\}\\ &= \left(\frac{1}{6}\right)\left[\frac{n}{4}+\left(\frac{n}{2}\right)^2\right]\\ &= \frac{n(n+1)}{24}.\end{aligned}$$

So,

$$\begin{aligned}\text{cov}(X_1, X_2) &= \text{E}(X_1X_2) - \text{E}(X_1)\text{E}(X_2)\\ &= \frac{n(n+1)}{24} - \left(\frac{n}{2}\right)\left(\frac{n}{12}\right)\\ &= \frac{n}{24},\end{aligned}$$

and hence

$$\begin{aligned}\text{corr}(X_1, X_2) &= \frac{\text{cov}(X_1, X_2)}{\sqrt{\text{V}(X_1)\text{V}(X_2)}}\\ &= \frac{(n/24)}{\sqrt{(n/4)(11n/144)}} = \frac{1}{\sqrt{11}} = 0.3015.\end{aligned}$$

Interestingly, $\text{corr}(X_1, X_2)$ does not depend on n.

(b) Using the results obtained in part (a), we have

$$\text{E}(S) = \text{E}(X_1) + \text{E}(X_2) = \frac{n}{2} + \frac{n}{12} = \frac{7n}{12},$$

and

$$\begin{aligned}\text{V}(S) &= \text{V}(X_1 + X_2) = \text{V}(X_1) + \text{V}(X_2) + (2)\text{cov}(X_1, X_2)\\ &= \frac{n}{4} + \frac{11n}{144} + 2\left(\frac{n}{24}\right) = \frac{59n}{144}.\end{aligned}$$

Solution 4.31. For $i = n+1, n+2, \ldots, N$, let the dichotomous random variable Y_i take the value 1 if a particular original member successfully recruits the i-th member, and let $Y_i = 0$ otherwise. Then, if the random variable T denotes the total number of members recruited by this particular original member, it follows that $T = \sum_{i=n+1}^{N} Y_i$.

Now, since $\text{pr}(Y_i = 1) = 1/(i-1)$, we have

$$\text{E}(T) = \sum_{i=n+1}^{N} \text{E}(Y_i) = \sum_{i=n+1}^{N} (i-1)^{-1}.$$

When $N = 200$ and $n = 190$, we have $\text{E}(T) \doteq 0.0514$.

Solution 4.33. Let X be the number of odd digits in any row; then, $X \sim \text{BIN}\left(n = 60, \pi = \frac{1}{2}\right)$. Then, the probability that any row contains between 25 and 35 odd digits is

$$\begin{aligned}\text{pr}(25 \leq X \leq 35) &= \sum_{x=25}^{35} \text{C}_x^{60} \left(\frac{1}{2}\right)^x \left(\frac{1}{2}\right)^{60-x}\\ &= \sum_{x=25}^{35} \text{C}_x^{60} \left(\frac{1}{2}\right)^{60}.\end{aligned}$$

For $i = 1, 2, \ldots, 60$, let X_i take the value 1 if the i-th position in a row contains an odd digit, and let X_i take the value 0 otherwise. Then, $X = \sum_{i=1}^{60} X_i$, where $X_1, X_2, \ldots, X_{60}$ constitute a set of 60 mutually independent and identically distributed random variables. Hence, we can approximate the desired probability using the Central Limit Theorem; in this situation, this approximation is often referred to as the "*normal approximation to the binomial distribution.*"

So, since $\mathrm{E}(X) = n\pi = (60)(1/2) = 30$ and $\mathrm{V}(X) = n\pi(1-\pi) = (60)(1/2)(1/2) = 15$, we have

$$\begin{aligned}\mathrm{pr}(25 \leq X \leq 35) &= \mathrm{pr}\left(\frac{25-\mathrm{E}(X)}{\sqrt{\mathrm{V}(X)}} \leq \frac{X-\mathrm{E}(X)}{\sqrt{\mathrm{V}(X)}} \leq \frac{35-\mathrm{E}(X)}{\sqrt{\mathrm{V}(X)}}\right) \\ &\approx \mathrm{pr}(-1.29 \leq Z \leq 1.29) = 0.803,\end{aligned}$$

since $Z \dot{\sim} \mathrm{N}(0,1)$ for large n.

Solution 4.35.

(a) The requirement that $S_N = k$ is the same as the requirement that exactly $(k-1) X_i$s take the value 1 among the first $(N-1) X_i$s sampled and that $X_N = 1$. In other words, $N \sim \mathrm{NEGBIN}(k, \pi)$, so that

$$\mathrm{p}_N(n) = \mathrm{C}_{k-1}^{n-1}\pi^k(1-\pi)^{n-k}, n = k, k+1, \ldots, \infty.$$

(b) Now, with $k^* = (k-1)$ and $n^* = (n-1)$, we have

$$\begin{aligned}\mathrm{E}\left(\frac{k-1}{N-1}\right) &= \sum_{n=k}^{\infty}\left(\frac{k-1}{n-1}\right)\mathrm{C}_{k-1}^{n-1}\pi^k(1-\pi)^{n-k} \\ &= \sum_{n=k}^{\infty}\mathrm{C}_{k-2}^{n-2}\pi^k(1-\pi)^{n-k} \\ &= \pi\sum_{n^*=k^*}^{\infty}\mathrm{C}_{k^*-1}^{n^*-1}\pi^{k^*}(1-\pi)^{n^*-k^*} \\ &= \pi.\end{aligned}$$

(c) If $\mathrm{g}(N) = [(N-1)!]^{-1}$, then

$$\begin{aligned}\mathrm{E}\left[\frac{1}{(N-1)!}\right] &= \sum_{n=1}^{\infty}\left[\frac{1}{(n-1)!}\right]\pi(1-\pi)^{n-1} \\ &= \pi\sum_{m=0}^{\infty}\frac{(1-\pi)^m}{m!} = \pi\mathrm{e}^{(1-\pi)}.\end{aligned}$$

Solution 4.37. Now,

$$\begin{aligned}\bar{Y}_1 &= n^{-1}\sum_{j=1}^{n} Y_{1j} = n^{-1}\sum_{j=1}^{n}(\mu_1 + X_1 + U_j) \\ &= \mu_1 + X_1 + n^{-1}\sum_{j=1}^{n} U_j = \mu_1 + X_1 + \bar{U},\end{aligned}$$

so that $\mathrm{V}(\bar{Y}_1) = \sigma_1^2 + \frac{\sigma_u^2}{n}$.

Analogously,

$$\bar{Y}_2 = \mu_2 + X_2 + n^{-1}\sum_{j=1}^{n} W_j = \mu_2 + X_2 + \bar{W},$$

so that $V(\bar{Y}_2) = \sigma_2^2 + \frac{\sigma_w^2}{n}$.

Also,

$$\begin{aligned}\mathrm{cov}(\bar{Y}_1, \bar{Y}_2) &= \mathrm{cov}\left(\mu_1 + X_1 + \bar{U}, \mu_2 + X_2 + \bar{W}\right) \\ &= \mathrm{cov}(X_1, X_2) = \rho\sigma_1\sigma_2.\end{aligned}$$

Thus,

$$\begin{aligned}\mathrm{corr}(\bar{Y}_1, \bar{Y}_2) &= \frac{\mathrm{cov}(\bar{Y}_1, \bar{Y}_2)}{\sqrt{\mathrm{V}(\bar{Y}_1)\mathrm{V}(\bar{Y}_2)}} \\ &= \frac{\rho\sigma_1\sigma_2}{\sqrt{\left(\sigma_1^2 + \frac{\sigma_u^2}{n}\right)\left(\sigma_2^2 + \frac{\sigma_w^2}{n}\right)}} \\ &= \left[\left(1 + \frac{\sigma_u^2}{n\sigma_1^2}\right)\left(1 + \frac{\sigma_w^2}{n\sigma_2^2}\right)\right]^{-1/2} \rho = \theta\rho,\end{aligned}$$

where $0 < \theta < 1$.

Solution 4.39.

(a)

$$\begin{aligned}\mathrm{p}_X(x) &= \int_0^\infty \mathrm{p}_X(x|\lambda)\mathrm{f}(\lambda)\mathrm{d}\lambda = \int_0^\infty \frac{\lambda^x \mathrm{e}^{-\lambda}}{x!}\frac{\lambda^{\beta-1}\mathrm{e}^{-\lambda/\alpha}}{\Gamma(\beta)\alpha^\beta}\mathrm{d}\lambda \\ &= \frac{1}{x!\Gamma^\beta\alpha^\beta}\int_0^\infty \lambda^{\beta+x-1}\mathrm{e}^{-\lambda\left(1+\frac{1}{\alpha}\right)}\mathrm{d}\lambda = \frac{\Gamma(\beta+x)\left(\frac{\alpha}{\alpha+1}\right)^{\beta+x}}{x!\Gamma(\beta)\alpha^\beta} \\ &= C_{\beta-1}^{\beta+x-1}\left(\frac{1}{\alpha+1}\right)^\beta\left(\frac{\alpha}{\alpha+1}\right)^x, x = 0, 1, \ldots, \infty.\end{aligned}$$

This distribution is known as the *negative binomial* distribution.

(b) Making use of conditional expectation theory, we have:

$$\mathrm{E}(X) = \mathrm{E}[\mathrm{E}(X|\lambda)] = \mathrm{E}(\lambda) = \alpha\beta,$$

and

$$\begin{aligned}\mathrm{V}(X) &= \mathrm{V}[\mathrm{E}(X|\lambda)] + \mathrm{E}[V(X|\lambda)] = \mathrm{V}(\lambda) + \mathrm{E}(\lambda) = \alpha^2\beta + \alpha\beta \\ &= \alpha(\alpha+1)\beta.\end{aligned}$$

There would be two reasons why one would expect the negative binomial distribution to be a better model than the Poisson distribution for laboratory animal tumor multiplicity data. First, for the negative binomial distribution, $\mathrm{V}(X) = (\alpha+1)\mathrm{E}(X) > \mathrm{E}(X)$, so that the negative binomial distribution would do a better job of modeling data where the variation in the data exceeds the mean. Second, the negative binomial distribution considered here involves two parameters (namely, α and β), while the Poisson distribution involves just one parameter (namely, λ); and a two-parameter model will always fit data better (although possibly not significantly better) than a one-parameter model. More generally, when modeling carcinogenic processes, there are more complex statistical models (e.g., multi-hit and multi-stage models) that would very often fit various types of tumor multiplicity data significantly better than the negative binomial distribution considered here.

Solution 4.41. The goal is to find the numerical value of $\mathrm{pr}(Y_1 = 3|T = \sum_{k=1}^{\infty} Y_k = 10)$ when $\lambda = 1$ and $\theta = 0.40$. Now, by the additivity property of mutually independent Poisson random variables, we know that

$$T = \sum_{k=1}^{\infty} Y_k \sim \mathrm{POI}\left[\sum_{k=1}^{\infty} \frac{\lambda\theta^k}{k!} = \lambda(\mathrm{e}^{\theta} - 1)\right],$$

and that

$$\sum_{k=2}^{\infty} Y_k \sim \mathrm{POI}\left[\sum_{k=2}^{\infty} \frac{\lambda\theta^k}{k!} = \lambda(\mathrm{e}^{\theta} - \theta - 1)\right].$$

So, in general,

$$\begin{aligned}&\mathrm{pr}(Y_1 = y_1|T = t) \\ &= \frac{\mathrm{pr}[(Y_1 = y_1) \cap (T = t)]}{\mathrm{pr}(T = t)} = \frac{\mathrm{pr}(Y_1 = y_1)\mathrm{pr}\left(\sum_{k=2}^{\infty} Y_k = t - y_1\right)}{\mathrm{pr}(T = t)} \\ &= \frac{\left[(\lambda\theta)^{y_1}\mathrm{e}^{-\lambda\theta}/y_1!\right]\left[[\lambda(\mathrm{e}^{\theta} - \theta - 1)]^{(t-y_1)}\mathrm{e}^{-\lambda(\mathrm{e}^{\theta}-\theta-1)}/(t-y_1)!\right]}{[\lambda(\mathrm{e}^{\theta}-1)]^t\,\mathrm{e}^{-\lambda(\mathrm{e}^{\theta}-1)}/t!} \\ &= C_{y_1}^t\left(\frac{\theta}{\mathrm{e}^{\theta}-1}\right)^{y_1}\left(\frac{\mathrm{e}^{\theta}-\theta-1}{\mathrm{e}^{\theta}-1}\right)^{t-y_1}, y_1 = 0, 1, \ldots, t.\end{aligned}$$

In other words, Y_1 given $T = t \sim \text{BIN}\left[t, \frac{\theta}{(e^{\theta}-1)}\right]$. Finally,

$$\begin{aligned}
&\text{pr}(Y_1 = 3|T = 10, \lambda = 1.0, \theta = 0.40) \\
&\qquad = C_3^{10}\left(\tfrac{0.40}{e^{0.40}-1}\right)^3\left(\tfrac{e^{0.40}-0.40-1}{e^{0.40}-1}\right)^7 \doteq 0.0005.
\end{aligned}$$

Solution 4.43.

(a)

$$\begin{aligned}
\text{P}_N(s) &= \text{E}\left(s^N\right) = \sum_{n=0}^{\infty} s^n\theta(1-\theta)^n = \theta\sum_{n=0}^{\infty}[s(1-\theta)]^n \\
&= \frac{\theta}{1-s(1-\theta)}, \quad |s(1-\theta)| < 1.
\end{aligned}$$

Now,

$$\begin{aligned}
\left[\frac{d\text{P}_N(s)}{ds}\right]_{s=1} &= \mu_{(1)} = \text{E}(N) \\
&= \left\{\theta[1-s(1-\theta)]^{-2}(1-\theta)\right\}_{s=1} \\
&= \theta(\theta^{-2})(1-\theta) = \frac{(1-\theta)}{\theta}; \text{ and,} \\
\left[\frac{d^2\text{P}_N(s)}{ds^2}\right]_{s=1} &= \mu_{(2)} = \text{E}[N(N-1)] \\
&= \left\{2\theta[1-s(1-\theta)]^{-3}(1-\theta)^2\right\}_{s=1} \\
&= 2\theta(\theta^{-3})(1-\theta)^2 = \frac{2(1-\theta)^2}{\theta^2}.
\end{aligned}$$

So,

$$\begin{aligned}
\text{E}[N(N+1)] &= \text{E}[N(N-1)] + 2\text{E}(N) = \mu_{(2)} + 2\mu_{(1)} \\
&= \frac{2(1-\theta)^2}{\theta^2} + 2\frac{(1-\theta)}{\theta} \\
&= = \frac{2(1-\theta)}{\theta}\left[\frac{(1-\theta)}{\theta} + 1\right] \\
&= \frac{2(1-\theta)}{\theta^2}.
\end{aligned}$$

(b) Now,

$$\begin{aligned}
\text{pr}(X=x) &= \text{pr}\left\{(X=x)\cap\left[\cup_{n=0}^{\infty}(N=n)\right]\right\} \\
&= \text{pr}\left\{\cup_{n=0}^{\infty}\left[(X=x)\cap(N=n)\right]\right\} \\
&= \sum_{n=0}^{\infty}\text{pr}[(X=x)\cap(N=n)] \\
&= \sum_{n=0}^{\infty}\text{pr}(X=x|N=n)\text{pr}(N=n) \\
&= \sum_{n=0}^{\infty}\text{C}_x^n\pi^x(1-\pi)^{n-x}\theta(1-\theta)^n \\
&= \left(\frac{\pi}{1-\pi}\right)^x\theta\sum_{n=x}^{\infty}\text{C}_x^n[(1-\pi)(1-\theta)]^n,
\end{aligned}$$

since we must have $n \geq x$.
Letting $y=(n-x)$, so that $n=(x+y)$, we have

$$\begin{aligned}
&\text{pr}(X=x) \\
&= \left(\frac{\pi}{1-\pi}\right)^x\theta\sum_{y=0}^{\infty}\text{C}_x^{x+y}[(1-\pi)(1-\theta)]^{x+y} \\
&= \frac{\left(\frac{\pi}{1-\pi}\right)^x\theta[(1-\pi)(1-\theta)]^x}{[1-(1-\pi)(1-\theta)]^{x+1}} \\
&\quad\times\sum_{y=0}^{\infty}\text{C}_x^{x+y}[1-(1-\pi)(1-\theta)]^{x+1}[(1-\pi)(1-\theta)]^y.
\end{aligned}$$

The expression to the right of the summation sign is a negative binomial probability distribution, and so the summation is equal to 1. Thus, we have

$$\begin{aligned}
&\text{pr}(X=x) \\
&= \frac{\left(\frac{\pi}{1-\pi}\right)^x\theta[(1-\pi)(1-\theta)]^x}{[1-(1-\pi)(1-\theta)]^{x+1}} \\
&= \frac{\theta[\pi(1-\theta)]^x}{[\theta+\pi(1-\theta)]^{x+1}} \\
&= \left[\frac{\theta}{\theta+\pi(1-\theta)}\right]\left[\frac{\pi(1-\theta)}{\theta+\pi(1-\theta)}\right]^x, \quad x=0,1,2,\ldots,\infty.
\end{aligned}$$

So, X has a geometric distribution with probability parameter equal to $\frac{\theta}{\theta+\pi(1-\theta)}$.

Solution 4.45.

(a)

$$
\begin{aligned}
\mathrm{F}_U(u) &= \int_0^{\infty} \int_y^{y+u} \lambda^2 \mathrm{e}^{-\lambda x} dx dy \\
&= \int_0^{u} \int_0^{x} \lambda^2 \mathrm{e}^{-\lambda x} dy dx + \int_u^{\infty} \int_{x-u}^{x} \lambda^2 \mathrm{e}^{-\lambda x} dy dx \\
&= 1 - \mathrm{e}^{-\lambda u}, 0 < u < \infty.
\end{aligned}
$$

Thus, $\mathrm{f}_U(u) = \lambda \mathrm{e}^{-\lambda u}, 0 < u < \infty$, so that U has a negative exponential distribution with $\mathrm{E}(U) = \lambda^{-1}$ and $\mathrm{V}(U) = \lambda^{-2}$.

(b)

$$
\begin{aligned}
\mathrm{M}_{X,Y}(s,t) &= \mathrm{E}\left(\mathrm{e}^{sX+tY}\right) \\
&= \int_0^{\infty} \int_0^{x} \mathrm{e}^{(sx+ty)} \lambda^2 \mathrm{e}^{-\lambda x} dy dx \\
&= \frac{\lambda^2}{(\lambda - s)(\lambda - s - t)}, s < \lambda \text{ and } (s+t) < \lambda.
\end{aligned}
$$

When $t = 0$, then $\mathrm{M}_{X,Y}(s, 0) = \lambda^2/(\lambda - s)^2$, so that $X \sim \mathrm{GAMMA}(\alpha = \lambda^{-1}, \beta = 2)$ with $\mathrm{E}(X) = 2\lambda^{-1}$ and $\mathrm{V}(X) = 2\lambda^{-2}$. When $s = 0$, then $\mathrm{M}_{X,Y}(0,t) = \lambda/(\lambda - t)$, so that $Y \sim \mathrm{GAMMA}(\alpha = \lambda^{-1}, \beta = 1)$ with $\mathrm{E}(Y) = \lambda^{-1}$ and $\mathrm{V}(Y) = \lambda^{-2}$. Now, since

$$
\frac{\partial^2 \mathrm{M}_{X,Y}(s,t)}{\partial s \partial t} = \lambda^2[(\lambda - s)^{-2}(\lambda - s - t)^{-2} + 2(\lambda - s)^{-1}(\lambda - s - t)^{-3}],
$$

$$
\left[\frac{\partial^2 \mathrm{M}_{X,Y}(s,t)}{\partial s \partial t}\right]_{|s=t=0} = \mathrm{E}(XY) = \lambda^2\left(\lambda^{-4} + 2\lambda^{-4}\right) = 3\lambda^{-2}.
$$

Finally,

$$
\mathrm{corr}(X,Y) = \frac{3\lambda^{-2} - (2\lambda^{-1})(\lambda^{-1})}{\sqrt{(2\lambda^{-2})(\lambda^{-2})}} = \frac{1}{\sqrt{2}} = 0.7071.
$$

(c) For a typical patient,

$$
\begin{aligned}
\mathrm{pr}[Y > (X - Y)] &= \mathrm{pr}\left(Y > \frac{X}{2}\right) \\
&= \int_0^{\infty} \int_{x/2}^{x} \lambda^2 \mathrm{e}^{-\lambda x} dy dx = \frac{1}{2}.
\end{aligned}
$$

So, using the binomial distribution, pr(*at least* 2 of 6 patients have waiting times that exceed their treatment times)

$$
= \sum_{j=2}^{6} C_j^6 \left(\frac{1}{2}\right)^j \left(\frac{1}{2}\right)^{6-j} = 0.8906.
$$

Solution 4.47.

(a) In general,

$$\mathrm{E}(X^r) = \int_0^\infty \frac{x^{r+2}}{2\beta^3}\mathrm{e}^{-x/\beta}\mathrm{d}x = \frac{\Gamma(r+3)}{2}\beta^r,\ (r+3) > 0.$$

So, $\mathrm{E}(X^{-1}) = 1/2\beta$, $\mathrm{E}(X^{-2}) = 1/2\beta^2$, and $\mathrm{V}(X^{-1}) = 1/2\beta^2 - (1/2\beta)^2 = 1/4\beta^2$.
So,

$$\mathrm{E}(Y) = \mathrm{E}_x[\mathrm{E}(Y|X=x)] = \mathrm{E}_x\left[\frac{1}{\alpha x}\right] = \frac{1}{\alpha}\left(\frac{1}{2\beta}\right) = \frac{1}{2\alpha\beta}.$$

And,

$$\begin{aligned}\mathrm{V}(Y) &= \mathrm{E}_x[\mathrm{V}(Y|X=x)] + \mathrm{V}_x[\mathrm{E}(Y|X=x)]\\ &= \mathrm{E}_x\left[\frac{1}{\alpha^2x^2}\right] + \mathrm{V}_x\left[\frac{1}{\alpha x}\right]\\ &= \frac{1}{\alpha^2}\left(\frac{1}{2\beta^2}\right) + \frac{1}{\alpha^2}\left(\frac{1}{4\beta^2}\right) = \frac{3}{4\alpha^2\beta^2}.\end{aligned}$$

(b) Since

$$\begin{aligned}\mathrm{E}(XY) &= \mathrm{E}_x[\mathrm{E}(XY|X=x)]\\ &= \mathrm{E}_x[x\mathrm{E}(Y|X=x)] = \mathrm{E}_x\left(\frac{x}{\alpha x}\right) = 1/\alpha,\end{aligned}$$

we have $\mathrm{cov}(X,Y) = \mathrm{E}(XY) - \mathrm{E}(X)\mathrm{E}(Y)$

$$= \frac{1}{\alpha} - \left(\frac{\Gamma(4)}{2}\beta\right)\left(\frac{1}{2\alpha\beta}\right) = -\frac{1}{2\alpha}.$$

$$\begin{aligned}\text{And, since } \mathrm{V}(X) &= \frac{\Gamma(5)}{2}\beta^2 - (3\beta)^2 = 3\beta^2,\\ \mathrm{corr}(X,Y) &= \frac{\mathrm{cov}(X,Y)}{\sqrt{\mathrm{V}(X)\mathrm{V}(Y)}} = \frac{-1/2\alpha}{\sqrt{(3\beta^2)[3/(4\alpha^2\beta^2)]}}\\ &= \frac{-1/2\alpha}{\sqrt{9/(4\alpha^2)}} = -\frac{1}{3}.\end{aligned}$$

Now, since $\mathrm{E}(Y|X=x) = 1/\alpha x = (1/\alpha)(1/x)$ is a *linear function* of $1/x$, it follows that

$$\begin{aligned}\mathrm{corr}\left(X^{-1}, Y\right) &= \left(\frac{1}{\alpha}\right)\sqrt{\frac{\mathrm{V}(X^{-1})}{\mathrm{V}(Y)}} = \left(\frac{1}{\alpha}\right)\sqrt{\frac{(1/4\beta^2)}{(3/4\alpha^2\beta^2)}}\\ &= \left(\frac{1}{\alpha}\right)\sqrt{\frac{\alpha^2}{3}} = \frac{1}{\sqrt{3}}.\end{aligned}$$

(c) First,

$$\begin{aligned} \mathrm{f}_Y(y) &= \int_0^\infty \mathrm{f}_{X,Y}(x,y)\mathrm{d}x = \int_0^\infty \mathrm{f}_X(x)\mathrm{f}_Y(y|X=x)\mathrm{d}x \\ &= \int_0^\infty \left(\frac{x^2}{2\beta^3}\mathrm{e}^{-x/\beta}\right)\left(\alpha x\mathrm{e}^{-\alpha xy}\right)\mathrm{d}x \\ &= \frac{3\alpha\beta}{(1+\alpha\beta y)^4},\ 0<y<\infty. \end{aligned}$$

So,

$$\begin{aligned} \mathrm{F}_Y(y) &= \mathrm{pr}(Y\le y) = \int_0^y \frac{3\alpha\beta}{(1+\alpha\beta u)^4}\mathrm{d}u \\ &= 1-(1+\alpha\beta y)^{-3},\ 0<y<\infty. \end{aligned}$$

Note that $\frac{\mathrm{dF}_Y(y)}{\mathrm{d}y} = \mathrm{f}_Y(y)$, $\mathrm{F}_Y(0)=0$, $\lim_{y\to\infty}\mathrm{F}_Y(y)=1$, and $\mathrm{F}_Y(y)$ is a monotonically increasing function of y.

(d) Now,

$$\mathrm{pr}[(X^2+Y^2)<1|X>Y] = \frac{\mathrm{pr}\{[(X^2+Y^2)<1]\cap(X>Y)\}}{\mathrm{pr}(X>Y)}$$

So,

$$\mathrm{pr}(X>Y) = \int_0^\infty\int_0^x \mathrm{f}_{X,Y}(x,y)\mathrm{d}y\mathrm{d}x;$$

and

$$\begin{aligned} &\mathrm{pr}\{[(X^2+Y^2)<1]\cap(X>Y)\} \\ &\quad= \int_0^{1/\sqrt{2}}\int_0^x \mathrm{f}_{X,Y}(x,y)\mathrm{d}y\mathrm{d}x + \int_{1/\sqrt{2}}^1\int_0^{\sqrt{1-x^2}} \mathrm{f}_{X,Y}(x,y)\mathrm{d}y\mathrm{d}x, \end{aligned}$$

where

$$\mathrm{f}_{X,Y}(x,y) = \mathrm{f}_X(x)\mathrm{f}_Y(y|X=x) = \left[(2\beta^3)^{-1}x^2\mathrm{e}^{-x/\beta}\right]\left[(\alpha x)\mathrm{e}^{-\alpha xy}\right],$$
$$x>0,\ y>0.$$

Solution 4.49.

(a)

$$\begin{aligned} \mathrm{f}_X(x) &= \int_x^\infty 2\theta^{-2}\mathrm{e}^{-(x+y)/\theta}\mathrm{d}y \\ &= 2\theta^{-2}\mathrm{e}^{-x/\theta}\int_x^\infty \mathrm{e}^{-y/\theta}\mathrm{d}y \\ &= 2\theta^{-2}\mathrm{e}^{-x/\theta}\left[-\theta\mathrm{e}^{-y/\theta}\right]_x^\infty \\ &= 2\theta^{-1}\mathrm{e}^{-2x/\theta}, 0<x<\infty. \end{aligned}$$

So, $X \sim \text{GAMMA}\left(\alpha = \frac{\theta}{2}, \beta = 1\right)$, with $\text{E}(X) = \frac{\theta}{2}$ and $\text{V}(X) = \frac{\theta^2}{4}$. Now,

$$\begin{aligned} \text{f}_Y(y|X=x) &= \frac{\text{f}_{X,Y}(x,y)}{\text{f}_X(x)} \\ &= \frac{2\theta^{-2}\text{e}^{-(x+y)/\theta}}{2\theta^{-1}\text{e}^{-2x/\theta}} \\ &= \theta^{-1}\text{e}^{-(y-x)/\theta}, 0 < x < y < \infty. \end{aligned}$$

So, for r a non-negative integer,

$$\begin{aligned} \text{E}(Y^r|X=x) &= \int_x^\infty \theta^{-1}y^r\text{e}^{-(y-x)/\theta}\text{d}y \\ &= \theta^{-1}\text{e}^{x/\theta}\int_x^\infty y^r\text{e}^{-y/\theta}\text{d}y. \end{aligned}$$

Using the change of variable $u = (y - x)$, so that $\text{d}u = \text{d}y$, we have

$$\begin{aligned} \text{E}(Y^r|X=x) &= \theta^{-1}\text{e}^{x/\theta}\int_0^\infty (u+x)^r\text{e}^{-(u+x)/\theta}\text{d}u \\ &= \theta^{-1}\text{e}^{x/\theta}\text{e}^{-x/\theta}\int_0^\infty \left(\sum_{j=0}^r C_j^r u^j x^{r-j}\right)\text{e}^{-u/\theta}\text{d}u \\ &= \theta^{-1}\sum_{j=0}^r C_j^r x^{r-j}\int_0^\infty u^{(j+1)-1}\text{e}^{-u/\theta}\text{d}u \\ &= \theta^{-1}\sum_{j=0}^r C_j^r x^{r-j}\Gamma(j+1)\theta^{j+1} \\ &= \sum_{j=0}^r C_j^r x^{r-j}\Gamma(j+1)\theta^j. \end{aligned}$$

(b) From part (a),

$$\text{E}(Y|X=x) = \sum_{j=0}^1 C_j^1 x^{1-j}\Gamma(j+1)\theta^j = (\theta + x);$$

so, $\text{E}(Y|X=x) = \beta_0 + \beta_1 x$, where $\beta_0 = \theta$ and $\beta_1 = 1$. Hence,

$$\rho = \sqrt{\frac{\text{V}(X)}{\text{V}(Y)}}(\beta_1) = \sqrt{\frac{\text{V}(X)}{\text{V}(Y)}}$$

since $\beta_1 = 1$.
Now,

$$\text{V}(Y) = \text{V}_x[\text{E}(Y|X=x)] + \text{E}_x[\text{V}(Y|X=x)].$$

Since

$$\mathrm{E}(Y^2|X=x) = \sum_{j=0}^{2} C_j^2 x^{2-j}\Gamma(j+1)\theta^j = (x^2 + 2x\theta + 2\theta^2),$$

we have

$$\mathrm{V}(Y|X=x) = (x^2 + 2x\theta + 2\theta^2) - (\theta + x)^2 = \theta^2;$$

thus,

$$\mathrm{V}(Y) = \mathrm{V}_x(\theta + x) + \mathrm{E}_x(\theta^2) = \frac{\theta^2}{4} + \theta^2 = \frac{5\theta^2}{4}.$$

Finally,

$$\rho = \sqrt{\frac{\mathrm{V}(X)}{\mathrm{V}(Y)}} = \sqrt{\frac{\theta^2/4}{5\theta^2/4}} = \frac{1}{\sqrt{5}}.$$

Alternatively,

$$\begin{aligned}
\mathrm{E}(XY) &= \mathrm{E}_x[E(XY|X=x)] = E_x[x\mathrm{E}(Y|X=x)] = \mathrm{E}_x[x(\theta + x)] \\
&= \theta\mathrm{E}(X) + \mathrm{E}(X^2) = \theta\left(\frac{\theta}{2}\right) + \frac{\theta^2}{4} + \left(\frac{\theta}{2}\right)^2 = \theta^2.
\end{aligned}$$

So,

$$\begin{aligned}
\rho = \mathrm{cor}(X,Y) &= \frac{\mathrm{E}(XY) - \mathrm{E}(X)\mathrm{E}(Y)}{\sqrt{\mathrm{V}(X)\mathrm{V}(Y)}} \\
&= \frac{\theta^2 - (\frac{\theta}{2}) \cdot \mathrm{E}_x[\mathrm{E}(Y|X=x)]}{\sqrt{\left(\frac{\theta^2}{4}\right)\left(\frac{5\theta^2}{4}\right)}} \\
&= \frac{\theta^2 - \frac{\theta}{2}\mathrm{E}_x(\theta + x)}{\sqrt{5\theta^4/16}} \\
&= \frac{\theta^2 - \frac{\theta}{2}(\theta + \frac{\theta}{2})}{(\sqrt{5}\theta^2/4)} \\
&= \frac{\theta^2 - 3\theta^2/4}{\sqrt{5}(\theta^2/4)} \\
&= \frac{1}{\sqrt{5}}.
\end{aligned}$$

(c) Clearly, $0 < P = X/Y < 1$. Now,

$$\begin{aligned}
\mathrm{F}_P(p) &= \mathrm{pr}(P \leq p) = \mathrm{pr}(\frac{X}{Y} \leq p) = \mathrm{pr}\left(\frac{X}{p} \leq Y\right) \\
&= \int_0^\infty \int_{x/p}^\infty 2\theta^{-2}\mathrm{e}^{-(x+y)/\theta}\mathrm{d}y\mathrm{d}x \\
&= 2\theta^{-2}\int_0^\infty \mathrm{e}^{-x/\theta}\left[-\theta\mathrm{e}^{-y/\theta}\right]_{x/p}^\infty \mathrm{d}x \\
&= 2\theta^{-1}\int_0^\infty \mathrm{e}^{-\left(\frac{1}{\theta}+\frac{1}{p\theta}\right)x}\mathrm{d}x \\
&= 2\theta^{-1}\left[-\left(\frac{1}{\theta}+\frac{1}{p\theta}\right)^{-1}\mathrm{e}^{-\left(\frac{1}{\theta}+\frac{1}{p\theta}\right)x}\right]_0^\infty \\
&= \frac{2}{\theta\left(\frac{1}{\theta}+\frac{1}{p\theta}\right)} \\
&= \frac{2p}{(1+p)}, 0 < p < 1.
\end{aligned}$$

So,

$$\mathrm{f}_P(p) = \frac{\mathrm{dF}_P(p)}{\mathrm{d}p} = 2\left[\frac{(1)(1+p) - p(1)}{(1+p)^2}\right] = 2(1+p)^{-2}, 0 < p < 1.$$

Hence,

$$\mathrm{E}(P) = \int_0^1 p \cdot 2(1+p)^{-2}\mathrm{d}p = 2\int_0^1 \frac{p}{(1+p)^2}\mathrm{d}p.$$

Using the change of variables $u = (1+p)$, so that $du = dp$, we have

$$\begin{aligned}
\mathrm{E}(P) &= 2\int_1^2 \frac{(u-1)}{u^2}\mathrm{d}u = 2\left[\ln u + u^{-1}\right]_1^2 \\
&= 2\left(\ln 2 + \frac{1}{2} - \ln 1 - \frac{1}{1}\right) = 0.3862.
\end{aligned}$$

Solution 4.51.

(a)

$$\begin{aligned}
\text{pr}(X_1 = X_2) &= \sum_{x=1}^{\infty} \text{pr}[(X_1 = x) \cap (X_2 = x)] \\
&= \sum_{x=1}^{\infty} [\text{pr}(X_1 = x) \cdot \text{pr}(X_2 = x)] \\
&= \sum_{x=1}^{\infty} [\theta_1(1-\theta_1)^{x-1} \cdot \theta_2(1-\theta_2)^{x-1}] \\
&= \frac{\theta_1\theta_2}{(1-\theta_1)(1-\theta_2)} \sum_{x=1}^{\infty} [(1-\theta_1)(1-\theta_2)]^x \\
&= \frac{\theta_1\theta_2}{(1-\theta_1)(1-\theta_2)} \left[\frac{(1-\theta_1)(1-\theta_2)}{1-(1-\theta_1)(1-\theta_2)}\right] \\
&= \frac{\theta_1\theta_2}{1-(1-\theta_1-\theta_2+\theta_1\theta_2)} = \frac{\theta_1\theta_2}{(\theta_1+\theta_2-\theta_1\theta_2)}.
\end{aligned}$$

(b) When $\theta_1 = \theta_2 = \theta$, then

$$\text{pr}(X_1 = X_2) = \frac{\theta^2}{(2\theta - \theta^2)} = \frac{\theta}{(2-\theta)}.$$

Thus, $\text{pr}(T_j = 0) = \theta/(2-\theta)$ and $\text{pr}(T_j = 1) = 1 - \theta/(2-\theta)$ $= 2(1-\theta)/(2-\theta)$. So,

$$\text{p}_{T_j}(t_j) = \left[\frac{2(1-\theta)}{(2-\theta)}\right]^{t_j} \left[\frac{\theta}{(2-\theta)}\right]^{1-t_j}, \quad t_j = 0, 1.$$

(c) Note that: (1) the outcomes (0 or 1) on each day are mutually independent; (2) $\text{pr}(T_j = 1) = 2(1-\theta)/(2-\theta)$ is the same for each day; and 3) $T = \sum_{j=1}^{4} T_j$ is the number of days out of 4 when $X_1 \neq X_2$.

Clearly, T has a binomial distribution with $n = 4$ and $\pi = 2(1-\theta)/(2-\theta)$, so that

$$\text{p}_T(t) = \text{C}_t^4 \left[\frac{2(1-\theta)}{(2-\theta)}\right]^{t} \left[\frac{\theta}{(2-\theta)}\right]^{1-t}, \quad t = 0, 1, 2, 3, 4.$$

(d) Now,

$$\text{pr}\left\{\bigcap_{j=1}^{4} (T_j = 1)\right\} = \left[\frac{2(1-\theta)}{(2-\theta)}\right]^4.$$

So, we want to find a range of values for θ such that

$$\frac{1}{2} \leq \left[\frac{2(1-\theta)}{(2-\theta)}\right]^4.$$

Solving this inequality gives $\theta \leq 0.2745$. So, the set $\{\theta : 0 \leq \theta \leq 0.2745\}$ will be such that

$$\frac{1}{2} \leq \left[\frac{2(1-\theta)}{(2-\theta)}\right]^4.$$

Solution 4.53.

(a) The random variables X_1, X_2, X_3, and X_4 have a *multinomial* distribution, namely,

$$\begin{aligned}
&\mathrm{p}_{X_1,X_2,X_3,X_4}(x_1, x_2, x_3, x_4) = \\
&\quad \frac{n!}{x_1!x_2!x_3!x_4!}\left(\frac{2+\theta}{4}\right)^{x_1}\left(\frac{1-\theta}{4}\right)^{x_2}\left(\frac{1-\theta}{4}\right)^{x_3}\left(\frac{\theta}{4}\right)^{x_4} \\
&\quad 0 \leq x_i \leq n\,\forall i, \sum_{i=1}^{4} x_i = n.
\end{aligned}$$

(b) If Y is the number of homosexuals in the random sample of size n, then $Y = (X_2 + X_4)$. Since a member of this random sample is a homosexual with probability $\left[\frac{(1-\theta)}{4} + \frac{\theta}{4}\right] = \frac{1}{4}$,

$$Y \sim \mathrm{BIN}\left(n, \frac{1}{4}\right),$$

namely

$$\mathrm{p}_Y(y) = C_y^n \left(\frac{1}{4}\right)^y \left(\frac{3}{4}\right)^{n-y}, y = 0, 1, \ldots, n.$$

(c) Since

$$\begin{aligned}
&\mathrm{pr}(\text{homosexual}|\text{intravenous drug user}) \\
&\quad = \frac{\mathrm{pr}[(\text{homosexual}) \cap (\text{intravenous drug user})]}{\mathrm{pr}(\text{intravenous drug user})} \\
&\quad = \frac{\frac{\theta}{4}}{\left[\frac{(1-\theta)}{4} + \frac{\theta}{4}\right]} \\
&\quad = \theta,
\end{aligned}$$

we would expect, on average, $k\theta$ homosexuals among these k intravenous drug users.

(d)

$$
\begin{aligned}
&\mathrm{E}\left(L_1\right)\\
&= n^{-1}\left[n\left(\frac{2+\theta}{4}\right) - n\left(\frac{1-\theta}{4}\right) - n\left(\frac{1-\theta}{4}\right) + n\left(\frac{\theta}{4}\right)\right]\\
&= \theta.
\end{aligned}
$$

And

$$
\begin{aligned}
&\mathrm{E}\left(L_2\right)\\
&= (2n)^{-1}\left[n\left(\frac{2+\theta}{4}\right) - n\left(\frac{1-\theta}{4}\right) - n\left(\frac{1-\theta}{4}\right) + 5n\left(\frac{\theta}{4}\right)\right]\\
&= \theta.
\end{aligned}
$$

(e) Now,

$$
\begin{aligned}
&\mathrm{V}(L_1)\\
&= n^{-2}\Big[(1)^2\mathrm{V}(X_1) + (-1)^2\mathrm{V}(X_2) + (-1)^2\mathrm{V}(X_3) + (1)^2\mathrm{V}(X_4)\\
&\quad + 2(1)(-1)\mathrm{cov}(X_1, X_2) + 2(1)(-1)\mathrm{cov}(X_1, X_3)\\
&\quad + 2(1)(1)\mathrm{cov}(X_1, X_4) + 2(-1)(-1)\mathrm{cov}(X_2, X_3)\\
&\quad + 2(-1)(1)\mathrm{cov}(X_2, X_4) + 2(-1)(1)\mathrm{cov}(X_3, X_4)\Big]\\
&= n^{-2}\left[n\left(\frac{2+\theta}{4}\right)\left(\frac{2-\theta}{4}\right) + n\left(\frac{1-\theta}{4}\right)\left(\frac{3+\theta}{4}\right)\right.\\
&\quad + n\left(\frac{1-\theta}{4}\right)\left(\frac{3+\theta}{4}\right) + n\left(\frac{\theta}{4}\right)\left(\frac{4-\theta}{4}\right)\\
&\quad + 2n\left(\frac{2+\theta}{4}\right)\left(\frac{1-\theta}{4}\right) + 2n\left(\frac{2+\theta}{4}\right)\left(\frac{1-\theta}{4}\right)\\
&\quad - 2n\left(\frac{2+\theta}{4}\right)\left(\frac{\theta}{4}\right) - 2n\left(\frac{1-\theta}{4}\right)\left(\frac{1-\theta}{4}\right)\\
&\quad \left. + 2n\left(\frac{1-\theta}{4}\right)\left(\frac{\theta}{4}\right) + 2n\left(\frac{1-\theta}{4}\right)\left(\frac{\theta}{4}\right)\right]\\
&= \frac{(1-\theta^2)}{n}.
\end{aligned}
$$

And

$$
\begin{aligned}
\mathrm{V}(L_2) &= (2n)^{-2}\left[(1)^2\mathrm{V}(X_1) + (-1)^2\mathrm{V}(X_2) + (-1)^2\mathrm{V}(X_3)\right. \\
&\quad +(5)^2\mathrm{V}(X_4) + 2(1)(-1)\mathrm{cov}(X_1, X_2) \\
&\quad +2(1)(-1)\mathrm{cov}(X_1, X_3) + 2(1)(5)\mathrm{cov}(X_1, X_4) \\
&\quad +2(-1)(-1)\mathrm{cov}(X_2, X_3) + 2(-1)(5)\mathrm{cov}(X_2, X_4) \\
&\quad \left. +2(-1)(5)\mathrm{cov}(X_3, X_4)\right] \\
&= (2n)^{-2}\left[n\left(\frac{2+\theta}{4}\right)\left(\frac{2-\theta}{4}\right) + n\left(\frac{1-\theta}{4}\right)\left(\frac{3+\theta}{4}\right)\right. \\
&\quad + n\left(\frac{1-\theta}{4}\right)\left(\frac{3+\theta}{4}\right) + 25n\left(\frac{\theta}{4}\right)\left(\frac{4-\theta}{4}\right) \\
&\quad + 2n\left(\frac{2+\theta}{4}\right)\left(\frac{1-\theta}{4}\right) + 2n\left(\frac{2+\theta}{4}\right)\left(\frac{1-\theta}{4}\right) \\
&\quad - 10n\left(\frac{2+\theta}{4}\right)\left(\frac{\theta}{4}\right) - 2n\left(\frac{1-\theta}{4}\right)\left(\frac{1-\theta}{4}\right) \\
&\quad \left. + 10n\left(\frac{1-\theta}{4}\right)\left(\frac{\theta}{4}\right) + 10n\left(\frac{1-\theta}{4}\right)\left(\frac{\theta}{4}\right)\right] \\
&= \frac{(1 + 6\theta - 4\theta^2)}{4n}.
\end{aligned}
$$

Now,

$$
\begin{aligned}
\mathrm{V}(L_1) - \mathrm{V}(L_2) &= \frac{(1-\theta^2)}{n} - \frac{(1 + 6\theta - 4\theta^2)}{4n} \\
&= \left(\frac{3}{2n}\right)\left(\frac{1}{2} - \theta\right).
\end{aligned}
$$

So, $\mathrm{V}(L_1) > \mathrm{V}(L_2)$ when $0 < \theta < \frac{1}{2}$, $\mathrm{V}(L_1) = \mathrm{V}(L_2)$ when $\theta = \frac{1}{2}$, and $\mathrm{V}(L_1) < \mathrm{V}(L_2)$ when $\frac{1}{2} < \theta < 1$.

Solution 4.55.

(a) We have

$$\begin{aligned}
\mathrm{p}_Y(y) &= \mathrm{pr}(Y=y)\\
&= \mathrm{pr}(X_1=y)\mathrm{pr}(X_2<y)\\
&\quad +\mathrm{pr}(X_1<y)\mathrm{pr}(X_2=y)\\
&\quad +\mathrm{pr}(X_1=y)\mathrm{pr}(X_2=y)\\
&= 2\theta(1-\theta)^{y-1}\sum_{j=1}^{y-1}\theta(1-\theta)^{j-1}+\left[\theta(1-\theta)^{y-1}\right]^2\\
&= 2\theta^2(1-\theta)^{y-1}\sum_{j=1}^{y-1}(1-\theta)^{j-1}+\theta^2(1-\theta)^{2(y-1)}\\
&= 2\theta^2(1-\theta)^{y-1}\left[\frac{1-(1-\theta)^{y-1}}{1-(1-\theta)}\right]+\theta^2(1-\theta)^{2(y-1)}\\
&= 2\theta(1-\theta)^{y-1}-\left[1-(1-\theta)^2\right]\left[(1-\theta)^2\right]^{y-1}\\
&= 2\theta(1-\theta)^{y-1}-(2\theta-\theta^2)\left[1-(2\theta-\theta^2)\right]^{y-1},
\end{aligned}$$

so that

$$\mathrm{p}_Y(y)=2\mathrm{p}_1(y)-\mathrm{p}_2(y), y=1,2,\ldots,\infty,$$

where $\mathrm{p}_1(y)$ is a geometric distribution with probability parameter θ and where $\mathrm{p}_2(y)$ is a geometric distribution with probability parameter $(2\theta-\theta^2)$.

(b) Now, if $U\sim$GEOM(π), then

$$\mathrm{E}(U)=\frac{1}{\pi}$$

and

$$\mathrm{E}(U^2)=\mathrm{V}(U)+[\mathrm{E}(U)]^2=\frac{(1-\pi)}{\pi^2}+\frac{1}{\pi^2}=\frac{2}{\pi^2}-\frac{1}{\pi}.$$

So,

$$\begin{aligned}
\mathrm{E}(Y) &= \sum_{y=1}^{\infty} y\mathrm{p}_Y(y)\\
&= 2\left(\frac{1}{\theta}\right)-\frac{1}{(2\theta-\theta^2)}=\frac{(3-2\theta)}{\theta(2-\theta)}.
\end{aligned}$$

And, since

$$\begin{aligned} \mathrm{E}(Y^2) &= \sum_{y=1}^{\infty} y^2 \mathrm{p}_Y(y) \\ &= 2\left(\frac{2}{\theta^2} - \frac{1}{\theta}\right) - \left[\frac{2}{(2\theta-\theta^2)^2} - \frac{1}{(2\theta-\theta^2)}\right] \\ &= \frac{(14 - 22\theta + 11\theta^2 - 2\theta^3)}{\theta^2(2-\theta)^2}, \end{aligned}$$

we obtain

$$\mathrm{V}(Y) = \mathrm{E}(Y^2) - [\mathrm{E}(Y)]^2 = \frac{(5 - 10\theta + 7\theta^2 - 2\theta^3)}{\theta^2(2-\theta)^2}.$$

As expected, when $\theta = 1, \mathrm{E}(Y) = \mathrm{E}(Y^2) = 1$, and $\mathrm{V}(Y) = 0$.

Solution 4.57. Since $Z \equiv (X - \mu)/\sigma \sim N(0, 1)$, we can write

$$\mathrm{pr}(X \geq L) \;=\; \mathrm{pr}\left[Z \geq \frac{L-\mu}{\sigma}\right] = 1 - \Phi\left(\frac{L-\mu}{\sigma}\right),$$

where

$$\Phi(c) \;=\; \int_{-\infty}^{c} \frac{1}{\sqrt{2\pi}} \mathrm{e}^{-z^2/2} \mathrm{d}z.$$

So,

$$\begin{aligned} &\mathrm{E}(X|X \geq L) \\ &= \left[1 - \Phi\left(\frac{L-\mu}{\sigma}\right)\right]^{-1} \int_{L}^{\infty} x \frac{1}{\sqrt{2\pi}\sigma} \mathrm{e}^{-(x-\mu)^2/(2\sigma^2)} \mathrm{d}x \\ &= \left[1 - \Phi\left(\frac{L-\mu}{\sigma}\right)\right]^{-1} \int_{\left(\frac{L-\mu}{\sigma}\right)}^{\infty} (\mu + \sigma z) \frac{1}{\sqrt{2\pi}} \mathrm{e}^{-z^2/2} \mathrm{d}z \\ &= \left[1 - \Phi\left(\frac{L-\mu}{\sigma}\right)\right]^{-1} \left\{\mu\left[1 - \Phi\left(\frac{L-\mu}{\sigma}\right)\right] + \frac{\sigma}{\sqrt{2\pi}} \int_{\left(\frac{L-\mu}{\sigma}\right)}^{\infty} z e^{-z^2/2} \mathrm{d}z\right\} \\ &= \mu + \frac{\frac{\sigma}{\sqrt{2\pi}}\left[-\mathrm{e}^{-z^2/2}\right]_{\left(\frac{L-\mu}{\sigma}\right)}^{\infty}}{\left[1 - \Phi\left(\frac{L-\mu}{\sigma}\right)\right]} \\ &= \mu + \frac{\frac{\sigma}{\sqrt{2\pi}}\left[\mathrm{e}^{-\left(\frac{L-\mu}{\sigma}\right)^2/2}\right]}{\left[1 - \Phi\left(\frac{L-\mu}{\sigma}\right)\right]}. \end{aligned}$$

This expression is the mean of a normal distribution that is left-truncated at the value L. If $\mu = 3$, $\sigma^2 = 1$, and $L = 1.60$, then

$$\begin{aligned}
\mathrm{E}(X|X \geq 1.60) &= 3 + \frac{\frac{1}{\sqrt{2\pi}}\left[\mathrm{e}^{-\left(\frac{1.60-3}{1}\right)^2/2}\right]}{\left[1 - \Phi\left(\frac{1.60-3}{1}\right)\right]} \\
&= 3 + \frac{0.1497}{0.9192} \\
&= 3.1629.
\end{aligned}$$

Solution 4.59. First,

$$\mathrm{pr}(T > 1) = \int_1^\infty \mathrm{e}^{-t}\mathrm{d}t = \left[-\mathrm{e}^{-t}\right]_1^\infty = 1/\mathrm{e} = 0.368.$$

For $i = 1, 2, \ldots, n$, let

$$X_i = \begin{cases} 1 & \text{if } T_i > 1 \\ 0 & \text{otherwise} \end{cases},$$

where T_i is the time to failure for the i-th component.

Then, consider $S = \sum_{i=1}^n X_i$, where $\{X_i\}_{i=1}^n$ constitute a set of i.i.d. random variables. Hence, by the Central Limit Theorem, it follows that, for large n,

$$\frac{S - \mathrm{E}(S)}{\sqrt{\mathrm{V}(S)}} \,\dot{\sim}\, \mathrm{N}(0, 1),$$

where $\mathrm{E}(S) = 0.368n$ and $\mathrm{V}(S) = 0.368(0.632)n = 0.233n$.

We want to find the smallest value of n, say n^*, such that $\mathrm{pr}(0.30n \leq S) \geq 0.95$. Now, with $Z \dot{\sim} \mathrm{N}(0, 1)$ for large n, we have

$$\begin{aligned}
\mathrm{pr}(0.30n \leq S) &\doteq \mathrm{pr}\left\{\frac{0.30n - \mathrm{E}(S)}{\sqrt{\mathrm{V}(S)}} \leq \frac{S - \mathrm{E}(S)}{\sqrt{\mathrm{V}(S)}}\right\} \\
&= \mathrm{pr}\left\{\frac{0.30n - 0.368n}{\sqrt{0.233n}} \leq Z\right\} \\
&= \mathrm{pr}\left[-0.141\sqrt{n} \leq Z\right] \geq 0.95.
\end{aligned}$$

So, we need $-0.141\sqrt{n} \leq -1.645 \Rightarrow n = 136.11$. Thus, we require $n^* = 137$.

Note that, since $S \sim \mathrm{BIN}(n, \pi = 0.368)$, we are using the classical "normal approximation to the binomial distribution."

Solution 4.61.

(a) Since

$$\begin{aligned} \mathrm{p}_{X,Y}(x,y|\theta) &= \mathrm{p}_X(x|\theta)\mathrm{p}_Y(y|\theta) = \left(\frac{\theta^x \mathrm{e}^{-\theta}}{x!}\right)\left(\frac{\theta^y \mathrm{e}^{-\theta}}{y!}\right), \\ &\quad \text{for } x = 0,1,\ldots,\infty \text{ and } y = 0,1,\ldots,\infty, \end{aligned}$$

it follows that

$$\begin{aligned} \mathrm{p}_{X,Y}(x,y) &= \int_0^\infty \mathrm{p}_{X,Y}(x,y|\theta)\mathrm{f}(\theta)\mathrm{d}\theta \\ &= \int_0^\infty \frac{\theta^{x+y}\mathrm{e}^{-2\theta}}{x!y!}\left[\Gamma(\alpha)\right]^{-1}\theta^{\alpha-1}\mathrm{e}^{-\theta}\mathrm{d}\theta \\ &= \frac{1}{x!y!\Gamma(\alpha)}\int_0^\infty \theta^{(x+y+\alpha)-1}\mathrm{e}^{-3\theta}\mathrm{d}\theta \\ &= \frac{\Gamma(x+y+\alpha)}{x!y!\Gamma(\alpha)}\left(\frac{1}{3}\right)^{x+y+\alpha} \\ &= \frac{(x+y+\alpha-1)!}{x!y!(\alpha-1)!}\left(\frac{1}{3}\right)^{x+y+\alpha}, \\ &\quad \text{for } x = 0,1,\ldots,\infty \text{ and } y = 0,1,\ldots,\infty. \end{aligned}$$

Clearly, $\mathrm{p}_{X,Y}(x,y) \geq 0$ for all permissible values of x and y. And, appealing to properties of the negative binomial distribution, we have

$$\begin{aligned} &\sum_{x=0}^\infty\sum_{y=0}^\infty \mathrm{p}_{X,Y}(x,y) \\ &= \sum_{x=0}^\infty\sum_{y=0}^\infty \frac{(x+y+\alpha-1)!}{x!y!(\alpha-1)!}\left(\frac{1}{3}\right)^{x+y+\alpha} \\ &= \sum_{x=0}^\infty \frac{1}{x!(\alpha-1)!}\left(\frac{1}{3}\right)^{x+\alpha}\sum_{y=0}^\infty \frac{(x+y+\alpha-1)!}{y!}\left(\frac{1}{3}\right)^{y} \\ &= \sum_{x=0}^\infty \frac{(x+\alpha-1)!}{x!(\alpha-1)!}\left(\frac{1}{3}\right)^{x+\alpha}\left(\frac{3}{2}\right)^{x+\alpha} \\ &\qquad \times \sum_{y=0}^\infty \mathrm{C}_{x+\alpha-1}^{y+x+\alpha-1}\left(\frac{2}{3}\right)^{x+\alpha}\left(\frac{1}{3}\right)^{y} \\ &= \sum_{x=0}^\infty \mathrm{C}_{\alpha-1}^{x+\alpha-1}\left(\frac{1}{2}\right)^{\alpha}\left(\frac{1}{2}\right)^{x} = 1, \end{aligned}$$

so that $\mathrm{p}_{X,Y}(x,y)$ is a valid bivariate discrete probability distribution.

(b) Since

$$\begin{aligned}\sum_{x=0}^{\infty}\sum_{y=0}^{\infty}\mathrm{p}_{X,Y}(x,y) &= \sum_{x=0}^{\infty}\sum_{y=0}^{\infty}\mathrm{p}_X(x)\mathrm{p}_Y(y|X=x)\\ &= \sum_{x=0}^{\infty}\mathrm{p}_X(x)\sum_{y=0}^{\infty}\mathrm{p}_Y(y|X=x),\end{aligned}$$

it follows from part (a) that

$$\mathrm{p}_X(x) = \mathrm{C}_{\alpha-1}^{x+\alpha-1}\left(\frac{1}{2}\right)^{\alpha}\left(\frac{1}{2}\right)^{x}, x = 0,1,\ldots,\infty,$$

and that

$$\mathrm{p}_Y(y|X=x) = \mathrm{C}_{x+\alpha-1}^{y+x+\alpha-1}\left(\frac{2}{3}\right)^{x+\alpha}\left(\frac{1}{3}\right)^{y}, y = 0,1,\ldots,\infty.$$

Thus, both the marginal distribution of X and the conditional distribution of Y given $X = x$ are negative binomial distributions. And, by symmetry, the marginal distribution of Y is the same as the marginal distribution of X.

Now, in general, if $U \sim$NEGBIN(k,π), then $\mathrm{E}(U) = k/\pi$. And, if $W = (U-k)$, then

$$\mathrm{p}_W(w) = \mathrm{C}_{k-1}^{w+k-1}\pi^k(1-\pi)^w, w = 0,1,\ldots,\infty,$$

and $\mathrm{E}(W) = \mathrm{E}(U) - k = \frac{k}{\pi} - k = k\left(\frac{1-\pi}{\pi}\right)$.

Thus, since $\mathrm{p}_Y(y|X=x)$ has the same structure as $\mathrm{p}_W(w)$ with $k = (x+\alpha)$ and $\pi = 2/3$, it follows that

$$\mathrm{E}(Y|X=x) = (x+\alpha)\left(\frac{1-\frac{2}{3}}{2/3}\right) = \frac{\alpha}{2} + \frac{x}{2} = \beta_0 + \beta_1 x,$$

where $\beta_0 = \alpha/2$ and $\beta_1 = 1/2$.

Finally, since $\mathrm{V}(X) = \mathrm{V}(Y)$, we have

$$\mathrm{corr}(X,Y) = \beta_1\sqrt{\frac{\mathrm{V}(X)}{\mathrm{V}(Y)}} = \beta_1 = \frac{1}{2}.$$

Solution 4.63. For $k \geq 0$, note that $|x_1 - x_2|^k$ equals $(x_1-x_2)^k$ if $x_1 \geq x_2$ and equals $(x_2-x_1)^k$ if $x_2 \geq x_1$. Then, with

$$\mathrm{f}_{X_1,X_2}(x_1,x_2) = \mathrm{f}_{X_1}(x_1)\mathrm{f}_{X_2}(x_2) = (1)(1) = 1, 0 < x_1 < 1, 0 < x_2 < 1,$$

we have

$$\begin{aligned}
\mathrm{E}\left(|X_1 - X_2|^k\right) &= \int_0^1 \int_0^1 |x_1 - x_2|^k (1) \mathrm{d}x_1 \mathrm{d}x_2 \\
&= \int_0^1 \int_0^{x_1} (x_1 - x_2)^k \mathrm{d}x_2 \mathrm{d}x_1 + \int_0^1 \int_0^{x_2} (x_2 - x_1)^k \mathrm{d}x_1 \mathrm{d}x_2 \\
&= \int_0^1 \left[\frac{-(x_1 - x_2)^{k+1}}{(k+1)} \right]_0^{x_1} \mathrm{d}x_1 + \int_0^1 \left[\frac{-(x_2 - x_1)^{k+1}}{(k+1)} \right]_0^{x_2} \mathrm{d}x_2 \\
&= \int_0^1 \frac{x_1^{k+1}}{(k+1)} \mathrm{d}x_1 + \int_0^1 \frac{x_2^{k+1}}{(k+1)} \mathrm{d}x_2 \\
&= \frac{1}{(k+1)} \left[\frac{x_1^{k+2}}{(k+2)} \right]_0^1 + \frac{1}{(k+1)} \left[\frac{x_2^{k+2}}{(k+2)} \right]_0^1 \\
&= \frac{2}{(k+1)(k+2)}, k \geq 0.
\end{aligned}$$

Solution 4.65.

(a) Now,

$$\begin{aligned}
&\mathrm{p}_{X_1}(x_1 | S = s) \\
&= \mathrm{pr}(X_1 = x_1 | S = s) = \frac{\mathrm{pr}\left[(X_1 = x_1) \cap (S = s)\right]}{\mathrm{pr}(S = s)} \\
&= \frac{\mathrm{pr}(X_1 = x_1)\mathrm{pr}(X_2 = s - x_1)}{\mathrm{pr}(S = s)} \\
&= \frac{\left[\mathrm{C}_{x_1}^{n_1} \pi_1^{x_1} (1 - \pi_1)^{n_1 - x_1}\right] \left[\mathrm{C}_{s-x_1}^{n_2} \pi_2^{s-x_1} (1 - \pi_2)^{n_2 - (s - x_1)}\right]}{\mathrm{pr}(S = s)}.
\end{aligned}$$

And, with $a = \max(0, s - n_2)$ and $b = \min(n_1, s)$, we have

$$\begin{aligned}
&\mathrm{pr}(S = s) \\
&= \sum_{u=a}^{b} \mathrm{pr}(X_1 = u)\mathrm{pr}(X_2 = s - u) \\
&= \sum_{u=a}^{b} \left[\mathrm{C}_u^{n_1} \pi_1^u (1 - \pi_1)^{n_1 - u}\right] \left[\mathrm{C}_{s-u}^{n_2} \pi_2^{s-u} (1 - \pi_2)^{n_2 - (s-u)}\right].
\end{aligned}$$

Finally,

$$\begin{aligned}
&\mathrm{p}_{X_1}(x_1 | S = s) \\
&= \frac{\left[\mathrm{C}_{x_1}^{n_1} \pi_1^{x_1} (1 - \pi_1)^{n_1 - x_1}\right] \left[\mathrm{C}_{s-x_1}^{n_2} \pi_2^{s-x_1} (1 - \pi_2)^{n_2 - (s - x_1)}\right]}{\sum_{u=a}^{b} \left[\mathrm{C}_u^{n_1} \pi_1^u (1 - \pi_1)^{n_1 - u}\right] \left[\mathrm{C}_{s-u}^{n_2} \pi_2^{s-u} (1 - \pi_2)^{n_2 - (s-u)}\right]} \\
&= \frac{\mathrm{C}_{x_1}^{n_1} \mathrm{C}_{s-x_1}^{n_2} \theta^{x_1}}{\sum_{u=a}^{b} \mathrm{C}_u^{n_1} \mathrm{C}_{s-u}^{n_2} \theta^u}, a \leq x_1 \leq b.
\end{aligned}$$

(b) When $n_1 = 3, n_2 = 2$, and $s = 4$, then $a = \max(0, 4-2) = 2$, and $b = \min(3,4) = 3$, so that the permissible values of x_1 are 2 and 3. So,

$$\begin{aligned}
\mathrm{pr}(X_1 = 2|S = 4) &= \frac{\mathrm{C}_2^3\mathrm{C}_2^2\theta^2}{\sum_{u=2}^{3}\mathrm{C}_u^3\mathrm{C}_{4-u}^2\theta^u} \\
&= \frac{3\theta^2}{(3\theta^2 + 2\theta^3)} = \frac{3}{(3+2\theta)},
\end{aligned}$$

and

$$\mathrm{pr}(X_1 = 3|S = 4) = 1 - \frac{3}{(3+2\theta)} = \frac{2\theta}{(3+2\theta)}.$$

Finally,

$$\mathrm{E}(X_1|S = 4) = 2\left(\frac{3}{3+2\theta}\right) + 3\left(\frac{2\theta}{3+2\theta}\right) = \frac{6(1+\theta)}{(3+2\theta)}.$$

Solution 4.67. For $i = 1, 2, \ldots, n$, let X_i be the discrete random variable denoting the number of cards correctly identified by subject i, and let G be the event that all n subjects are making purely random guesses about the numbers on the backs of the cards. Then, with $X_{(n)} = \max\{X_1, X_2 \ldots, X_n\}$, we are interested in determining a reasonable value for $\mathrm{pr}\left[X_{(n)} \geq 30|\mathrm{G}\right]$.

So,

$$\begin{aligned}
\mathrm{pr}\left[X_{(n)} \geq 30|\mathrm{G}\right] &= 1 - \mathrm{pr}\left[X_{(n)} < 30|\mathrm{G}\right] \\
&= 1 - \mathrm{pr}\left[\cap_{i=1}^{n}(X_i < 30|\mathrm{G})\right] \\
&= 1 - \prod_{i=1}^{n}\mathrm{pr}(X_i < 30|\mathrm{G}) \\
&= 1 - \left[\mathrm{pr}(X_i < 30|\mathrm{G})\right]^n.
\end{aligned}$$

Now, given G, $X_i \sim \mathrm{BIN}(100, 1/5)$, so that

$$\mathrm{pr}(X_i < 30|G) = \sum_{x=0}^{29}\mathrm{C}_x^{100}\left(\frac{1}{5}\right)^x\left(\frac{4}{5}\right)^{100-x}.$$

We can use the Central Limit Theorem to obtain an accurate approximate value for this probability. In particular, since $\mathrm{E}(X_i) = 100(1/5) = 20$ and $\mathrm{V}(X_i) = 100(1/5)(4/5) = 16$, we have

$$\begin{aligned}
\mathrm{pr}(X_i < 30|G) &= \mathrm{pr}(0 \leq X_i \leq 29|G) \\
&= \mathrm{pr}\left[\left(\frac{0-20}{4}\right) \leq \left(\frac{X_i - 20}{4}\right) \leq \left(\frac{29-20}{4}\right)\Big|\mathrm{G}\right] \\
&\doteq \mathrm{pr}(Z_i \leq 2.25) = 0.9878,
\end{aligned}$$

where $Z_i = (X_i - 20)/4 \dot{\sim} \mathrm{N}(0,1)$ for large n.

So, since $n = 50$,

$$\mathrm{pr}\left[X_{(n)} \geq 30|G\right] \doteq 1 - (0.9878)^{50} = 0.4587,$$

so that this result strongly supports the hypothesis that all 50 subjects participating in this ESP experiment have made purely random guesses about the numbers on the backs of the cards.

Solution 4.69. Now,

$$\begin{aligned}
\mathrm{pr}(\bar{X} > k|\mu > 46) &\geq \mathrm{pr}(\bar{X} > k|\mu = 46) = \mathrm{pr}\left(\frac{\bar{X} - 46}{\sqrt{10/n}} > \frac{k - 46}{\sqrt{10/n}}\right) \\
&= \mathrm{pr}\left(Z > \frac{k - 46}{\sqrt{10/n}}\right) \geq 0.95,
\end{aligned}$$

where $Z \sim \mathrm{N}(0,1)$. Hence, we require

$$\frac{k - 46}{\sqrt{10/n}} \leq -1.645, \text{ or } k \leq 46 - \frac{5.202}{\sqrt{n}}.$$

Similarly, we have

$$\begin{aligned}
\mathrm{pr}(\bar{X} \leq k|\mu < 42) &\geq \mathrm{pr}(\bar{X} \leq k|\mu = 42) = \mathrm{pr}\left(\frac{\bar{X} - 42}{\sqrt{10/n}} \leq \frac{k - 42}{\sqrt{10/n}}\right) \\
&= \mathrm{pr}\left(Z \leq \frac{k - 42}{\sqrt{10/n}}\right) \geq 0.98,
\end{aligned}$$

where $Z \sim \mathrm{N}(0,1)$. Hence, we require

$$\frac{k - 42}{10/\sqrt{n}} \geq 2.054, \text{ or } k \geq 42 + \frac{6.495}{\sqrt{n}}.$$

Equating these two inequality expressions for k gives

$$42 + \frac{6.495}{\sqrt{n}} = 46 - \frac{5.202}{\sqrt{n}},$$

so that $\sqrt{n} = 2.9243$. Thus, $n = 8.5512$, and hence we take $n^* = 9$.

Now, for $n^* = 9$, we have

$$k \geq 42 + \frac{6.495}{\sqrt{9}} = 44.165 \text{ and } k \leq 46 - \frac{5.202}{\sqrt{n}} = 44.266,$$

so that one reasonable (but not unique) choice for k^* is

$$k^* = \frac{(44.165 + 44.266)}{2} = 44.216.$$

Finally, for $n^* = 9$ and $k^* = 44.216$, we have

$$\text{pr}(\bar{X} > k^*|\mu > 46) \geq \text{pr}\left(Z > \frac{44.216 - 46}{\sqrt{10/9}}\right) = \text{pr}(Z > -1.6924) > 0.95,$$

and

$$\text{pr}(\bar{X} \leq k^*|\mu < 42) \geq \text{pr}\left(Z \leq \frac{44.216 - 42}{\sqrt{10/9}}\right) = \text{pr}(Z \leq 2.1023) > 0.98,$$

so that the requirements are met when $n^* = 9$ and $k^* = 44.216$.

Solution 4.71. First, we know that $X_2 \sim \text{BIN}\left[n; \pi(1-\theta)\right], X_3 \sim \text{BIN}(n, \pi\theta)$, and that $\text{cov}(X_2, X_3) = -n[\pi(1-\theta)](\pi\theta) = -n\pi^2\theta(1-\theta)$. Now,

$$\text{E}(U) = \frac{\text{E}(X_2) + \text{E}(X_3)}{n} = \frac{n\pi(1-\theta) + n\pi\theta}{n} = \pi,$$

and

$$\begin{aligned} \text{V}(U) &= n^{-2}\left[\text{V}(X_2) + \text{V}(X_3) + 2\text{cov}(X_2, X_3)\right] \\ &= n^{-2}\left\{n\pi(1-\theta)\left[1 - \pi(1-\theta)\right] + n\pi\theta(1-\pi\theta) - 2n\pi^2\theta(1-\theta)\right\} \\ &= \frac{\pi(1-\pi)}{n}. \end{aligned}$$

Note that these two results follow more directly by noting that $(X_2 + X_3) \sim \text{BIN}(n, \pi)$. And,

$$\begin{aligned} \text{E}(W) &= \text{E}\left[\text{E}(W|X_2 + X_3 = x_2 + x_3\right] \\ &= \text{E}\left[(x_2 + x_3)^{-1}\text{E}(X_3|X_2 + X_3 = x_2 + x_3)\right], \end{aligned}$$

so that we need the conditional distribution of X_3 given that $(X_2 + X_3) = (x_2 + x_3)$. So,

$$\begin{aligned} &\text{pr}(X_3 = x_3|X_2 + X_3 = x_2 + x_3) \\ &= \frac{\text{pr}\left[(X_3 = x_3) \cap (X_2 + X_3 = x_2 + x_3)\right]}{\text{pr}(X_2 + X_3 = x_2 + x_3)} \\ &= \frac{\text{pr}\left[(X_2 = x_2) \cap (X_3 = x_3)\right]}{\text{pr}(X_2 + X_3 = x_2 + x_3)} \\ &= \frac{\frac{n!}{(n-x_2-x_3)!x_2!x_3!}(1-\pi)^{(n-x_2-x_3)}\left[\pi(1-\theta)\right]^{x_2}(\pi\theta)^{x_3}}{\text{C}^n_{(x_2+x_3)}\pi^{(x_2+x_3)}(1-\pi)^{n-(x_2+x_3)}} \\ &= \frac{(x_2 + x_3)!}{x_3!x_2!}\theta^{x_3}(1-\theta)x_2, \end{aligned}$$

so that the conditional distribution of X_3 given that $(X_2 + X_3) = (x_2 + x_3)$ is $\text{BIN}(x_2 + x_3, \theta)$.

It then follows that

$$\text{E}(W) = \text{E}\left[(x_2 + x_3)^{-1}(x_2 + x_3)\theta\right] = \theta.$$

Finally,

$$\begin{aligned}
\text{cov}(U, W) &= \text{E}(UW) - \text{E}(U)\text{E}(W) \\
&= \text{E}\left[\left(\frac{X_2 + X_3}{n}\right)\left(\frac{X_3}{X_2 + X_3}\right)\right] - (\pi)(\theta) \\
&= \text{E}\left(\frac{X_3}{n}\right) - \pi\theta \\
&= \frac{n\pi\theta}{n} - \pi\theta = 0,
\end{aligned}$$

so that $\text{corr}(U, W) = 0$.

Solution 4.73.

(a) Let $T_{(1)} = \min\{T_1, T_2, \ldots, T_s\}$. Then, for $w_1 > 0$, we have

$$\begin{aligned}
\text{F}_{W_1}(w_1) &= \text{pr}(W_1 \leq w_1) = \text{pr}\left[T_{(1)} \leq w_1\right] \\
&= 1 - \text{pr}\left[T_{(1)} > w_1\right] = 1 - \text{pr}\left[\cap_{i=1}^{s}(T_i > w_1)\right] \\
&= 1 - \prod_{i=1}^{s}\left[\int_{w_1}^{\infty} \frac{1}{\mu}\text{e}^{-t/\mu}\text{d}t\right] \\
&= 1 - \prod_{i=1}^{s} \text{e}^{-w_1/\mu} = 1 - \text{e}^{-sw_1/\mu}.
\end{aligned}$$

So,

$$\text{f}_{W_1}(w_1) = \frac{\text{dF}_{W_1}(w_1)}{\text{d}w_1} = \frac{s}{\mu}\text{e}^{-sw_1/\mu}, w_1 > 0,$$

so that $W_1 \sim \text{NEGEXP}(\alpha = \mu/s)$. Thus, $\text{E}(W_1) = \mu/s$ and $\text{V}(W_1) = \mu^2/s^2$.

(b) As soon as the first customer in the queue leaves the queue to receive service, the second customer in the queue now becomes the first customer in the queue, and so W_1 will be the random variable representing the additional time that this customer will have to wait for service. In other words, the total waiting time W_2 for the second customer in the queue can be expressed as $W_2 = (W_{11} + W_{12})$, where W_{11} and W_{12} are independent random variables, each following the $\text{NEGEXP}(\alpha = \mu)$ distribution. In general, then,

$$W_n = \sum_{j=1}^{n} W_{1j}$$

is the total time that the n-th customer in the queue will have to wait for service, where $W_{1j} \sim \text{NEGEXP}(\alpha = \mu), j = 1, 2, \ldots, n$, and where the $\{W_{1j}\}_{j=1}^{n}$ constitute a set of mutually independent random variables.

Now, the moment generating function $\text{M}_{W_n}(t)$ for the random variable W_n is

$$\begin{aligned} \text{M}_{W_n}(t) &= \text{E}\left(\text{e}^{W_n}\right) = \text{E}\left(\text{e}^{\sum_{j=1}^{n} W_{1j}}\right) \\ &= \text{E}\left(\prod_{j=1}^{n} \text{e}^{W_{1j}}\right) \\ &= \prod_{j=1}^{n} \text{E}\left(\text{e}^{W_{1j}}\right) \\ &= \prod_{j=1}^{n} (1 - \mu t)^{-1} \\ &= (1 - \mu t)^{-n}, \end{aligned}$$

so that $W_n \sim \text{GAMMA}(\alpha = \mu, \beta = n), \text{E}(W_n) = n\mu$, and $\text{V}(W_n) = n\mu^2$.

Solution 4.75.

(a) We have

$$\begin{aligned} \text{pr}(a \le X \le b) &= k \sum_{x=a}^{b} (1 - \pi)^{x-1} \pi \\ &= k \left[\text{pr}(X \ge a) - \text{pr}(X > b)\right] \\ &= k \left[\sum_{x=a}^{\infty} (1 - \pi)^{x-1} \pi - \sum_{x=b+1}^{\infty} (1 - \pi)^{x-1} \pi\right] \\ &= k\pi \left[\frac{(1 - \pi)^{a-1}}{1 - (1 - \pi)} - \frac{(1 - \pi)^{b}}{1 - (1 - \pi)}\right] \\ &= k \left[(1 - \pi)^{a-1} - (1 - \pi)^{b}\right], \end{aligned}$$

so that $k = \left[(1 - \pi)^{a-1} - (1 - \pi)^{b}\right]^{-1}$.

(b) Now, since $\text{E}(Y) = \text{E}_x\left[\text{E}(Y|X = x)\right] = \text{E}_x(x\theta) = \theta\text{E}(X)$, we need to find an explicit expression for $\text{E}(X)$.

So,

$$\begin{aligned}
\mathrm{E}(X) &= k\sum_{x=a}^{b} x(1-\pi)^{x-1}\pi \\
&= k\pi\sum_{x=a}^{b}\left[-\frac{\mathrm{d}}{\mathrm{d}\pi}(1-\pi)^{x}\right] \\
&= -k\pi\frac{\mathrm{d}}{\mathrm{d}\pi}\left[\sum_{x=a}^{b}(1-\pi)^{x}\right] \\
&= -k\pi\frac{\mathrm{d}}{\mathrm{d}\pi}\left[\sum_{x=a}^{\infty}(1-\pi)^{x} - \sum_{x=b+1}^{\infty}(1-\pi)^{x}\right] \\
&= -k\pi\frac{\mathrm{d}}{\mathrm{d}\pi}\left[\frac{(1-\pi)^{a}}{1-(1-\pi)} - \frac{(1-\pi)^{b+1}}{1-(1-\pi)}\right] \\
&= -k\pi\frac{\mathrm{d}}{\mathrm{d}\pi}\left[\frac{(1-\pi)^{a}-(1-\pi)^{b+1}}{\pi}\right] \\
&= -k\pi\left\{\frac{\pi\left[-a(1-\pi)^{a-1}+(b+1)(1-\pi)^{b}\right]-\left[(1-\pi)^{a}-(1-\pi)^{b+1}\right]}{\pi^{2}}\right\}.
\end{aligned}$$

With some algebraic simplification, we finally obtain

$$\mathrm{E}(Y) = \theta\mathrm{E}(X) = \theta\left\{\frac{(1-\pi)^{a-1}\left[1+(a-1)\pi\right]-(1-\pi)^{b}(1+b\pi)}{\pi\left[(1-\pi)^{a-1}-(1-\pi)^{b}\right]}\right\}.$$

And, since $\lim_{b\to\infty}(1-\pi)^{b} = \lim_{b\to\infty}\left[(1-\pi)^{b}(1+b\pi)\right] = 0$, we obtain

$$\lim_{b\to\infty}\mathrm{E}(Y) = \theta\left[\frac{1+(a-1)\pi}{\pi}\right].$$

Solution 4.77.

(a) Note that Y_{ij} can be written in the form

$$Y_{ij} = (\beta_0 + \beta_1 x_{1ij} + \sum_{l=2}^{p}\beta_l x_{lij}) + (U_{0j} + x_{1ij}U_{1j} + e_{ij}).$$

Since Y_{ij} is a linear combination of mutually independent normally distributed random variables, it follows that Y_{ij} has a normal distribution with

$$\mathrm{E}(Y_{ij}) = \beta_0 + \beta_1 x_{1ij} + \sum_{l=2}^{p}\beta_l x_{lij},$$

and

$$\begin{aligned}
\mathrm{V}(Y_{ij}) &= \mathrm{V}\left(U_{0j} + x_{1ij}U_{1j} + e_{ij}\right) \\
&= \mathrm{V}(U_{0j}) + x_{1ij}^2 \mathrm{V}(U_{1j}) + 2x_{1ij}\mathrm{cov}(U_{0j}, U_{1j}) + \mathrm{V}(e_{ij}) \\
&= \sigma_0^2 + x_{1ij}^2\sigma_1^2 + 2x_{1ij}\sigma_{01} + \sigma_e^2.
\end{aligned}$$

(b) Now,

$$\begin{aligned}
\mathrm{cov}(Y_{ij}, Y_{i'j}) &= \mathrm{cov}\left[U_{0j} + x_{1ij}U_{1j} + e_{ij}, U_{0j} + x_{1i'j}U_{1j} + e_{i'j}\right] \\
&= \mathrm{cov}(U_{0j}, U_{0j}) + \mathrm{cov}(U_{0j}, x_{1i'j}U_{1j}) + \mathrm{cov}(x_{1ij}U_{1j}, U_{0j}) + \\
&\quad \mathrm{cov}(x_{1ij}U_{1j}, x_{1i'j}U_{1j}) \\
&= \sigma_0^2 + x_{1i'j}\sigma_{01} + x_{1ij}\sigma_{01} + x_{1ij}x_{1i'j}\sigma_1^2 \\
&= \sigma_0^2 + (x_{1ij} + x_{1i'j})\sigma_{01} + x_{1ij}x_{1i'j}\sigma_1^2.
\end{aligned}$$

(c) When $\sigma_1^2 = 0$, then $\sigma_{01} = 0$. Thus,

$$\mathrm{V}(Y_{ij}) = \mathrm{V}(Y_{i'j}) = (\sigma_0^2 + \sigma_e^2), \text{ and } \mathrm{cov}(Y_{ij}, Y_{i'j}) = \sigma_0^2.$$

Thus, in this special case,

$$\begin{aligned}
\mathrm{corr}(Y_{ij}, Y_{i'j}) &= \frac{\mathrm{cov}(Y_{ij}, Y_{i'j})}{\sqrt{\mathrm{V}(Y_{ij})\mathrm{V}(Y_{i'j})}} \\
&= \frac{\sigma_0^2}{(\sigma_0^2 + \sigma_e^2)}.
\end{aligned}$$

Since two students in the same school would be expected to perform more similarly than would two students from different schools, it is reasonable to believe that $\mathrm{corr}(Y_{ij}, Y_{i'j})$ would be positive. Furthermore, ignoring such intra-school correlations among student responses can lead to incorrect statistical conclusions. Multilevel statistical analyses are one way to account appropriately for intra-cluster correlations.

Solution 4.79. Since the sample is selected randomly without replacement from a finite population, it follows that Y has a *hypergeometric* distribution, namely

$$\mathrm{p}_Y(y) = \mathrm{pr}(Y = y) = \frac{\mathrm{C}_y^{\theta}\mathrm{C}_{n-y}^{N-\theta}}{\mathrm{C}_n^N}, \max[0, n - (N - \theta)] \leq y \leq \min(n, \theta),$$

so that

$$\mathrm{E}(Y) = n\left(\frac{\theta}{N}\right) \text{ and } \mathrm{V}(Y) = n\left(\frac{\theta}{N}\right)\left(1 - \frac{\theta}{N}\right)\left(\frac{N-n}{N-1}\right).$$

Also, the conditional distribution of X_1 given $Y = y$ is $\mathrm{BIN}(y, \pi_1)$, the conditional distribution of X_0 given $Y = y$ is $\mathrm{BIN}(n - y, 1 - \pi_0)$, and X_1 and X_0 are independent given $Y = y$.

Thus,

$$\begin{aligned}
\mathrm{E}(X) &= \mathrm{E}_y[\mathrm{E}(X|Y=y)] = \mathrm{E}_y[\mathrm{E}(X_1|Y=y)] + \mathrm{E}_y[\mathrm{E}(X_0|Y=y)] \\
&= \mathrm{E}_y[y\pi_1 + (n-y)(1-\pi_0)] = n(1-\pi_0) + (\pi_1+\pi_0-1)\mathrm{E}(Y) \\
&= n(1-\pi_0) + (\pi_1+\pi_0-1)\left[n\left(\frac{\theta}{N}\right)\right] \\
&= n\left[\frac{(\pi_1+\pi_0-1)\theta}{N} + (1-\pi_0)\right].
\end{aligned}$$

And,

$$\begin{aligned}
\mathrm{V}(X) &= \mathrm{E}_y[\mathrm{V}(X|Y=y)] + \mathrm{V}_y[\mathrm{E}(X|Y=y)] \\
&= \mathrm{E}_y[\mathrm{V}(X_1|Y=y)+\mathrm{V}(X_0|Y=y)]+\mathrm{V}_y[\mathrm{E}(X_1|Y=y)+\mathrm{E}(X_0|Y=y)] \\
&= \mathrm{E}_y[y\pi_1(1-\pi_1) + (n-y)\pi_0(1-\pi_0)] + \mathrm{V}_y[y\pi_1 + (n-y)(1-\pi_0)] \\
&= \mathrm{E}_y\left\{n\pi_0(1-\pi_0) + [\pi_1(1-\pi_1) - \pi_0(1-\pi_0)]y\right\} + \mathrm{V}_y[n(1-\pi_0) + \\
&\quad (\pi_1+\pi_0-1)y] \\
&= n\pi_0(1-\pi_0) + [\pi_1(1-\pi_1) - \pi_0(1-\pi_0)]\left[n\left(\frac{\theta}{N}\right)\right] \\
&+ (\pi_1+\pi_0-1)^2\left[n\left(\frac{\theta}{N}\right)\left(1-\frac{\theta}{N}\right)\left(\frac{N-n}{N-1}\right)\right].
\end{aligned}$$

Solution 4.81.

(a) Clearly, $\mathrm{f}_Y(y) > 0, 0 < \alpha^{-1} < y < \infty$. And, with $w = (\alpha y - 1)$, so that $\mathrm{d}w = \alpha \mathrm{d}y$ and $y = (1+w)/\alpha$, we have

$$\begin{aligned}
\int_{\alpha^{-1}}^{\infty} \mathrm{f}_Y(y)\mathrm{d}y &= \int_{\alpha^{-1}}^{\infty} \frac{\left(y-\frac{1}{\alpha}\right)^{-\beta}\alpha^{-\beta}}{y\Gamma(\beta)\Gamma(1-\beta)}\mathrm{d}y \\
&= \int_0^{\infty} \frac{(w/\alpha)^{-\beta}\alpha^{-\beta}}{\left[\frac{(1+w)}{\alpha}\right]\Gamma(\beta)\Gamma(1-\beta)}\left(\frac{\mathrm{d}w}{\alpha}\right) \\
&= \int_0^{\infty} \frac{w^{-\beta}(1+w)^{-1}}{\Gamma(\beta)\Gamma(1-\beta)}\mathrm{d}w.
\end{aligned}$$

Now, with $v = w/(1+w)$, so that $\mathrm{d}w = (1-v)^{-2}\mathrm{d}v$, we obtain

$$\begin{aligned}
\int_{\alpha^{-1}}^{\infty} \mathrm{f}_Y(y)\mathrm{d}y &= \int_0^1 \frac{\left(\frac{v}{1-v}\right)^{-\beta}\left[(1-v)^{-1}\right]^{-1}}{\Gamma(\beta)\Gamma(1-\beta)}(1-v)^{-2}\mathrm{d}v \\
&= \int_0^1 \frac{v^{-\beta}(1-v)^{\beta-1}}{\Gamma(\beta)\Gamma(1-\beta)}\mathrm{d}v = \int_0^1 \mathrm{f}_V(v)\mathrm{d}v = 1,
\end{aligned}$$

since $\mathrm{f}_V(v)$ is a beta distribution.

(b) Now, with $u = x(y - \alpha^{-1})$, so that $y = \left(\frac{u}{x} + \frac{1}{\alpha}\right)$ and $\mathrm{d}y = \mathrm{d}u/x$, we have

$$\begin{aligned}
\mathrm{f}_X(x) &= \int_{\alpha^{-1}}^{\infty} \mathrm{f}_X(x|Y=y)\mathrm{f}_Y(y)\mathrm{d}y \\
&= \int_{\alpha^{-1}}^{\infty} \left[y\mathrm{e}^{-yx}\right] \frac{\left(y-\frac{1}{\alpha}\right)^{-\beta}\alpha^{-\beta}}{y\Gamma(\beta)\Gamma(1-\beta)}\mathrm{d}y \\
&= \int_0^{\infty} \left[\mathrm{e}^{-\left(\frac{u}{x}+\frac{1}{\alpha}\right)x}\right] \frac{(u/x)^{-\beta}\alpha^{-\beta}}{\Gamma(\beta)\Gamma(1-\beta)}\left(\frac{\mathrm{d}u}{x}\right) \\
&= \frac{x^{\beta-1}\mathrm{e}^{-x/\alpha}}{\Gamma(\beta)\alpha^{\beta}} \int_0^{\infty} \frac{u^{-\beta}\mathrm{e}^{-u}}{\Gamma(1-\beta)}\mathrm{d}u \\
&= \frac{x^{\beta-1}\mathrm{e}^{-x/\alpha}}{\Gamma(\beta)\alpha^{\beta}}, 0 < x < \infty,
\end{aligned}$$

since, with $\gamma = (1 - \beta)$,

$$\int_0^{\infty} \frac{u^{-\beta}\mathrm{e}^{-u}}{\Gamma(1-\beta)}\mathrm{d}u = \int_0^{\infty} \frac{u^{\gamma-1}\mathrm{e}^{-u}}{\Gamma(\gamma)}\mathrm{d}u = 1.$$

So, X has a gamma distribution.

Solution 4.83. First, we have

$$\begin{aligned}
\mathrm{cov}(U_n, U_m) &= \mathrm{cov}\left(\sum_{i=1}^{n} a_i X_i, \sum_{i=1}^{m} a_i X_i\right) \\
&= \mathrm{cov}\left(\sum_{i=1}^{m} a_i X_i + \sum_{i=m+1}^{n} a_i X_i, \sum_{i=1}^{m} a_i X_i\right) \\
&= \mathrm{cov}\left(\sum_{i=1}^{m} a_i X_i, \sum_{i=1}^{m} a_i X_i\right) \\
&= \mathrm{V}\left(\sum_{i=1}^{m} a_i X_i\right) \\
&= \mathrm{V}(U_m) = \sigma^2 \sum_{i=1}^{m} a_i^2.
\end{aligned}$$

So, we have

$$\begin{aligned}\operatorname{corr}(U_n, U_m) &= \frac{\operatorname{cov}(U_n, U_m)}{\sqrt{\operatorname{V}(U_n)\operatorname{V}(U_m)}} \\ &= \sqrt{\frac{\operatorname{V}(U_m)}{\operatorname{V}(U_n)}} \\ &= \sqrt{\frac{\sigma^2 \sum_{i=1}^{m} a_i^2}{\sigma^2 \sum_{i=1}^{n} a_i^2}} \\ &= \sqrt{\frac{\sum_{i=1}^{m} a_i^2}{\sum_{i=1}^{n} a_i^2}}.\end{aligned}$$

When $a_i = n^{-1}, i = 1, 2, \ldots, n$, we obtain

$$\begin{aligned}\operatorname{corr}(U_n, U_m) &= \operatorname{corr}(\bar{X}_n, \bar{X}_m) \\ &= \sqrt{\frac{m(n^{-2})}{n(n^{-2})}} \\ &= \sqrt{\frac{m}{n}}.\end{aligned}$$

Solution 4.85.

a) We have

$$\begin{aligned}\operatorname{pr}(U = 1) &= \operatorname{E}(U) = \operatorname{E}_t[\operatorname{E}(U|T = t)] \\ &= \operatorname{E}_t[\operatorname{pr}(U = 1|T = t)] = \operatorname{E}_t[\operatorname{pr}(C \geq T|T = t)] \\ &= \operatorname{E}_t[\operatorname{pr}(C \geq t)] = \int_0^{\infty} \operatorname{pr}(C \geq t)\operatorname{f}_T(t)\mathrm{d}t \\ &= \int_0^{1} \operatorname{pr}(C \geq t)\operatorname{f}_T(t)\mathrm{d}t,\end{aligned}$$

since $\int_1^{\infty} \operatorname{pr}(C \geq t)\operatorname{f}_T(t)\mathrm{d}t = 0$.

b) Since $\operatorname{pr}(C \geq t) = (1 - t), 0 < t < 1$, and using integration by parts, it follows that

$$\begin{aligned}
\text{pr}(U=1) &= \int_0^1 (1-t)\frac{1}{\alpha}e^{-t/\alpha}dt \\
&= \int_0^1 \frac{1}{\alpha}\text{e}^{-t/\alpha}\text{d}t - \int_0^1 \frac{t}{\alpha}\text{e}^{-t/\alpha}\text{d}t \\
&= \left[-\text{e}^{-t/\alpha}\right]_0^1 - \left\{\left[-t\text{e}^{-t/\alpha}\right]_0^1 - \int_0^1 \left(-\text{e}^{-t/\alpha}\right)\text{d}t\right\} \\
&= 1-\text{e}^{-1/\alpha}+\text{e}^{1/\alpha}+\left[\alpha\text{e}^{-t/\alpha}\right]_0^1 \\
&= 1-\alpha\left(1-\text{e}^{-1/\alpha}\right).
\end{aligned}$$

First, note that $\alpha = \text{E}(T)$ is the expected time to death for a patient receiving this experimental drug treatment. So, as anticipated, $\text{pr}(U = 1) \to 1$ as $\alpha \to 0$, and $\text{pr}(U = 1) \to 0$ as $\alpha \to \infty$.

Solution 4.87*.

(a) Let $\text{p}_k(t+\Delta t) = \text{pr}[X(0, t+\Delta t) = k]$ denote the probability of k hits during the time interval $(0, t+\Delta t)$. Note that the event $\{X(0, t+\Delta t) = k\}$ can occur in one of the following mutually exclusive ways:
$X(0, t) = k$ and $X(t, t+\Delta t) = 0$; or,
$X(0, t) = (k-1)$ and $X(t, t+\Delta t) = 1$; or,
$X(0, t) = (k-j)$ and $X(t, t+\Delta t) = j, j = 2, \ldots, k$.
Now, by conditions (2)–(4),

$$\begin{aligned}
\text{pr}\{[X(0,t)=k]\cap[X(t,t+\Delta t)=0]\} &= \text{pr}[X(0,t)=k]\text{pr}[X(t,t+\Delta t)=0] \\
&= \text{p}_k(t)(1-\lambda\Delta t).
\end{aligned}$$

Likewise,

$$\text{pr}\{[X(0,t)=(k-1)]\cap[X(t,t+\Delta t)=1]\} = \text{p}_{k-1}(t)(\lambda\Delta t).$$

And finally,

$$\text{pr}\{[X(0,t)=(k-j)]\cap[X(t,t+\Delta t)=j]\} = 0, j=2,3,\ldots,k.$$

Thus,

$$\text{p}_k(t+\Delta t) = \text{p}_k(t)(1-\lambda\Delta t)+\text{p}_{k-1}(t)(\lambda\Delta t),$$

so that

$$\frac{\text{p}_k(t+\Delta t)-\text{p}_k(t)}{\Delta t} = -\lambda[\text{p}_k(t)-\text{p}_{k-1}(t)].$$

Now, taking the limit as $\Delta t \to 0$, we obtain

$$\frac{\mathrm{dp}_k(t)}{\mathrm{d}t} = -\lambda[\mathrm{p}_k(t) - \mathrm{p}_{k-1}(t)].$$

Since $\mathrm{p}_{-1}(t) = 0$, it follows that

$$\frac{\mathrm{dp}_0(t)}{\mathrm{d}t} = -\lambda \mathrm{p}_0(t), \text{ giving } \frac{\mathrm{dlnp}_0(t)}{\mathrm{d}t} = -\lambda.$$

So,

$$\mathrm{lnp}_0(t) = -\lambda t + \mathrm{c}_0, \text{ giving } \mathrm{p}_0(t) = \mathrm{e}^{-\lambda t + \mathrm{c}_0}.$$

Since

$$\mathrm{p}_0(0) = 1 = \mathrm{e}^{-\lambda(0)+\mathrm{c}_0} = \mathrm{e}^{\mathrm{c}_0}, \text{ so that } \mathrm{c}_0 = 0,$$

it follows that

$$\mathrm{p}_0(t) = \mathrm{e}^{-\lambda t}, t > 0.$$

Note that $\mathrm{p}_0(t)$ is the probability that a Poisson random variable with mean $\mu = \lambda t$ takes the value 0.

When $k = 1$,

$$\begin{aligned}
\frac{\mathrm{dp}_1(t)}{\mathrm{d}t} &= -\lambda[\mathrm{p}_1(t) - \mathrm{p}_0(t)] \\
&= -\lambda \mathrm{p}_1(t) + \lambda \mathrm{e}^{-\lambda t} \\
\Rightarrow \quad \mathrm{e}^{\lambda t}\left[\frac{\mathrm{dp}_1(t)}{\mathrm{d}t} + \lambda \mathrm{p}_1(t)\right] &= \lambda \\
\Rightarrow \quad \frac{\mathrm{d}[\mathrm{e}^{\lambda t}\mathrm{p}_1(t)]}{\mathrm{d}t} &= \lambda \\
\Rightarrow \quad \mathrm{e}^{\lambda t}\mathrm{p}_1(t) &= \lambda t + \mathrm{c}_1 \\
\Rightarrow \quad \mathrm{p}_1(t) &= (\lambda t + \mathrm{c}_1)\mathrm{e}^{-\lambda t}.
\end{aligned}$$

Since $\mathrm{p}_1(0) = \mathrm{pr}[X(0,0) = 1] = 0 = \mathrm{c}_1$, it follows that

$$\mathrm{p}_1(t) = \lambda t \mathrm{e}^{-\lambda t}, t > 0.$$

Note that $\mathrm{p}_1(t)$ is the probability that a Poisson random variable with mean $\mu = \lambda t$ takes the value 1.

To obtain a general expression for $\mathrm{p}_k(t)$, we apply mathematical induction on k. Suppose,

$$\mathrm{p}_k(t) = \frac{(\lambda t)^k \mathrm{e}^{-\lambda t}}{k!},$$

which is the probability that a Poisson random variable with mean $\mu = \lambda t$ takes the value k. We wish to prove that

$$\mathrm{p}_{k+1}(t) = \frac{(\lambda t)^{k+1} \mathrm{e}^{-\lambda t}}{(k+1)!}.$$

Now,

$$
\begin{aligned}
\frac{\mathrm{dp}_{k+1}(t)}{\mathrm{d}t} &= -\lambda[\mathrm{p}_{k+1}(t) - \mathrm{p}_k(t)] \\
&= -\lambda \mathrm{p}_{k+1}(t) + \lambda\left[\frac{(\lambda t)^k \mathrm{e}^{-\lambda t}}{k!}\right] \\
\Rightarrow \quad \frac{\mathrm{d}[\exp(\lambda t)\mathrm{p}_{k+1}(t)]}{\mathrm{d}t} &= \lambda\frac{(\lambda t)^k}{k!} \\
\Rightarrow \quad \mathrm{e}^{\lambda t}\mathrm{p}_{k+1}(t) &= \lambda\left[\frac{(\lambda t)^{k+1}}{k!(k+1)\lambda}\right] + \mathrm{c}_2 \\
\Rightarrow \quad \mathrm{p}_{k+1}(t) &= \frac{(\lambda t)^{k+1}\mathrm{e}^{-\lambda t}}{(k+1)!} + \mathrm{c}_2\mathrm{e}^{-\lambda t}.
\end{aligned}
$$

Since $\mathrm{p}_{k+1}(0) = 0 = \mathrm{c}_2$, the proof by induction is completed.

Thus, for $k = 0, 1, \ldots, \infty$, $\mathrm{p}_k(t) = (\lambda t)^k \mathrm{e}^{-\lambda t}/k!$, and hence $X(0,t) \sim \mathrm{POI}(\lambda t), t > 0$. The sequence of random variables $\{X(0,t)\}, t > 0$, is said to form a *Poisson process* with *intensity parameter* λ. When λ is independent of t (as in this example), the sequence $\{X(0,t)\}$ forms a *homogeneous* or *stationary* Poisson process. More generally, $\lambda(t)$ may be a function of t, leading to a *non-homogeneous* or *time-dependent* Poisson process.

(b) Now, using a result in part (a), we have

$$
\begin{aligned}
\frac{\mathrm{dP}_X(s;t)}{\mathrm{d}t} &= \sum_{k=0}^{\infty} \frac{\mathrm{dp}_k(t)}{\mathrm{d}t} s^k \\
&= \sum_{k=0}^{\infty} -\lambda[\mathrm{p}_k(t) - \mathrm{p}_{k-1}(t)]s^k \\
&= -\lambda \sum_{k=0}^{\infty} \mathrm{p}_k(t)s^k + \lambda \sum_{k=1}^{\infty} \mathrm{p}_{k-1}(t)s^k \\
&= -\lambda \mathrm{e}^{-\lambda t} \sum_{k=0}^{\infty} \frac{(\lambda ts)^k}{k!} + \lambda \mathrm{e}^{-\lambda t} s \sum_{k=1}^{\infty} \frac{(\lambda ts)^{k-1}}{(k-1)!} \\
&= -\lambda \mathrm{e}^{\lambda t(s-1)} + \lambda s \mathrm{e}^{\lambda t(s-1)} = \lambda(s-1)\mathrm{e}^{\lambda t(s-1)}
\end{aligned}
$$

so that

$$
\mathrm{P}_X(s;t) = \mathrm{e}^{\lambda t(s-1)},
$$

which is the probability generating function of a Poisson random variable with mean λt.

Solution 4.89*.

$$\begin{aligned}
M_{U_n}(t) &= \mathrm{E}\left(\mathrm{e}^{tU_n}\right) = \mathrm{E}\left\{\mathrm{e}^{t\left[\frac{\sum_{i=1}^{n} X_i - n\theta}{\sqrt{2n}}\right]}\right\} \\
&= \mathrm{E}\left\{\mathrm{e}^{\frac{t}{\sqrt{2n}}\sum_{i=1}^{n} X_i}\mathrm{e}^{-t\theta\sqrt{\frac{n}{2}}}\right\} \\
&= \mathrm{e}^{-t\theta\sqrt{\frac{n}{2}}}\mathrm{E}\left\{\prod_{i=1}^{n}\mathrm{e}^{\frac{t}{\sqrt{2n}}X_i}\right\} \\
&= \mathrm{e}^{-t\theta\sqrt{\frac{n}{2}}}\prod_{i=1}^{n}\mathrm{E}\left(\mathrm{e}^{\frac{t}{\sqrt{2n}}X_i}\right) \\
&= \mathrm{e}^{-t\theta\sqrt{\frac{n}{2}}}\left[\mathrm{E}\left(\mathrm{e}^{t'X_i}\right)\right]^n,
\end{aligned}$$

where $t' = \frac{t}{\sqrt{2n}}$. Now,

$$\begin{aligned}
E\left(\mathrm{e}^{t'X_i}\right) &= \int_{-\infty}^{\infty}\mathrm{e}^{t'x_i}\frac{1}{2}\mathrm{e}^{-|x_i-\theta|}\mathrm{d}x_i \\
&= \frac{1}{2}\left\{\int_{-\infty}^{\theta}\mathrm{e}^{t'x_i}\mathrm{e}^{-(\theta-x_i)}\mathrm{d}x_i + \int_{\theta}^{\infty}\mathrm{e}^{t'x_i}\mathrm{e}^{-(x_i-\theta)}\mathrm{d}x_i\right\} \\
&= \frac{1}{2}\left\{\mathrm{e}^{-\theta}\int_{-\infty}^{\theta}\mathrm{e}^{(1+t')x_i}\mathrm{d}x_i + \mathrm{e}^{\theta}\int_{\theta}^{\infty}\mathrm{e}^{-(1-t')x_i}\mathrm{d}x_i\right\} \\
&= \frac{1}{2}\left\{\frac{\mathrm{e}^{-\theta}\mathrm{e}^{(1+t')\theta}}{(1+t')} + \frac{\mathrm{e}^{\theta}\mathrm{e}^{-(1-t')\theta}}{(1-t')}\right\} \\
&= \frac{\mathrm{e}^{t'\theta}}{2}\left[\frac{1}{(1+t')} + \frac{1}{(1-t')}\right] \\
&= \frac{\mathrm{e}^{t'\theta}}{[1-(t')^2]},\ |t'| < 1.
\end{aligned}$$

So, with $t' = \frac{t}{\sqrt{2n}}$, we have:

$$\begin{aligned}
M_{U_n}(t) &= \mathrm{e}^{-t\theta\sqrt{\frac{n}{2}}}\left[\frac{\mathrm{e}^{t\theta/\sqrt{2n}}}{(1-\frac{t^2}{2n})}\right]^n \\
&= \frac{\mathrm{e}^{-t\theta\sqrt{\frac{n}{2}}}\mathrm{e}^{t\theta\sqrt{\frac{n}{2}}}}{\left(1-\frac{t^2}{2n}\right)^n} \\
&= \left[\left(1-\frac{t^2}{2n}\right)^n\right]^{-1}.
\end{aligned}$$

Since

$$\lim_{n\to\infty} M_{U_n}(t) = \left[\lim_{n\to\infty}\left(1-\frac{t^2}{2n}\right)^n\right]^{-1} = \left(\mathrm{e}^{\frac{-t^2}{2}}\right)^{-1} = \mathrm{e}^{\frac{t^2}{2}},$$

it follows that U_n is asymptotically $N(0,1)$. Since

$$\mathrm{E}\left(\sum_{i=1}^{n} X_i\right) = \sum_{i=1}^{n} \mathrm{E}(X_i) = n\theta,$$

and since

$$\mathrm{V}\left(\sum_{i=1}^{n} X_i\right) = \sum_{i=1}^{n} \mathrm{V}(X_i) = 2n,$$

so that

$$U_n = \frac{\sum_{i=1}^{n} X_i - \mathrm{E}\left(\sum_{i=1}^{n} X_i\right)}{\sqrt{\mathrm{V}\left(\sum_{i=1}^{n} X_i\right)}} = \frac{\sum_{i=1}^{n} X_i - n\theta}{\sqrt{2n}},$$

the $N(0,1)$ asymptotic distribution for U_n follows from the Central Limit Theorem.

Solution 4.91*.

(a)

$$\begin{aligned}
\mathrm{pr}(Y \leq X) &= \mathrm{pr}[(Y \leq X) \cap (X=0)] + \mathrm{pr}[(Y \leq X) \cap (X=1)] \\
&= \mathrm{pr}[(Y \leq X)|X=0]\mathrm{pr}(X=0) + \mathrm{pr}[(Y \leq X)|X=1]\mathrm{pr}(X=1) \\
&= \mathrm{pr}[(Y=0)|X=0]\mathrm{pr}(X=0) + \mathrm{pr}[(Y \leq 1)|X=1]\mathrm{pr}(X=1) \\
&= e^{-\alpha}(1-\pi) + \left[e^{-(\alpha+\beta)} + (\alpha+\beta)e^{-(\alpha+\beta)}\right]\pi
\end{aligned}$$

(b) Note that

$$L = c\bar{X} + (1-c)\bar{Y} = \frac{1}{n}\sum_{i=1}^{n}[cX_i + (1-c)Y_i] = \frac{1}{n}\sum_{i=1}^{n} L_i,$$

where $L_i = cX_i + (1-c)Y_i$, $i = 1, 2, \ldots, n$. Since the $\{L_i\}_{i=1}^{n}$ are mutually independent and identically distributed, $\mathrm{V}(L) = \mathrm{V}(L_i)/n$, and so minimizing $\mathrm{V}(L_i)$ is equivalent to minimizing $\mathrm{V}(L)$. Now,

$$\begin{aligned}
\mathrm{V}(L_i) &= \mathrm{V}_{x_i}[\mathrm{E}(L_i|X_i = x_i)] + \mathrm{E}_{x_i}[\mathrm{V}(L_i|X_i = x_i)] \\
&= \mathrm{V}_{x_i}[cx_i + (1-c)\mathrm{E}(Y_i|X_i = x_i)] + \mathrm{E}_{x_i}[(1-c)^2(\alpha + \beta x_i)] \\
&= \mathrm{V}_{x_i}[cx_i + (1-c)(\alpha + \beta x_i)] + (1-c)^2(\alpha + \beta\pi) \\
&= \mathrm{V}_{x_i}\{(1-c)\alpha + [c + (1-c)\beta]x_i\} + (1-c)^2(\alpha + \beta\pi) \\
&= [c + (1-c)\beta]^2\pi(1-\pi) + (1-c)^2(\alpha + \beta\pi) \\
&= Q, \text{ say.}
\end{aligned}$$

So,

$$\begin{aligned}
\frac{\partial Q}{\partial c} &= 2[c+(1-c)\beta](1-\beta)\pi(1-\pi)+2(1-c)(-1)(\alpha+\beta\pi) \\
&= 2c[(1-\beta)\pi(1-\pi)-\beta(1-\beta)\pi(1-\pi)+(\alpha+\beta\pi)] \\
&\qquad +2[\beta(1-\beta)\pi(1-\pi)-(\alpha+\beta\pi)] \\
&= 2c[(1-\beta)^2\pi(1-\pi)+(\alpha+\beta\pi)]+2[\beta(1-\beta)\pi(1-\pi)-(\alpha+\beta\pi)].
\end{aligned}$$

The equation $\frac{\partial Q}{\partial c}=0$ implies that

$$c^* = \frac{(\alpha+\beta\pi)-\beta(1-\beta)\pi(1-\pi)}{(1-\beta)^2\pi(1-\pi)+(\alpha+\beta\pi)}.$$

Since

$$\frac{\partial^2 Q}{\partial c^2} = 2\left[(1-\beta)^2\pi(1-\pi)+(\alpha+\beta\pi)\right] > 0,$$

c^* minimizes $\mathrm{V}(L)$. When $\beta=0$, X and Y are independent, and so $\bar{X}$ and $\bar{Y}$ are independent. Also, when $\beta=0$,

$$c^* = \frac{\alpha}{\pi(1-\pi)+\alpha} = \frac{\alpha/n}{\frac{\pi(1-\pi)}{n}+\frac{\alpha}{n}} = \frac{\mathrm{V}(\bar{Y})}{\mathrm{V}(\bar{X})+\mathrm{V}(\bar{Y})}.$$

Thus, when $\beta=0$, c^* gives more weight to $\bar{X}$ when $\mathrm{V}(\bar{X})<\mathrm{V}(\bar{Y})$, equal weight to $\bar{X}$ and $\bar{Y}$ when $\mathrm{V}(\bar{X})=\mathrm{V}(\bar{Y})$, and less weight to $\bar{X}$ when $\mathrm{V}(\bar{X})>\mathrm{V}(\bar{Y})$, for the linear combination

$$L = c^*\bar{X}+(1-c^*)\bar{Y} = \left[\frac{\mathrm{V}(\bar{Y})}{\mathrm{V}(\bar{X})+\mathrm{V}(\bar{Y})}\right]\bar{X} + \left[\frac{V(\bar{X})}{V(\bar{X})+\mathrm{V}(\bar{Y})}\right]\bar{Y}.$$

(c)

$$\begin{aligned}
\mathrm{E}\left(\mathrm{e}^{tL}\right) &= \mathrm{E}\left\{\mathrm{e}^{t[c\bar{X}+(1-c)\bar{Y}]}\right\} \\
&= \mathrm{E}\left\{\mathrm{e}^{\frac{t}{n}\sum_{i=1}^{n}[cX_i+(1-c)Y_i]}\right\} \\
&= \mathrm{E}\left\{\prod_{i=1}^{n}\mathrm{e}^{\frac{t}{n}[cX_i+(1-c)Y_i]}\right\} \\
&= \prod_{i=1}^{n}\mathrm{E}\left\{\mathrm{e}^{\frac{t}{n}[cX_i+(1-c)Y_i]}\right\} \\
&= \left\{\mathrm{E}\left[\mathrm{e}^{\frac{t}{n}[cX_i+(1-c)Y_i]}\right]\right\}^n.
\end{aligned}$$

Now,

$$\begin{aligned}
\mathrm{E}\left\{\mathrm{e}^{\frac{t}{n}[cX_i+(1-c)Y_i]}\right\} &= \mathrm{E}_{x_i}\left\{\mathrm{E}\left[\mathrm{e}^{\frac{t}{n}[cX_i+(1-c)Y_i]}|X_i=x_i\right]\right\} \\
&= \mathrm{E}_{x_i}\left\{\mathrm{e}^{\frac{tc}{n}x_i}\mathrm{E}\left[\mathrm{e}^{\frac{t(1-c)Y_i}{n}}\middle| X_i=x_i\right]\right\} \\
&= \mathrm{E}_{x_i}\left\{\mathrm{e}^{\frac{tc}{n}x_i}\mathrm{e}^{(\alpha+\beta x_i)[\mathrm{e}^{\frac{t(1-c)}{n}}-1]}\right\} \\
&= \mathrm{e}^{\alpha[\mathrm{e}^{\frac{t(1-c)}{n}}-1]}\mathrm{E}\left\{\mathrm{e}^{kX_i}\right\},
\end{aligned}$$

where

$$k = \frac{tc}{n} + \beta\left[\mathrm{e}^{\frac{t(1-c)}{n}} - 1\right].$$

Thus,

$$\mathrm{E}\left(\mathrm{e}^{kX_i}\right) = (1-\pi) + \pi\mathrm{e}^{k}.$$

So,

$$\mathrm{E}\left(\mathrm{e}^{tL}\right) = \mathrm{e}^{n\alpha\left[\mathrm{e}^{\frac{t(1-c)}{n}}-1\right]}\left\{(1-\pi)+\pi\mathrm{e}^{\frac{tc}{n}+\beta\left[\mathrm{e}^{\frac{t(1-c)}{n}}-1\right]}\right\}^{n}.$$

As desired, note that $\mathrm{E}\left(\mathrm{e}^{tL}\right) = 1$ when $t = 0$, and that

$$\mathrm{E}\left(\mathrm{e}^{tL}\right) = \left\{(1-\pi)+\pi\mathrm{e}^{\frac{t}{n}}\right\}^{n}$$

when $c = 1$ (so that $L = \bar{X}$).

Solution 4.93*.

(a) Using Markov's Inequality, we have

$$\begin{aligned}
\mathrm{pr}\left[(|U_1|>k)\cup(|U_2|>k)\right] &= \mathrm{pr}\left[(U_1^2>k^2)\cup(U_2^2>k^2)\right] \\
&\leq \mathrm{pr}(U_1^2>k^2)+\mathrm{pr}(U_2^2>k^2) \\
&\leq \frac{\mathrm{E}(U_1^2)}{k^2}+\frac{\mathrm{E}(U_2^2)}{k^2} \\
&= \frac{1}{k^2}+\frac{1}{k^2}=\frac{2}{k^2}.
\end{aligned}$$

(b) Since

$$\max(U_1^2, U_2^2) = \frac{1}{2}\left[|U_1^2-U_2^2|+|U_1^2+U_2^2|\right],$$

it follows that

$$\begin{aligned}
\mathrm{E}\left[\max(U_1^2, U_2^2)\right] &= \frac{1}{2}\mathrm{E}\left(|U_1^2 - U_2^2|\right) + \frac{1}{2}\mathrm{E}\left(|U_1^2 + U_2^2|\right) \\
&= \frac{1}{2}\mathrm{E}\left(|U_1 - U_2||U_1 + U_2|\right) + \frac{1}{2}\mathrm{E}\left(U_1^2 + U_2^2\right) \\
&\leq \frac{1}{2}\sqrt{\mathrm{E}\left[(U_1 - U_2)^2\right]\mathrm{E}\left[(U_1 + U_2)^2\right]} + \frac{1}{2}(1+1),
\end{aligned}$$

by the Cauchy-Schwartz Inequality.

Finally, since

$$\mathrm{E}\left[(U_1 - U_2)^2\right] = \mathrm{E}(U_1^2) - 2\mathrm{E}(U_1U_2) + \mathrm{E}(U_2^2) = 1 - 2\rho + 1 = 2(1-\rho),$$

and since

$$\mathrm{E}\left[(U_1 + U_2)^2\right] = \mathrm{E}(U_1^2) + 2\mathrm{E}(U_1U_2) + \mathrm{E}(U_2^2) = 1 + 2\rho + 1 = 2(1+\rho),$$

we obtain

$$\mathrm{E}\left[\max(U_1^2, U_2^2)\right] \leq 1 + \frac{1}{2}\sqrt{[2(1-\rho)][2(1+\rho)]} = 1 + \sqrt{1-\rho^2}.$$

(c) Using the result in part (b) and Markov's Inequality, we have

$$\begin{aligned}
\mathrm{pr}\left[(|U_1| > k) \cup (|U_2| > k)\right] &= \mathrm{pr}\left[(U_1^2 > k^2) \cup (U_2^2 > k^2)\right] \\
&= \mathrm{pr}\left[\max(U_1^2, U_2^2) > k^2\right] \\
&\leq \frac{\mathrm{E}\left[\max(U_1^2, U_2^2)\right]}{k^2} \\
&\leq \frac{1 + \sqrt{1-\rho^2}}{k^2}.
\end{aligned}$$

(d) For $i = 1, 2$, it follows that $U_i = (Y_i - \mu_i)/\sigma_i$ has mean 0 and variance 1, and that $\mathrm{corr}(U_1, U_2) = \mathrm{corr}(Y_1, Y_2) = \rho$. Then, using the result in part (c), we have

$$\begin{aligned}
\mathrm{pr}\left[(|U_1| \leq k) \cap (|U_2| \leq k)\right] &= \mathrm{pr}\left[(U_1^2 \leq k^2) \cap (U_2^2 \leq k^2)\right] \\
&= \mathrm{pr}\left[\max(U_1^2, U_2^2) \leq k^2\right] \\
&= 1 - \mathrm{pr}\left[\max(U_1^2, U_2^2) > k^2\right] \\
&\geq 1 - \left[\frac{1 + \sqrt{1-\rho^2}}{k^2}\right].
\end{aligned}$$

Then, solving the equation

$$1 - \left[\frac{1 + \sqrt{1-\rho^2}}{k^2}\right] = \alpha$$

gives

$$k = \sqrt{\frac{1+\sqrt{1-\rho^2}}{(1-\alpha)}}.$$

As expected, k increases as α increases, and k decreases as $|\rho|$ increases.

Solution 4.95*.

(a) Since $Y_i = T_i + U_i, i = 1, 2, \ldots, k$, so that $\bar{Y} = \bar{T} + \bar{U}$, where $\bar{U} = k^{-1}\sum_{i=1}^{k} U_i$, it follows that

$$\text{cov}(\bar{Y}, \bar{T}) = \text{cov}(\bar{T} + \bar{U}, \bar{T}) = \text{cov}(\bar{T}, \bar{T}) + \text{cov}(\bar{U}, \bar{T}) = \text{V}(\bar{T}) + 0 = \text{V}(\bar{T}).$$

Thus,

$$\begin{aligned} \theta &= \left[\frac{\text{V}(\bar{T})}{\sqrt{\text{V}(\bar{T})\text{V}(\bar{Y})}}\right]^2 = \frac{\text{V}(\bar{T})}{\text{V}(\bar{Y})} \\ &= \frac{n^{-2}\text{V}\left(\sum_{i=1}^{k} T_i\right)}{n^{-2}\text{V}\left(\sum_{i=1}^{k} Y_i\right)} = \frac{\text{V}\left(\sum_{i=1}^{k} T_i\right)}{\text{V}\left(\sum_{i=1}^{k} Y_i\right)}. \end{aligned}$$

(b) Now, $\sum_{i=1}^{k-1}\sum_{j=i+1}^{k} \text{V}(T_j - T_i)$ can be written as

$$\begin{aligned} &= \sum_{i=1}^{k-1}\sum_{j=i+1}^{k} [\text{V}(T_j) + \text{V}(T_i) - 2\text{cov}(T_j, T_i)] \\ &= \sum_{i=1}^{k-1}\sum_{j=i+1}^{k} \text{V}(T_j) + \sum_{i=1}^{k-1}\sum_{j=i+1}^{k} \text{V}(T_i) - \sum_{\text{all } i\neq j} \text{cov}(T_i, T_j) \\ &= \sum_{i=1}^{k-1} [\text{V}(T_{i+1}) + \text{V}(T_{i+2}) + \cdots + \text{V}(T_k)] \\ &\quad + \sum_{i=1}^{k-1}(k-i)\text{V}(T_i) - \sum_{\text{all } i\neq j} \text{cov}(T_i, T_j) \\ &= \sum_{i=2}^{k}(i-1)\text{V}(T_i) + \sum_{i=1}^{k-1}(k-i)\text{V}(T_i) - \sum_{\text{all } i\neq j} \text{cov}(T_i, T_j) \\ &= \sum_{i=2}^{k-1}(i-1)\text{V}(T_i) + (k-1)\text{V}(T_k) + (k-1)\text{V}(T_1) \\ &\quad + \sum_{i=2}^{k-1}(k-i)\text{V}(T_i) - \sum_{\text{all } i\neq j} \text{cov}(T_i, T_j) \\ &= (k-1)\sum_{i=1}^{k} \text{V}(T_i) - \sum_{\text{all } i\neq j} \text{cov}(T_i, T_j) \geq 0, \end{aligned}$$

which gives the desired inequality.

(c) Now, using the inequality derived in part (b), we have

$$\begin{aligned} \mathrm{V}\left(\sum_{i=1}^{k} T_i\right) &= \sum_{i=1}^{k} \mathrm{V}(T_i) + \sum_{\text{all } i \neq j} \operatorname{cov}(T_i, T_j) \\ &\geq \frac{1}{(k-1)} \sum_{\text{all } i \neq j} \operatorname{cov}(T_i, T_j) + \sum_{\text{all } i \neq j} \operatorname{cov}(T_i, T_j) \\ &= \left(\frac{k}{k-1}\right) \sum_{\text{all } i \neq j} \operatorname{cov}(T_i, T_j). \end{aligned}$$

And, since $\operatorname{cov}(Y_i, Y_j) = \operatorname{cov}(T_i + U_i, T_j + U_j) = \operatorname{cov}(T_i, T_j)$, we have

$$\begin{aligned} \theta &= \frac{\mathrm{V}\left(\sum_{i=1}^{k} T_i\right)}{\mathrm{V}\left(\sum_{i=1}^{k} Y_i\right)} \geq \frac{\left(\frac{k}{k-1}\right) \sum_{\text{all } i \neq j} \operatorname{cov}(T_i, T_j)}{\mathrm{V}\left(\sum_{i=1}^{k} Y_i\right)} \\ &= \left(\frac{k}{k-1}\right) \left[\frac{\sum_{\text{all } i \neq j} \operatorname{cov}(Y_i, Y_j)}{\mathrm{V}\left(\sum_{i=1}^{k} Y_i\right)}\right] \\ &= \left(\frac{k}{k-1}\right) \left[\frac{\mathrm{V}\left(\sum_{i=1}^{k} Y_i\right) - \sum_{i=1}^{k} \mathrm{V}(Y_i)}{\mathrm{V}\left(\sum_{i=1}^{k} Y_i\right)}\right] \\ &= \left(\frac{k}{k-1}\right) \left[1 - \frac{\sum_{i=1}^{k} \mathrm{V}(Y_i)}{\mathrm{V}\left(\sum_{i=1}^{k} Y_i\right)}\right] = \alpha. \end{aligned}$$

(d) Now,

$$\begin{aligned} \alpha &= \left(\frac{k}{k-1}\right) \left[1 - \frac{\sum_{i=1}^{k} \mathrm{V}(Y_i)}{\mathrm{V}\left(\sum_{i=1}^{k} Y_i\right)}\right] \\ &= \left(\frac{k}{k-1}\right) \left[1 - \frac{\sum_{i=1}^{k} \mathrm{V}(Y_i)}{\sum_{i=1}^{k} \mathrm{V}(Y_i) + \sum_{\text{all } i \neq j} \operatorname{cov}(Y_i, Y_j)}\right] \\ &= \left(\frac{k}{k-1}\right) \left[\frac{\sum_{\text{all } i \neq j} \operatorname{cov}(Y_i, Y_j)}{\sum_{i=1}^{k} \mathrm{V}(Y_i) + \sum_{\text{all } i \neq j} \operatorname{cov}(Y_i, Y_j)}\right] \\ &= \frac{k\left[\sum_{\text{all } i \neq j} \operatorname{cov}(Y_i, Y_j)/k(k-1)\right]}{\left[\frac{\sum_{i=1}^{k} \mathrm{V}(Y_i)}{k} + (k-1)\frac{\sum_{\text{all } i \neq j} \operatorname{cov}(Y_i, Y_j)}{k(k-1)}\right]} \\ &= \frac{k\bar{C}}{\bar{V} + (k-1)\bar{C}}, \end{aligned}$$

where

$$\bar{C} = \frac{1}{k(k-1)} \sum_{\text{all } i \neq j} \text{cov}(Y_i, Y_j) \text{ and } \bar{V} = \frac{1}{k} \sum_{i=1}^{k} \text{V}(Y_i).$$

(e) Since $\text{cov}(Y_i, Y_j) = (\sigma_Y^2) \text{corr}(Y_i, Y_j)$ for $i \neq j$, we obtain

$$\begin{aligned} \alpha &= \frac{k\left[\frac{\sigma_Y^2}{k(k-1)} \sum_{\text{all } i \neq j} \text{corr}(Y_i, Y_j)\right]}{\sigma_Y^2 + (k-1)\left[\frac{\sigma_Y^2}{k(k-1)} \sum_{\text{all } i \neq j} \text{corr}(Y_i, Y_j)\right]} \\ &= \frac{\sum_{\text{all } i \neq j} \text{corr}(Y_i, Y_j)}{(k-1) + \left(\frac{k-1}{k}\right) \sum_{\text{all } i \neq j} \text{corr}(Y_i, Y_j)}. \end{aligned}$$

Thus, a sufficient condition for which $\alpha = 1$ is $\text{corr}(Y_i, Y_j) = 1$ for all $i \neq j$, since then

$$\alpha = \frac{k(k-1)}{(k-1) + \left(\frac{k-1}{k}\right)[k(k-1)]} = 1.$$

Solution 4.97*. First, with $\bar{X}^* = m^{-1} \sum_{i=1}^{m} X_i^*$, we have

$$\begin{aligned} U &= \bar{Y} + \hat{\beta}\bar{X}' - \hat{\beta}\bar{X} = \bar{Y} + (n+m)^{-1}\hat{\beta}(n\bar{X} + m\bar{X}^*) - \hat{\beta}\bar{X} \\ &= \bar{Y} + \left(\frac{m}{n+m}\right)\hat{\beta}\bar{X}^* + \left(\frac{n}{n+m}\right)\hat{\beta}\bar{X} - \hat{\beta}\bar{X} \\ &= \bar{Y} + \left(\frac{m}{n+m}\right)\hat{\beta}\bar{X}^* - \left(\frac{m}{n+m}\right)\hat{\beta}\bar{X}. \end{aligned}$$

Now, we know that $\text{E}(\bar{Y}) = \mu_y$, that $\text{E}(\bar{X}^*) = \mu_x$, and that $\hat{\beta}$ and $\bar{X}^*$ are independent random variables. So,

$$\text{E}(U) = \mu_y + \left(\frac{m}{n+m}\right)\mu_x \text{E}(\hat{\beta}) - \left(\frac{m}{n+m}\right)\text{E}(\hat{\beta}\bar{X}).$$

Thus, we need to determine $\text{E}(\hat{\beta})$ and $\text{E}(\hat{\beta}\bar{X})$.

Now, from the properties of the bivariate normal distribution, we know that

$$\text{E}(Y_i|X_i = x_i) = \alpha + \left(\rho\frac{\sigma_y}{\sigma_x}\right)x_i = \alpha + \beta x_i, i = 1, 2, \ldots, n.$$

Then, with $\boldsymbol{X} = (X_1, X_2, \ldots, X_n)$ and $\boldsymbol{x} = (x_1, x_2, \ldots, x_n)$, we have

$$\text{E}(\hat{\beta}) = \text{E}[\text{E}(\hat{\beta}|\boldsymbol{X} = \boldsymbol{x})],$$

where

$$\begin{aligned}
\mathrm{E}(\hat{\beta}|\boldsymbol{X}=\boldsymbol{x}) &= \frac{\sum_{i=1}^{n}(x_i-\bar{x})\mathrm{E}(Y_i|\boldsymbol{X}=\boldsymbol{x})}{\sum_{i=1}^{n}(x_i-\bar{x})^2} = \frac{\sum_{i=1}^{n}(x_i-\bar{x})\left[\alpha+\left(\rho\frac{\sigma_y}{\sigma_x}\right)x_i\right]}{\sum_{i=1}^{n}(x_i-\bar{x})^2} \\
&= \frac{\alpha\sum_{i=1}^{n}(x_i-\bar{x})}{\sum_{i=1}^{n}(x_i-\bar{x})^2} + \frac{\left(\rho\frac{\sigma_y}{\sigma_x}\right)\sum_{i=1}^{n}(x_i-\bar{x})x_i}{\sum_{i=1}^{n}(x_i-\bar{x})^2} = \rho\frac{\sigma_y}{\sigma_x},
\end{aligned}$$

so that

$$\mathrm{E}(\hat{\beta}) = \rho\frac{\sigma_y}{\sigma_x}.$$

And,

$$\begin{aligned}
\mathrm{E}(\hat{\beta}\bar{X}) &= \mathrm{E}[\mathrm{E}(\hat{\beta}\bar{X}|\boldsymbol{X}=\boldsymbol{x})] = \mathrm{E}[\bar{x}\mathrm{E}(\hat{\beta}|\boldsymbol{X}=\boldsymbol{x})] \\
&= \mathrm{E}\left(\rho\frac{\sigma_y}{\sigma_x}\bar{x}\right) = \rho\frac{\sigma_y}{\sigma_x}\mu_x.
\end{aligned}$$

Finally,

$$\mathrm{E}(U) = \mu_y + \left(\frac{m}{n+m}\right)\mu_x\left(\rho\frac{\sigma_y}{\sigma_x}\right) - \left(\frac{m}{n+m}\right)\left(\rho\frac{\sigma_y}{\sigma_x}\mu_x\right) = \mu_y.$$

Solution 4.99*.

(a) Let E_i be the event that subject i is a terrorist. Then,

$$\begin{aligned}
\mathrm{E}(T) &= \sum_{i=1}^{n}\mathrm{pr}(\mathrm{E}_i)\mathrm{E}(Y_i|\mathrm{E}_i) \\
&= \sum_{i=1}^{n}\pi_i\left(\frac{1}{\theta_i}\right) = \sum_{i=1}^{n}\frac{\pi_i}{\theta_i}.
\end{aligned}$$

Clearly, $\mathrm{E}(T)$ is minimized by choosing $\theta_i = 1, i = 1, 2, \ldots, n$. In other words, $\mathrm{E}(T)$ is minimized by selecting every single one of the n subjects for secondary screening, which would guarantee that any terrorist would be identified the first time that he or she attempted to pass through a checkpoint. However, such a screening strategy is clearly impractical due to logistical constraints.

(b) Consider the function

$$Q = \sum_{i=1}^{n}\frac{\pi_i}{\theta_i} + \lambda\left(\sum_{i=1}^{n}\theta_i - K\right).$$

Then,

$$\frac{\partial Q}{\partial \theta_i} = \frac{-\pi_i}{\theta_i^2} + \lambda = 0 \text{ gives } \theta_i = \frac{\sqrt{\pi_i}}{\sqrt{\lambda}}, i = 1, 2, \ldots, n.$$

Also,

$$\frac{\partial Q}{\partial \theta_i} = 0 \text{ gives } \lambda = \frac{\pi_i}{\theta_i^2};$$

so, since $\partial Q/\partial\lambda = 0$ gives $\sum_{i=1}^{n} \theta_i = K$, we have

$$\sum_{i=1}^{n} \lambda\theta_i = \lambda \sum_{i=1}^{n} \theta_i = \lambda K = \sum_{i=1}^{n} \frac{\pi_i}{\theta_i}.$$

Thus,

$$\lambda = \frac{1}{K} \sum_{i=1}^{n} \frac{\pi_i}{\theta_i} = \frac{1}{K} \sum_{i=1}^{n} \frac{\pi_i}{\left(\sqrt{\pi_i}/\sqrt{\lambda}\right)},$$

so that

$$\sqrt{\lambda} = \frac{\sum_{i=1}^{n} \sqrt{\pi_i}}{K}.$$

Finally, we obtain

$$\theta_i = \frac{\sqrt{\pi_i}}{\sqrt{\lambda}} = \frac{K\sqrt{\pi_i}}{\sum_{l=1}^{n} \sqrt{\pi_l}}, i = 1, 2, \ldots, n.$$

For these optimal values of $\theta_1, \theta_2, \ldots, \theta_n$, the minimized value of $\text{E}(T)$ is

$$\begin{aligned}\sum_{i=1}^{n} \frac{\pi_i}{\theta_i} &= \sum_{i=1}^{n} \frac{\pi_i}{\left(K\sqrt{\pi_i}/\sum_{l=1}^{n}\sqrt{\pi_l}\right)} \\ &= \frac{\left(\sum_{i=1}^{n}\sqrt{\pi_i}\right)^2}{K}.\end{aligned}$$

Note that $\sqrt{\pi_i}$ instead of π_i appears in these optimal expressions for θ_i and for the minimized value of $E(T)$. This is known as *square root sampling*, and it has been studied and utilized in other scientific settings.

Solution 4.101*.

(a) Let T be the event that a tumor is present, and let γ denote the probability of a detection error. Then,

$$\begin{aligned}
\gamma &= \text{pr}[(V \leq D) \cap \text{T}] + \text{pr}[(V > D) \cap \bar{\text{T}}] \\
&= \text{pr}(\text{T})\text{pr}(V \leq D|\text{T}) + \text{pr}(\bar{\text{T}})\text{pr}(V > D|\bar{\text{T}}] \\
&= (\theta)\text{pr}(U \leq \text{ln}D|\text{T}) + (1-\theta)\text{pr}(U > \text{ln}D|\bar{\text{T}}] \\
&= \theta \int_{-\infty}^{\text{ln}D} \frac{1}{\sqrt{2\pi}\sigma}\text{e}^{-(u-\mu_1)^2/2\sigma^2}du + (1-\theta)\int_{\text{ln}D}^{\infty} \frac{1}{\sqrt{2\pi}\sigma}\text{e}^{-(u-\mu_0)^2/2\sigma^2}du \\
&= \theta \int_{-\infty}^{(\text{ln}D-\mu_1)/\sigma} \frac{1}{\sqrt{2\pi}}\text{e}^{-z^2/2}\text{d}z + (1-\theta)\int_{(\text{ln}D-\mu_0)/\sigma}^{\infty} \frac{1}{\sqrt{2\pi}}\text{e}^{-z^2/2}\text{d}z \\
&= \theta\text{F}_Z\left[\frac{(\text{ln}D-\mu_1)}{\sigma}\right] + (1-\theta)\left\{1 - \text{F}_Z\left[\frac{(\text{ln}D-\mu_0)}{\sigma}\right]\right\},
\end{aligned}$$

where $\text{F}_Z(z) = \text{pr}(Z \leq z)$ when $Z \sim \text{N}(0,1)$.
Now,

$$\frac{\partial\gamma}{\partial(\text{ln}D)} = \left(\frac{\theta}{\sigma}\right)\frac{1}{\sqrt{2\pi}}\text{e}^{-\frac{1}{2}\left(\frac{\text{ln}D-\mu_1}{\sigma}\right)^2} - \left[\frac{(1-\theta)}{\sigma}\right]\frac{1}{\sqrt{2\pi}}\text{e}^{-\frac{1}{2}\left(\frac{\text{ln}D-\mu_0}{\sigma}\right)^2} = 0,$$

which gives

$$\text{ln}\theta - \frac{1}{2}\left(\frac{\text{ln}D-\mu_1}{\sigma}\right)^2 = \text{ln}(1-\theta) - \frac{1}{2}\left(\frac{\text{ln}D-\mu_0}{\sigma}\right)^2,$$

or

$$\text{ln}\theta - \frac{(\text{ln}D)^2 - 2\mu_1\text{ln}D + \mu_1^2)}{2\sigma^2} - \text{ln}(1-\theta) + \frac{(\text{ln}D)^2 - 2\mu_0\text{ln}D + \mu_0^2)}{2\sigma^2} = 0.$$

So, solving the above expression for $\text{ln}D$ gives

$$\begin{aligned}
\text{ln}D &= \frac{(\mu_1^2 - \mu_0^2) - 2\sigma^2\text{ln}\left(\frac{\theta}{1-\theta}\right)}{2(\mu_1 - \mu_0)} \\
&= \frac{(\mu_0 + \mu_1)}{2} + \left(\frac{\sigma^2}{\mu_1 - \mu_0}\right)\text{ln}\left(\frac{1-\theta}{\theta}\right).
\end{aligned}$$

Thus, the optimal choice D^* that minimizes γ is equal to

$$D^* = \left(\frac{1-\theta}{\theta}\right)^{\left(\frac{\sigma^2}{\mu_1-\mu_0}\right)}\text{e}^{\left(\frac{\mu_0+\mu_1}{2}\right)}.$$

(b) As $\theta \to 1$, so that a tissue sample has probability 1 of containing a tumor, then $D^* \to 0$, so that all tissue samples will be correctly classified as containing a tumor. And, as $\theta \to 0$, so that a tissue sample has probability 0 of containing a tumor, then $D^* \to +\infty$, so that all tissue samples will be correctly classified as being tumor-free.

Solution 4.103*. Clearly,

$$\mathrm{V}(S|\mathrm{RA}) = \mathrm{V}(X_1|\mathrm{RA}) + \mathrm{V}(X_2|\mathrm{RA}) = 1 + 1 = 2.$$

Now, if ξ denotes the median of the density function for X, we have

$$\int_0^{\xi} \mathrm{e}^{-x}\mathrm{d}x = 1 - \mathrm{e}^{-\xi} = \frac{1}{2},$$

so that $\xi = \ln 2$.

So, under SA, the appropriate truncated distributions for X_1 and X_2 are

$\mathrm{f}_{X_1}(x_1|x_1 \leq \xi) = 2\mathrm{e}^{-x_1}, 0 < x_1 < \xi$ and $\mathrm{f}_{X_2}(x_2|x_2 > \xi) = 2\mathrm{e}^{-x_2}, \xi < x_2 < \infty$.

Thus, using integration by parts, it follows directly that

$$\mathrm{E}(X_1|X_1 \leq \xi) = \int_0^{\xi} (x_1)2\mathrm{e}^{-x_1}\mathrm{d}x_1 = (1 - \ln 2).$$

And, using integration by parts with $u = x_1^2$ and $\mathrm{d}v = \mathrm{e}^{-x_1}\mathrm{d}x_1$, we have

$$\begin{aligned}
\mathrm{E}(X_1^2|X_1 \leq \xi) &= \int_0^{\xi} (x_1^2)2\mathrm{e}^{-x_1}\mathrm{d}x_1 \\
&= 2\left[-x_1^2\mathrm{e}^{-x_1}\right]_0^{\xi} + 4\int_0^{\xi} x_1\mathrm{e}^{-x_1}\mathrm{d}x_1 \\
&= -2\xi^2\mathrm{e}^{-\xi} + 2\mathrm{E}(X_1|X_1 \leq \xi) \\
&= -(\ln 2)^2 + 2(1 - \ln 2),
\end{aligned}$$

so that

$$\mathrm{V}(X_1|X_1 \leq \xi) = -(\ln 2)^2 + 2(1 - \ln 2) - (1 - \ln 2)^2 = 1 - 2(\ln 2)^2.$$

Now, since $\mathrm{f}_{X_2}(x_2|x_2 > \xi) = 2\mathrm{e}^{-x_2}, \xi < x_2 < \infty$, consider the transformation $Y = (X_2 - \xi)$, so that $\mathrm{d}Y = \mathrm{d}X_2$. Then, it follows that

$$\mathrm{f}_Y(y) = 2\mathrm{e}^{-(y+\xi)} = \mathrm{e}^{-y}, 0 < y < \infty, \text{ so that } \mathrm{V}(Y) = 1 = \mathrm{V}(X_2|X_2 > \xi).$$

Finally,

$$\mathrm{V}(S|\mathrm{SA}) = \mathrm{V}(X_1|\mathrm{SA}) + \mathrm{V}(X_2|\mathrm{SA}) = \left[1 - 2(\ln 2)^2\right] + 1 = 2 - 2(\ln 2)^2.$$

Hence,

$$\mathrm{V}(S|\mathrm{RA}) - \mathrm{V}(S|\mathrm{SA}) = 2 - \left[2 - 2(\ln 2)^2\right] = 2(\ln 2)^2 = 0.961,$$

so that stratified assembly produces a lower variance for S than does random assembly.

Solution 4.105*. For $i = 2, 3, \ldots, n$, let $X_i = 1$ if the individual wins both the $(i-1)$-th game and the i-th game, so that $Y_n = \sum_{i=2}^{n} X_i$. Now,

$$\begin{aligned} \mathrm{E}(X_i) &= \mathrm{pr}(X_i = 1) \\ &= \mathrm{pr}[\text{individual wins game } (i-1)] \times \mathrm{pr}[\text{individual wins game } i] \\ &= \pi^2, \end{aligned}$$

so that $\mathrm{E}(Y_n) = \sum_{i=2}^{n} \mathrm{E}(X_i) = (n-1)\pi^2$.

Also, $\mathrm{V}(X_i) = \mathrm{E}(X_i^2) - [\mathrm{E}(X_i)]^2 = \pi^2 - (\pi^2)^2 = \pi^2(1-\pi^2)$.

Now, for $|i - i'| \geq 2$, it follows that $\mathrm{cov}(X_i, X_{i'}) = 0$ since X_i and $X_{i'}$ have no games in common. And, for $i = 2, 3, \ldots, n-1$, we have

$$\begin{aligned} \mathrm{cov}(X_i, X_{i+1}) &= \mathrm{E}(X_i X_{i+1}) - \mathrm{E}(X_i)\mathrm{E}(X_{i+1}) \\ &= \mathrm{pr}\left[(X_i = 1) \cap (X_{i+1} = 1)\right] - (\pi^2)(\pi^2) \\ &= \mathrm{pr}\left[\text{individual wins games } (i-1), i, \text{ and } (i+1)\right] - \pi^4 \\ &= \pi^3 - \pi^4 = \pi^3(1-\pi). \end{aligned}$$

So, we have

$$\begin{aligned} \mathrm{V}(Y_n) &= \sum_{i=2}^{n} \mathrm{V}(X_i) + 2\sum_{i=2}^{n-1} \sum_{i'=i+1}^{n} \mathrm{cov}(X_i, X_{i'}) \\ &= (n-1)\pi^2(1-\pi^2) + 2\sum_{i=2}^{n-1} \mathrm{cov}(X_i, X_{i+1}) \\ &= (n-1)\pi^2(1-\pi^2) + 2(n-2)\pi^3(1-\pi) \\ &= \pi^2(1-\pi)\left[(n-1) + (3n-5)\pi\right]. \end{aligned}$$

As a simple check on the validity of these general formulas for $\mathrm{E}(Y_n)$ and $\mathrm{V}(Y_n)$ when $n = 2$, we obtain

$$\mathrm{E}(Y_2) = \mathrm{E}(X_2) = (2-1)\pi^2 = \pi^2,$$

and

$$\mathrm{V}(Y_2) = \mathrm{V}(X_2) = \pi^2(1-\pi)\left[(2-1) + (6-5)\pi\right] = \pi^2(1-\pi^2),$$

which are the desired results.

Solution 4.107*.

(a) We have

$$\begin{aligned} \mathrm{pr}(T > t) &= 1 - \mathrm{pr}(T \leq t) = 1 - \mathrm{F}_T(t) \\ &= \mathrm{pr}(N = 0) = \mathrm{e}^{-\lambda t} \text{ since } N \sim \mathrm{POI}(\lambda t). \end{aligned}$$

Hence, $F_T(t) = 1 - e^{-\lambda t}$, so that

$$f_T(t) = \frac{dF_T(t)}{dt} = \lambda e^{-\lambda t}, t > 0.$$

(b) It follows easily that

$$\begin{aligned} pr(T > t + s | T > t) &= \frac{pr(T > t + s)}{pr(T > t)} = \frac{e^{-\lambda(t+s)}}{e^{-\lambda t}} \\ &= e^{-\lambda s} = pr(T > s). \end{aligned}$$

(c) Using integration by parts with $u = t$ and $dv = \lambda e^{-\lambda t}dt$, we have

$$\begin{aligned} E(T|T \leq t^*) &= \int_0^{t^*} (t) \frac{\lambda e^{-\lambda t}}{F_T(t^*)} dt = (1 - e^{-\lambda t^*})^{-1} \int_0^{t^*} (t) \lambda e^{-\lambda t} dt \\ &= (1 - e^{-\lambda t^*})^{-1} \left\{ \left[-te^{-\lambda t} \right]_0^{t^*} + \int_0^{t^*} e^{-\lambda t} dt \right\} \\ &= (1 - e^{-\lambda t^*})^{-1} \left\{ -t^* e^{-\lambda t^*} + \left[-\lambda^{-1} e^{-\lambda t} \right]_0^{t^*} \right\} \\ &= (1 - e^{-\lambda t^*})^{-1} \left\{ -t^* e^{-\lambda t^*} + \left[\lambda^{-1} - \lambda^{-1} e^{-\lambda t^*} \right] \right\} \\ &= \frac{1}{\lambda} - \frac{t^*}{(e^{\lambda t^*} - 1)}. \end{aligned}$$

(d) Let $X(\geq 1)$ be the number of time intervals required until the first time interval longer than t^* occurs. Then, given the memoryless property, $X \sim$ GEOM(π), where $\pi = pr(T > t^*) = e^{-\lambda t^*}$; so, $E(X) = \pi^{-1} = e^{\lambda t^*}$.
Now, let $W = \sum_{i=1}^{X} T_{i-1}$, where $T_i(0 < T_i \leq t^*)$ is the length of the i-th time interval and where $T_0 \equiv 0$. Thus, if $X = 1$, then $W = 0$; if $X = 2$, then $W = T_1$; if $X = 3$, then $W = (T_1 + T_2)$, etc.
So, it follows that

$$\begin{aligned} E(W) &= E_x \left[E(W|X = x) \right] = E_x \left[E\left(\sum_{i=1}^{x} T_{i-1} \right) \right] \\ &= E_x \left[\sum_{i=2}^{x} E\left(T_{i-1}\right) \right] = E(T|T \leq t^*)E(X - 1) \\ &= \left[\frac{1}{\lambda} - \frac{t^*}{(e^{\lambda t^*} - 1)} \right] \left(e^{\lambda t^*} - 1 \right) \\ &= \frac{(e^{\lambda t^*} - \lambda t^* - 1)}{\lambda}. \end{aligned}$$

When $\lambda = 10$ cars per minute and when $t^* = 0.25$ minutes, then

$$E(W) = \frac{\left[e^{10(0.25)} - 10(0.25) - 1\right]}{10} = 0.8682 \text{ minutes},$$

so that the average waiting time is about 52 seconds.

Solution 4.109*. First, note that

$$\mathrm{E}(X+Y|Z=z) = z = \mathrm{E}(X|Z=z) + \mathrm{E}(Y|Z=z),$$

so that we need only to find one of these two conditional expectations, say, $\mathrm{E}(X|Z=z)$.

Now, we have

$$\begin{aligned}
\mathrm{p}_Z(z) &= \mathrm{pr}(Z=z) = \mathrm{pr}(X+Y=z) = \sum_{x=0}^{z} \mathrm{pr}\left[(X=x)\cap(Y=z-x)\right] \\
&= \sum_{x=0}^{z} \mathrm{pr}(X=x)\mathrm{pr}(Y=z-x) = \sum_{x=0}^{z} \left[\alpha(1-\alpha)^x\right]\left[\beta(1-\beta)^{z-x}\right] \\
&= \alpha\beta(1-\beta)^z \sum_{x=0}^{z} \left(\frac{1-\alpha}{1-\beta}\right)^x \\
&= \alpha\beta(1-\beta)^z \left[\frac{1-\left(\frac{1-\alpha}{1-\beta}\right)^{z+1}}{1-\left(\frac{1-\alpha}{1-\beta}\right)}\right] \\
&= \frac{\alpha\beta}{(\alpha-\beta)}\left[(1-\beta)^{z+1}-(1-\alpha)^{z+1}\right], z=0,1,\ldots,\infty.
\end{aligned}$$

Thus, $\mathrm{p}_X(x|Z=z)$, the conditional distribution of X given $Z=z$, has the structure

$$\begin{aligned}
\mathrm{p}_X(x|Z=z) &= \frac{\mathrm{pr}\left[(X=x)\cap(Z=z)\right]}{\mathrm{pr}(Z=z)} \\
&= \frac{\mathrm{pr}(X=x)\mathrm{pr}(Y=z-x)}{\mathrm{pr}(Z=z)} \\
&= \frac{[\alpha(1-\alpha)^x][\beta(1-\beta)^{z-x}]}{\frac{\alpha\beta}{(\alpha-\beta)}\left[(1-\beta)^{z+1}-(1-\alpha)^{z+1}\right]} \\
&= k\left(\frac{1-\alpha}{1-\beta}\right)^x, x=0,1,\ldots,z,
\end{aligned}$$

where

$$k = \frac{(\alpha-\beta)(1-\beta)^z}{\left[(1-\beta)^{z+1}-(1-\alpha)^{z+1}\right]}.$$

A tractable way to develop an explicit expression for $\mathrm{E}(X|Z=z)$ is to first find $\mathrm{E}(\mathrm{e}^{tX}|Z=z)$, the moment generating function (MGF) of X given $Z=z$. Then,

$$\mathrm{E}(X|Z=z) = \left[\frac{\partial \mathrm{E}(\mathrm{e}^{tX}|Z=z)}{\partial t}\right]_{|t=0}.$$

So,

$$\begin{aligned}
\mathrm{E}(\mathrm{e}^{tX}|Z=z) &= \sum_{x=0}^{z} \mathrm{e}^{tx}\mathrm{p}_X(x|Z=z)\\
&= k\sum_{x=0}^{z} \mathrm{e}^{tx}\mathrm{e}^{x\ln\left(\frac{1-\alpha}{1-\beta}\right)}\\
&= k\sum_{x=0}^{z} \mathrm{e}^{\left[t+\ln\left(\frac{1-\alpha}{1-\beta}\right)\right]x}\\
&= k\left\{\frac{1-\mathrm{e}^{\left[t+\ln\left(\frac{1-\alpha}{1-\beta}\right)\right](z+1)}}{1-\mathrm{e}^{\left[t+\ln\left(\frac{1-\alpha}{1-\beta}\right)\right]}}\right\}.
\end{aligned}$$

Now, $\partial\mathrm{E}(\mathrm{e}^{tX}|Z=z)/\partial t$ is equal to

$$k\left\{\frac{-\mathrm{e}^{\left[t+\ln\left(\frac{1-\alpha}{1-\beta}\right)\right](z+1)}(z+1)\left(1-\mathrm{e}^{\left[t+\ln\left(\frac{1-\alpha}{1-\beta}\right)\right]}\right)-\left(1-\mathrm{e}^{\left[t+\ln\left(\frac{1-\alpha}{1-\beta}\right)\right](z+1)}\right)\left(-\mathrm{e}^{\left[t+\ln\left(\frac{1-\alpha}{1-\beta}\right)\right]}\right)}{\left(1-\mathrm{e}^{\left[t+\ln\left(\frac{1-\alpha}{1-\beta}\right)\right]}\right)^2}\right\}.$$

Then, setting $t=0$, using the expression for k, and doing some algebraic manipulations, we finally obtain

$$\mathrm{E}(X|Z=z) = \frac{(1-\beta)^{z+2}}{(\alpha-\beta)}\left[\frac{z\left(\frac{1-\alpha}{1-\beta}\right)^{z+2}-(z+1)\left(\frac{1-\alpha}{1-\beta}\right)^{z+1}+\left(\frac{1-\alpha}{1-\beta}\right)}{(1-\beta)^{z+1}-(1-\alpha)^{z+1}}\right],$$

so that $\mathrm{E}(Y|Z=z)=z-\mathrm{E}(X|Z=z)$.

When $z=2, \alpha=0.30$, and $\beta=0.60$, $\mathrm{E}(X|Z=z)=1.3543$ and $\mathrm{E}(Y|Z=z)$ $=2-1.3543=0.6457$.

Chapter 5

Estimation Theory

5.1 Exercises

Exercise 5.1. The sulfur dioxide (SO_2) concentration Y (in parts per million, or ppm) in a certain city is postulated to have the gamma-type density function

$$\mathrm{f}_Y(y;\sigma) = (2\pi\sigma^2 y)^{-1/2}\mathrm{e}^{-y/2\sigma^2}, 0 < y < +\infty, \sigma > 0.$$

Let $Y_1, Y_2, \ldots, Y_n$ constitute a random sample of size n from $\mathrm{f}_Y(y;\sigma)$, so that $y_1, y_2, \ldots, y_n$ are the corresponding realizations (or observed particulate concentrations).

Use the Central Limit Theorem to derive an appropriate $100(1-\alpha)\%$ large-sample confidence interval for the parameter σ. If $n = 50$ and $\bar{y} = n^{-1}\sum_{i=1}^{n} y_i$ $= 10$, use the derived confidence interval to compute an appropriate 95% confidence interval for σ.

Exercise 5.2. Let $X_1, X_2, \ldots, X_n$ constitute a random sample of size n from the parent population

$$\mathrm{f}_X(x;\theta) = \mathrm{e}^{-(x-\theta)}, 0 < \theta < x < +\infty.$$

(a) Prove rigorously that $X_{(1)}$, the smallest order statistic, is a sufficient statistic for θ.

(b) Prove that $X_{(1)}$ is a consistent estimator of θ.

Exercise 5.3. Let $Y_1, Y_2, \ldots, Y_n$ constitute a random sample of size $n\,(\geq 2)$ from a $\mathrm{N}(\mu, \sigma^2)$ population.

(a) Find an *explicit expression* for k such that the estimator $\hat{\sigma} = kS$ is an unbiased estimator of σ.

(b) Find the Cramér-Rao (C-R) lower bound for the variance of any unbiased estimator of σ. Ascertain whether $\hat{\sigma}$ is the minimum variance bound unbiased estimator (MVBUE) of σ.

Exercise 5.4. An occupational hygienist postulates that the concentration Y (in parts per million) of a certain airborn pollutant in a particular industrial setting varies according to the density function

$$\mathrm{f}_Y(y;\theta) = \theta^k [\Gamma(k)]^{-1} y^{(k-1)} \mathrm{e}^{-\theta y}, y > 0, \theta > 0, k > 0,$$

where k is a known positive constant and where θ is an unknown parameter. Given that there is available a data set consisting of a random sample $Y_1, Y_2, \ldots, Y_n$ of size n from $\mathrm{f}_Y(y;\theta)$, the goal is to use the information in this data set to help this occupational hygienist make appropriate statistical inferences about the unknown parameter θ.

(a) Prove that the statistic $\overline{Y} = n^{-1}\sum_{i=1}^{n} Y_i$ is a sufficient statistic for the parameter θ.

(b) Construct a function $\hat{\theta}$ of $\overline{Y}$ that is an unbiased estimator of θ. Is $\hat{\theta}$ a consistent estimator of θ?

(c) If $k = 2, n = 50$, and the observed (or realized) value of $\overline{Y}$ is $\overline{y} = 3$, compute what you believe to be an appropriate 95% confidence interval (CI) for the unknown parameter θ.

Exercise 5.5. Let $X_1, X_2, \ldots, X_{n_1}$ constitute a random sample of size n_1 from a $\mathrm{N}(\mu_1, \sigma_1^2)$ parent population, and let $Y_1, Y_2, \ldots, Y_{n_2}$ constitute a random sample of size n_2 from a $\mathrm{N}(\mu_2, \sigma_2^2)$ parent population. Assume that μ_1 and μ_2 are *unknown* population parameters, and that σ_1^2 and σ_2^2 are *known* population parameters.

(a) Using basic principles, develop an exact $100(1-\alpha)\%$ confidence interval $(0 < \alpha \leq 0.10)$ for the unknown parameter $\theta = (3\mu_1 - 5\mu_2)$ that is a function of the sample means $\bar{X} = n_1^{-1}\sum_{i=1}^{n_1} X_i$ and $\bar{Y} = n_2^{-1}\sum_{i=1}^{n_2} Y_i$.

(b) Suppose that an epidemiologist can only afford to select a *total* sample size $n = (n_1+n_2)$ equal to 100 in order to make statistical inferences about the unknown parameter θ. If $\alpha = 0.05, \sigma_1^2 = 2$, and $\sigma_2^2 = 3$, find specific numerical values for n_1 and n_2, subject to the constraint $(n_1+n_2) = 100$, that *minimize* the width of the confidence interval developed in part (a) of this problem.

Exercise 5.6. Let X_1 and X_2 constitute a random sample of size $n = 2$ from the discrete distribution

$$\mathrm{p}_X(x;\theta) = \mathrm{C}_x^2 \theta^x (1-\theta)^{2-x}, \quad x = 0, 1, 2 \text{ and } 0 < \theta < 1.$$

(a) Prove that the probability distribution of $\mathrm{p}_Y(y;\theta)$ of $Y = \max(X_1, X_2)$ is

y	$\mathrm{p}_Y(y;\theta)$
0	$(1-\theta)^4$
1	$4\theta(1-\theta)^2$
2	$\theta^2(2-\theta^2)$

(b) In a random sample of size n from $\mathrm{p}_Y(y;\theta)$, suppose we observe n_0 zeros and n_1 ones, where $(n_0 + n_1) = n$. Express the maximum likelihood estimator (MLE) of θ as an explicit function of n_0 and n_1.

Exercise 5.7. A toxicologist postulates that the time Y (in seconds) to respiratory distress for rats after intravenous exposure to a certain potentially toxic chemical follows the uniform density

$$\mathrm{f}_Y(y;\theta) = \theta^{-1}, 0 < y < \theta < \infty,$$

where θ is an unknown parameter. This toxicologist wants help in finding the minimum variance unbiased estimators (MVUEs) of both $\mu = \mathrm{E}(Y)$ and $\sigma^2 = \mathrm{V}(Y)$ using the information contained in a random sample $Y_1, Y_2, \ldots, Y_n$ of size $n(> 1)$ from $\mathrm{f}_Y(y;\theta)$. Develop explicit expressions for $\hat{\mu}$ and $\hat{\sigma}^2$, the MVUEs of μ and σ^2.

Exercise 5.8. Let $X_1, X_2, \ldots, X_n$ constitute a random sample of size n from the parent population

$$\mathrm{f}_X(x;\theta) = (2\theta)^{-1}\mathrm{e}^{-|x|/\theta}, \quad -\infty < x < +\infty,\ \theta > 0.$$

(a) Derive an *explicit expression* for the maximum likelihood estimator (MLE) $\hat{\theta}$ of θ, and then find $\mathrm{E}(\hat{\theta})$ and $\mathrm{V}(\hat{\theta})$.

(b) Find an *explicit expression* for the Cramér-Rao (C-R) lower bound for the variance of any unbiased estimator of θ. Is the MLE $\hat{\theta}$ the minimum-variance bound unbiased estimator (MVBUE) of θ [i.e., does $\mathrm{V}(\hat{\theta})$ equal the C-R bound]?

Exercise 5.9. For a certain rubber manufacturing process, the random variable Y_x (the amount in kilograms manufactured per day) has mean $\mathrm{E}(Y_x) = \alpha x + \beta x^2$ and *known* variance $\mathrm{V}(Y_x) = \sigma^2$, where x is the known amount of raw material in kilograms used per day in the manufacturing process. The n mutually independent data pairs $(x, Y_x), x = 1, 2, \ldots, n$, are available to estimate the unknown parameters of interest. Thus, $Y_1, Y_2, \ldots, Y_n$ are a set of n mutually independent random variables. Further, let $S_k = \sum_{x=1}^{n} x^k, k = 1, 2, 3 \ldots$; again, note that the S_ks are non-stochastic quantities with known values.

(a) Derive explicit expressions for the unweighted least squares estimators $\hat{\alpha}$ and $\hat{\beta}$ of the unknown parameters α and β.

(b) Derive explicit expressions for $\mathrm{E}(\hat{\beta})$ and $\mathrm{V}(\hat{\beta})$, the mean and variance of $\hat{\beta}$.

(c) If $Y_x \sim \mathrm{N}(\alpha x + \beta x^2, \sigma^2), x = 1, 2, \ldots, n$, with the Y_xs being mutually independent random variables, compute an exact 95% confidence interval for β if $n = 4, \hat{\beta} = 2$, and $\sigma^2 = 1$.

Exercise 5.10. Let $Y_1, Y_2, \ldots, Y_n$ constitute a random sample of size n from the parent population

$$\mathrm{p}_Y(y; \pi) = \mathrm{C}_y^2 \pi^y (1-\pi)^{2-y}, y = 0, 1, 2; 0 < \pi < 1.$$

It is of interest to find the "best" estimator $\hat{\theta}$ of the parameter $\theta = \pi^2$ under the restriction that $\mathrm{E}(\hat{\theta}) = \theta$ for any finite value of n (≥ 1).

(a) Find an explicit expression for the Cramér-Rao lower bound for the variance of any unbiased estimator of θ.

(b) Develop an explicit expression for what you consider to be the "best" unbiased estimator $\hat{\theta}$ of the parameter $\theta = \pi^2$.

(c) Find the asymptotic efficiency of the estimator $\hat{\theta}$ found in part (b) relative to the Cramèr-Rao lower bound found in part (a).

Exercise 5.11. Two continuous random variables have the joint density function

$$\mathrm{f}_{X,Y}(x, y; \theta) = \mathrm{e}^{-\left(\theta x + \theta^{-1} y\right)}, x > 0, y > 0, \theta > 0.$$

Let (X_i, Y_i), $i = 1, 2, \ldots, n$, constitute a random sample of size n from $\mathrm{f}_{X,Y}(x, y; \theta)$. Consider estimating θ using

$$\hat{\theta} = \left(\bar{Y}/\bar{X}\right)^{1/2},$$

where $\bar{X} = n^{-1} \sum_{i=1}^n X_i$ and $\bar{Y} = n^{-1} \sum_{i=1}^n Y_i$.

Find an *explicit expression* for $\mathrm{E}(\hat{\theta})$. Is $\hat{\theta}$ an unbiased estimator of θ? Is $\hat{\theta}$ an asymptotically unbiased estimator of θ?

Exercise 5.12. Let $Y_1, Y_2, \ldots, Y_n$ constitute a random sample of size n from the continuous parent population

$$\mathrm{f}_Y(y; \theta) = (1 + \theta y)/2, \quad -1 < y < +1; \ -1 < \theta < +1.$$

Using the random sample $Y_1, Y_2, \ldots, Y_n$ $(n > 1)$, find that *unbiased estimator*

$\hat{\theta}$ of θ that is a linear function of $\bar{Y} = n^{-1}\sum_{i=1}^{n} Y_i$. Then, *prove* that no unbiased estimator of θ has a smaller variance than $\hat{\theta}$ when $\theta = 0$. More generally, when $\theta \neq 0$, would you use $\hat{\theta}$ to estimate θ? Why or why not? Do you notice any undesirable properties of $\hat{\theta}$ as an estimator for θ?

Exercise 5.13. In various laboratory-based chemical research investigations carried out over extended periods of time, experimenters often obtain accurate information concerning the *population coefficient of variation* (namely, $\theta = \sigma/\mu$ = population standard deviation/population mean) of a quantitative characteristic under study. Since random variation is a typical feature of most chemical systems, knowledge of the value of σ/μ can be used to obtain an asymptotically unbiased estimator of an important population parameter (e.g., the population mean μ), this estimator having smaller variance than the best linear unbiased estimator (BLUE) of that parameter.

In particular, suppose that a certain characteristic Y (such as the amount of product produced via a chemical reaction) is under investigation, where $\mathrm{E}(Y) = \mu$ and $\mathrm{V}(Y) = \sigma^2$ are unknown parameters. Let $Y_1, Y_2, \ldots, Y_n$ constitute a random sample of n observations from the underlying density $\mathrm{f}_Y(y; \mu, \sigma^2)$ of the continuous random variable Y, where $\mathrm{f}_Y(y; \mu, \sigma^2)$ is of unspecified structure.

Consider the following two estimators of μ:

$$\bar{Y} = n^{-1}\sum_{i=1}^{n} Y_i, \text{ the BLUE of } \mu;$$

$$T = C\sum_{i=1}^{n} Y_i, \text{ where } C \text{ is a constant to be determined.}$$

(a) *Prove rigorously* that the value of C, which minimizes the mean squared error (MSE) of T as an estimator of μ, namely the quantity $\mathrm{E}[(T-\mu)^2]$, is equal to

$$C^* = (n + \theta^2)^{-1}.$$

(b) Show that $T^* = C^* \sum_{i=1}^{n} Y_i$ has a smaller variance than $\bar{Y}$.

(c) Define the *efficiency* of T^* relative to $\bar{Y}$ to be the ratio of their mean squared errors, namely,

$$\frac{\mathrm{MSE}(\bar{Y}, \mu)}{\mathrm{MSE}(T^*, \mu)}.$$

Find a general expression for this ratio. Then, find the limit of this ratio as $n \to \infty$ to determine the *asymptotic efficiency* of T^* relative to $\bar{Y}$.

(d) Are T^* and $\bar{Y}$ MSE-consistent estimators of the parameter μ?

Exercise 5.14. For the estimation of the unknown true mean radius γ (> 0) of circular-shaped land masses on a distant planet viewed through the Hubble telescope, it is reasonable to assume that the error E in estimating γ has a normal distribution with expected value of 0 and *known variance* σ^2. More specifically, if R_i is the i-th of n measurements of γ, then the model relating R_i to γ is assumed to be of the form

$$R_i = \gamma + E_i, \quad i = 1, 2, \ldots, n,$$

where $E_i \sim \text{N}(0, \sigma^2)$ and where $E_1, E_2, \ldots, E_n$ are mutually independent random variables.

It is of interest to use the available data to make statistical inferences about the true average area $\alpha = \pi\gamma^2$ of such circular-shaped land masses.

If $n = 100$, $\hat{\alpha} = 5$, and $\sigma^2 = 2$, develop an appropriate large-sample 95% confidence interval for the unknown parameter α.

Exercise 5.15. The performance rating Y (scaled so that $0 < Y < 1$) of individual subjects on a certain manual dexterity task is assumed to follow the beta density function

$$\text{f}_Y(y; \theta) = (\theta + 1)(\theta + 2)y^{\theta}(1 - y), 0 < y < 1,$$

where θ (> 0) is an unknown parameter.

(a) Suppose that a random sample of $n = 50$ subjects from $\text{f}_Y(y; \theta)$ supplies data $y_1, y_2, \ldots, y_{50}$, producing the value $\hat{\theta} = 3.00$ for the maximum likelihood estimate (MLE) of θ. In other words, if $Y_1, Y_2, \ldots Y_{50}$ constitute a random sample of size $n = 50$ from $\text{f}_Y(y; \theta)$, then $y_1, y_2, \ldots y_{50}$ are the corresponding observed values (or "realizations") of the random variables $Y_1, Y_2, \ldots, Y_{50}$. Use the available information to construct an appropriate large-sample 95% confidence interval for the unknown parameter θ.

(b) In planning for the future sampling of subjects, suppose that it is desired to find the minimum sample size n^* such that the width of a 95% confidence interval for θ will be no wider than 1.00 in value. Using the available information given in part (a), provide a reasonable value for n^*.

Exercise 5.16. For residents in a certain city in the United States, suppose that it is reasonable to assume that the distribution of the *proportion* X of a certain antibody in a cubic centimeter of blood taken from a randomly chosen resident follows the density function

$$\text{f}_X(x; \theta) = \theta x^{\theta - 1}, \quad 0 < x < 1, \ \theta > 0.$$

Let $X_1, X_2, \ldots, X_n$ constitute a random sample of size $n\,(> 2)$ from $\mathrm{f}_X(x;\theta)$.

Develop an *explicit expression* for the *maximum likelihood estimator* (MLE) $\hat{\theta}$ of θ, and then use this result to develop an *explicit expression* for an unbiased estimator $\tilde{\theta}$ of θ that is a function of a sufficient statistic for θ. Does $\tilde{\theta}$ achieve the Cramér-Rao (C-R) lower bound for the variance of any unbiased estimator of θ? Does $\tilde{\theta}$ achieve this lower bound asymptotically?

Exercise 5.17. Let $Y_1, Y_2, \ldots, Y_n$ constitute a set of mutually independent random variables with $\mathrm{E}(Y_i) = \theta$ and $\mathrm{V}(Y_i) = \sigma^2$ for $i = 1, 2, \ldots, n$. Let

$$\hat{\theta} = \sum_{i=1}^{n} c_i Y_i$$

be an estimator of θ, where $c_1, c_2, \ldots, c_n$ are constants independent of the $\{Y_i\}_{i=1}^{n}$.

Find choices for $c_1, c_2, \ldots, c_n$ so that $\hat{\theta}$ has the smallest variance among all unbiased estimators of θ.

Exercise 5.18. Let $Y_1, Y_2, \ldots, Y_n$ constitute a random sample of size n from the parent population

$$\mathrm{f}_Y(y;\alpha) = \alpha \mathrm{e}^{-\alpha y},\ y > 0, \alpha > 0.$$

For a particular sample of size $n = 100$, suppose that $\bar{y} = n^{-1}\sum_{i=1}^{n} y_i = 4.00$. Compute an appropriate large-sample 95% confidence interval for the unknown parameter $\theta_k = \mathrm{E}(Y^k) = \alpha^{-k}\Gamma(k+1)$ when $k = 2$.

Exercise 5.19. Let $X_1, X_2, \ldots, X_n$ constitute a random sample of size $n\,(> 3)$ from $\mathrm{f}_X(x)$, a density function of *unknown* functional form, with $\mathrm{E}(X) = \mu$ and $\mathrm{V}(X) = \sigma^2$. With $S^2 = (n-1)^{-1}\sum_{i=1}^{n}(X_i - \bar{X})^2$, where $\bar{X} = n^{-1}\sum_{i=1}^{n} X_i$, recall that $\mathrm{E}(S^2) = \sigma^2$ and that $\mathrm{V}(S^2) = n^{-1}\left[\mu_4 - \left(\frac{n-3}{n-1}\right)\sigma^4\right]$, where $\mu_4 = \mathrm{E}[(X-\mu)^4]$.

Consider using cS^2 to estimate σ^2, where c is a positive constant to be specified. Derive an explicit expression for the value of c, say c^*, such that the mean-squared error (MSE) of cS^2 as an estimator of σ^2 is *minimized.* What can you say about the choices of c^* when $\mu_4 < 3\sigma^4$ and when $\mu_4 > 3\sigma^4$?

Exercise 5.20. For $i = 1, 2, 3$, let $X_{i1}, X_{i2}, \ldots, X_{in_i}$ constitute a random sample of size n_i from an $\mathrm{N}(\mu_i, 1)$ population, and let $\bar{X}_i = n_i^{-1}\sum_{j=1}^{n_i} X_{ij}$. Observations from different populations can be assumed to be independent of one another.

Using $\bar{X}_1, \bar{X}_2$, and $\bar{X}_3$, develop a general formula for an *exact* $100(1-\alpha)\%$ confidence interval for the parameter $\theta = (2\mu_1 - 3\mu_2 + \mu_3)$. Then, use your formula to calculate a 95% confidence interval for θ based on the following data: $n_1 = n_2 = n_3 = 25$, $\bar{x}_1 = 10$, $\bar{x}_2 = 5$, and $\bar{x}_3 = 8$.

Exercise 5.21. Let $X_1, X_2, \ldots, X_n$ constitute a random sample of size n from a normal distribution with *known* mean μ_0 and *unknown* variance σ^2. Derive an explicit expression for the minimum variance unbiased estimator (MVUE) $\hat{\theta}$ of $\theta = \sigma^4$. For a particular set of data, if $n = 4$, $\mu_0 = 2$, $\sum_{i=1}^{4} x_i = 9$, and $\sum_{i=1}^{4} x_i^2 = 25$, determine the numerical value of $\hat{\theta}$, compute an *exact* 98% confidence interval for θ, and then comment on your findings.

Exercise 5.22. Suppose that the conditional distribution of Y_i given that $X_i = x_i$ is *normal* with conditional mean $\mathrm{E}(Y_i|X_i = x_i) = \theta x_i$ and conditional variance $\mathrm{V}(Y_i|X_i = x_i) = \theta^2 x_i^2$, where $\theta\,(> 0)$ is an unknown parameter. It is desired to use the n *mutually independent* pairs (x_i, Y_i), $i = 1, 2, \ldots, n$, to estimate θ. Here, for the i-th person, $x_i\,(> 0)$ is a person's reported level of strenuous physical activity (as measured by responses to a questionnaire) and $Y_i\,(> 0)$ is the concentration (in parts per million) of a certain exercise-producing chemical in a cubic centimeter of that person's blood.

(a) Show that the estimator

$$\hat{\theta} = \sum_{i=1}^{n} x_i Y_i \Big/ \sum_{i=1}^{n} x_i^2$$

minimizes (with respect to θ) the "sum of squares"

$$Q = \sum_{i=1}^{n} [Y_i - \mathrm{E}(Y_i|X_i = x_i)]^2.$$

(b) Given that $X_i = x_i, i = 1, 2, \ldots, n$, find the *exact* distribution of the estimator $\hat{\theta}$. Then, use this result to find random variables L and U such that $\mathrm{pr}(L < \theta < U) = (1-\alpha), 0 < \alpha \leq 0.20$. [Note: Assume that $\mathrm{pr}(\hat{\theta} \leq 0) = 0$.]

Exercise 5.23. Let $X_1, X_2, \ldots, X_n$ constitute a random sample of size n from the Bernoulli parent population $\mathrm{p}_X(x;\pi) = \pi^x(1-\pi)^{1-x}$, $x = 0, 1$ and $0 < \pi < 1$. Let $\theta = \pi^k$, where k is a known constant satisfying $1 \leq k \leq n$. Using the Rao-Blackwell Theorem, develop an explicit expression for the minimum variance unbiased estimator (MVUE) $\hat{\theta}$ of the unknown parameter θ. Further, by finding $\mathrm{E}(\hat{\theta})$ directly, demonstrate that $\hat{\theta}$ is an unbiased estimator of θ.

Exercise 5.24. Suppose $X_1, X_2, \ldots, X_n$ constitute a random sample from a $\mathrm{N}(\mu_0, \sigma^2)$ parent population, where μ_0 is a *known constant* and where σ^2 is an *unknown* parameter. Using all n observations, develop an *exact* $100(1-\alpha)\%$ confidence interval for the population standard deviation σ. If $n=5$, $x_1=1$, $x_2=2$, $x_3=3$, $x_4=4$, $x_5=5$, and $\mu_0=2$, compute an exact 95% confidence interval for σ.

Exercise 5.25. For $i=1,2$, let $X_{i1}, X_{i2}, \ldots, X_{in_i}$ constitute a sample of size n_i from a parent population having unknown mean μ and known variance σ_i^2. With $\overline{X}_i = n_i^{-1}\sum_{j=1}^{n_i} X_{ij}, i=1,2$, a biostatistician suggests that an "optimal" estimator of the unknown parameter μ is $\hat{\mu} = (w_1\overline{X}_1 + w_2\overline{X}_2)$, where the "weights" w_1 and w_2 are chosen so that the following two conditions are simultaneously satisfied: (1) $\mathrm{E}(\hat{\mu}) = \mu$ (namely, $\hat{\mu}$ is an unbiased estimator of μ); and (2) $\mathrm{V}(\hat{\mu})$, the variance of the estimator $\hat{\mu}$, is a minimum. Derive explicit expressions for the particular choices of w_1 and w_2 that simultaneously satisfy these two conditions, and then interpret your findings.

Exercise 5.26. The distance X (in centiMorgans) between mutations along a DNA strand can often be reasonably approximated by the negative exponential density function

$$\mathrm{f}_X(x;\theta) = \theta \mathrm{e}^{-\theta x}, \ x>0, \ \theta>0.$$

Suppose that $x_1, x_2, \ldots, x_n$ constitute $n(>2)$ realizations (i.e., observed distances) based on a random sample $X_1, X_2, \ldots, X_n$ from $\mathrm{f}_X(x;\theta)$.

(a) Develop an explicit expression for the minimum variance unbiased estimator (MVUE) $\hat{\theta}$ of θ.

(b) Derive an explicit expression for the *efficiency* $\mathrm{EFF}(\hat{\theta}, \theta)$ of the MVUE $\hat{\theta}$ of θ relative to the Cramér-Rao lower bound (CRLB) for the variance of any unbiased estimator of θ. What is the value of the asymptotic efficiency of $\hat{\theta}$?

Exercise 5.27. Let $X_1, X_2, \ldots, X_n$ constitute a random sample of size n from some population with unknown mean μ and unknown variance σ^2.

(a) Show that there exists a set of values for $k, 0<k<1$, such that the estimator $k\bar{X}$ has smaller mean-squared error (MSE) as an estimator of μ than does the sample mean $\bar{X} = n^{-1}\sum_{i=1}^{n} X_i$.

(b) Are there any possible drawbacks associated with the use of $k\bar{X}$ as an estimator of μ?

Exercise 5.28. Let $X_1, X_2, \ldots, X_n$ constitute a random sample of size $n(>1)$

from an $\mathrm{N}(\mu, \sigma^2)$ population, and let

$$\bar{X} = n^{-1} \sum_{i=1}^{n} X_i \text{ and } S^2 = (n-1)^{-1} \sum_{i=1}^{n} (X_i - \bar{X})^2.$$

It is of interest to estimate the unknown parameter $\theta = \mu^2$. Biostatistician #1 suggests using the *biased* estimator $\hat{\theta}_1 = (\bar{X})^2$. Biostatistician #2 suggests using an *unbiased* estimator $\hat{\theta}_2$ of θ that is a function of *both* $\bar{X}$ and S^2. Compare the mean-squared errors of these two estimators of θ, and describe under what circumstances you would select one of these estimators over the other using mean-squared error as the sole selection criterion.

Exercise 5.29. The lognormal density function is a popular and appropriate choice for describing the distribution of the concentration X of an air pollutant, either in an occupational environment or in the ambient atmosphere [e.g., see Rappaport and Kupper (2008)].

Suppose that the random variable $Y = \ln X \sim \mathrm{N}(\mu, \sigma^2)$, so that X has a lognormal distribution. Suppose that $X_1, X_2, \ldots, X_n$ constitute a random sample from the lognormal distribution for X, and suppose that it is of interest to consider the maximum likelihood estimator (MLE) $\hat{\xi}$ of $\xi = \mathrm{E}(X)$, the true mean concentration of the pollutant X.

(a) Develop an explicit expression for the MLE $\hat{\xi}$ of the parameter ξ.

(b) Develop explicit expressions for $\mathrm{E}(\hat{\xi})$ and for $\lim_{n \to \infty} \mathrm{E}(\hat{\xi})$, and then comment on your findings.

Exercise 5.30. The concentration X (in parts per million or ppm) of styrene in the air in a certain styrene manufacturing plant has a lognormal distribution; in particular, $Y = \ln X \sim \mathrm{N}(\mu, \sigma^2)$. Suppose that $x_1, x_2, \ldots, x_n$ represent n measurements of the airborne styrene concentration in this plant; these n measured concentration values can be considered to be realized values of a random sample $X_1, X_2, \ldots, X_n$ of size n from the lognormal distribution for X. With $y_i = \ln x_i, i = 1, 2, \ldots, n$, suppose that $n = 30, \bar{y} = n^{-1} \sum_{i=1}^{n} y_i = 3.00$ and $s_y^2 = (n-1)^{-1} \sum_{i=1}^{n} (y_i - \bar{y})^2 = 2.50$. Using these data, construct a maximum likelihood (ML)-based large-sample 95% confidence interval for $\mathrm{E}(X)$, the true mean concentration of airborne styrene in this styrene manufacturing plant.

Exercise 5.31. For $i = 1, 2$, let $X_{i1}, X_{i2}, \ldots, X_{in}$ constitute a random sample of size $n (\geq 2)$ from a $\mathrm{N}(\mu, \sigma^2)$ population, and let $\bar{X}_i = n^{-1} \sum_{j=1}^{n} X_{ij}$.

(a) With $S_1^2 = (n-1)^{-1} \sum_{j=1}^{n} (X_{1j} - \bar{X}_1)^2$, show that the probability θ_n that

$\bar{X}_2$ falls in the interval $\left(\bar{X}_1 - t_{n-1,1-\frac{\alpha}{2}} \frac{S_1}{\sqrt{n}}, \bar{X}_1 + t_{n-1,1-\frac{\alpha}{2}} \frac{S_1}{\sqrt{n}}\right)$ involves the CDF of a random variable having a t-distribution with $(n-1)$ degrees of freedom.

(b) Find the limiting value of θ_n as $n \to \infty$. If $\alpha = 0.05$, what is the numerical value of this limiting value of θ_n?

Exercise 5.32. In a very large population of N people, suppose that N_i of these people all belong to exactly one of k mutually exclusive and exhaustive categories, so that $\sum_{i=1}^{k} N_i = N$. For $i = 1, 2, \ldots, k$, let $\pi_i = N_i/N$ denote the proportion of people in this population who belong to category i, so that $\sum_{i=1}^{k} \pi_i = 1$.

The values of $N_1, N_2, \ldots, N_k$ are unknown, and it is of interest to find an appropriate 95% confidence interval for the unknown parameter $\theta_{ij} = (\pi_i - \pi_j), i \neq j$, based on a random sample of size n from this population.

In what follows, assume that the sampling fraction n/N is small. Thus, with x_i denoting the observed number of people in the sample who belong to category i, so that $\sum_{i=1}^{k} x_i = n$, it is a reasonable strategy to assume that the joint distribution of the random variables $X_1, X_2, \ldots, X_k$ (with respective realizations $x_1, x_2, \ldots, x_k$) is $\text{MULT}(n; \pi_1, \pi_2 \ldots, \pi_k)$.

(a) Let the estimator of θ_{ij} be

$$\hat{\theta}_{ij} = \hat{\pi}_i - \hat{\pi}_j = \frac{X_i}{n} - \frac{X_j}{n}.$$

Find explicit expressions for $\text{E}(\hat{\theta}_{ij})$ and $\text{V}(\hat{\theta}_{ij})$.

(b) If $n = 100, k = 3, x_1 = 50, x_2 = 20$, and $x_3 = 30$, compute an appropriate large-sample 95% confidence interval for the unknown parameter $\theta_{12} = (\pi_1 - \pi_2)$ using the estimator $\hat{\theta}_{12}$.

Exercise 5.33. In Bayesian inference, model parameters are treated as random variables and assigned *prior distributions* that quantify uncertainty about their values prior to collecting data. These prior distributions are then updated via Bayes' Theorem to obtain *posterior distributions* given the observed data. For example, if θ denotes a model parameter and y is an observed realization of a random variable Y, then the posterior distribution of θ, given that $Y = y$, is obtained from the following application of Bayes' Theorem:

$$\pi(\theta|Y = y) = \frac{\text{f}_Y(y|\theta)\pi(\theta)}{\int_{\Theta} \text{f}_Y(y|\theta)\pi(\theta)\text{d}\theta},$$

where $\pi(\theta)$ is the prior distribution of θ and where Θ denotes the parameter space (i.e., the domain of θ).

Note that $\pi(\theta|Y = y) = [\mathrm{h}(y)]^{-1}\mathrm{f}_Y(y|\theta)\pi(\theta)$, where the "normalizing constant" $\mathrm{h}(y) = \int_\Theta \mathrm{f}_Y(y|\theta)\pi(\theta)\mathrm{d}\theta$ can often be determined indirectly to satisfy the requirement $\int_\Theta \pi(\theta|Y = y)\mathrm{d}\theta = 1$.

Suppose that the random variable $Y \sim \mathrm{N}(\mu, \sigma^2)$, and that interest lies in making Bayesian-type inferences about the unknown parameters μ and σ^2. In particular, consider the following prior distributions for μ and σ^2:

$$\begin{aligned}\pi(\mu) &= \mathrm{N}(\mu_0, \sigma_0^2), \ -\infty < \mu_0 < \infty, \ \sigma_0^2 > 0 \\ \pi(\sigma^2) &= \mathrm{IG}(a, b), a > 0, b > 0\end{aligned}$$

where $\mathrm{IG}(a, b)$ denotes the *inverse-gamma* (IG) distribution with shape parameter a and scale parameter b; that is,

$$\pi(\sigma^2) = \frac{b^a(\sigma^2)^{-a-1}}{\Gamma(a)}\mathrm{e}^{-b/\sigma^2}, 0 < \sigma^2 < \infty.$$

(a) Assuming *prior independence* between μ and σ^2, namely, assuming that $\pi(\mu, \sigma^2) = \pi(\mu)\pi(\sigma^2)$, derive an explicit expression for $\pi(\mu|Y = y, \sigma^2)$, the posterior distribution of μ conditional on both $Y = y$ and σ^2. Comment on the form of this posterior distribution.

(b) Assuming prior independence between μ and σ^2, derive an explicit expression for $\pi(\sigma^2|Y = y, \mu)$, the posterior distribution for σ^2 conditional on $Y = y$ and μ, and comment on its distributional form.

(c) Assuming prior independence between μ and σ^2, derive an explicit expression for $\mathrm{f}_Y(y|\sigma^2)$, which is obtained by integrating over (i.e., eliminating) μ.

For further information about Bayesian inference, see Gelman et al. (2004) and Hoff (2009).

Exercise 5.34. Let $\boldsymbol{Y} = (Y_1, \ldots, Y_n)'$ constitute a random sample from an $\mathrm{N}(\mu, \sigma^2)$ parent population, and suppose that interest lies in making Bayesian inferences about μ and σ^2. (For further details on Bayesian inference, see Exercise 5.33.)

Consider the following *diffuse* (i.e., infinite-variance) prior distributions for μ and σ^2:

$$\begin{aligned}\pi(\mu) &\propto 1, \ -\infty < \mu < \infty \\ \pi[\ln(\sigma)] &\propto 1, \ -\infty < \ln(\sigma) < \infty\end{aligned}$$

Distributions such as these are called *improper* because, unlike standard (or "proper") distributions, they do not integrate to 1. Strictly speaking, then, *improper prior distributions* are not probability densities. However, in many cases, the resulting posterior distributions will be proper even when the prior distributions are not.

(a) Assuming prior independence between μ and σ^2, that is, assuming that $\pi(\mu, \sigma^2) = \pi(\mu)\pi(\sigma^2)$, find an explicit expression for $\pi(\mu|\boldsymbol{Y} = \mathbf{y})$, the posterior distribution of μ given $\boldsymbol{Y} = \mathbf{y}$. Then, derive an explicit expression for the posterior distribution of the parameter $\psi = \frac{(\mu - \bar{y})}{s/\sqrt{n}}$ given $\boldsymbol{Y} = \mathbf{y}$, where $\bar{y} = \frac{1}{n}\sum_{i=1}^{n} y_i$ and $s^2 = \frac{1}{n-1}\sum_{i=1}^{n}(y_i - \bar{y})^2$. Do you see any connections to "frequentist" (i.e., non-Bayesian) inference?

(b) Again assuming prior independence between μ and σ^2, find the structure of $\pi(\sigma^2|\boldsymbol{Y} = \mathbf{y})$, the posterior distribution of σ^2 given $\boldsymbol{Y} = \mathbf{y}$.

Exercise 5.35. Environmental exposures may affect human reproduction by many diverse mechanisms. Sperm production may be suppressed in the male, or subtle abnormalities in the spermatozoa may impair their ability to fertilize the ovum. The exposed female may experience anovulation, or may produce ova that are nonviable. All such mechanisms lead to a common observable effect: longer time is required, on average, for such affected couples to achieve pregnancy. Information on the number of menstrual cycles required for conception can be gathered with little inconvenience or embarrassment to couples under study. Consequently, evaluation of such readily available data provides a useful epidemiological screening method for detecting harmful effects of human exposures to reproductive toxins.

Suppose that the per-menstrual cycle conception probability is $\pi(0 < \pi < 1)$, and that, for now, π is assumed not to vary from couple to couple. In other words, the probability of a woman becoming pregnant in any particular menstrual cycle is π. Let X denote the number of menstrual cycles required for conception, and assume that X has the geometric distribution

$$p_X(x; \pi) = \pi(1-\pi)^{x-1}, \quad x = 1, 2, \ldots, \infty; \quad 0 < \pi < 1.$$

Let $X_1, X_2, \ldots, X_n$ constitute a random sample of size n from this geometric distribution; in other words, information is obtained on the number of cycles required for conception for each of n independently selected couples.

(a) Find a sufficient statistic for π.

(b) Find the maximum likelihood estimator $\hat{\pi}$ of π.

(c) If $\hat{\pi} = 0.20$ when $n = 50$, construct an appropriate 95% confidence interval for the parameter π.

(d) Derive an explicit expression for the minimum variance unbiased estimator (MVUE) of π.

(e) It is, in fact, unreasonable to assume that the per-cycle conception probability π is the same for all couples. As an alternative statistical model, suppose that π is assumed to vary from couple to couple according to some probability distribution. In particular, assume that π has the distribution

$$\mathrm{f}(\pi) = \theta\pi^{\theta-1}, \quad 0 < \pi < 1, \quad \theta > 2.$$

Find the unweighted least squares estimator $\hat{\theta}$ of θ based on a random sample of size n from the compound distribution of X based on unconditionalizing $\mathrm{p}_X(x;\pi)$ with respect to $\mathrm{f}(\pi)$. In addition, prove that $\hat{\theta}$ is also the method of moments estimator of θ, and develop an appropriate approximation for $\mathrm{V}(\hat{\theta})$.

Exercise 5.36. Let $X_1, X_2, \ldots, X_n$ constitute a random sample of size n from an $\mathrm{N}(\mu, \sigma^2)$ parent population, where σ^2 has a *known* value.

(a) Develop an explicit expression for the minimum variance unbiased estimator (MVUE) $\hat{\theta}$ of the parameter $\theta = \mathrm{e}^{\mu}$.

(b) Develop an explicit expression for the relative efficiency (RE) of $\hat{\theta}$, namely, the Cramér-Rao lower bound (CRLB) for the variance of any unbiased estimator of θ divided by $\mathrm{V}(\hat{\theta})$, the variance of $\hat{\theta}$. Does the variance of $\hat{\theta}$ equal the CRLB (i.e., is $\hat{\theta}$ a fully efficient estimator of θ for all finite values of n)? What is the limiting value of this RE as $n \to +\infty$ [i.e., what is the asymptotic relative efficiency (ARE) of $\hat{\theta}$]?

Exercise 5.37. Suppose that $Y_1, Y_2, \ldots, Y_n$ constitute a random sample of size n from the population

$$\mathrm{f}_Y(y;\theta) = \theta^{-1}, 0 < y < \theta < +\infty.$$

Develop an exact $100(1-\alpha)\%$ lower confidence limit for the unknown parameter θ. In other words, find an explicit expression for a random variable L (which is a function of the available data) such that $\mathrm{pr}(L \leq \theta) = (1-\alpha)$. What is the numerical value of L when $n = 10$, $y_{(10)} = 2.10$, and $\alpha = 0.05$? HINT: Consider $\theta^{-1}Y_{(n)}$, where $Y_{(n)} = \max\{Y_1, Y_2, \ldots, Y_n\}$.

Exercise 5.38. Suppose that a model for a certain biological process specifies that a random variable W can be described as the product of two other *independent* random variables X and Y, namely,

$$W = XY.$$

Assume that it is possible to directly observe (or measure) X and Y, but *not* W.

A parameter of particular interest to a biologist studying this process is the *coefficient of variation* of W, namely,

$$\text{CV}_w = \sigma_w/\mu_w, \; \mu_w \neq 0,$$

where $\text{E}(W) = \mu_w$ and $\text{V}(W) = \sigma_w^2$.

Since CV_w cannot be estimated directly (because W is not observable), this biologist wants to know if CV_w can be estimated indirectly using estimates of CV_x and CV_y, where $\text{CV}_x = \sigma_x/\mu_x$ and $\text{CV}_y = \sigma_y/\mu_y$, and where $\mu_x \neq 0$ and $\mu_y \neq 0$.

(a) Prove that

$$\text{CV}_w = \left[(\text{CV}_x)^2 (\text{CV}_y)^2 + (\text{CV}_x)^2 + (\text{CV}_y)^2\right]^{1/2}.$$

(b) Assume that the independent random variables X and Y both have the same GAMMA($\alpha = 1, \beta$) distribution, so that $\text{E}(X)$=$\text{E}(Y)$=$\text{V}(X)$ =$\text{V}(Y) = \beta$. In particular, let $x_1, x_2, \ldots, x_n$ and $y_1, y_2, \ldots, y_n$ represent sets of observed values of X and Y that have been randomly selected from this underlying gamma distribution. Further, suppose that these $2n$ observed values are used to produce $\hat{\beta} = 2.00$ as the maximum likelihood (ML) estimate of β and $\hat{\text{V}}(\hat{\beta}) = 0.04$ as the ML estimate of the variance of $\hat{\beta}$. Use this information, the gamma distribution assumption, and the result in part (a) to compute the ML estimate of CV_w, and then to compute an ML-based 95% confidence interval for CV_w.

Exercise 5.39. Let $X_1, X_2, \ldots, X_n$ constitute a random sample of size n from

$$\text{f}_X(x; \alpha, \beta) = \frac{\alpha^\beta}{\Gamma(\beta)} x^{\beta - 1} \text{e}^{-\alpha x}, \quad x > 0; \; \alpha > 0, \; \beta > 0.$$

Here, β is a *known positive constant*, and α is an *unknown parameter.*

(a) Find a sufficient statistic for the unknown parameter α.

(b) Let $\hat{\alpha} = \beta/\bar{X}$ be a proposed estimator of α. Develop explicit expressions for $\text{E}(\hat{\alpha})$ and $\text{V}(\hat{\alpha})$, the mean and variance of $\hat{\alpha}$. Is $\hat{\alpha}$ a consistent estimator of the parameter α?

(c) If $\beta = 2$, $n = 50$, and $\bar{x} = 3$, use the Central Limit Theorem to develop an appropriate large-sample approximate 95% confidence interval for the unknown parameter α.

Exercise 5.40. Let X_1 and X_2 constitute a random sample of size $n = 2$ from the discrete parent population

$$\mathrm{p}_X(x;\theta) = (-\ln\theta)^{-1}\frac{(1-\theta)^x}{x}, \quad x = 1, 2, \ldots, \infty \text{ and } 0 < \theta < 1.$$

(a) Prove rigorously that the probability distribution of $S = (X_1 + X_2)$ is

$$\mathrm{p}_S(s;\theta) = \frac{2(1-\theta)^s}{s(\ln\theta)^2}\sum_{l=1}^{s-1} l^{-1}, \quad s = 2, 3, \ldots, \infty.$$

(b) Find the minimum variance unbiased estimator (MVUE) of

$$\pi = \mathrm{pr}(X = 1) = (\theta - 1)/\ln\theta.$$

You may assume that $S = (X_1 + X_2)$ is a complete sufficient statistic for π. Also, what is the numerical value of your MVUE if $X_1 = 2$ and $X_2 = 3$?

Exercise 5.41. Suppose that the time X to death of non-smoking heart transplant patients follows the distribution

$$\mathrm{f}_X(x;\alpha) = \alpha^{-1}\mathrm{e}^{-x/\alpha}, \ 0 < x < +\infty, \ \alpha > 0.$$

Further, suppose that the time Y to death of smoking heart transplant patients follows the density function

$$\mathrm{f}_Y(y;\beta) = \beta^{-1}\mathrm{e}^{-y/\beta}, \ 0 < y < +\infty, \ \beta > 0.$$

It is of interest to make statistical inferences about the parameter $\theta = (\alpha - \beta)$.

(a) Let $X_1, X_2 \ldots, X_n$ be a random sample of size n (n large) from $\mathrm{f}_X(x;\alpha)$, and let $Y_1, Y_2 \ldots, Y_n$ be a random sample of size n (n large) from $\mathrm{f}_Y(y;\beta)$. Let $X_{(1)} = \min\{X_1, X_2 \ldots, X_n\}$ and let $Y_{(1)} = \min\{Y_1, Y_2 \ldots, Y_n\}$. Find the constant k such that the estimator $k[X_{(1)} - Y_{(1)}]$ is an unbiased estimator of $\theta = (\alpha - \beta)$. Find an explicit expression for the variance of this unbiased estimator for θ, and then show that this unbiased estimator of θ is not a *consistent* estimator of θ. Do you notice any other undesirable properties of this unbiased estimator of θ?

(b) Using an estimator that is a function of sufficient statistics for α and β, develop an explicit expression for what you deem to be the most appropriate large-sample $100(1-\alpha)\%$ confidence interval for the parameter θ.

Exercise 5.42. In a certain population of teenagers, the number Y of times that a member of this population has "unprotected sex" is assumed to have the distribution

$$\mathrm{p}_Y(y;\pi) = \pi(1-\pi)^y,\ y = 0, 1, \ldots, \infty;\ 0 < \pi < 1.$$

An epidemiologist is interested in making statistical inferences about the parameter $\theta = \mathrm{pr}(Y > 1|Y > 0)$, namely, the conditional probability that a member of this population of teenagers has unprotected sex more than once, given that this teenager has unprotected sex at least once.

(a) Suppose that n teenagers are randomly selected from this population and are interviewed, giving observed responses $y_1, y_2, \ldots, y_n$. These observed values can be considered to be realizations for a random sample $Y_1, Y_2, \ldots, Y_n$ of size n selected from $\mathrm{p}_Y(y;\pi)$. Use these data to find an explicit expression for the minimum variance unbiased estimator (MVUE) of θ.

(b) Compute an appropriate 95% confidence interval for θ if $n = 100$ and $\sum_{i=1}^{100} y_i = 30$.

Exercise 5.43. Racing car windshields made of a newly developed impact-resistant glass are tested for breaking strength by striking them repeatedly with a mechanical device that simulates the stresses caused by high-speed crashes in automobile races. If each windshield has a constant probability θ of surviving a particular strike, independently of the number of previous strikes received, then the number of strikes X required to break a windshield will have the geometric distribution

$$\mathrm{p}_X(x;\theta) = (1-\theta)\theta^{x-1}, x = 1, 2, \ldots, \infty.$$

Suppose that the results of tests on $n = 200$ independently produced windshields are as follows: 112 windshields broke on the first strike (i.e., $x = 1$ for each of these 112 windshields), 36 windshields broke on the second strike (i.e., $x = 2$ for each of these 36 windshields), 22 windshields broke on the third strike (i.e., $x = 3$ for each of these 22 windshields), and 30 windshields each required at least four strikes before breaking (i.e., $x \geq 4$ for each of these 30 windshields). For these data, compute the numerical value of the maximum likelihood estimator $\hat{\theta}$ of the unknown parameter θ, and then compute appropriate large-sample 95% confidence intervals for θ using both *observed* and *expected* information.

Exercise 5.44. For the i-th of k cities ($i = 1, 2, \ldots, k$), suppose that the systolic blood pressure Y_{ij} of the j-th randomly chosen resident of city i has a normal distribution with mean μ_i and variance σ^2, $j = 1, 2, \ldots, n\,(n > 1)$, where $N = kn$. Define $\bar{Y}_i = n^{-1}\sum_{j=1}^{n} Y_{ij}$ and $S_i^2 = (n-1)^{-1}\sum_{j=1}^{n}(Y_{ij} - \bar{Y}_i)^2$, and assume that the set $\{Y_{ij}\}$ constitutes a set of N mutually independent random variables.

(a) Derive from basic principles what you consider to be the best *exact* $100(1-\alpha)\%$ confidence interval for the parameter $\theta = \sum_{i=1}^{k} c_i\mu_i$, where $c_1, c_2, \ldots, c_k$ are known constants.

(b) Now, suppose that k is a large, even, and positive integer. Further, for $i = 1, 2, \ldots, \frac{k}{2}$ and $j = 1, 2, \ldots, n$, suppose that $Y_{ij} \sim \mathrm{N}(\mu_i, \sigma_1^2)$; and for $i = (\frac{k}{2} + 1, \ldots, k)$ and $j = 1, 2, \ldots, n$, $Y_{ij} \sim \mathrm{N}(\mu_i, \sigma_2^2)$. Derive from basic principles what you consider to be an appropriate $100(1-\alpha)\%$ confidence interval for the parameter $\gamma = (\sigma_1^2 - \sigma_2^2)$.

Exercise 5.45. The joint distribution of the concentrations X and Y (in milligrams per liter) of two enzymes in a certain biological system is assumed to be adequately described by the bivariate density function

$$\mathrm{f}_{X,Y}(x, y; \theta) = 3\theta^{-3}(x + y),\ 0 < x < \theta,\ 0 < y < \theta,\ \text{ and } 0 < (x + y) < \theta,$$

where θ (> 0) is an unknown parameter. A biologist is interested in making statistical inferences about the unknown parameter θ, but is only able to measure values of the random variable $S = (X + Y)$. In other words, this biologist is not able to measure values of X and Y separately, but can only measure the sum of X and Y, namely, $S = (X + Y)$.

Suppose that we have available n mutually independent observations of the random variable S. More specifically, we have available observed values of the n mutually independent and identically distributed random variables $S_i = (X_i + Y_i)$, $i = 1, 2, \ldots, n$. Using $S_1, S_2, \ldots, S_n$, develop an exact $100(1-\alpha)\%$ upper one-sided confidence interval $(0, U)$ for the unknown parameter θ, where U is a function of $S_{(n)} = \max\{S_1, S_2, \ldots, S_n\}$; in particular, find U such that $\mathrm{pr}(0 < \theta < U) = (1 - \alpha)$. If $n = 25$ and the observed value of $S_{(n)}$ is equal to $s_{(n)} = 8.20$, compute the exact 95% upper one-sided confidence interval for θ.

Exercise 5.46. In a certain population of teenagers, the number Y of times that a member of this population has smoked marijuana is assumed to have the geometric distribution

$$\mathrm{p}_Y(y; \pi) = \pi(1 - \pi)^y,\ y = 0, 1, \ldots, \infty;\ 0 < \pi < 1.$$

An epidemiologist is interested in estimating the population parameter $\theta = \mathrm{pr}(Y > 1 | Y > 0)$, namely, the *conditional probability* that a member of this population of teenagers has smoked marijuana more than once given that he or she has smoked marijuana at least once. Suppose that n teenagers are randomly selected from this population, and suppose that each teenager reports (with anonymity) the number of times that he or she has smoked marijuana. The observed responses $y_1, y_2, \ldots, y_n$ of these n teenagers can be considered to be realizations of a random sample $Y_1, Y_2, \ldots, Y_n$ of size n selected from $\mathrm{p}_Y(y;\pi)$. Show directly that $\mathrm{E}(\hat{\theta}) = \theta$, where $\hat{\theta}$ is the minimum variance unbiased estimator (MVUE) of θ.

Exercise 5.47. The geometric distribution is often used to model the number of trials required before a mechanical system (or a component of that system) breaks down. In particular, if X has the geometric distribution

$$\mathrm{p}_X(x;\theta) = (1-\theta)\theta^{x-1},\ x = 1, 2, \ldots, \infty \text{ and } 0 < \theta < 1,$$

a system could be considered reliable if $\mathrm{pr}(X > k)$, its so-called "reliability probability," is large, where k is some specified positive integer.

a) For the geometric distribution given above, prove that $\mathrm{pr}(X > k) = \theta^k$.

b) Using a random sample of $X_1, X_2, \ldots, X_n$ from $\mathrm{p}_X(x;\theta)$, derive a formula for the minimum variance unbiased estimator (MVUE) of the "reliability probability" θ^k. If $n = 5$, $k = 6$, $x_1 = 4$, $x_2 = 3$, $x_3 = 2$, $x_4 = 1$, and $x_5 = 5$, use the MVUE to compute a numerical estimate of θ^k.

Exercise 5.48. The joint distribution of the concentrations X and Y (in milligrams per liter) of two enzymes in a certain biological system is assumed to be adequately described by the bivariate density function

$$\mathrm{f}_{X,Y}(x,y;\theta) = 3\theta^{-3}(x+y),\ 0 < x < \theta,\ 0 < y < \theta, \text{ and } 0 < (x+y) < \theta,$$

where θ (> 0) is an unknown parameter. A biologist is interested in making statistical inferences about the unknown parameter θ, but is only able to measure values of the random variable $S = (X+Y)$. In other words, this biologist is not able to measure values of X and Y separately, but can only measure the sum of X and Y, namely, $S = (X+Y)$.

Suppose that we have available n mutually independent observations of the random variable S. More specifically, we have available the observed values of the n mutually independent and identically distributed random variables $S_i = (X_i + Y_i)$, $i = 1, 2, \ldots, n$. For $0 < \alpha \leq 0.10$, develop a large-sample $100(1-\alpha)\%$ confidence interval for the unknown parameter θ that utilizes the observed values of the n random variables $S_1, S_2, \ldots, S_n$. If $n = 100$ and if the observed value of $\overline{S} = n^{-1}\sum_{i=1}^{n} S_i$ is equal to $\bar{s} = 6$, compute a 95% confidence interval for θ.

Exercise 5.49. Let $X_1, X_2, \ldots, X_n$ constitute a random sample of size n from the Poisson parent population

$$\mathrm{p}_X(x) = \frac{\lambda^x \mathrm{e}^{-\lambda}}{x!}, x = 0, 1, \ldots, +\infty \text{ and } \lambda > 0.$$

(a) If $n = 3, \bar{X} = n^{-1}\sum_{i=1}^{n} X_i$, and $\lambda = 0.25$, find the numerical value of $\mathrm{pr}(\bar{X} > 0.50)$.

(b) To estimate $\mathrm{p}_X(x) = \mathrm{pr}(X = x) = \pi_x$, a non-statistician suggests using the estimator

$$\hat{\pi}_x = \frac{(\bar{X})^x \mathrm{e}^{-\bar{X}}}{x!},$$

his motivation being that $\mathrm{E}(\bar{X}) = \lambda$. Develop an expression for $\mathrm{E}(\hat{\pi}_x)$ that is an explicit function of $\mathrm{E}(Y^x)$, where $Y \sim \mathrm{POI}(\theta)$ with $\theta = n\lambda \mathrm{e}^{-1/n}$.

(c) If $x = 0$, derive an explicit expression for $\lim_{n\to\infty} \hat{\pi}_0$, and then comment on your finding. More generally, is $\hat{\pi}_x$ a consistent estimator of π_x?

Exercise 5.50. For any family in the United States with exactly k children (k is a known positive integer), the number X out of k teenagers in such a family who are overweight relative to the National Center for Health Statistics (NCHS) guidelines is postulated to have the binomial distribution

$$\mathrm{p}_X(x; \pi) = \mathrm{C}_x^k \pi^x (1-\pi)^{k-x}, \; x = 0, 1, \ldots k; \; 0 < \pi < 1.$$

It is of interest to find an explicit expression for the minimum variance unbiased estimator (MVUE) of the unknown parameter $\theta = \mathrm{pr}(X = 1)$, namely, the probability that a U.S. family with k teenage children has exactly one teenager who is overweight based on NCHS guidelines.

(a) If $X_1, X_2, \ldots, X_n$ constitute a random sample of size n from $\mathrm{p}_X(x; \pi)$, develop an explicit expression for the MVUE $\hat{\theta}$ of θ using the fact that $S = \sum_{i=1}^{n} X_i$ is a complete sufficient statistic for θ. Also, show directly that $\mathrm{E}(\hat{\theta}) = \theta$.

(b) Comment on the appropriateness of the use of the binomial model $\mathrm{p}_X(x; \pi)$ for the distribution of the random variable X.

Exercise 5.51. The concentration Y (in milligrams per cubic centimeter) of lead in the blood of children of age x is postulated to have the (conditional) density

$$\mathrm{f}_Y(y; \beta | X = x) = (\beta x)^{-1} \mathrm{e}^{-y/\beta x}, \; y > 0, \; x > 0, \; \beta > 0.$$

Let (x_i, Y_i) be a randomly chosen observation from $\mathrm{f}_Y(y; \beta | X = x_i), i =$

$1, 2, \ldots, n$, and assume that the n pairs $(x_1, Y_1), (x_2, Y_2) \ldots, (x_n, Y_n)$ are mutually independent.

(a) Derive explicit expressions for the *maximum likelihood* (ML) estimator $\hat{\beta}_{\text{ML}}$ of β and for its expected value and variance given $x_1, x_2, \ldots, x_n$.

(b) Derive explicit expressions for the (unweighted) *least squares* (LS) estimator $\hat{\beta}_{\text{LS}}$ of β and for its expected value and variance given $x_1, x_2, \ldots, x_n$.

(c) Derive explicit expressions for the *method of moments* estimator $\hat{\beta}_{\text{MM}}$ of β and for its expected value and variance given $x_1, x_2, \ldots, x_n$.

(d) On statistical grounds, which of these three estimators would you recommend? (Be precise and thorough in your statistical reasoning.) For the estimator that you recommend, provide a formula for an appropriate 95% confidence interval for β which can be computed using the available data (x_i, y_i), $i = 1, 2, \ldots, n$.

Exercise 5.52. Let $Y_1, Y_2, \ldots, Y_n$ constitute a random sample of size n from the parent population

$$\mathrm{f}_Y(y) = C^{-1}\mathrm{e}^{-(y-\mu)/C},\ 0 < \mu < y < +\infty,$$

where C is a known constant and where μ is an unknown parameter. Consider the following two estimators of μ:

1) $\hat{\mu}_1 = \bar{Y} - C$, where $\bar{Y} = n^{-1}\sum_{i=1}^{n} Y_i$;
2) $\hat{\mu}_2 = Y_{(1)} - n^{-1}C$, where $Y_{(1)} = \min\{Y_1, Y_2, \ldots, Y_n\}$.

Which of these two estimators of μ do you prefer, and why?

Exercise 5.53. Suppose that $X_1, X_2, \ldots, X_n$ constitute a random sample of size n (> 2) from $\mathrm{f}_X(x; \theta_1) = \theta_1^{-1}, 0 < x < \theta_1$. Further, suppose that $Y_1, Y_2, \ldots, Y_n$ constitute a random sample of size n (> 2) from $\mathrm{f}_Y(y; \theta_2) = \theta_2^{-1}, 0 < y < \theta_2$. A statistician proposes using $R = X_{(n)}/Y_{(n)}$, the ratio of the largest order statistics in the two independent random samples, as a possible point estimator of the unknown ratio parameter $\rho = \theta_1/\theta_2$.

(a) Derive an explicit expression for $\mathrm{f}_R(r; \rho)$, the density function of the estimator R.

(b) Find explicit expressions for $\mathrm{E}(R)$ and $\mathrm{V}(R)$. Is R a consistent estimator of ρ? Comment on this statistician's suggestion to use R as a point estimator of ρ.

Exercise 5.54. Let $X_1, X_2, \ldots, X_n$ constitute a random sample of size n (> 1) from an $\mathrm{N}(\mu, \sigma^2)$ population, where σ^2 has a *known* value.

(a) Find a sufficient statistic U for the parameter $\theta = \mathrm{e}^{\mu}$.

(b) Find a function $\hat{\theta} = \mathrm{g}(U)$, an explicit function of U, that can serve as an unbiased estimator of the parameter θ under the stated assumptions. Is $\hat{\theta}$ a consistent estimator of θ?

Exercise 5.55. Researchers have theorized that monozygotic twins separated at birth will tend, as adults, to be more alike than different with regard to their exercise habits. To examine this theory, a random sample of n sets of such adult monozygotic twins are interviewed regarding their current exercise habits. For $i = 1, 2, \ldots, n$, suppose that the random variable Y_i takes the value 0 if neither member of the i-th set of twins exercises on a regular basis, that Y_i takes the value 1 if one twin in the i-th set exercises on a regular basis and the other does not, and that Y_i takes the value 2 if both twins in the i-th set exercise on a regular basis.

Further, for $i = 1, 2, \ldots, n$, assume that the random variable Y_i has the probability distribution

$$\mathrm{p}_{Y_i}(y_i) = \left[\frac{1 + y_i(2 - y_i)}{2}\right] \frac{\theta^{y_i(2-y_i)}}{(1+\theta)}, y_i = 0, 1, 2 \text{ and } \theta > 0.$$

For a data set involving $n = 100$ sets of monozygotic twins, suppose that there are no regular exercisers for each of 50 sets of these twins, that there is one regular exerciser and one non-regular exerciser for each of 20 sets of these twins, and that there are two regular exercisers for each of 30 sets of these twins. Using both *observed* information and *expected* information, compute appropriate large-sample 95% confidence intervals for the unknown parameter θ, and then comment on your findings with regard to the stated theory.

Exercise 5.56. Suppose that a certain company has developed a new type of electric car battery that is designed to last at least k years before maintenance is required, where k is a known positive integer. Let X be the discrete random variable denoting the number of years that such a battery lasts before requiring maintenance, and assume that X has the geometric distribution

$$\mathrm{p}_X(x) = \theta^{x-1}(1-\theta), x = 1, 2, \ldots, \infty \text{ and } 0 < \theta < 1.$$

Suppose that the proposed battery warranty guarantees free maintenance only if such maintenance is required during the first k years of battery life. Thus, when company scientists test a battery, they decide to record the exact number of years before required maintenance only if $X \leq k$, and otherwise they simply note that the battery lasted at least $(k+1)$ years before requiring maintenance. In other words, instead of X, consider a new discrete random variable Y, defined as follows: $Y = X$ if $X \leq k$ and $Y = (k+1)$ if $X \geq (k+1)$.

(a) Find the probability distribution $\mathrm{p}_Y(y)$ of the random variable Y, and then show that

$$\mathrm{E}(Y) = \frac{(1-\theta^{k+1})}{(1-\theta)}.$$

(b) Suppose that $Y_1, Y_2, \ldots, Y_n$ constitute a random sample of size n from $\mathrm{p}_Y(y)$. Show that the maximum likelihood estimator (MLE) $\hat{\theta}$ of θ is equal to

$$\hat{\theta} = \frac{\sum_{i=1}^{n} Y_i - n}{\sum_{i=1}^{n} Y_i - T},$$

where $T, 0 \leq T \leq n$, is the total number of the n Y_i values that take the value $(k+1)$. Also, if $n = 30, k = 4, \sum_{i=1}^{n} y_i = 120$, and if the observed value t of T is equal to 20, compute an appropriate large-sample 95% confidence interval for the unknown parameter θ.

Exercise 5.57*. To estimate the unknown proportion $\pi (0 < \pi \leq 0.50)$ of a particular species of fish inhabiting the Pacific Ocean, the following sampling plan will be implemented. Each of $n (> 1)$ fishing boats will catch fish until exactly $k (> 1)$ fish of the particular species of interest are caught, and the total number of fish caught by each fishing boat will be recorded. All caught fish will be returned unharmed to the Pacific Ocean.

(a) For $i = 1, 2, \ldots, n$, let X_i denote the total number of fish caught by the i-th fishing boat. Assuming that $X_1, X_2, \ldots, X_n$ constitute a random sample from an appropriately specified probability distribution, develop an explicit expression for the maximum likelihood estimator (MLE) $\hat{\pi}$ of π. If $n = 25, k = 5$, and $\hat{\pi} = 0.40$, use expected information to compute an appropriate large-sample 95% confidence interval for the unknown parameter π.

(b) Under the assumptions that $0 < \pi \leq 0.50$ and that $n = 25$, provide a value for k, say k^*, such that the large-sample 95% confidence interval for π developed in part (a) will *never* have a width larger than 0.05.

Exercise 5.58*. Suppose that $Y_1 \sim \mathrm{BIN}(n_1, \pi_1)$, that $Y_2 \sim \mathrm{BIN}(n_2, \pi_2)$, and that Y_1 and Y_2 are independent random variables. Let $\hat{\pi}_1 = Y_1/n_1$ and let $\hat{\pi}_2 = Y_2/n_2$. Further, let

$$\theta = \ln\left[\frac{\pi_1/(1-\pi_1)}{\pi_2/(1-\pi_2)}\right]$$

be the *log odds ratio*, and let

$$\hat{\theta} = \ln\left[\frac{\hat{\pi}_1/(1-\hat{\pi}_1)}{\hat{\pi}_2/(1-\hat{\pi}_2)}\right]$$

be an estimator of θ.

Under the constraint $(n_1 + n_2) = N$, where N is a fixed positive integer, find expressions for n_1 and n_2 (as a function of N and θ) that minimize a large-sample approximation to $\text{V}(\hat{\theta})$ based on the delta method.

If $N = 100$ and $\theta = 2$, what are the numerical values of n_1 and n_2?

Exercise 5.59. For $i = 1, 2, 3$, suppose that U_i has the Bernoulli distribution

$$\text{p}_{U_i}(u_i) = \pi^{u_i}(1-\pi)^{1-u_i}, u_i = 0, 1 \text{ and } 0 < \pi < 1.$$

Now, let

$$X = WU_1 + (1-W)U_2 \text{ and } Y = WU_1 + (1-W)U_3,$$

where the random variable W has the Bernoulli distribution

$$\text{p}_W(w) = \theta^w(1-\theta)^{1-w}, w = 0, 1 \text{ and } 0 < \theta < 1.$$

Further, assume that U_1, U_2, U_3 and W are mutually independent random variables.

(a) Suppose that $(X_1, Y_1), (X_2, Y_2), \ldots, (X_n, Y_n)$ constitute a random sample of size $n(\geq 1)$ from the joint distribution of X and Y. Then, consider the following two estimators of π, namely,

$$\hat{\pi}_1 = \bar{X} \text{ and } \hat{\pi}_2 = \frac{1}{2}\left(\bar{X} + \bar{Y}\right),$$

where $\bar{X} = n^{-1}\sum_{i=1}^{n} X_i$ and $\bar{Y} = n^{-1}\sum_{i=1}^{n} Y_i$.

Which of these two estimators should be preferred and why?

(b) If π has a known value, provide an explicit expression for an unbiased estimator of θ.

Exercise 5.60*. Assume that, given $X = x$, the conditional density function of Y is normal with conditional mean $\text{E}(Y|X = x) = (\alpha + \beta x)$ and conditional variance $\text{V}(Y|X = x) = \sigma^2$. Let $(x_i, Y_i), i = 1, 2, \ldots, n$, constitute a random sample of size n from this conditional density.

A statistician proposes using $\widehat{Y_0} = \overline{Y} + \widehat{\beta}(x_0 - \overline{x})$ as an estimator of $\text{E}(Y|X = x_0)$, the true mean of Y when $X = x_0 (x_1 \leq x_0 \leq x_n)$, where $\overline{Y} = n^{-1}\sum_{i=1}^{n} Y_i, \overline{x} = n^{-1}\sum_{i=1}^{n} x_i$, and where

$$\widehat{\beta} = \sum_{i=1}^{n}(x_i - \overline{x})Y_i \bigg/ \sum_{i=1}^{n}(x_i - \overline{x})^2.$$

In other words, this statistician is interested in estimating the true mean of Y when $X = x_0$ using a simple linear regression (i.e., straight-line) statistical model. In the questions that follow, keep in mind that $x_1, x_2, \ldots, x_n$ are fixed, known constants.

(a) Prove that $\widehat{\beta}$ is normally distributed with mean β and variance $\sigma^2 / \sum_{i=1}^{n}(x_i - \overline{x})^2$.

(b) Assuming that $\widehat{\beta}$ and $\overline{Y}$ are independent random variables, determine the exact distribution of $\widehat{Y}_0$.

(c) Define $\widehat{Y}_i = \overline{Y} + \widehat{\beta}(x_i - \overline{x}), i = 1, 2, \ldots, n$, and let $\text{SSE} = \sum_{i=1}^{n}(Y_i - \widehat{Y}_i)^2$. Assuming that $\text{SSE}/\sigma^2 \sim \chi^2_{(n-2)}$, and that SSE and $\widehat{Y}_0$ are independent random variables, derive an exact $100(1-\alpha)\%$ confidence interval for $\text{E}(Y|X = x_0)$ based on Student's t-distribution. Find a 90% confidence interval for $\text{E}(Y|X = 2)$ if $n = 10, \overline{Y} = 1, \overline{x} = 3, \widehat{\beta} = 4, \sum_{i=1}^{n}(x_i - \overline{x})^2 = 3$, and $\text{SSE} = 4$.

Exercise 5.61*. A certain rare cancer can be classified as being one of four types. A random sample of $n = 100$ subjects with this rare cancer contains $n_1 = 70$ subjects who have type 1, $n_2 = 10$ subjects who have type 2, $n_3 = 15$ subjects who have type 3, and $n_4 = 5$ subjects who have type 4. Based on genetic models, researchers who study the causes of this rare cancer have determined that a subject with this rare cancer has probability $(2 + \theta)/4$ of having type 1, has probability $(1 - \theta)/2$ of having type 2, has probability $(1 - \theta)/2$ of having type 3, and has probability $\theta/4$ of having type 4, where $\theta(0 < \theta < 1)$ is an unknown parameter.

Use the available data to compute an appropriate large-sample 95% confidence interval for θ.

Exercise 5.62*. Let X_1, X_2, X_3, X_4 constitute a random sample of size 4 from an N(0,1) parent population. Consider the following two random variables:

$$U = \frac{(X_1 + X_2 + X_3)}{\left[\frac{1}{2}(X_1 - X_3)^2 + \frac{1}{3}(X_1 - X_2 + X_3)^2 + X_4^2\right]^{1/2}};$$

and

$$V = \frac{\sum_{i=1}^{2} X_i^2}{\sum_{i=1}^{4} X_i^2}.$$

Find constants k_1 and k_2 such that $\text{pr}(|U| > k_1) = 0.05$ and $\text{pr}(V > k_2) = 0.05$.

Exercise 5.63*. In a small U.S. college town containing a number of homeless people, let the random variable Y be the nightly number of homeless people who have no shelter, and assume that $Y \sim \text{POI}(\lambda)$. Information concerning $\text{E}(Y) = \lambda$ would be helpful to town planners for assessing requirements for new homeless shelters.

Suppose that town employees attempt to count the number of homeless people without shelter on any particular night, and further suppose that each homeless person without nighttime shelter has probability $\pi (0 < \pi < 1)$ of being counted. Also, assume that whether a particular homeless person is counted is not affected by whether any other homeless person is counted. Let the random variable X denote the number of homeless persons without nighttime shelter who are *actually* counted on any particular night.

Now, suppose that these town employees attempt to count the number of homeless people without shelter on each of $n = 50$ randomly selected nights during a particular time period. For $i = 1, 2, \ldots, 50$, let x_i be the observed count on the i-th night, and suppose that these 50 observed counts produce a sample mean equal to $\bar{x} = (50)^{-1} \sum_{i=1}^{50} x_i = 24.80$. Using the fact that $\pi = 0.75$ from prior experience, compute an appropriate large-sample 95% confidence interval for the parameter $\text{E}(Y) = \lambda$.

Exercise 5.64*. Let $X_1, X_2, \ldots, X_n$ constitute a random sample of size n from the geometric distribution

$$\text{p}_X(x) = (1 - \pi)\pi^x, x = 0, 1, \ldots, \infty \text{ and } 0 < \pi < 1.$$

Find an explicit expression for the minimum variance unbiased estimator (MVUE) $\hat{\pi}$ of π, and then show directly that $\text{E}(\hat{\pi}) = \pi$.

Exercise 5.65*. Suppose that θ (> 0) is an unknown parameter and that $\hat{\theta}$ is an *unbiased* estimator of θ with unknown variance $\text{V}(\hat{\theta}) > 0$.

(a) For $k > 0$, find an explicit expression for $\text{MSE}(k\hat{\theta}, \theta) = \text{E}\left[(k\hat{\theta} - \theta)^2\right]$, the *mean-squared error* of $k\hat{\theta}$ as an estimator of θ. For what value k^* of k is $\text{MSE}(k\hat{\theta}, \theta)$ minimized? Also, discuss any problems that you see regarding the use of $k^*\hat{\theta}$ as an estimator of θ.

(b) Show that $\text{MSE}(k^*\hat{\theta}, \theta) < \text{MSE}(\hat{\theta}, \theta)$. For a related discussion, see Copas (1983).

(c) If $X_1, X_2, \ldots, X_n$ constitute a random sample of size n (≥ 1) from an $\text{N}(0, \theta)$ population, and $\hat{\theta} = n^{-1} \sum_{i=1}^{n} X_i^2$, find an explicit expression for k^*.

Exercise 5.66*. For product quality control purposes, a certain manufacturing company periodically inspects for defects consecutively chosen items as they come off a production line. Suppose that $n(> 1)$ consecutively chosen items are inspected. For $i = 1, 2, \ldots, n$ (where $i = 1$ pertains to the first item chosen for inspection, $i = 2$ pertains to the second item chosen for inspection, etc.), let the random variable $X_i = 1$ if the i-th item is found to be defective, and let $X_i = 0$ otherwise; also, let $\pi = \text{pr}(X_i = 1), 0 \leq \pi \leq 1$.

A common statistical assumption for such a quality control inspection plan is that the dichotomous random variables $X_1, X_2, \ldots, X_n$ constitute a set of *mutually independent* random variables. However, a company statistician theorizes that items chosen for inspection after a defective item is found may have a probability (say, θ) larger than π of being defective, since the occurrence of a defective item suggests that the production process itself may be experiencing problems.

To support his theory, this statistician suggests a possibly more appropriate probabilistic model, namely, one that does not assume mutual independence among the random variables $X_1, X_2, \ldots, X_n$. More specifically, this statistician's suggested probabilistic model has the following structure:

$$\text{pr}(X_i = 1) = \pi, i = 1, 2, \ldots, n, \text{ and } \text{pr}(X_i = 1 | X_{i-1} = 1) = \theta, i = 1, 2, \ldots, n.$$

(a) Based on this statistician's probabilistic model, show that the joint distribution of the n random variables $X_1, X_2, \ldots, X_n$ [i.e., the likelihood function $\mathcal{L} = \text{p}_{X_1, X_2, \ldots, X_n}(x_1, x_2, \ldots, x_n)$] can be written in the form

$$\mathcal{L} = \pi^{(x_1+a-c)}(1-\pi)^{(2-x_1-n+b)}\theta^c(1-\theta)^{(a+b-2c)}(1-2\pi+\pi\theta)^{(n-1-a-b+c)},$$

where $a = \sum_{i=2}^n x_i, b = \sum_{i=2}^n x_{i-1}$, and $c = \sum_{i=2}^n x_{i-1}x_i$.

(b) Suppose that $n = 50$ consecutively chosen items are inspected for defects, producing a data set $x_1, x_2, \ldots, x_{50}$ that gives $\hat{\pi} = 0.03$ and $\hat{\theta} = 0.05$ as the maximum likelihood estimates of π and θ, respectively. Assuming, for large n, that

$$\text{V}(\hat{\theta}) = \frac{\theta(1-\theta)}{n\pi}, \text{V}(\hat{\pi}) = \frac{\pi(1-\pi)(1-2\pi+\theta)}{n(1-\theta)}, \text{ and } \text{cov}(\hat{\theta}, \hat{\pi}) = \frac{(1-\pi)\theta}{n},$$

compute an appropriate large-sample 95% for the parameter $(\theta - \pi)$, and then comment on how this computed confidence interval supports (or not) the conjecture by the company statistician that $\theta > \pi$.

Exercise 5.67*. Consider the kn random variables

$$Y_{ij} = \alpha + \beta x_{ij} + U_i + E_{ij}, i = 1, 2, \ldots, k \text{ and } j = 1, 2, \ldots, n,$$

where α and β are unknown parameters, where U_i and E_{ij} are random variables, and where, for all i and j, x_{ij} is a known constant. Further, assume that $\mathrm{E}(U_i) = \mathrm{E}(E_{ij}) = 0$, that $\mathrm{V}(U_i) = \sigma_u^2$, that $\mathrm{V}(E_{ij}) = \sigma_e^2$, that U_i and E_{ij} are independent for all i and j, that the set $\{U_1, U_2, \ldots, U_k\}$ consists of k mutually independent random variables, and that the set $\{E_{11}, E_{12}, \ldots, E_{kn}\}$ consists of kn mutually independent random variables.

Now, consider estimating the parameter β with the estimator

$$\hat{\beta} = \frac{\sum_{i=1}^{k}\sum_{j=1}^{n}(x_{ij} - \bar{x}_i)Y_{ij}}{\sum_{i=1}^{k}\sum_{j=1}^{n}(x_{ij} - \bar{x}_i)^2}, \text{ where } \bar{x}_i = n^{-1}\sum_{j=1}^{n} x_{ij}.$$

Develop explicit expressions for $\mathrm{E}(\hat{\beta})$ and $\mathrm{V}(\hat{\beta})$.

Exercise 5.68*. Let $X_1, X_2, \ldots, X_n$ constitute a random sample from

$$\mathrm{f}_X(x;\theta) = \theta^{-1}\mathrm{e}^{-x/\theta},\ x > 0,\ \theta > 0.$$

Suppose that it is of interest to find the minimum variance unbiased estimator (MVUE) of $\mathrm{V}(X) = \theta^2$ using the Rao-Blackwell Theorem and involving $S_n = \sum_{i=1}^{n} X_i$, a complete sufficient statistic for θ^2.

(a) Starting with the joint distribution of $S_{n-1} = \sum_{i=1}^{n-1} X_i$ and X_n, first find (by the method of transformations) the joint distribution of S_n and $Y_n = X_n^2$, and then prove that

$$\mathrm{f}_{Y_n}(y_n|S_n = s_n) = \frac{(n-1)(s_n - \sqrt{y_n})^{n-2}}{2\sqrt{y_n}s_n^{n-1}},\ 0 < y_n < s_n^2.$$

(b) Use the result in part (a) to prove rigorously that the MVUE of θ^2 is

$$S_n^2/n(n+1).$$

(c) Verify the finding in part (b) by employing a much simpler approach that does not involve the use of the Rao-Blackwell Theorem.

Exercise 5.69*. Suppose that a discrete random variable X has the following probability distribution:

$$\mathrm{p}_X(x;\pi,\lambda) = \begin{cases} (1-\pi) & \text{for } x = 0; \\ \frac{\pi\lambda^x}{x!(\mathrm{e}^\lambda - 1)} & \text{for } x = 1, 2, \ldots, \infty. \end{cases}$$

Let $X_1, X_2, \ldots, X_n$ constitute a random sample of size n from $\mathrm{p}_X(x;\pi,\lambda)$; in particular, let N_x be the discrete random variable denoting the number of the n observations that take the specific value x. So, the random variable N_x can take one of the set of possible values $\{0, 1, \ldots, n\}$ subject to the restriction $\sum_{x=0}^{\infty} N_x = n$. If $n = 50$, and if the maximum likelihood (ML) estimates of π and λ are $\hat{\pi} = 0.30$ and $\hat{\lambda} = 2.75$, compute appropriate large-sample 95% confidence intervals for both π and λ.

Exercise 5.70*. Let $X_1, X_2, \ldots, X_{2n+1}$ constitute a random sample of size $(2n+1)$ from the density function $\mathrm{f}_X(x)$, $-\infty < x < \infty$. Let θ be the *population median* of the density function for X, so that θ satisfies the equation

$$\int_{-\infty}^{\theta} \mathrm{f}_X(x)dx = \frac{1}{2}.$$

Further, let $X_{(n+1)}$ denote the *sample median* [i.e., the $(n+1)$-th order statistic for the ordered observations $-\infty < X_{(1)} < X_{(2)} < \ldots < X_{(2n)} < X_{(2n+1)} < +\infty$].

(a) Under the assumption that the maximum value of $\mathrm{f}_X(x)$ occurs at $x = \theta$ [so that θ is also the *population mode* for the density function $\mathrm{f}_X(x)$], prove that the mean-squared error of $X_{(n+1)}$ as an estimator of θ, namely $\mathrm{E}[(X_{(n+1)} - \theta)^2]$, satisfies the inequality

$$\mathrm{E}[(X_{(n+1)} - \theta)^2] \geq \frac{1}{4(2n+3)[\mathrm{f}_X(\theta)]^2}.$$

(b) If $n = 20$ and if $X \sim \mathrm{N}(\mu, \sigma^2)$, find an explicit expression for the lower bound derived in part (a), and then comment on how this lower bound varies with σ^2.

Exercise 5.71*. For various reasons, individuals participating in a survey may prefer *not* to reveal to an interviewer the correct answers to certain sensitive or stigmatizing questions about their personal lives (e.g., about whether they have ever used certain drugs, about whether they have ever stolen anything, etc.). To combat this interview response problem, Warner (1965) introduced a technique for estimating the proportion π of a human population having a sensitive or stigmatizing attribute A. The method, which Warner called *randomized response*, is designed to eliminate untruthful responses that would result in a biased estimate of π. Following this initial work by Warner, there have been many extensions and applications of this randomized response methodology.

This randomized response procedure works as follows. A random sample of n people is selected from the population of interest. Before a particular sensitive issue is discussed (e.g., like whether an individual has or does not have a sensitive or stigmatizing attribute A), the interviewer gives each interviewee a spinner with a face marked so that the spinner points to the letter A with probability θ and not to the letter A (i.e., to the complementary outcome $\overline{\mathrm{A}}$)

with probability $(1-\theta), 0 < \theta < 1$ and $\theta \neq \frac{1}{2}$; here, θ has a *known* value. Each of the n interviewees in the sample spins the spinner (while unobserved by the interviewer) and reports *only* whether the spinner points to the letter representing the group (either A or $\overline{\text{A}}$) to which the interviewee truly belongs (with Group A members having the attribute of interest and Group $\overline{\text{A}}$ members not having the attribute of interest). That is, the interviewee is required *only* to say either "yes" or "no" according to whether the spinner points to the correct group; the interviewee does *not* report the actual letter (or, equivalently, the group) to which the spinner points.

For $i = 1, 2, \ldots, n$, let the random variable X_i take the value 1 if the i-th person in the sample responds "yes," and let X_i take the value 0 if the i-th person in the sample responds "no."

(a) Prove that $\text{pr}(X_i = 1) = \pi(2\theta - 1) + (1 - \theta)$.

(b) Derive an explicit expression for $\hat{\pi}$, the maximum likelihood estimator (MLE) of π. Is $\hat{\pi}$ an unbiased estimator of the unknown parameter π?

(c) Develop an explicit expression for $\text{V}(\hat{\pi})$, the variance of the MLE $\hat{\pi}$. Then, construct an appropriate large-sample 95% confidence interval for π when $n = 100, \theta = 0.20$, and $\hat{\pi} = 0.10$.

(d) If $\theta = 0.20$, develop an expression that can be used to provide an approximate value for the smallest sample size n^* so that

$$\text{pr}(|\hat{\pi} - \pi| < \delta) \geq 0.95,$$

where $\delta(> 0)$ is a known positive constant. Note that your answer will be a function of the unknown value of π; for what value of π does n^* take its largest value?

Exercise 5.72*. Consider the parent population

$$\text{p}_Y(y; \lambda) = (e^{\lambda} - 1)^{-1}\lambda^y/y!, \ y = 1, 2, \ldots, +\infty; \lambda > 0.$$

Suppose that $Y_1, Y_2, \ldots, Y_n$ constitute a random sample of size n from $\text{p}_Y(y; \lambda)$. Define the random variable N_r to be the number of the n observations taking the integer value $r, r = 1, 2, \ldots, +\infty$. Hence, $0 \leq N_r < +\infty$ for each r, and $\sum_{r=1}^{\infty} N_r = n$.

(a) Prove that $\hat{\lambda} = n^{-1}\sum_{r=2}^{\infty} rN_r$ is an unbiased estimator of λ.

(b) Given that

$$\text{V}(\hat{\lambda}) = n^{-1}\lambda[1 + \lambda(e^{\lambda} - 1)^{-1}],$$

find an explicit expression for the efficiency of $\hat{\lambda}$ relative to the Cramèr-Rao lower bound for the variance of any unbiased estimator of λ. If $\lambda = 2$, what is the numerical value of this efficiency?

Exercise 5.73*. For $i = 1, 2, \ldots, n$, assume that $\mathrm{E}(Y_i) = \alpha + \beta x_i$, that $\mathrm{V}(Y_i) = \sigma^2$, and that $\mathrm{corr}(Y_i, Y_{i'}) = \rho \ (\neq 0)$ for all $i \neq i'$. Further, assume that $\{x_1, x_2, \ldots, x_n\}$ constitute a set of n known constants. Recall that the unweighted least squares estimator of β is

$$\hat{\beta} = \frac{\sum_{i=1}^{n}(x_i - \bar{x})(Y_i - \bar{Y})}{(n-1)s_x^2},$$

where $\bar{x} = n^{-1}\sum_{i=1}^{n} x_i$, $\bar{Y} = n^{-1}\sum_{i=1}^{n} Y_i$, and $s_x^2 = (n-1)^{-1}\sum_{i=1}^{n}(x_i - \bar{x})^2$. Given the stated assumptions, derive explicit expressions for $\mathrm{E}(\hat{\beta})$ and $\mathrm{V}(\hat{\beta})$.

Exercise 5.74*. Suppose that

$$(X_1, X_2, X_3, X_4) \sim \mathrm{MULT}\left[n, \alpha(1-\alpha), \alpha\beta, \alpha^2, (1 - \alpha - \alpha\beta)\right],$$

where $0 < \alpha < 1, 0 < \beta < 1$, and $\alpha(1 + \beta) < 1$.

For a particular research study, suppose that the realized (or observed) values of X_1, X_2, X_3, and X_4 are $x_1 = 25, x_2 = 15, x_3 = 10$, and $x_4 = 50$. Use maximum likelihood theory and the delta method to compute an appropriate large-sample 95% confidence interval for the unknown parameter α.

Exercise 5.75*. Group testing is a method designed to save resources when an attribute of interest (e.g., the presence of a particular virus in a human blood sample) rarely occurs in some population under study, and when it is of interest to estimate the proportion of the population having the particular attribute of interest (i.e., the population *prevalence* of the attribute). In particular, group testing refers to the process of combining individual specimens (e.g., individual blood samples) together and then testing whether the attribute of interest is present in the combined sample. If the combined sample tests positively for the presence of the attribute, then it is known that at least one of the individual specimens in the combined sample possesses the attribute; if the combined sample tests negatively for the presence of the attribute, then it is known that none of the individual specimens in the combined sample possesses the attribute. For a discussion of one particular group testing strategy, see Hepworth (2005).

Consider the following scenario, designed to provide data that can be used to compute a maximum likelihood-based confidence interval for the unknown prevalence $\pi, 0 < \pi < 1$, of a certain blood virus in the population under study. Suppose that a total of N subjects are randomly sampled from the population under study (with each such subject having probability π of possessing the blood virus), and that each of these N subjects independently contributes a blood specimen. Further, for $i = 1, 2, \ldots, k$, suppose that g_i groups of combined specimens are formed, with each of these g_i groups containing n_i individual blood specimens mixed together. Hence, it follows that $N = \sum_{i=1}^{k} g_i n_i$.

Let the random variable $Y_i, i = 1, 2, \ldots, k$, be the number of the g_i groups that test positively for the blood virus.

(a) Using the observed realizations $\{y_1, y_2, \ldots, y_k\}$ of the random variables $\{Y_1, Y_2, \ldots, Y_k\}$, develop an equation that can be used to solve iteratively for the maximum likelihood estimator $\hat{\pi}$ of the unknown prevalence π.

(b) Using expected information, show that the large-sample variance $\mathrm{V}(\hat{\pi})$ of $\hat{\pi}$ is equal to

$$\left[\sum_{i=1}^{k} \frac{n_i^2 g_i (1-\pi)^{n_i - 2}}{1 - (1-\pi)^{n_i}}\right]^{-1}.$$

(c) Now, suppose that there are 10 groups (i.e., 10 combined samples), with each group involving the merging of 20 individual blood specimens. If 6 of these 10 groups test positively for the blood virus, develop an appropriate large-sample 95% confidence interval for π.

Exercise 5.76*. For elderly people with early signs of dementia, the time Y (in minutes) to complete a certain written test is assumed to have the negative exponential distribution

$$\mathrm{f}_Y(y) = \theta \mathrm{e}^{-\theta y}, 0 < y < \infty, \theta > 0.$$

Using a random sample $Y_1, Y_2, \ldots, Y_n$ from $\mathrm{f}_Y(y)$, it is of interest to find the minimum variance unbiased estimator (MVUE) of the probability π that an elderly person with early signs of dementia takes no longer than t minutes to complete this written test, where $t, 0 < t < \infty$, is a specified positive number.

(a) Develop an explicit expression for the MVUE $\hat{\pi}$ of the parameter π.

(b) Show directly that $\mathrm{E}(\hat{\pi}) = \pi$.

(c) Discuss any connection that you see between $\hat{\pi}$ and the maximum likelihood estimator (MLE) of π.

Exercise 5.77*. Suppose that the continuous random variable X has the distribution

$$\mathrm{f}_X(x) = \frac{m}{\Gamma\left(\frac{1}{2m}\right)} \mathrm{e}^{-(x-\theta)^{2m}}, -\infty < x < \infty,$$

where $-\infty < \theta < \infty$ and where m is a known positive integer.

Let $X_1, X_2, \ldots, X_n$ constitute a random sample of size n from $\mathrm{f}_X(x)$, and consider using the sample mean $\bar{X} = n^{-1} \sum_{i=1}^{n} X_i$ as an estimator of the parameter θ.

(a) Prove that $\eta_m = \text{EFF}(\bar{X}, \theta)$, the efficiency of $\bar{X}$ relative to the Cramér-Rao lower bound (CRLB) for the variance of any unbiased estimator of θ, is a function only of m.

(b) Compute numerical values for η_1, η_2, η_3, and η_4, and then comment on your findings. Also, find $\lim_{m\to\infty} \eta_m$.

Exercise 5.78*. The standardized score Y on a particular manual dexterity test ranges from -1 (indicating poor manual dexterity) to $+1$ (indicating excellent manual dexterity). Suppose that the density function of the random variable Y is

$$\text{f}_Y(y) = \frac{1}{2}(1 + \alpha y), -1 < y < +1 \text{ and } -1 < \alpha < +1.$$

Let $Y_1, Y_2, \ldots, Y_n$ constitute a random sample of size $n(> 1)$ from $\text{f}_Y(y)$.

(a) Using $Y_1, Y_2, \ldots, Y_n$, develop an explicit expression for $\hat{\alpha}$, the unweighted least squares (ULS) estimator of α. Also, find explicit expressions for $\text{E}(\hat{\alpha})$ and $\text{V}(\hat{\alpha})$. Do you notice any obvious undesirable properties of $\hat{\alpha}$?

(b) Develop an explicit expression for the efficiency of $\hat{\alpha}$ relative to the Cramér-Rao lower bound (CRLB) for the variance of any unbiased estimator of α. Comment on the behavior of this efficiency expression for values of α satisfying $0 \leq \alpha \leq 1$. The following indefinite integral may be useful:

$$\int \frac{x^2}{(a + bx)} \text{d}x = b^{-3} \left[\frac{1}{2}(a + bx)^2 - 2a(a + bx) + a^2 \ln(a + bx) \right], b \neq 0.$$

Exercise 5.79*. It is well-documented that U.S. office workers spend a significant amount of time each workday using the Internet for non-work-related purposes. Suppose that the proportion X of an 8-hour workday that a typical U.S. office worker spends using the Internet for non-work-related purposes is assumed to have the distribution

$$\text{f}_X(x) = 2(1 - \theta)x + 2\theta(1 - x), 0 < x < 1 \text{ and } 0 < \theta < 1.$$

Suppose that a large number n of randomly selected U.S. office workers complete a questionnaire, with the i-th worker providing a value x_i of the random variable $X, i = 1, 2, \ldots, n$. The values $x_1, x_2, \ldots, x_n$ can be considered to be realizations of a random sample $X_1, X_2, \ldots, X_n$ of size n from $\text{f}_X(x)$.

(a) Find an explicit expression for the Cramér-Rao lower bound (CRLB) for the variance of any unbiased estimator of θ involving $X_1, X_2, \ldots, X_n$.

As a hint, for the integration required, consider the transformation $u = (1-\theta)x + \theta(1-x)$, and then evaluate the integral separately for $0 < \theta < \frac{1}{2}, \theta = \frac{1}{2}$, and $\frac{1}{2} < \theta < 1$.

(b) Find an explicit expression for an unbiased estimator $\hat{\theta}$ of θ that is a function of $\bar{X} = n^{-1}\sum_{i=1}^{n} X_i$. Describe how EFF$(\hat{\theta}, \theta)$=CRLB/V$(\hat{\theta})$ varies as a function of θ.

Exercise 5.80*. Suppose that a randomized clinical trial is conducted to compare two new experimental drugs (denoted drug 1 and drug 2) designed to prolong the lives of patients with advanced metastatic colorectal cancer. For $i = 1, 2$, it is assumed that the survival time X_i (in years) for a patient using drug i can be described by the cumulative distribution function (CDF)

$$\mathrm{F}_{X_i}(x_i) = 1 - \mathrm{e}^{-(x_i/\theta_i)^m}, 0 < x_i < \infty, 0 < \theta_i < \infty,$$

where $m(\geq 1)$ is a *known* positive constant.

Suppose that n patients are randomly allocated to each of these two drug therapies. For $i = 1, 2$, let $x_{i1}, x_{i2}, \ldots, x_{in}$ be the n observed survival times (in years) for the n patients receiving drug i. For $i = 1, 2$, these n observed survival times can be considered to be the realizations of a random sample $X_{i1}, X_{i2}, \ldots, X_{in}$ of size n from the CDF $\mathrm{F}_{X_i}(x_i)$.

Suppose that $n = 50, m = 2, \sum_{j=1}^{n} x_{1j}^2 = 350$, and $\sum_{j=1}^{n} x_{2j}^2 = 200$. Use these data to compute an appropriate large-sample 95% confidence interval for $\gamma = \mathrm{E}(X_1) - \mathrm{E}(X_2)$, the difference in mean survival times for these two drug therapies. Based on your computations, do these data provide statistical evidence that favors one drug over the other?

Exercise 5.81*. Ear infections are quite common in infants. To estimate the prevalence of ear infections in infants in a certain area of the United States, and to assess whether ear infections tend to occur in both ears rather than in just one ear, the following statistical model is proposed.

For a random sample of n infants whose parents reside in this U.S. area, suppose, for $i = 1, 2, \ldots, n$, that the random variable X_i equals 0 with probability $(1-\pi)$ if the i-th infant does not have an ear infection, that X_i equals 1 with probability $\pi(1-\theta)$ if the i-th infant has an ear infection in only one ear, and that X_i equals 2 with probability $\pi\theta$ if the i-th infant has ear infections in both ears. Here, $\pi(0 < \pi < 1)$ is the probability that an infant has an infection in at least one ear; that is, π is the *prevalence* in this U.S. area of children with an infection in at least one ear. Since

$$\begin{aligned}\text{pr}(X_i = 2|X_i \geq 1) &= \frac{\text{pr}\left[(X_i = 2) \cap (X_i \geq 1)\right]}{\text{pr}(X_i \geq 1)} \\ &= \frac{\text{pr}(X_i = 2)}{\text{pr}(X_i \geq 1)} = \frac{\pi\theta}{\pi} = \theta,\end{aligned}$$

it follows that $\theta(0 < \theta < 1)$ is the conditional probability that an infant has ear infections in both ears given that this infant has at least one ear that is infected.

(a) Suppose that $n = 100$, that $n_0 = 20$ is the number of infants with no ear infections, that $n_1 = 30$ is the number of infants with an ear infection in only one ear, and that $n_2 = 50$ is the number of infants with ear infections in both ears. Use these data to compute appropriate large-sample 95% confidence intervals for the unknown parameters π and θ, and then comment on your findings.

(b) Using the available data, compute an appropriate large-sample 95% confidence interval for the prevalence γ of ear infections in this U.S. area.

Exercise 5.82*. For $i = 1, 2, \ldots, k$, let $X_{i1}, X_{i2}, \ldots, X_{in_i}$ constitute a random sample of size n_i from a population with unknown mean μ_i and known variance σ_i^2.

(a) Suppose that the goal is to estimate the parameter $\mu = k^{-1}\sum_{i=1}^{k}\mu_i$ using the *unbiased* estimator $\bar{X} = k^{-1}\sum_{i=1}^{k}\bar{X}_i$, where $\bar{X}_i = n_i^{-1}\sum_{j=1}^{n_i}X_{ij}$. Subject to the restriction that $\sum_{i=1}^{k}n_i = N$, where N is a fixed positive integer, two options are suggested for the sizes $n_1, n_2, \ldots, n_k$ of the samples to be selected from these k populations. Determine analytically which of the following two options is the better one:
Option 1: $n_i = N\left(\frac{\sigma_i}{\sum_{i=1}^{k}\sigma_i}\right), i = 1, 2, \ldots, k;$
Option 2: $n_i = N\left(\frac{\sigma_i^2}{\sum_{i=1}^{k}\sigma_i^2}\right), i = 1, 2, \ldots, k.$

(b) Using analytical arguments, decide whether it is possible to find a better option than the ones given in part (a).

Exercise 5.83*. Assessments concerning the presence or absence of a particular disease in human beings cannot always be made with certainty, and so imperfect diagnostic tests are often used to help make such assessments. Suppose that the epidemiologic goal is to estimate the unknown proportion $\pi(0 < \pi < 1)$ of individuals having a difficult-to-diagnose disease in a certain human population; the unknown parameter π is known as the *prevalence* of this disease in the population.

To achieve this goal, suppose that a large number n of individuals is randomly selected from this population, and that each of these individuals is given a certain diagnostic test. More specifically, for the i-th such individual, let the random variable Y_i equal 1 if the i-th individual produces a positive result when given this diagnostic test, and let Y_i equal 0 if not, $i = 1, 2, \ldots, n$. Further, let the random variable D_i equal 1 if the i-th individual actually has the disease in question and let D_i equal 0 if not. Then, define the *sensitivity* of this diagnostic test to be $\gamma = \text{pr}(Y_i = 1|D_i = 1)$, define the *specificity* of this diagnostic test to be $\delta = \text{pr}(Y_i = 0|D_i = 0)$, and assume that the values of γ and δ are *known*.

If $n = 100, s = \sum_{i=1}^{n} y_i = 20, \gamma = 0.90$, and $\delta = 0.85$, develop an appropriate large-sample 95% confidence interval for the prevalence $\pi = \text{pr}(D_i = 1)$.

Note that the random variables $D_1, D_2, \ldots, D_n$ (which *classify* the n individuals as having the disease or not) are unobservable (or *latent*), and that the only data available are the observed values $y_1, y_2, \ldots, y_n$ of the random variables $Y_1, Y_2, \ldots, Y_n$. In this situation, the development of an appropriate large-sample confidence interval for π falls under the general heading of *latent class analysis*. For detailed information about latent class analysis of diagnostic test performance, see Pepe and Janes (2007).

Exercise 5.84*. Suppose that a certain clinical trial involves comparing two different drugs (designated Drug 1 and Drug 2) designed to help patients who suffer from migraine headaches. For patient enrollment, treatment, and follow-up, and for other important activities (e.g., data collection and statistical analysis, salaries for personnel, etc.), suppose that it will cost c_1 dollars for each patient randomly assigned to Drug 1 and that it will cost c_2 dollars for each patient randomly assigned to Drug 2.

Let the dichotomous random variable Y_{ij} take the value 1 if the j-th patient $(j = 1, 2, \ldots, n_i)$ assigned to take the i-th drug $(i = 1, 2)$ reports having *fewer* migraine headaches during a one-month period involving daily consumption of a fixed dose (in pill form) of drug i, and let Y_{ij} take the value 0 otherwise.

Further, for $i = 1, 2$, let $\bar{Y}_i = n_i^{-1} \sum_{j=1}^{n_i} Y_{ij}$, and assume that $\text{E}(Y_{ij}) = \pi_i$ so that $\text{E}(\bar{Y}_i) = \pi_i$ and that $\text{V}(\bar{Y}_i) = \pi_i(1 - \pi_i)/n_i$.

Since cost is an important consideration in any research study, it is desired to find values for n_1 and n_2 (say, n_1^* and n_2^*) that *minimize* the cost C $(= \text{c}_1 n_1 + \text{c}_2 n_2)$ of this clinical trial, subject to the constraint that the large-sample 95% confidence interval for the important parameter $(\pi_1 - \pi_2)$, namely

$$(\bar{Y}_1 - \bar{Y}_2) \pm 1.96\sqrt{\frac{\bar{Y}_1(1 - \bar{Y}_1)}{n_1} + \frac{\bar{Y}_2(1 - \bar{Y}_2)}{n_2}},$$

has a width W no larger than w, where w is a specified positive real number.

Find explicit expressions for n_1^* and n_2^* that hold for all possible values of $\bar{Y}_1$ and $\bar{Y}_2$. Also, find an explicit expression for the ratio n_1^*/n_2^*, and then comment on all these findings.

Exercise 5.85*. In many important practical data analysis situations, the statistical models being used involve several parameters, only a few of which are relevant for directly addressing the research questions of interest. The irrelevant parameters, generally referred to as "nuisance parameters", are typically employed to ensure that the statistical models make scientific sense, but are generally unimportant otherwise. One method for eliminating the need to estimate these nuisance parameters, and hence to improve both statistical validity and precision, is to employ a *conditional inference* approach, whereby a conditioning argument is used to produce a conditional likelihood function that involves only the relevant parameters. For an excellent discussion of methods of conditional inference, see McCullagh and Nelder (1989).

As an example, consider the matched-pairs case-control study design often used in epidemiologic research to examine the association between a potentially harmful exposure and a particular disease. In such a design, a case (i.e., a diseased person, denoted D) is matched (on covariates such as age, race, and sex) to a control (i.e., a non-diseased person, denoted $\overline{\text{D}}$). Each member of the pair is then categorized with regard to the presence (E) or absence ($\overline{\text{E}}$) of a history of exposure to some potentially harmful substance (e.g., cigarette smoke, asbestos, benzene, etc.). The data from such a study involving n case-control pairs can be presented in tabular form, as follows:

		$\overline{\text{D}}$		
		E	$\overline{\text{E}}$	
D	E	Y_{11}	Y_{10}	
	$\overline{\text{E}}$	Y_{01}	Y_{00}	
				n

Here, Y_{11} is the number of pairs for which *both* the case and the control are exposed (i.e., both have a history of exposure to the potentially harmful agent under study), Y_{10} is the number of pairs for which the case is exposed but the control is not, and so on. Clearly, $\sum_{i=0}^{1}\sum_{j=0}^{1} Y_{ij} = n$.

In what follows, assume that the $\{Y_{ij}\}$ have a multinomial distribution with sample size n and associated cell probabilities $\{\pi_{ij}\}$, where $\sum_{i=0}^{1}\sum_{j=0}^{1} \pi_{ij} = 1$. For example, π_{10} is the probability of obtaining a pair in which the case is exposed and its matched control is not, and π_{01} is the probability of obtaining a pair in which the case is not exposed but the control is.

Now, let $\alpha = \text{pr}(\text{E}|\text{D})$ and let $\beta = \text{pr}(\text{E}|\overline{\text{D}})$, so that $\pi_{10} = \alpha(1-\beta)$ and $\pi_{01} = (1-\alpha)\beta$. A parameter used to quantify the association between exposure status and disease status in a matched-pairs case-control study is the *exposure odds ratio* ψ, namely,

$$\begin{aligned}\psi &= \frac{\text{pr}(\text{E}|\text{D})/\text{pr}(\overline{\text{E}}|\text{D})}{\text{pr}(\text{E}|\overline{\text{D}})/\text{pr}(\overline{\text{E}}|\overline{\text{D}})} \\ &= \frac{\alpha/(1-\alpha)}{\beta/(1-\beta)} = \frac{\alpha(1-\beta)}{(1-\alpha)\beta} \\ &= \frac{\pi_{10}}{\pi_{01}}.\end{aligned}$$

(a) Let $S = (Y_{10} + Y_{01})$ and $s = (y_{10} + y_{01})$. Show that the *conditional* distribution $\text{p}_{Y_{10}}(y_{10}|S = s)$ of Y_{10} given $S = s$ can be expressed as a function of the exposure odds ratio ψ, and not of the parameters α and β.

(b) If $\text{p}_{Y_{10}}(y_{10}|S = s) = \mathcal{L}$ is taken as the *conditional likelihood function*, use $\mathcal{L}$ to develop an appropriate large-sample 95% confidence interval for ψ. Note that conditioning eliminates the need to estimate the two probability parameters α and β.

(c) For a particular matched-pairs case-control study, suppose that the observed value of Y_{10} equals $y_{10} = 26$ and the observed value of Y_{01} equals $y_{01} = 10$. Compute an appropriate large-sample 95% confidence interval for ψ. Do these data provide evidence of an exposure-disease association?

(d) Research has shown that the estimator $\hat{\psi} = Y_{10}/Y_{01}$ has a distribution that is highly skewed to the right. To overcome this distributional problem, researchers have suggested working with the MLE estimator $\ln\hat{\psi}$ of $\ln\psi$, first computing a large-sample confidence interval for $\ln\psi$, and then converting this confidence interval into one for ψ. Use the available data to compute an appropriate large-sample 95% confidence interval for ψ based on this alternative approach, and then compare your numerical result with that obtained in part (c).

For further discussion about these issues and about the design and analysis of matched case-control studies, see the books by Breslow and Day (1980) and by Kleinbaum, Kupper, and Morgenstern (1982).

Exercise 5.86*. Suppose that an urn contains $N(\geq 1)$ balls, numbered individually from 1 to N, where N is an *unknown* positive integer. Suppose that $n(1 \leq n \leq N)$ balls are selected randomly *without replacement* from this urn, and let X_i be the number observed on the i-th ball selected, $i = 1, 2, \ldots, n$.

(a) If $U = \max\{X_1, X_2, \ldots, X_n\}$, show that

$$\hat{N} = \left(\frac{n+1}{n}\right)U - 1$$

is an unbiased estimator of the unknown value of N.

(b) Show that

$$\mathrm{V}(\hat{N}) = \frac{(N+1)(N-n)}{n(n+2)}.$$

Exercise 5.87*. Obesity rates for children in the United States have been increasing fairly steadily in recent years. In particular, research has shown that blood triglyceride levels are positively correlated with sugar consumption levels in U.S. children, and high blood triglyceride levels are associated with an increased risk of heart disease.

For the i-th subject in a random sample of size n from a population of obese teenagers in a certain urban area of the United States, suppose that Y_i is the measured blood triglyceride level (in milligrams per deciliter, or mg/dl), that μ_i is the true (but *unobservable*) average weekly amount (in grams) of sugar consumed, and that Y_i and μ_i are related by the straight-line equation

$$Y_i = \theta_0 + \theta_1\mu_i, i = 1, 2, \ldots, n,$$

where $\mu_i = (\mu + \beta_i)$, where $\beta_i \sim \mathrm{N}(0, \sigma_\beta^2)$, and where the $\{\beta_i\}$ constitute a set of n mutually independent random variables. Here, μ is the true average weekly amount (in grams) of sugar consumed by subjects in this population, and μ_i is the average weekly amount (in grams) of sugar consumed by a randomly chosen subject in this population. Note that σ_β^2 is quantifying inter-subject variation in the amount of weekly sugar consumption in this population.

To obtain information about the unobservable random variables $\mu_1, \mu_2, \ldots, \mu_n$, each subject in the random sample completes a weekly food-frequency questionnaire for k (> 1) weeks. For the i-th subject, let X_{ij} be the reported weekly sugar consumption amount for the j-th week, where it is assumed that

$$X_{ij} = \mu_i + \epsilon_{ij}, j = 1, 2, \ldots, k;$$

here, $\epsilon_{ij} \sim \mathrm{N}(0, \sigma_\epsilon^2)$, and the nk random variables $\{\epsilon_{ij}\}$ are mutually independent and are independent of the $\{\beta_i\}$.

To estimate the slope parameter θ_1 using unweighted least squares, it is proposed to use $\bar{X}_i = k^{-1}\sum_{j=1}^{k} X_{ij}$ as a *surrogate* measure of μ_i. Then, based on the data set $\{(\bar{X}_i, Y_i), i = 1, 2, \ldots, n\}$, the unweighted least squares estimator of θ_1 is

$$\hat{\theta}_1 = \frac{\sum_{i=1}^{n}(\bar{X}_i - \bar{X})Y_i}{\sum_{i=1}^{n}(\bar{X}_i - \bar{X})^2}, \text{ where } \bar{X} = n^{-1}\sum_{i=1}^{n}\bar{X}_i.$$

Prove that

$$\mathrm{E}(\hat{\theta}_1) = \gamma\theta_1, \text{ where } 0 < \gamma < 1.$$

This result demonstrates that using $\bar{X}_i$ as a surrogate for μ_i leads to *underestimation* of θ_1 (i.e., $\hat{\theta}_1$ is said to be *attenuated*). Comment on how the degree of *attenuation* varies with n, k, σ^2_β, and σ^2_ϵ.

In your proof, assume that $(\bar{X}_i, \mu_i)$ have a bivariate normal distribution, and employ the equality

$$\mathrm{E}(Y_i|\mu_i, \bar{X}_i = \bar{x}_i) = \mathrm{E}(Y_i|\mu_i), i = 1, 2, \ldots, n.$$

This equality is known as the *nondifferential error assumption* and states that $\bar{X}_i$ contributes no further information about Y_i if the value of μ_i is known. For an analysis of measurement error in the field of nutritional epidemiology, see Willett (1990); for an application in the field of environmental health, see Rappaport et al. (1995).

Exercise 5.88*. Research studies where a pair of correlated Bernoulli (i.e., dichotomous) responses are recorded for each study subject are quite common. For example, each infant enrolled in a study about ear infections could have none, one, or two ears infected, where a dichotomous response (1 if infected, 0 if not) is recorded for each ear. More generally, a pair of dichotomous responses would tend to be correlated when measured at two sites (e.g., ears, eyes, hands) on the same subject. Ignoring such intra-subject response correlation can lead to invalid statistical inferences. For more information on this topic, see Agresti (2012) and Diggle, Liang, and Zeger (1994).

(a) Consider the following statistical model developed for the analysis of correlated Bernoulli response data. For $i = 1, 2, \ldots, n$ subjects and $j = 1, 2$ sites, let X_{ij} be the dichotomous response for subject i at site j, and let $\mathrm{pr}(X_{ij} = 1) = \alpha, 0 < \alpha < 1$. Further, let

$$\mathrm{pr}(X_{i1} = 1|X_{i2} = 1) = \mathrm{pr}(X_{i2} = 1|X_{i1} = 1) = \beta, 0 < \beta < 1.$$

Develop an explicit expression for $\mathrm{corr}(X_{i1}, X_{i2})$. How does the sign of $\mathrm{corr}(X_{i1}, X_{i2})$ vary as a function of α and β?

(b) For a study about ear infections involving n infants, suppose that we have n realizations $\{(x_{i1}, x_{i2})\}_{i=1}^n$ of the n mutually independent pairs of Bernoulli response random variables $\{(X_{i1}, X_{i2})\}_{i=1}^n$. Let the random variable Y_0 denote the number of infants with no ear infections, let Y_1 denote the number of infants with exactly one ear infection, and let Y_2 denote the number of infants with two ears infected; then, let y_0, y_1, and y_2 denote the corresponding realizations of these three random variables, where $(y_0 + y_1 + y_2) = n$. Under the statistical model described in part (a), show that the maximum likelihood estimates (MLEs) $\hat{\alpha}$ and $\hat{\beta}$ of α and β are

$$\hat{\alpha} = \frac{(y_1 + 2y_2)}{2n} \text{ and } \hat{\beta} = \frac{2y_2}{(y_1 + 2y_2)}.$$

(c) Suppose that $n = 100, y_0 = 60, y_1 = 15$, and $y_2 = 25$. Using expected information, compute an appropriate large-sample 95% confidence interval for the parameter $\theta = (\beta - \alpha)$, and then interpret your findings.

Exercise 5.89*. Suppose that a small double-blind randomized clinical trial is conducted to compare the short-term efficacies of two experimental antihypertension drugs (say, Drug 1 and Drug 2) designed to reduce diastolic blood pressure (DBP). Suppose that n subjects are randomly assigned to receive Drug 1 and that n subjects are randomly assigned to receive Drug 2. Subjects take a pill (containing either Drug 1 or Drug 2) each morning for six weeks. At the end of the six-week period, two DBP readings (taken 10 minutes apart) are obtained for each patient and are then averaged. The reason for using an average DBP for each patient is that individual DBP measurements can be highly variable.

Suppose that Y_{ijk} represents the k-th DBP measurement for the j-th patient receiving Drug $i, i = 1, 2, j = 1, 2, \ldots, n$, and $k = 1, 2$. Let $\overline{Y}_{ij} = \frac{1}{2}\sum_{k=1}^{2} Y_{ijk}$, and let $\bar{Y}_i = \frac{1}{n}\sum_{j=1}^{n} \overline{Y}_{ij} = \frac{1}{2n}\sum_{j=1}^{n}\sum_{k=1}^{2} Y_{ijk}$.

(a) Suppose it is assumed that $Y_{ijk} = \mu_i + \epsilon_{ijk}$, that $\epsilon_{ijk} \sim \text{N}(0, \sigma_\epsilon^2)$, that the $\{\epsilon_{ijk}\}$ constitute a set of mutually independent random variables, and that $\sigma_\epsilon^2 = 21$. Use these assumptions and the estimator $(\bar{Y}_1 - \overline{Y}_2)$ to develop an exact 95% confidence interval for the parameter $(\mu_1 - \mu_2)$, which is a measure of the true difference in the efficacies of these two experimental antihypertensive drugs. If $n = 10$ and if the observed values of $\bar{Y}_1$ and $\overline{Y}_2$ are $\bar{y}_1 = 90$ and $\bar{y}_2 = 87$, respectively, compute an exact 95% confidence interval for $(\mu_1 - \mu_2)$, and then comment on your findings.

(b) A biostatistician suggests that the model assumed in part (a) is not correct, because DBP measurements taken close together in time on the same subject are not independent and would be expected to be positively correlated. This biostatistician suggests that a better model would be one where the pair (Y_{ij1}, Y_{ij2}) is assumed to have a bivariate normal distribution and, with the same assumptions about the $\{\epsilon_{ijk}\}$ as stated in part (a), that

$$Y_{ijk} = \mu_i + \beta_{ij} + \epsilon_{ijk},$$

where $\beta_{ij} \sim \text{N}(0, \sigma_\beta^2)$, where the $\{\beta_{ij}\}$ constitute a set of mutually independent random variables, where the $\{\beta_{ij}\}$ are independent of the $\{\epsilon_{ijk}\}$, and where $\sigma_\beta^2 = 7$.
Under this model, find the numerical value of $\text{corr}(Y_{ij1}, Y_{ij2})$. Also, under

this model, and for the data given in part (a), use the estimator $(\bar{Y}_1 - \bar{Y}_2)$ to develop an exact 95% confidence interval for $(\mu_1 - \mu_2)$. Comment on your findings.

For detailed discussions concerning the analysis of correlated data, see Agresti (2012), Kleinbaum et al. (2008, Chapters 25 and 26), and Diggle, Liang, and Zeger (1994).

Exercise 5.90*. For research studies in most areas of science, a typical issue is whether it is worthwhile to increase the number of observations from n to $(n+k)$, the goal being to produce a more accurate estimator of a parameter of interest.

To quantify this issue in a simple situation, let $X_1, X_2, \ldots, X_n, X_{n+1}, \ldots, X_{n+k}$ constitute a random sample of size $(n+k)$ from a $\mathrm{N}(\mu, \sigma^2)$ population, and let

$$\bar{X}_n = n^{-1}\sum_{i=1}^{n} X_i \text{ and } \bar{X}_{n+k} = (n+k)^{-1}\sum_{i=1}^{n+k} X_i.$$

Suppose that the ratio

$$\frac{\left(\bar{X}_{n+k} - \mu\right)^2}{\left(\bar{X}_n - \mu\right)^2}$$

is to be used as a relative measure comparing $\bar{X}_{n+k}$ to $\bar{X}_n$ as an estimator of μ.

Further, suppose that it is desired to choose the smallest value of k, say k^*, such that

$$\mathrm{pr}\left[\frac{\left(\bar{X}_{n+k} - \mu\right)^2}{\left(\bar{X}_n - \mu\right)^2} < \delta\right] \geq \pi,$$

where $n(>1), \delta(0 < \delta < 1)$, and $\pi(0 < \pi < 1)$ have specified values.

Find explicit expressions for A and B (where A and B are each functions of n, k, and π) such that k^* is the smallest value of k satisfying the probability inequality

$$\mathrm{pr}\,(A < T_1 < B) \geq \pi,$$

where T_1 has a t-distribution with 1 degree of freedom (i.e., $T_1 \sim t_1$).

HINT: Use the fact that $Z_1/Z_2 \sim t_1$ when $Z_1 \sim \mathrm{N}(0,1), Z_2 \sim \mathrm{N}(0,1)$, and Z_1 and Z_2 are independent random variables.

For a more detailed discussion of this topic, see Webb, Smith, and Firag (2010).

5.2 Solutions to Odd-Numbered Exercises

Solution 5.1. For $i = 1, 2, \ldots, n$, since $Y_i \sim$GAMMA$(\alpha = 2\sigma^2, \beta = 1/2)$, it follows that $\mathrm{E}(Y_i) = \alpha\beta = \sigma^2$ and $\mathrm{V}(Y_i) = \alpha^2\beta = 2\sigma^4$. Thus, $\mathrm{E}(\bar{Y}) = \sigma^2$ and $\mathrm{V}(\bar{Y}) = 2\sigma^4/n$. So, by the Central Limit Theorem,

$$\frac{\bar{Y} - \sigma^2}{\sqrt{2\sigma^4/n}} = \sqrt{\frac{n}{2}}\left(\frac{\bar{Y}}{\sigma^2} - 1\right) \dot{\sim} \mathrm{N}(0, 1)$$

for large n.

So, if $\mathrm{pr}[Z < Z_{1-\frac{\alpha}{2}}] = (1 - \frac{\alpha}{2})$ when $Z \sim$N(0,1), then, for large n,

$$\begin{aligned}
(1-\alpha) &\approx \mathrm{pr}\left[-Z_{1-\frac{\alpha}{2}} < \sqrt{\frac{n}{2}}\left(\frac{\bar{Y}}{\sigma^2} - 1\right) < Z_{1-\frac{\alpha}{2}}\right] \\
&= \mathrm{pr}\left[1 - \sqrt{\frac{2}{n}}Z_{1-\frac{\alpha}{2}} < \frac{\bar{Y}}{\sigma^2} < 1 + \sqrt{\frac{2}{n}}Z_{1-\frac{\alpha}{2}}\right] \\
&= \mathrm{pr}\left[\frac{\bar{Y}}{1 + \sqrt{\frac{2}{n}}Z_{1-\frac{\alpha}{2}}} < \sigma^2 < \frac{\bar{Y}}{1 - \sqrt{\frac{2}{n}}Z_{1-\frac{\alpha}{2}}}\right] \\
&= \mathrm{pr}\left\{\left[\frac{\bar{Y}}{1 + \sqrt{\frac{2}{n}}Z_{1-\frac{\alpha}{2}}}\right]^{1/2} < \sigma < \left[\frac{\bar{Y}}{1 - \sqrt{\frac{2}{n}}Z_{1-\frac{\alpha}{2}}}\right]^{1/2}\right\},
\end{aligned}$$

where, since n is large, $(1 - \sqrt{\frac{2}{n}}Z_{1-\frac{\alpha}{2}}) > 0$ for any typically chosen value of α (say, $0.01 \leq \alpha \leq 0.10$).

Finally, for $n = 50, \alpha = 0.05$, and $\bar{y} = 10$, the computed 95% confidence interval for σ is (2.68, 4.06).

Solution 5.3.

(a) Since

$$U = \frac{(n-1)S^2}{\sigma^2} \sim \chi^2_{n-1},$$

it follows that

$$\begin{aligned}
\mathrm{E}\left(U^{1/2}\right) &= \frac{\sqrt{n-1}}{\sigma}\mathrm{E}(S) = \int_0^\infty u^{1/2}\frac{u^{\left(\frac{n-1}{2}\right)-1}\mathrm{e}^{-u/2}}{\Gamma\left(\frac{n-1}{2}\right)2^{\left(\frac{n-1}{2}\right)}}\mathrm{d}u \\
&= \frac{1}{\Gamma\left(\frac{n-1}{2}\right)\cdot 2^{\left(\frac{n-1}{2}\right)}}\int_0^\infty u^{\left(\frac{n}{2}\right)-1}\mathrm{e}^{-u/2}\mathrm{d}u \\
&= \frac{\Gamma\left(\frac{n}{2}\right)2^{n/2}}{\Gamma\left(\frac{n-1}{2}\right)2^{\left(\frac{n-1}{2}\right)}} = \frac{\sqrt{2}\Gamma\left(\frac{n}{2}\right)}{\Gamma\left(\frac{n-1}{2}\right)}; \\
\Rightarrow \quad \mathrm{E}(S) &= \sqrt{\frac{2}{(n-1)}}\frac{\Gamma\left(\frac{n}{2}\right)}{\Gamma\left(\frac{n-1}{2}\right)}\sigma, \\
\text{so that} \quad \mathrm{E}(\hat{\sigma}) &= \mathrm{E}(kS) = \sigma \text{ when } k = \sqrt{\frac{(n-1)}{2}}\frac{\Gamma\left(\frac{n-1}{2}\right)}{\Gamma\left(\frac{n}{2}\right)}.
\end{aligned}$$

(b) The likelihood function $\mathcal{L}$ is

$$\begin{aligned}
\mathcal{L} &= \prod_{i=1}^{n}\frac{1}{(2\pi\sigma^2)^{1/2}}\,\mathrm{e}^{-\frac{1}{2\sigma^2}(y_i-\mu)^2} \\
&= (2\pi)^{-\frac{n}{2}}\sigma^{-n}\,\mathrm{e}^{-\frac{1}{2\sigma^2}\sum_{i=1}^{n}(y_i-\mu)^2} \\
\Rightarrow \quad \ln\mathcal{L} &= -\frac{n}{2}\ln(2\pi) - n\ln\sigma - \frac{1}{2\sigma^2}\sum_{i=1}^{n}(y_i-\mu)^2 \\
\Rightarrow \quad \frac{\partial\ln\mathcal{L}}{\partial\sigma} &= \frac{-n}{\sigma} + \frac{1}{\sigma^3}\sum_{i=1}^{n}(y_i-\mu)^2 \\
\Rightarrow \quad \frac{\partial^2\ln\mathcal{L}}{\partial\sigma^2} &= \frac{n}{\sigma^2} - \frac{3}{\sigma^4}\sum_{i=1}^{n}(y_i-\mu)^2 \\
\Rightarrow \quad -\mathrm{E}\left(\frac{\partial^2\ln\mathcal{L}}{\partial\sigma^2}\right) &= \frac{-n}{\sigma^2} + \frac{3}{\sigma^4}\sum_{i=1}^{n}\mathrm{E}\left[(y_i-\mu)^2\right] \\
&= \frac{-n}{\sigma^2} + \frac{3}{\sigma^4}(n\sigma^2) = \frac{2n}{\sigma^2}.
\end{aligned}$$

Note that

$$\frac{\partial^2\ln\mathcal{L}}{\partial\sigma\partial\mu} = -\frac{2}{\sigma^3}\sum_{i=1}^{n}(y_i-\mu), \quad \text{so that } -\mathrm{E}\left(\frac{\partial^2\ln\mathcal{L}}{\partial\sigma\partial\mu}\right) = 0.$$

So, the C-R lower bound for the variance of any unbiased estimator of σ based on a random sample $Y_1, Y_2, \ldots, Y_n$ of size n from $\mathrm{N}(\mu, \sigma^2)$ is

$$\left[-\mathrm{E}\left(\frac{\partial^2\ln\mathcal{L}}{\partial\sigma^2}\right)\right]^{-1} = \frac{\sigma^2}{2n}.$$

Now,

$$\begin{aligned}
\mathrm{V}(\hat{\sigma}) &= \mathrm{V}(kS) = k^2\mathrm{V}(S) = k^2\left\{\mathrm{E}\left(S^2\right) - [\mathrm{E}(S)]^2\right\} \\
&= k^2\left[\sigma^2 - \frac{\sigma^2}{k^2}\right] = \left(k^2 - 1\right)\sigma^2.
\end{aligned}$$

For $n = 3$,

$$\begin{aligned}
k^2 &= \left(\frac{3-1}{2}\right)\frac{\Gamma^2\left(\frac{3-1}{2}\right)}{\Gamma^2\left(\frac{3}{2}\right)} \\
&= \Gamma^{-2}\left(\frac{3}{2}\right) = \left(\frac{\sqrt{\pi}}{2}\right)^{-2} = \frac{4}{\pi} \doteq 1.2732.
\end{aligned}$$

So, $\mathrm{V}(\hat{\sigma}) = (1.2732 - 1)\sigma^2 = 0.2732\,\sigma^2$, whereas the C-R lower bound is given by $\sigma^2/(2 \cdot 3) = 0.1667\,\sigma^2$. Hence, $\hat{\sigma}$ is *not* the MVBUE of σ.

Solution 5.5.

(a) Let $\hat{\theta} = (3\bar{X} - 5\bar{Y})$. Then, since $\bar{X} \sim \mathrm{N}\left(\mu_1, \sigma_1^2/n_1\right)$, $\bar{Y} \sim \mathrm{N}(\mu_2, \sigma_2^2/n_2)$, and $\bar{X}$ and $\bar{Y}$ are independent, it follows that

$$\hat{\theta} \sim \mathrm{N}\left(\theta, \frac{9\sigma_1^2}{n_1} + \frac{25\sigma_2^2}{n_2}\right), \text{ so that } \frac{\hat{\theta} - \theta}{\sqrt{\frac{9\sigma_1^2}{n_1} + \frac{25\sigma_2^2}{n_2}}} \sim \mathrm{N}(0, 1).$$

If $\mathrm{pr}\left(Z > Z_{1-\frac{\alpha}{2}}\right) = \frac{\alpha}{2}$ when $Z \sim \mathrm{N}(0, 1)$, then

$$\begin{aligned}
(1-\alpha) &= \mathrm{pr}(-Z_{1-\frac{\alpha}{2}} < Z < Z_{1-\frac{\alpha}{2}}) \\
&= \mathrm{pr}\left[-Z_{1-\frac{\alpha}{2}} < \frac{\hat{\theta} - \theta}{\sqrt{\frac{9\sigma_1^2}{n_1} + \frac{25\sigma_2^2}{n_2}}} < Z_{1-\frac{\alpha}{2}}\right] \\
&= \mathrm{pr}\left[\hat{\theta} - Z_{1-\frac{\alpha}{2}}\sqrt{\frac{9\sigma_1^2}{n_1} + \frac{25\sigma_2^2}{n_2}} < \theta < \hat{\theta} + Z_{1-\frac{\alpha}{2}}\sqrt{\frac{9\sigma_1^2}{n_1} + \frac{25\sigma_2^2}{n_2}}\right].
\end{aligned}$$

So, the appropriate $100(1-\alpha)\%$ confidence interval is:

$$(3\bar{X} - 5\bar{Y}) \pm Z_{1-\frac{\alpha}{2}}\sqrt{\frac{9\sigma_1^2}{n_1} + \frac{25\sigma_2^2}{n_2}}.$$

(b) When $\alpha = 0.05, \sigma_1^2 = 2$, and $\sigma_2^2 = 3$, then the width W of the confidence interval developed in part (a) is

$$2(1.96)\sqrt{\frac{9(2)}{n_1} + \frac{25(3)}{n_2}} = 3.92\sqrt{\frac{18}{n_1} + \frac{75}{n_2}}.$$

Minimizing W subject to $(n_1 + n_2) = 100$ is equivalent to minimizing $Q = 18/n_1 + 75/(100 - n_1)$ with respect to n_1.
So,

$$\begin{aligned}\frac{dQ}{dn_1} &= \frac{-18}{n_1^2} + \frac{75}{(100-n_1)^2} = 0\\ &\Rightarrow 75n_1^2 - 18(100-n_1)^2 = 0\\ &\Rightarrow 57n_1^2 + 3,600n_1 - 180,000 = 0.\end{aligned}$$

Applying the quadratic formula and choosing the positive root gives $n_1 = 33$ and $n_2 = 67$ as the sample size values that minimize W.

Solution 5.7. We know that $\mu = \mathrm{E}(Y) = \frac{\theta}{2}$ and $\sigma^2 = \mathrm{V}(Y) = \frac{\theta^2}{12}$. First, we need to find a sufficient statistic for θ. Let $I_E(x)$ be the indicator function for the set E, so that $I_E(x)$ equals 1 if $x \in E$ and $I_E(x)$ equals 0 otherwise. Then, letting $A = (0, \theta)$ and $B = (0, \infty)$, we have

$$\begin{aligned}\mathrm{f}_{Y_1,Y_2,\ldots,Y_n}(y_1, y_2, \ldots, y_n; \theta) &= \prod_{i=1}^{n} \{\theta^{-1} I_A(y_i)\}\\ &= \left[\left(\theta^{-n}\right) I_A(y_{(n)})\right] \cdot \left[\prod_{i=1}^{n} I_B(y_i)\right]\\ &= \mathrm{g}(u;\theta) \cdot \mathrm{h}(y_1, y_2, \ldots, y_n),\end{aligned}$$

where $y_{(n)} = \max\{y_1, y_2, \ldots, y_n\}$. Thus, it follows that $Y_{(n)}$ is sufficient for θ since, given $Y_{(n)} = y_{(n)}$, $\mathrm{h}(y_1, y_2, \ldots, y_n)$ does not in any way depend on θ.

Now, since $\mathrm{F}_Y(y;\theta) = \frac{y}{\theta}, 0 < y < \theta$, it follows that

$$\begin{aligned}\mathrm{f}_{Y_{(n)}}(y_{(n)};\theta) &= n\left[\frac{y_{(n)}}{\theta}\right]^{n-1}\theta^{-1}\\ &= n\theta^{-n}y_{(n)}^{n-1}, 0 < y_{(n)} < \theta.\end{aligned}$$

For r a non-negative integer,

$$\begin{aligned}\mathrm{E}\left(Y_{(n)}^r\right) &= \int_0^\theta y_{(n)}^r n\theta^{-n} y_{(n)}^{n-1} dy_{(n)}\\ &= n\theta^{-n}\int_0^\theta y_{(n)}^{(n+r)-1} dy_{(n)}\\ &= n\theta^{-n}\left[\frac{y_{(n)}^{n+r}}{(n+r)}\right]_0^\theta\\ &= \left(\frac{n}{n+r}\right)\theta^r.\end{aligned}$$

So, since $\mathrm{E}[Y_{(n)}] = \left(\frac{n}{n+1}\right)\theta$ and $\mathrm{E}(Y) = \frac{\theta}{2}$, the candidate MVUE for $\mathrm{E}(Y)$ is

$$\hat{\mu} = \left(\frac{n+1}{2n}\right) Y_{(n)}.$$

Since $\mathrm{E}[Y_{(n)}^2] = \left(\frac{n}{n+2}\right)\theta^2$ and $\mathrm{V}(Y) = \sigma^2 = \frac{\theta^2}{12}$, the candidate MVUE for V(Y) is

$$\hat{\sigma}^2 = \left(\frac{n+2}{12n}\right) Y_{(n)}^2.$$

To conclude that these candidate MVUEs are actually the unique MVUEs of μ and σ^2, we need to show that $Y_{(n)}$ is a complete sufficient statistic for θ. Now, with $U = Y_{(n)}$, let $\mathrm{g}(U)$ be any function of U. If we can show that $\mathrm{E}[\mathrm{g}(U)] = 0$ for all $\theta > 0$ implies that $\mathrm{g}(U) = 0$ for $U > 0$, then $Y_{(n)}$ is a complete sufficient statistic for θ. Now, with

$$\mathrm{E}[\mathrm{g}(U)] = \int_0^\theta \mathrm{g}(u) n\theta^{-n} u^{n-1} \mathrm{d}u = 0,$$

we have

$$\begin{aligned}
\frac{\mathrm{dE}[\mathrm{g}(U)]}{\mathrm{d}\theta} &= 0 = \frac{\mathrm{d}}{\mathrm{d}\theta}\left[\theta^{-n}\int_0^\theta n\mathrm{g}(u)u^{n-1}\mathrm{d}u\right] \\
&= \theta^{-n}\frac{\mathrm{d}}{\mathrm{d}\theta}\left[\int_0^\theta n\mathrm{g}(u)u^{n-1}\mathrm{d}u\right] + \left[\int_0^\theta n\mathrm{g}(u)u^{n-1}\mathrm{d}u\right]\frac{\mathrm{d}}{\mathrm{d}\theta}\left(\theta^{-n}\right) \\
&= \theta^{-n} n\mathrm{g}(\theta)\theta^{n-1} + (0)(n\theta^{n-1}) = \theta^{-1} n\mathrm{g}(\theta).
\end{aligned}$$

So, since $\theta > 0$ and $n > 1$, the equation $\theta^{-1}n\mathrm{g}(\theta) = 0$ implies $\mathrm{g}(\theta) = 0$ for all $\theta > 0$, so that $U = Y_{(n)}$ is a complete sufficient statistic for θ.

Solution 5.9.

(a) By definition, the unweighted least squares estimators are the values of α and β that minimize the function

$$Q = \sum_{x=1}^{n}[Y_x - (\alpha x + \beta x^2)]^2.$$

The equation

$$\frac{\partial Q}{\partial \alpha} = -2\sum_{x=1}^{n} x[Y_x - (\alpha x + \beta x^2)] = 0$$

implies that

$$\sum_{x=1}^{n} xY_x - \alpha S_2 - \beta S_3 \quad = \quad 0,$$

or

$$\hat{\alpha} = \frac{\sum_{x=1}^{n} xY_x - \hat{\beta} S_3}{S_2}.$$

Similarly,

$$\frac{\partial Q}{\partial \beta} \quad = \quad -2\sum_{x=1}^{n} x^2[Y_x - (\alpha x + \beta x^2)] = 0$$

implies that

$$\sum_{x=1}^{n} x^2 Y_x - \alpha S_3 - \beta S_4 = 0,$$

or

$$\sum_{x=1}^{n} x^2 Y_x - \left[\frac{\sum_{x=1}^{n} xY_x - \hat{\beta} S_3}{S_2}\right] S_3 - \hat{\beta} S_4 = 0,$$

or

$$\begin{aligned}\hat{\beta} &= \frac{\sum_{x=1}^{n} x^2 Y_x - S_2^{-1} S_3 \sum_{x=1}^{n} xY_x}{(S_4 - S_2^{-1} S_3^2)} \\ &= \frac{S_2 \sum_{x=1}^{n} x^2 Y_x - S_3 \sum_{x=1}^{n} xY_x}{(S_2 S_4 - S_3^2)}.\end{aligned}$$

(b)

$$\begin{aligned}\mathrm{E}(\hat{\beta}) &= \frac{S_2 \sum_{x=1}^{n} x^2(\alpha x + \beta x^2) - S_3 \sum_{x=1}^{n} x(\alpha x + \beta x^2)}{(S_2 S_4 - S_3^2)} \\ &= \frac{\alpha S_2 S_3 + \beta S_2 S_4 - \alpha_2 S_2 S_3 - \beta S_3^2}{(S_2 S_4 - S_3^2)} = \beta.\end{aligned}$$

And,

$$\begin{aligned}
\mathrm{V}(\hat{\beta}) &= \mathrm{V}\left\{\frac{\sum_{x=1}^{n}(S_2x^2 - S_3x)Y_x}{(S_2S_4 - S_3^2)}\right\} \\
&= \frac{\sigma^2\sum_{x=1}^{n}(S_2x^2 - S_3x)^2}{(S_2S_4 - S_3^2)^2} \\
&= \frac{\sigma^2\sum_{x=1}^{n}(S_2^2x^4 - 2S_2S_3x^3 + S_3^2x^2)}{(S_2S_4 - S_3^2)^2} \\
&= \frac{\sigma^2[S_2^2S_4 - 2S_2S_3^2 + S_2S_3^2]}{(S_2S_4 - S_3^2)^2} \\
&= \frac{\sigma^2(S_2^2S_4 - S_2S_3^2)}{(S_2S_4 - S_3^2)^2} \\
&= \frac{\sigma^2S_2(S_2S_4 - S_3^2)}{(S_2S_4 - S_3^2)^2} \\
&= \frac{S_2\sigma^2}{(S_2S_4 - S_3^2)}.
\end{aligned}$$

(c) Since $\hat{\beta} = \sum_{x=1}^{n} C_xY_x$, where $C_x = (S_2x^2 - S_3x)/(S_2S_4 - S_3^2)$, $\hat{\beta}$ is a linear combination of mutually independent normal variates and so is itself normally distributed. Thus, since $\frac{\hat{\beta}-\beta}{\sqrt{\mathrm{V}(\hat{\beta})}} \sim \mathrm{N}(0,1)$ since $\mathrm{E}(\hat{\beta}) = \beta$, this gives an exact 95% confidence interval for β of the form:

$$\hat{\beta} \pm 1.96\sqrt{\frac{S_2\sigma^2}{(S_2S_4 - S_3^2)}}.$$

Since $n = 4$, $S_2 = (1)^2 + (2)^2 + (3)^2 + (4)^2 = 30$, $S_3 = (1)^3 + (2)^3 + (3)^3 + (4)^3 = 100$, and $S_4 = (1)^4 + (2)^4 + (3)^4 + (4)^4 = 354$. Thus, since $\sigma^2 = 1$, the confidence interval is:

$$2 \pm 1.96\sqrt{\frac{(30)(1)}{(30)(354) - (100)^2}},$$

or $(1.57, 2.43)$.

Solution 5.11. Note that

$$\mathrm{f}_{X,Y}(x, y; \theta) = \left(\theta \mathrm{e}^{-\theta x}\right)\left(\theta^{-1}\mathrm{e}^{-\theta^{-1}y}\right), x > 0, y > 0, \theta > 0.$$

So, $X \sim$GAMMA$(\alpha = \theta^{-1}, \beta = 1)$, $Y \sim$GAMMA$(\alpha = \theta, \beta = 1)$, and X and Y are independent random variables. So, $\sum_{i=1}^{n} X_i \sim$GAMMA$(\alpha = \theta^{-1}, \beta = n)$ and $\sum_{i=1}^{n} Y_i \sim$GAMMA$(\alpha = \theta, \beta = n)$.

Now, if $U \sim$GAMMA(α, β), then

$$\mathrm{E}\left(U^r\right) = \int_0^\infty u^r \frac{u^{\beta-1}\mathrm{e}^{-u/\alpha}}{\Gamma(\beta)\alpha^\beta}\mathrm{d}u = \frac{\Gamma(\beta+r)}{\Gamma(\beta)}\alpha^r, \quad (\beta+r) > 0.$$

So, since $\sum\limits_{i=1}^{n} X_i$ and $\sum\limits_{i=1}^{n} Y_i$ are independent random variables, we have

$$\begin{aligned}
\mathrm{E}(\hat{\theta}) &= \mathrm{E}\left[\frac{\left(\sum\limits_{i=1}^{n} Y_i\right)^{1/2}}{\left(\sum\limits_{i=1}^{n} X_i\right)^{1/2}}\right] = \mathrm{E}\left[\left(\sum_{i=1}^{n} Y_i\right)^{1/2}\right] \cdot \mathrm{E}\left[\left(\sum_{i=1}^{n} X_i\right)^{-1/2}\right] \\
&= \frac{\Gamma\left(n+\frac{1}{2}\right)}{\Gamma(n)}\theta^{1/2} \cdot \frac{\Gamma\left(n-\frac{1}{2}\right)}{\Gamma(n)}\left(\theta^{-1}\right)^{-1/2} = \frac{\Gamma\left(n+\frac{1}{2}\right)\Gamma\left(n-\frac{1}{2}\right)}{\Gamma^2(n)}\theta,
\end{aligned}$$

so that $\hat{\theta}$ is *not* an unbiased estimator of θ. However, since

$$\lim_{n\to\infty}\left[\frac{\Gamma\left(n+\frac{1}{2}\right)\Gamma\left(n-\frac{1}{2}\right)}{\Gamma^2(n)}\right] = 1,$$

$\hat{\theta}$ is an *asymptotically unbiased* estimator of θ.

Solution 5.13.

(a) Using the fact that $\mathrm{E}\left[C\sum_{i=1}^{n} Y_i\right] = Cn\mu$, that Y_i and Y_j are independent random variables for all $i \neq j$, and that $\mathrm{E}(Y_i - \mu) = \mathrm{E}(Y_j - \mu) = 0\,\forall i$ and j, we have

$$\begin{aligned}
\mathrm{E}\left[(T-\mu)^2\right] &= \mathrm{E}\left\{\left[C\sum_{i=1}^{n} Y_i - \mu\right]^2\right\} \\
&= \mathrm{E}\left\{\left[C\sum_{i=1}^{n} Y_i - Cn\mu + Cn\mu - \mu\right]^2\right\} \\
&= \mathrm{E}\left\{\left[C\sum_{i=1}^{n}(Y_i - \mu) + \mu(Cn-1)\right]^2\right\} \\
&= \mathrm{E}\left\{C^2\left[\sum_{i=1}^{n}(Y_i-\mu)\right]^2 + 2C\mu(Cn-1)\sum_{i=1}^{n}(Y_i-\mu)\right. \\
&\quad \left. + \, \mu^2(Cn-1)^2\right\}
\end{aligned}$$

$$\begin{aligned}
&= C^2\mathrm{E}\left[\sum_{i=1}^{n}(Y_i-\mu)^2 + 2\sum_{\text{all } i<j}(Y_i-\mu)(Y_j-\mu)\right]\\
&\quad + 2C\mu(Cn-1)\sum_{i=1}^{n}\mathrm{E}(Y_i-\mu) + \mu^2(Cn-1)^2\\
&= C^2\sum_{i=1}^{n}\mathrm{E}\left[(Y_i-\mu)^2\right] + C^2\cdot 2\sum_{\text{all } i<j}\mathrm{E}(Y_i-\mu)\mathrm{E}(Y_j-\mu)\\
&\quad + \mu^2(Cn-1)^2\\
&= C^2 n\sigma^2 + \mu^2(Cn-1)^2 = Q, \text{ say.}
\end{aligned}$$

Now,

$$\begin{aligned}
\frac{\mathrm{d}Q}{\mathrm{d}C} &= 2Cn\sigma^2 + 2\mu^2(Cn-1)n = 0\\
\Rightarrow &\quad Cn\sigma^2 + C\mu^2 n^2 - n\mu^2 = 0\\
\Rightarrow &\quad C(\sigma^2 + n\mu^2) = \mu^2\\
\Rightarrow &\quad C^* = \frac{\mu^2}{(n\mu^2+\sigma^2)} = \frac{1}{(n+\sigma^2/\mu^2)}\\
&\quad = (n+\theta^2)^{-1}.
\end{aligned}$$

This is a minimum since $\frac{\mathrm{d}^2Q}{\mathrm{d}C^2} = 2n\sigma^2 + 2n^2\mu^2 > 0$.

(b)

$$\begin{aligned}
\mathrm{V}(T^*) &= (C^*)^2(n\sigma^2) = \frac{n\sigma^2}{(n+\theta^2)^2}\\
&= \frac{\sigma^2}{\left(\frac{n}{\sqrt{n}} + \frac{\theta^2}{\sqrt{n}}\right)^2}\\
&= \frac{\sigma^2}{\left(n + 2\theta^2 + \frac{\theta^4}{n}\right)} < \frac{\sigma^2}{n} = \mathrm{V}(\bar{Y}).
\end{aligned}$$

(c) Now, $\mathrm{MSE}(\bar{Y},\mu) = \mathrm{V}(\bar{Y}) = \sigma^2/n$. And,

$$\begin{aligned}
\mathrm{MSE}(T^*,\mu) &= \mathrm{E}\left[(T^*-\mu)^2\right] = \mathrm{V}(T^*) + [\mathrm{E}(T^*)-\mu]^2\\
&= \frac{n\sigma^2}{(n+\theta^2)^2} + \left[\frac{n\mu}{(n+\theta^2)} - \mu\right]^2 = \frac{n\sigma^2 + [n\mu - \mu(n+\theta^2)]^2}{(n+\theta^2)^2}\\
&= \frac{n\sigma^2 + \mu^2\theta^4}{(n+\theta^2)^2} = \frac{n\sigma^2 + \mu^2(\sigma^2/\mu^2)\theta^2}{(n+\theta^2)^2}\\
&= \frac{\sigma^2(n+\theta^2)}{(n+\theta^2)^2} = \frac{\sigma^2}{(n+\theta^2)}.
\end{aligned}$$

So,

$$\begin{aligned}
\frac{\text{MSE}(\bar{Y},\mu)}{\text{MSE}(T^*,\mu)} &= \frac{\sigma^2/n}{\sigma^2/(n+\theta^2)} = \frac{1/n}{1/(n+\theta^2)} \\
&= \frac{(n+\theta^2)}{n} = 1 + \frac{\theta^2}{n}.
\end{aligned}$$

For finite n, $\text{MSE}(\bar{Y},\mu) > \text{MSE}(T^*,\mu)$. However, since

$$\lim_{n\to\infty}\left(1+\frac{\theta^2}{n}\right) = 1,$$

the two estimators are *asymptotically equally efficient*.

(d) Since $\lim_{n\to\infty}\text{MSE}(T^*,\mu) = \lim_{n\to\infty}\text{MSE}(\bar{Y},\mu) = 0$, both T^* and $\bar{Y}$ are MSE-consistent estimators of the parameter μ.

Solution 5.15.

(a) With $\boldsymbol{y} = (y_1, y_2, \ldots, y_n)$, we have

$$\begin{aligned}
\mathcal{L}(\boldsymbol{y};\theta) &= \prod_{i=1}^{n}\left[(\theta+1)(\theta+2)y_i^{\theta}(1-y_i)\right] \\
&= (\theta+1)^n(\theta+2)^n\left(\prod_{i=1}^{n} y_i\right)^{\theta}\prod_{i=1}^{n}(1-y_i); \\
\ln\mathcal{L}(\boldsymbol{y};\theta) &= n\ln(\theta+1) + n\ln(\theta+2) + \theta\sum_{i=1}^{n}\ln y_i + \sum_{i=1}^{n}\ln(1-y_i); \\
\frac{\partial\ln\mathcal{L}(\boldsymbol{y};\theta)}{\partial\theta} &= \frac{n}{(\theta+1)} + \frac{n}{(\theta+2)} + \sum_{i=1}^{n}\ln y_i; \\
\frac{-\partial^2\ln\mathcal{L}(\boldsymbol{y};\theta)}{\partial\theta^2} &= \frac{n}{(\theta+1)^2} + \frac{n}{(\theta+2)^2} \\
&= \frac{n(\theta^2+4\theta+4+\theta^2+2\theta+1)}{(\theta+1)^2(\theta+2)^2} \\
&= \frac{n(2\theta^2+6\theta+5)}{(\theta+1)^2(\theta+2)^2}.
\end{aligned}$$

Since, for large n,

$$\frac{(\hat{\theta}-\theta)}{\sqrt{\frac{(\hat{\theta}+1)^2(\hat{\theta}+2)^2}{n(2\hat{\theta}^2+6\hat{\theta}+5)}}} \,\dot{\sim}\, \text{N}(0,1),$$

the 95% large-sample confidence interval for θ given $\hat{\theta} = 3.00$ and $n = 50$

is:

$$3.00 \pm 1.96\sqrt{\frac{(3.00+1)^2(3.00+2)^2}{50[2(3.00)^2+6(3.00)+5]}} = 3.00 \pm 1.96\sqrt{0.195}$$
$$= (2.134, 3.866).$$

(b) We require the minimum sample size n^* such that

$$\text{width} = 2\left[1.96\sqrt{\frac{(\hat{\theta}+1)^2(\hat{\theta}+2)^2}{n(2\hat{\theta}^2+6\hat{\theta}+5)}}\right] \leq 1.00,$$

where $\hat{\theta}$ is an estimated value of θ. From part (a), *a* reasonable value to use for $\hat{\theta}$ is 3.00. So, we have:

$$n \geq \frac{(1.96)^2(3.00+1)^2(3.00+2)^2}{[2(3.00)^2+6(3.00)+5](0.50)^2} = 149.92;$$

thus, a reasonable value for n^* is $n^* = 150$.

Solution 5.17. Since

$$\mathrm{E}(\hat{\theta}) = \mathrm{E}\left(\sum_{i=1}^{n} c_i Y_i\right) = \sum_{i=1}^{n} c_i \mathrm{E}(Y_i) = \theta \sum_{i=1}^{n} c_i,$$

we require the condition $\sum_{i=1}^{n} c_i = 1$ for $\hat{\theta}$ to be an unbiased estimator of θ.

Now, using the condition $\sum_{i=1}^{n} c_i = 1$, we have

$$\begin{aligned}
\mathrm{V}(\hat{\theta}) &= \sum_{i=1}^{n} a_i^2 \mathrm{V}(Y_i) = \sigma^2 \sum_{i=1}^{n} a_i^2 \\
&= \sigma^2 \sum_{i=1}^{n}\left(a_i^2 - \frac{2a_i}{n} + \frac{1}{n^2}\right) + \frac{\sigma^2}{n} \\
&= \sigma^2 \sum_{i=1}^{n}\left(a_i - \frac{1}{n}\right)^2 + \frac{\sigma^2}{n} \geq \frac{\sigma^2}{n}.
\end{aligned}$$

Thus, the required specification is $c_i = 1/n, i = 1, 2, \ldots, n$, giving $\hat{\theta} = \bar{Y} = n^{-1}\sum_{i=1}^{n} Y_i$.

Solution 5.19.

$$\begin{aligned}
\text{MSE}(cS^2, \sigma^2) &= \text{V}(cS^2) + [\text{E}(cS^2) - \sigma^2]^2 \\
&= \frac{c^2}{n}\left[\mu_4 - \left(\frac{n-3}{n-1}\right)\sigma^4\right] + \sigma^4(c-1)^2 = Q, \text{ say.} \\
\text{Now, } \frac{\text{d}Q}{\text{d}c} &= \frac{2c}{n}\left[\mu_4 - \left(\frac{n-3}{n-1}\right)\sigma^4\right] + 2\sigma^4(c-1) = 0 \\
\Rightarrow \quad & \frac{c}{n}\left[\frac{\mu_4}{\sigma^4} - \left(\frac{n-3}{n-1}\right) + n\right] = 1 \\
\Rightarrow \quad c^* &= \frac{n}{\left(\frac{\mu_4}{\sigma^4}\right) - \left(\frac{n-3}{n-1}\right) + n} = \frac{n}{\left(\frac{\mu_4}{\sigma^4}\right) + \frac{n^2-2n+3}{n-1}}.
\end{aligned}$$

Since $\frac{\text{d}^2Q}{\text{d}c^2} = 2\text{V}(S^2) + 2\sigma^4 > 0$, c^* minimizes Q.

When $\mu_4/\sigma^4 < 3$, $c^* > n/\left[3 + \frac{n^2-2n+3}{n-1}\right] = \left(\frac{n-1}{n+1}\right)$ and when $\mu_4/\sigma^4 > 3$, $c^* < \left(\frac{n-1}{n+1}\right)$.

Note: When $X \sim \text{N}(\mu, \sigma^2)$, then $\mu_4 = 3\sigma^4$, so that $\left(\frac{n-1}{n+1}\right) S^2$ has the smallest MSE of all estimators of σ^2 of the form cS^2.

Solution 5.21. Since

$$\prod_{i=1}^{n}\left\{\frac{1}{\sqrt{2\pi}\sigma}\text{e}^{-(x_i-\mu_0)^2/2\sigma^2}\right\} = (2\pi)^{-n/2}\theta^{-n/4}\text{e}^{-\frac{1}{2\sqrt{\theta}}\sum_{i=1}^{n}(x_i-\mu_0)^2},$$

it follows from the Factorization Theorem that $U = \sum_{i=1}^{n}(X_i - \mu_0)^2$ is a sufficient statistic for θ. And, from exponential family theory, U is a *complete* sufficient statistic for θ.

Since

$$\frac{U}{\sigma^2} = \sum_{i=1}^{n}\left(\frac{X_i - \mu_0}{\sigma}\right)^2 \sim \chi^2_n \equiv \text{GAMMA}(\alpha = 2, \beta = n/2),$$

so that

$$\text{E}\left(\frac{U^r}{\sigma^{2r}}\right) = \frac{\Gamma\left(\frac{n}{2}+r\right)}{\Gamma\left(\frac{n}{2}\right)}2^r \text{ for } \left(\frac{n}{2}+r\right) > 0,$$

we have

$$\text{E}\left(\frac{U^2}{\sigma^4}\right) = \frac{\Gamma\left(\frac{n}{2}+2\right)}{\Gamma\left(\frac{n}{2}\right)}2^2 = n(n+2).$$

Thus,

$$\hat{\theta} = \frac{U^2}{n(n+2)} = \frac{\left[\sum_{i=1}^n (X_i - \mu_0)^2\right]^2}{n(n+2)}$$

is the MVUE of $\theta = \sigma^4$.

Now,

$$\begin{aligned}
(1-\alpha) &= \text{pr}\left[\chi^2_{n,\frac{\alpha}{2}} < \frac{U}{\sigma^2} < \chi^2_{n,1-\frac{\alpha}{2}}\right] \\
&= \text{pr}\left[\frac{1}{\chi^2_{n,1-\frac{\alpha}{2}}} < \frac{\sigma^2}{U} < \frac{1}{\chi^2_{n,\frac{\alpha}{2}}}\right] \\
&= \text{pr}\left[\frac{U^2}{\left(\chi^2_{n,1-\frac{\alpha}{2}}\right)^2} < \sigma^4 < \frac{U^2}{\left(\chi^2_{n,\frac{\alpha}{2}}\right)^2}\right].
\end{aligned}$$

Since $U = \sum_{i=1}^n (X_i - \mu_0)^2 = \sum_{i=1}^n X_i^2 - 2\mu_0 \sum_{i=1}^n X_i + n\mu_0^2$, the available data yield $u = 5$ and $\hat{\theta} = \frac{(5)^2}{4(4+2)} = 1.042$. For $\alpha = 0.02$, since $\chi^2_{4,\,0.01} = 0.297$, and $\chi^2_{4,\,0.99} = 13.277$, the computed exact 98% confidence interval for $\theta = \sigma^4$ is (0.142, 283.418). The computed confidence interval is extremely wide because the sample size is very small.

Solution 5.23. Since

$$\prod_{i=1}^n \pi^{x_i}(1-\pi)^{1-x_i} = \pi^{\sum_{i=1}^n x_i}(1-\pi)^{n-\sum_{i=1}^n x_i} = (\theta^{1/k})^u(1-\theta^{1/k})^{n-u},$$

where $u = \sum_{i=1}^n x_i$, it follows by the Factorization Theorem that $U = \sum_{i=1}^n X_i$ is a sufficient statistic for θ. And, from exponential family theory, U is a *complete* sufficient statistic for θ.

Let $U^* = \prod_{i=1}^k X_i$. Then, $\text{E}(U^*) = \pi^k$. So, by the Rao-Blackwell Theorem, $\hat{\theta} = \text{E}(U^*|U = u)$. Since U^* can only take the values zero and one, it follows that

$$\begin{aligned}
\hat{\theta} &= \text{pr}(U^* = 1|U = u) = \frac{\text{pr}\left[\left(\prod\limits_{i=1}^k X_i = 1\right) \cap \left(\sum\limits_{i=1}^n X_i = u\right)\right]}{\text{pr}\left(\sum_{i=1}^n X_i = u\right)} \\
&= \frac{\text{pr}\left[\left(\prod\limits_{i=1}^k X_i = 1\right) \cap \left(\sum\limits_{i=(k+1)}^n X_i = u - k\right)\right]}{\text{pr}\left(\sum_{i=1}^n X_i = u\right)}.
\end{aligned}$$

Note that $\hat{\theta}$ equals zero when $u < k$. And, for $u \geq k$, since $\sum_{i=(k+1)}^n X_i \sim$

$\text{BIN}(n-k,\pi)$ and $\sum_{i=1}^{n} X_i \sim \text{BIN}(n,\pi)$, we have

$$\hat{\theta} = \frac{(\pi^k)\cdot \text{C}_{u-k}^{n-k}\pi^{u-k}(1-\pi)^{n-u}}{\text{C}_u^n \pi^u (1-\pi)^{n-u}} = \frac{\text{C}_{u-k}^{n-k}}{\text{C}_u^n}.$$

So, the MVUE of θ is:

$$\hat{\theta} = \begin{cases} 0, & \text{if } u < k; \\ \frac{\text{C}_{u-k}^{n-k}}{\text{C}_u^n}, & \text{if } u \geq k. \end{cases}$$

As an example, if $k=1$, so that $\theta = \pi$, then

$$\hat{\theta} = \begin{cases} 0, & \text{if } u < 1; \\ \frac{u}{n}, & \text{if } u \geq 1. \end{cases}$$

In other words, $\hat{\theta} = \sum_{i=1}^{n} X_i/n$, the usual estimator of θ.

Now, to show directly that $\text{E}(\hat{\theta}) = \theta$, we have

$$\begin{aligned} \text{E}(\hat{\theta}) &= \sum_{u=k}^{n} \frac{\text{C}_{u-k}^{n-k}}{\text{C}_u^n}\text{C}_u^n \pi^u (1-\pi)^{n-u} = \sum_{u=k}^{n} \text{C}_{u-k}^{n-k}\pi^u(1-\pi)^{n-u} \\ &= \sum_{s=0}^{n-k} \text{C}_s^{n-k}\pi^{s+k}(1-\pi)^{n-(s+k)} \\ &= \pi^k \sum_{s=0}^{n-k} \text{C}_s^{n-k}\pi^s(1-\pi)^{(n-k)-s} = \pi^k = \theta. \end{aligned}$$

Solution 5.25. Clearly, $\text{E}(\overline{X}_i) = \mu$ and $\text{V}(\overline{X}_i) = \sigma_i^2/n_i$, $i = 1, 2$. So,

$$\text{E}(\hat{\mu}) = \text{E}(w_1\overline{X}_1 + w_2\overline{X}_2) = w_1\mu + w_2\mu = (w_1+w_2)\mu,$$

so that we require $(w_1 + w_2) = 1$. And,

$$\text{V}(w_1\overline{X}_1 + w_2\overline{X}_2) = w_1^2\left(\frac{\sigma_1^2}{n_1}\right) + w_2^2\left(\frac{\sigma_2^2}{n_2}\right).$$

Thus, we want to minimize $\text{V}(w_1\overline{X}_1 + w_2\overline{X}_2) = \text{V}(\hat{\mu})$ subject to the restriction that $(w_1 + w_2) = 1$. So,

$$\begin{aligned} 0 = \frac{\partial \text{V}(\hat{\mu})}{\partial w_1} &= \frac{\partial}{\partial w_1}\left[w_1^2\left(\frac{\sigma_1^2}{n_1}\right) + (1-w_1)^2\left(\frac{\sigma_2^2}{n_2}\right)\right] \\ &= 2w_1\left(\frac{\sigma_1^2}{n_1}\right) - 2(1-w_1)\left(\frac{\sigma_2^2}{n_1}\right) \\ \implies & w_1 = \frac{\sigma_2^2/n_2}{(\sigma_1^2/n_1 + \sigma_2^2/n_2)}, \end{aligned}$$

and so

$$w_2 = (1 - w_1) = \frac{\sigma_1^2/n_1}{(\sigma_1^2/n_1 + \sigma_2^2/n_2)}.$$

Thus, if $V(\overline{X}_1)$ is much greater than $V(\overline{X}_2)$, then $\overline{X}_2$ gets more weight than $\overline{X}_1$ in the estimator $\hat{\mu}$. In general, we are weighting the two sample means inversely proportional to the sizes of their variances.

Solution 5.27.

(a) If $\hat{\theta}$ is a point estimator of a parameter θ, recall that the MSE of $\hat{\theta}$ as an estimator of θ is equal to

$$\text{MSE}(\hat{\theta}, \theta) = \text{V}(\hat{\theta}) + [\text{E}(\hat{\theta}) - \theta]^2.$$

Since $\text{E}(\bar{X}) = \mu$ and $\text{V}(\bar{X}) = \frac{\sigma^2}{n}$, it follows that

$$\text{MSE}(\bar{X}, \mu) = \text{V}(\bar{X}) = \frac{\sigma^2}{n}.$$

And, since $\text{E}(k\bar{X}) = k\mu$ and $\text{V}(k\bar{X}) = k^2\left(\frac{\sigma^2}{n}\right)$, it follows that

$$\text{MSE}(k\bar{X}, \mu) = k^2\left(\frac{\sigma^2}{n}\right) + (1-k)^2\mu^2.$$

So,

$$\text{MSE}(\bar{X}, \mu) - \text{MSE}(k\bar{X}, \mu) = (1-k^2)\left(\frac{\sigma^2}{n}\right) - (1-k)^2\mu^2.$$

So, for $0 < k < 1$, the previous expression is positive when

$$\frac{(1+k)}{(1-k)} > \frac{n\mu^2}{\sigma^2}.$$

Thus, for finite values of μ and σ^2, there will always exist a set of values for k, namely $\{k : 0 < c < k < 1\}$, with c appropriately chosen based on the value of $n\mu^2/\sigma^2$, for which $\text{MSE}(k\bar{X}, \mu) < \text{MSE}(\bar{X}, \mu)$.

(b) The obvious problem with the use of $k\bar{X}$ as a point estimator of μ is that a proper choice for the value of k will require knowledge about the ratio μ^2/σ^2, which is problematic since μ and σ^2 are unknown parameters. However, if we knew, for example, that $|\mu| < \text{A}$ and that $\sigma^2 > \text{B}$, then we would know to choose a value for k that satisfies the inequality

$$\frac{(1+k)}{(1-k)} > \frac{n\text{A}^2}{\text{B}}.$$

Specifying ranges of values for the parameters μ and σ^2 relates to Bayesian methods, where prior distributions for unknown parameters are often used.

Solution 5.29.

(a) First, since $\xi = \mathrm{E}(X) = \mathrm{e}^{\mu+\sigma^2/2}$, the MLE of ξ is $\hat{\xi} = \mathrm{e}^{\hat{\mu}+\hat{\sigma}^2/2}$, where $\hat{\mu}$ and $\hat{\sigma}^2$ are, respectively, the MLE's of μ and σ^2.

Now, with $y_i = \ln x_i$ for $i = 1, 2, \ldots, n$, and with $\boldsymbol{y} = (y_1, y_2, \ldots, y_n)$, the likelihood $\mathcal{L}(\boldsymbol{y}; \mu, \sigma^2) \equiv \mathcal{L}$ takes the form

$$\mathcal{L} = \prod_{i=1}^{n}(2\pi\sigma^2)^{-1/2}\mathrm{e}^{-(y_i-\mu)^2/2\sigma^2} = (2\pi\sigma^2)^{-n/2}\mathrm{e}^{-\sum_{i=1}^{n}(y_i-\mu)^2/2\sigma^2},$$

so that

$$\ln\mathcal{L} = -\frac{2}{n}\ln 2\pi - \frac{n}{2}\ln\sigma^2 - \frac{\sum_{i=1}^{n}(y_i-\mu)^2}{2\sigma^2}.$$

Thus, solving simultaneously the two equations

$$\frac{\partial\ln\mathcal{L}}{\partial\mu} = \frac{2}{2\sigma^2}\sum_{i=1}^{n}(y_i - \mu) = 0 \text{ and } \frac{\partial\ln\mathcal{L}}{\partial(\sigma^2)} = -\frac{n}{2\sigma^2} + \frac{\sum_{i=1}^{n}(y_i-\mu)^2}{2\sigma^4} = 0$$

gives

$$\hat{\mu} = n^{-1}\sum_{i=1}^{n}Y_i = \bar{Y} \text{ and } \hat{\sigma}^2 = n^{-1}\sum_{i=1}^{n}(Y_i - \bar{Y})^2 = \left(\frac{n-1}{n}\right)S^2,$$

so that

$$\hat{\xi} = \mathrm{e}^{\hat{\mu}+\hat{\sigma}^2/2} = \mathrm{e}^{\bar{Y}+\left(\frac{n-1}{2n}\right)S^2}.$$

(b) Since $Y_1, Y_2, \ldots, Y_n$ constitute a random sample of size n from a $\mathrm{N}(\mu, \sigma^2)$ population, we know that $\bar{Y} \sim \mathrm{N}(\mu, \frac{\sigma^2}{n})$, that $(n-1)S^2/\sigma^2 \sim \chi^2_{n-1} = \mathrm{GAMMA}\left(\alpha = 2, \beta = \frac{n-1}{2}\right)$, and that $\bar{Y}$ and S^2 are independent random variables. So, appealing to moment generating function theory results and letting $U = (n-1)S^2/\sigma^2$, we have

$$\begin{aligned}
\mathrm{E}(\hat{\xi}) &= \mathrm{E}\left(\mathrm{e}^{\bar{Y}}\right)\mathrm{E}\left[\mathrm{e}^{\left(\frac{n-1}{2n}\right)S^2}\right] \\
&= \left(\mathrm{e}^{\mu+\frac{\sigma^2}{2n}}\right)\mathrm{E}\left(\mathrm{e}^{\frac{\sigma^2 U}{2n}}\right) \\
&= \left(\mathrm{e}^{\mu+\frac{\sigma^2}{2n}}\right)\left[1 - 2\left(\frac{\sigma^2}{2n}\right)\right]^{-\left(\frac{n-1}{2}\right)} \\
&= \left(\mathrm{e}^{\mu+\frac{\sigma^2}{2n}}\right)\left(1 - \frac{\sigma^2}{n}\right)^{1/2}\left[\left(1 - \frac{\sigma^2}{n}\right)^n\right]^{-1/2},
\end{aligned}$$

so that

$$\begin{aligned}
\lim_{n\to\infty}\mathrm{E}(\hat{\xi}) &= (\mathrm{e}^{\mu})(1)\left(\mathrm{e}^{-\sigma^2}\right)^{-1/2} \\
&= \mathrm{e}^{\mu+\sigma^2/2} = \xi.
\end{aligned}$$

These results illustrate a general property of an MLE, namely, that it is not necessarily an unbiased estimator for finite n, but that it is asymptotically unbiased.

Solution 5.31.

(a) With $Z = \frac{(\bar{X}_1 - \bar{X}_2)}{\sqrt{2}\sigma/\sqrt{n}} \sim \mathrm{N}(0,1), U = (n-1)S_1^2/\sigma^2 \sim \chi^2_{n-1}$, and $T_{n-1} = Z/\sqrt{U/(n-1)} \sim t_{n-1}$ since Z and U are independent random variables, we have

$$\begin{aligned}
\theta_n &= \mathrm{pr}\left[|\bar{X}_1 - \bar{X}_2| \leq t_{n-1,1-\frac{\alpha}{2}} \frac{S_1}{\sqrt{n}}\right] \\
&= \mathrm{pr}\left[-t_{n-1,1-\frac{\alpha}{2}} \frac{S_1}{\sqrt{n}} \leq (\bar{X}_1 - \bar{X}_2) \leq t_{n-1,1-\frac{\alpha}{2}} \frac{S_1}{\sqrt{n}}\right] \\
&= \mathrm{pr}\left[\frac{-t_{n-1,1-\frac{\alpha}{2}} S_1/\sqrt{n}}{\sqrt{2}\sigma/\sqrt{n}} \leq Z \leq \frac{t_{n-1,1-\frac{\alpha}{2}} S_1/\sqrt{n}}{\sqrt{2}\sigma/\sqrt{n}}\right] \\
&= \mathrm{pr}\left[\frac{-t_{n-1,1-\frac{\alpha}{2}} \sqrt{U}}{\sqrt{2(n-1)}} \leq Z \leq \frac{t_{n-1,1-\frac{\alpha}{2}} \sqrt{U}}{\sqrt{2(n-1)}}\right] \\
&= \mathrm{pr}\left[\frac{-t_{n-1,1-\frac{\alpha}{2}}}{\sqrt{2}} \leq T_{n-1} \leq \frac{t_{n-1,1-\frac{\alpha}{2}}}{\sqrt{2}}\right] \\
&= 2\mathrm{F}_{T_{n-1}}\left(\frac{t_{n-1,1-\frac{\alpha}{2}}}{\sqrt{2}}\right) - 1,
\end{aligned}$$

where $\mathrm{F}_{T_{n-1}}(t) = \mathrm{pr}\,(T_{n-1} \leq t)$.

(b) Since T_{n-1} converges in distribution to a standard normal variate as $n \to \infty$, it follows that

$$\lim_{n\to\infty}\theta_n = 2\mathrm{F}_Z\left(\frac{z_{1-\frac{\alpha}{2}}}{\sqrt{2}}\right) - 1,$$

where $z_{1-\frac{\alpha}{2}}$ is defined such that $\mathrm{F}_Z\left(z_{1-\frac{\alpha}{2}}\right) = \mathrm{pr}\left(Z \leq z_{1-\frac{\alpha}{2}}\right) = \left(1 - \frac{\alpha}{2}\right)$ when $Z \sim \mathrm{N}(0,1)$. When $\alpha = 0.05$, $\lim_{n\to\infty}\theta_n = 2\mathrm{F}_Z\left(\frac{1.96}{\sqrt{2}}\right) - 1 = 0.834$.

Solution 5.33.

(a) Under prior independence of μ and σ^2, we have

$$\begin{aligned}
\pi(\mu|Y=y,\sigma^2) &\propto \mathrm{f}_Y(y|\mu,\sigma^2)\pi(\mu|\sigma^2) = \mathrm{f}_Y(y|\mu,\sigma^2)\pi(\mu) \\
&\propto K \times \exp\left\{-\frac{1}{2}\left[\tau(y-\mu)^2 + \tau_0(\mu-\mu_0)^2\right]\right\},
\end{aligned}$$

where $\tau = 1/\sigma^2$ is the *precision* of Y, $\tau_0 = 1/\sigma_0^2$ is the *prior precision* of μ, and $K = \sqrt{\tau\tau_0}/(2\pi)$.

So,

$$\begin{aligned}\pi(\mu|Y=y,\sigma^2) &\propto K \times \exp\left\{-\frac{1}{2}\left[\tau(y^2-2y\mu+\mu^2)+\tau_0(\mu^2-2\mu_0\mu+\mu_0^2]\right\}\right.\\ &\propto \mathrm{g}(y)\times\exp\left\{-\frac{1}{2}\left[(\tau+\tau_0)\mu^2-2(\tau y+\tau_0\mu_0)\mu\right]\right\},\end{aligned}$$

where $\mathrm{g}(y)$ now incorporates terms involving y (and μ_0). Completing the square, we obtain

$$\pi(\mu|Y=y,\sigma^2)=\mathrm{h}(y)\times\exp\left[-\frac{(\tau+\tau_0)}{2}\left(\mu-\frac{\tau y+\tau_0\mu_0}{\tau+\tau_0}\right)^2\right],$$

so that $\pi(\mu|Y=y,\sigma^2)$ is a normal density with posterior mean and posterior variance

$$\begin{aligned}\mathrm{E}(\mu|Y=y,\sigma^2) &= \left(\frac{\tau}{\tau+\tau_0}\right)y+\left(\frac{\tau_0}{\tau+\tau_0}\right)\mu_0 \text{ and}\\ \mathrm{V}(\mu|Y=y,\sigma^2) &= \frac{1}{(\tau+\tau_0)},\end{aligned}$$

and where $\mathrm{h}(y)=\sqrt{(\tau+\tau_0)/2\pi}$ does not, in this case, depend on y. Note that the *posterior precision* is $(\tau+\tau_0)$, and that the *posterior mean* is a weighted average of the data y and the prior mean μ_0; for example, if $\tau>\tau_0$, then y receives more weight in determining the posterior mean than does μ_0. Also, the posterior distribution is from the same distributional family as the prior distribution (namely, the normal family). When the posterior distribution is from the same family as the prior distribution, the prior is referred to as a *conjugate prior distribution.*

(b) Assuming prior independence of μ and σ^2,

$$\begin{aligned}\pi(\sigma^2|y,\mu) &\propto \mathrm{f}(y|\mu,\sigma^2)\pi(\sigma^2|\mu)=\mathrm{f}_Y(y|\mu,\sigma^2)\pi(\sigma^2)\\ &\propto (\sigma^2)^{-\frac{1}{2}}\mathrm{e}^{-\frac{1}{2\sigma^2}(y-\mu)^2}(\sigma^2)^{-a-1}\mathrm{e}^{-b/\sigma^2}\\ &\propto (\sigma^2)^{-\left(a+\frac{1}{2}\right)-1}\mathrm{e}^{-\frac{b+(y-\mu)^2/2}{\sigma^2}},\end{aligned}$$

which is proportional to (i.e., is the *kernel* for) an $\mathrm{IG}[a+1/2, b+(y-\mu)^2/2]$ distribution. Thus, conditional on $Y=y$ and μ, σ^2 is distributed IG with shape parameter $(a+1/2)$ and scale parameter $\left[b+(y-\mu)^2/2\right]$. Hence, the IG family is conjugate for σ^2.

(c) Under prior independence,

$$\mathrm{f}_Y(y|\sigma^2) = \int_{-\infty}^{\infty}\mathrm{f}_Y(y|\mu,\sigma^2)\pi(\mu)\mathrm{d}\mu.$$

So, we have

$$
\begin{aligned}
\mathrm{f}_Y(y|\sigma^2) &= \frac{\sqrt{\tau\tau_0}}{\sqrt{2\pi}}\int_{-\infty}^{\infty}\frac{1}{\sqrt{2\pi}}\mathrm{e}^{-\frac{1}{2}\left[\tau(y-\mu)^2+\tau_0(\mu-\mu_0)^2\right]}\mathrm{d}\mu \\
&= \frac{\sqrt{\tau\tau_0}}{\sqrt{2\pi}}\mathrm{e}^{-\frac{1}{2}\left(\tau y^2+\tau_0\mu_0^2\right)}\int_{-\infty}^{\infty}\frac{1}{\sqrt{2\pi}}\mathrm{e}^{-\frac{1}{2}\left[(\tau+\tau_0)\mu^2-2(\tau y+\tau_0\mu_0)\mu\right]}\mathrm{d}\mu \\
&= \frac{\sqrt{\tau\tau_0}}{\sqrt{2\pi}}\mathrm{e}^{-\frac{1}{2}\left(\tau y^2+\tau_0\mu_0^2\right)}\int_{-\infty}^{\infty}\frac{1}{\sqrt{2\pi}}\mathrm{e}^{-\frac{\tau+\tau_0}{2}\left[\mu^2-2\left(\frac{\tau y+\tau_0\mu_0}{\tau+\tau_0}\right)\mu\right]}\mathrm{d}\mu \\
&= \frac{\sqrt{\tau\tau_0}}{\sqrt{2\pi}}\mathrm{e}^{-\frac{1}{2}\left(\tau y^2+\tau_0\mu_0^2\right)}\frac{1}{\sqrt{\tau+\tau_0}}\mathrm{e}^{\frac{\tau+\tau_0}{2}\left(\frac{\tau y+\tau_0\mu_0}{\tau+\tau_0}\right)^2} \\
&\quad\times\underbrace{\int_{-\infty}^{\infty}\frac{\sqrt{\tau+\tau_0}}{\sqrt{2\pi}}\mathrm{e}^{-\frac{\tau+\tau_0}{2}\left[\mu-\left(\frac{\tau y+\tau_0\mu_0}{\tau+\tau_0}\right)\right]^2}\mathrm{d}\mu}_{=\int_{-\infty}^{\infty}\pi(\mu|y,\sigma^2)\mathrm{d}\mu=1\text{ (from part a)}} \\
&= \frac{1}{\sqrt{2\pi}}\sqrt{\frac{\tau\tau_0}{\tau+\tau_0}}\mathrm{e}^{-\frac{1}{2}\left(\frac{\tau\tau_0}{\tau+\tau_0}\right)(y-\mu_0)^2}.
\end{aligned}
$$

Thus, $\mathrm{f}_Y(y|\sigma^2)$ is a normal density with conditional mean $\mathrm{E}(Y|\sigma^2)=\mu_0$ and with conditional variance

$$\mathrm{V}(Y|\sigma^2)=\frac{1}{\tau}+\frac{1}{\tau_0}=\sigma^2+\sigma_0^2.$$

As a check, note that

$$\mathrm{E}(Y|\sigma^2)=\mathrm{E}_\mu\left[\mathrm{E}(Y|\mu,\sigma^2)\right]=\mathrm{E}(\mu)=\mu_0,$$

and that

$$\mathrm{V}(Y|\sigma^2)=\mathrm{E}_\mu\left[\mathrm{V}(Y|\mu,\sigma^2)\right]+\mathrm{V}_\mu\left[\mathrm{E}(Y|\mu,\sigma^2)\right]=\mathrm{E}(\sigma^2)+\mathrm{V}(\mu)=\sigma^2+\sigma_0^2.$$

Solution 5.35.

(a) With $\boldsymbol{x}=(x_1,x_2,\ldots,x_n)$,

$$
\begin{aligned}
\mathcal{L}(\boldsymbol{x};\pi) &= \prod_{i=1}^{n}\mathrm{p}_X(x_i;\pi) \\
&= \prod_{i=1}^{n}\{\pi(1-\pi)^{x_i-1}\} \\
&= \pi^n(1-\pi)^{(\sum_{i=1}^n x_i-n)} \\
&= \pi^n(1-\pi)^{n(\overline{x}-1)}.
\end{aligned}
$$

Hence, by the Factorization Theorem, $\sum_{i=1}^n X_i$ (or any 1-to-1 function of $\sum_{i=1}^n X_i$ such as $\overline{X}=n^{-1}\sum_{i=1}^n X_i$) is a sufficient statistic for π.

(b) First,

$$\ln\mathcal{L}(\boldsymbol{x};\pi) = n \ln \pi + n(\overline{x}-1) \ln (1-\pi).$$

Solving for π in the equation

$$\frac{\partial \ln\mathcal{L}(\boldsymbol{x};\pi)}{\partial \pi} = \frac{n}{\pi} - \frac{n(\overline{x}-1)}{(1-\pi)} = 0$$

yields the MLE

$$\hat{\pi} = \overline{x}^{-1}.$$

(c) First,

$$\frac{\partial^2 \ln\mathcal{L}(\boldsymbol{x};\pi)}{\partial \pi^2} = \frac{-n}{\pi^2} - \frac{n(\overline{x}-1)}{(1-\pi)^2}.$$

So,

$$\begin{aligned}
-\mathrm{E}\left[\frac{\partial^2 \ln\mathcal{L}(\boldsymbol{x};\pi)}{\partial \pi^2}\right] &= \frac{n}{\pi^2} + \frac{n}{(1-\pi)^2}\left[\mathrm{E}\left(\overline{X}\right) - 1\right] \\
&= \frac{n}{\pi^2} + \frac{n}{(1-\pi)^2}\left(\frac{1}{\pi} - 1\right) \\
&= \frac{n}{\pi^2} + \frac{n}{\pi(1-\pi)} \\
&= \frac{n}{\pi}\left[\frac{1}{\pi} + \frac{1}{(1-\pi)}\right] \\
&= \frac{n}{\pi^2(1-\pi)}.
\end{aligned}$$

Now, from maximum likelihood theory, for large n,

$$\frac{\hat{\pi} - \pi}{\sqrt{\hat{\pi}^2(1-\hat{\pi})/n}} \;\dot{\sim}\; \mathrm{N}(0,1).$$

So, an approximate large-sample 95% confidence interval for π is

$$\begin{aligned}
\hat{\pi} \pm 1.96\sqrt{\frac{\hat{\pi}^2(1-\hat{\pi})}{n}} &= 0.20 \pm 1.96\sqrt{\frac{(0.20)^2(0.80)}{50}} \\
&= 0.20 \pm 0.0496 \\
&= (0.1504, 0.2496).
\end{aligned}$$

(d) By the additivity of mutually independent and identically distributed geometric random variables, $U = \sum_{i=1}^{n} X_i$ has the negative binomial distribution

$$\mathrm{p}_U(u;\pi) = \mathrm{C}_{n-1}^{u-1}\pi^n(1-\pi)^{u-n}, u = n, n+1, \ldots, \infty.$$

Furthermore, from part (a), U is sufficient for π. Let

$$U^* = \begin{cases} 1 \text{ if } X_1 = 1, \\ 0 \text{ if } X_1 > 1 \end{cases}.$$

Then, $\mathrm{E}(U^*) = (1) \cdot \mathrm{pr}(X_1 = 1) = \pi$. So,

$$\begin{aligned}
\mathrm{E}(U^*|U = u) &= \mathrm{pr}(X_1 = 1|U = u) \\
&= \frac{\mathrm{pr}[(X_1 = 1) \cap (U = u)]}{\mathrm{pr}(U = u)} \\
&= \frac{\mathrm{pr}(X_1 = 1) \cdot \mathrm{pr}(\sum_{i=2}^{n} X_i = u - 1)}{\mathrm{pr}(\sum_{i=1}^{n} X_i = u)} \\
&= \frac{\pi \cdot \mathrm{C}_{n-2}^{u-2} \pi^{n-1}(1-\pi)^{u-n}}{\mathrm{C}_{n-1}^{u-1} \pi^{n}(1-\pi)^{u-n}} \\
&= \frac{\mathrm{C}_{n-2}^{u-2}}{\mathrm{C}_{n-1}^{u-1}} \\
&= \frac{(n-1)}{(u-1)}, u = n, n+1, \ldots, \infty.
\end{aligned}$$

Since the equation

$$\mathrm{E}[\mathrm{g}(U)] = \sum_{u=n}^{\infty} \mathrm{g}(u) \mathrm{C}_{n-1}^{u-1} \pi^{n}(1-\pi)^{u-n} = 0$$

is only true, for all $\pi \in (0, 1)$, if $\mathrm{g}(u) \equiv 0, u = n, n+1, \ldots, \infty$, we may conclude that $U = \sum_{i=1}^{n} X_i$ is a *complete* sufficient statistic for π. Hence,

$$\hat{\pi}^* = \frac{(n-1)}{(\sum_{i=1}^{n} X_i - 1)}$$

is the MVUE of π.

(e) Note that

$$\mathrm{E}(\pi^r) = \theta \int_0^1 \pi^r \pi^{\theta-1} \mathrm{d}\pi = \theta \left[\frac{\pi^{\theta+r}}{(\theta+r)} \right]_0^1 = \frac{\theta}{(\theta+r)}, (\theta+r) > 0.$$

So,

$$\mathrm{E}(\pi) = \frac{\theta}{(\theta+1)},$$

and

$$\mathrm{E}(\pi^2) = \frac{\theta}{(\theta+2)},$$

so that

$$V(\pi) = E(\pi^2) - [E(\pi)]^2 = \frac{\theta}{(\theta+2)} - \frac{\theta^2}{(\theta+1)^2} = \frac{\theta}{(\theta+1)^2(\theta+2)}.$$

So,

$$E(X) = E_\pi[E_x(X|\pi)] = E_\pi(1/\pi) = \frac{\theta}{(\theta-1)}, \theta > 1.$$

And

$$\begin{aligned}
V(X) &= E_\pi[V_x(X|\pi)] + V_\pi[E_x(X|\pi)] \\
&= E_\pi\left(\frac{1-\pi}{\pi^2}\right) + V_\pi\left(\frac{1}{\pi}\right) \\
&= E\left(\frac{1}{\pi^2}\right) - E\left(\frac{1}{\pi}\right) + E\left(\frac{1}{\pi^2}\right) - \left[E\left(\frac{1}{\pi}\right)\right]^2 \\
&= 2E\left(\pi^{-2}\right) - E\left(\pi^{-1}\right) - \left[E\left(\pi^{-1}\right)\right]^2 \\
&= \frac{2\theta}{(\theta-2)} - \frac{\theta}{(\theta-1)} - \frac{\theta^2}{(\theta-1)^2} \\
&= \frac{\theta^2}{(\theta-1)^2(\theta-2)}, \theta > 2.
\end{aligned}$$

We need to choose $\hat{\theta}$ to minimize the quantity

$$Q = \sum_{i=1}^{n}[X_i - E(X_i)]^2 = \sum_{i=1}^{n}\left[X_i - \frac{\theta}{(\theta-1)}\right]^2.$$

Solving for θ in the equation

$$\begin{aligned}
\frac{\partial Q}{\partial \theta} &= 2\sum_{i=1}^{n}\left[X_i - \frac{\theta}{(\theta-1)}\right] \cdot \frac{d}{d\theta}\left(-\frac{\theta}{\theta-1}\right) \\
&= 2n\left(\overline{X} - \frac{\theta}{(\theta-1)}\right) = 0
\end{aligned}$$

yields

$$\hat{\theta} = \frac{\overline{X}}{(\overline{X}-1)}$$

as the unweighted least squares estimator of θ. To obtain the method of moments estimator of θ, we need to solve the equation

$$\overline{X} = E(X_i) = \frac{\theta}{(\theta-1)},$$

giving

$$\hat{\theta}_{mm} = \frac{\overline{X}}{(\overline{X}-1)}.$$

Now,

$$\mathrm{V}(\hat{\theta}) \doteq \left\{[\mathrm{f}'(\overline{X})]_{\overline{X}=\mathrm{E}(\overline{X})}\right\}^2 \cdot \mathrm{V}\left(\overline{X}\right),$$

where $\mathrm{f}(\overline{X}) = \overline{X}/(\overline{X}-1)$ and $\mathrm{f}'\left(\overline{X}\right) = -1/\left(\overline{X}-1\right)^2$. So

$$\begin{aligned}\mathrm{V}(\hat{\theta}) &\doteq \left[\frac{-1}{\left(\frac{\theta}{\theta-1}-1\right)^2}\right]^2 \cdot \frac{\theta^2}{n(\theta-1)^2(\theta-2)} \\ &= \frac{(\theta-1)^4\theta^2}{n(\theta-1)^2(\theta-2)} \\ &= \frac{\theta^2(\theta-1)^2}{n(\theta-2)}, \theta > 2.\end{aligned}$$

Solution 5.37. We know that $\mathrm{f}_{Y_{(n)}}(y_{(n)};\theta) = n\theta^{-n}y_{(n)}^{n-1}$, $0 < y_{(n)} < \theta$. Let $U = \theta^{-1}Y_{(n)}$, so that $dU = \theta^{-1}dY_{(n)}$. Hence, $\mathrm{f}_U(u) = n\theta^{-n}(\theta u)^{n-1}(\theta) = nu^{n-1}, 0 < u < 1$. So,

$$\mathrm{pr}(U \leq u) = \int_0^u nw^{n-1}\mathrm{d}w = \mathrm{F_U}(u) = [w^n]_0^u = u^n, 0 < u < 1.$$

So,

$$\mathrm{pr}(U \leq u) = \mathrm{pr}\left[\frac{Y_{(n)}}{\theta} \leq u\right] = \mathrm{pr}\left[\frac{Y_{(n)}}{u} \leq \theta\right] = u^n.$$

We want $u^n = (1-\alpha)$, so that $u = (1-\alpha)^{1/n}$. So,

$$\mathrm{pr}\left[\frac{Y_{(n)}}{(1-\alpha)^{1/n}} \leq \theta\right] = (1-\alpha),$$

so that

$$L = \frac{Y_{(n)}}{(1-\alpha)^{1/n}}.$$

For the given data,

$$\hat{L} = \frac{y_{(10)}}{(1-0.05)^{1/10}} = \frac{2.10}{0.9949} = 2.1108.$$

Solution 5.39.

(a)

$$
\begin{aligned}
\prod_{i=1}^{n} \mathrm{f}_X(x_i;\alpha,\beta) &= \prod_{i=1}^{n} \left\{ \frac{\alpha^{\beta}}{\Gamma(\beta)} x_i^{\beta-1} \mathrm{e}^{-\alpha x_i} \right\} \\
&= \frac{\alpha^{n\beta}}{\Gamma^{n}(\beta)} \left(\prod_{i=1}^{n} x_i \right)^{\beta-1} \mathrm{e}^{-\alpha \sum_{i=1}^{n} x_i} \\
&= \left[\alpha^{n\beta} \mathrm{e}^{-\alpha \sum_{i=1}^{n} x_i} \right] \cdot \left[\Gamma^{-n}(\beta) \cdot \left(\prod_{i=1}^{n} x_i \right)^{\beta-1} \right] \\
&= g\left(\sum_{i=1}^{n} x_i; \alpha \right) \cdot h(x_1, x_2, \ldots, x_n),
\end{aligned}
$$

where $h(x_1, x_2, \ldots, x_n)$ *does not in any way* depend on α for every fixed value u of $U = \sum_{i=1}^{n} X_i$. Hence, $\sum_{i=1}^{n} X_i$ is a sufficient statistic for α.

(b) Since

$$
\begin{aligned}
\bar{X} &\sim \text{GAMMA}\left[(n\alpha)^{-1}, n\beta\right], \\
\mathrm{E}\left(\bar{X}^{r}\right) &= \int_{0}^{\infty} \bar{x}^{r} \frac{(n\alpha)^{n\beta-1}}{\Gamma(n\beta)} (\bar{x})^{n\beta-1} \mathrm{e}^{-n\alpha\bar{x}} \mathrm{d}\bar{x} \\
&= \int_{0}^{\infty} \frac{(n\alpha)^{n\beta-1}}{\Gamma(n\beta)} (\bar{x})^{(n\beta+r)-1} \mathrm{e}^{-n\alpha\bar{x}} \mathrm{d}\bar{x} \\
&= \frac{\Gamma(n\beta+r)}{\Gamma(n\beta)} (n\alpha)^{-r}, \quad (n\beta+r) > 0.
\end{aligned}
$$

So,

$$
\begin{aligned}
\mathrm{E}(\hat{\alpha}) &= \beta \mathrm{E}\left(\bar{X}^{-1}\right) = \beta \frac{\Gamma(n\beta-1)}{\Gamma(n\beta)} (n\alpha)^{-(-1)} \\
&= \frac{\beta}{(n\beta-1)} (n\alpha) = \frac{\alpha}{1-\frac{1}{n\beta}}, \quad n\beta > 1,
\end{aligned}
$$

so that $\lim_{n\to\infty} \mathrm{E}(\hat{\alpha}) = \alpha$.
And,

$$
\begin{aligned}
\mathrm{E}\left(\hat{\alpha}^{2}\right) &= \beta^{2} \mathrm{E}\left(\bar{X}^{-2}\right) = \beta^{2} \frac{\Gamma(n\beta-2)}{\Gamma(n\beta)} (n\alpha)^{-(-2)} \\
&= \frac{n^{2}\alpha^{2}\beta^{2}}{(n\beta-1)(n\beta-2)}, \quad n\beta > 2.
\end{aligned}
$$

So, for $\beta > 2$,

$$\begin{aligned}
\mathrm{V}(\hat{\alpha}) &= \frac{n^2\alpha^2\beta^2}{(n\beta-1)(n\beta-2)} - \left[\frac{n\alpha\beta}{n\beta-1}\right]^2 \\
&= \frac{n^2\alpha^2\beta^2}{(n\beta-1)}\left[\frac{1}{(n\beta-2)} - \frac{1}{(n\beta-1)}\right] \\
&= \frac{n^2\alpha^2\beta^2}{(n\beta-1)^2(n\beta-2)} = \frac{\alpha^2\beta^2}{(\beta-\frac{1}{n})^2(n\beta-2)} \rightarrow 0 \text{ as } n \rightarrow \infty.
\end{aligned}$$

Thus, $\hat{\alpha}$ is *consistent* for α.

(c) Since

$$\mathrm{E}\left(\bar{X}^r\right) = \frac{\Gamma(n\beta+r)}{\Gamma(n\beta)}(n\alpha)^{-r}, \quad (n\beta+r) > 0,$$

we have

$$\begin{aligned}
\mathrm{E}(\bar{X}) &= \frac{\Gamma(n\beta+1)}{\Gamma(n\beta)}(n\alpha)^{-1} = (n\beta)(n\alpha)^{-1} = \frac{\beta}{\alpha}, \\
\mathrm{E}\left(\bar{X}^2\right) &= \frac{\Gamma(n\beta+2)}{\Gamma(n\beta)}(n\alpha)^{-2} = \frac{n\beta(n\beta+1)}{n^2\alpha^2} = \frac{\beta(n\beta+1)}{n\alpha^2}, \text{ and} \\
\mathrm{V}(\bar{X}) &= \frac{\beta(n\beta+1)}{n\alpha^2} - \left(\frac{\beta}{\alpha}\right)^2 = \frac{\beta}{\alpha^2}\left[\frac{(n\beta+1)}{n} - \beta\right] = \frac{\beta}{n\alpha^2}.
\end{aligned}$$

So, from the Central Limit Theorem,

$$\frac{\bar{X}-\mathrm{E}(\bar{X})}{\sqrt{\mathrm{V}(\bar{X})}} = \frac{\bar{X}-\beta/\alpha}{\sqrt{\beta/n\alpha^2}} = \left[\sqrt{\frac{n}{\beta}}\alpha\bar{X} - \sqrt{n\beta}\right] \dot{\sim} \mathrm{N}(0,1) \text{ for large } n.$$

So, for large n,

$$\begin{aligned}
0.95 &\doteq \mathrm{pr}\left\{-1.96 < \sqrt{\frac{n}{\beta}}\alpha\bar{X} - \sqrt{n\beta} < +1.96\right\} \\
&= \mathrm{pr}\left\{\sqrt{n\beta} - 1.96 < \sqrt{\frac{n}{\beta}}\alpha\bar{X} < \sqrt{n\beta} + 1.96\right\} \\
&= \mathrm{pr}\left\{\bar{X}^{-1}\sqrt{\frac{\beta}{n}}\left(\sqrt{n\beta} - 1.96\right) < \alpha < \bar{X}^{-1}\sqrt{\frac{\beta}{n}}\left(\sqrt{n\beta} + 1.96\right)\right\}.
\end{aligned}$$

So, an appropriate 95% confidence interval for α is:

$$\bar{X}^{-1}\sqrt{\frac{\beta}{n}}\left(\sqrt{n\beta} \pm 1.96\right).$$

For $\beta = 2$, $n = 50$, and $\bar{x} = 3$, we obtain:

$$\begin{aligned}
(3)^{-1}\sqrt{\frac{2}{50}}(\sqrt{50\cdot 2} \pm 1.96) &= \frac{0.20}{3}(10 \pm 1.96), \\
&\text{or } (0.536, 0.797).
\end{aligned}$$

Solution 5.41.

(a) In general,

$$\mathrm{f}_{X_{(1)}}(x_{(1)}) = n[1 - \mathrm{F}_X(x_{(1)})]^{n-1}\mathrm{f}_X(x_{(1)}), \quad -\infty < x_{(1)} < +\infty.$$

Here, since $\mathrm{F}_X(x;\alpha) = 1 - \mathrm{e}^{-x/\alpha}$, $x > 0$, we have

$$\begin{aligned} \mathrm{f}_{X_{(1)}}(x_{(1)};\alpha) &= \frac{n}{\alpha}\mathrm{e}^{-nx_{(1)}/\alpha}, \ x_{(1)} > 0; \\ \text{hence, } \mathrm{E}[X_{(1)}] &= \frac{\alpha}{n} \text{ and } \mathrm{V}[X_{(1)}] = \frac{\alpha^2}{n^2}. \end{aligned}$$

Also,

$$\begin{aligned} \mathrm{f}_{Y_{(1)}}(y_{(1)};\beta) &= \frac{n}{\beta}\mathrm{e}^{-ny_{(1)}/\beta}, \ y_{(1)} > 0, \\ \text{so that } \mathrm{E}[Y_{(1)}] &= \frac{\beta}{n} \text{ and } \mathrm{V}[Y_{(1)}] = \frac{\beta^2}{n^2}. \end{aligned}$$

So, since $\mathrm{E}[X_{(1)} - Y_{(1)}] = (\alpha - \beta)/n$, choosing $k = n$ means that

$$\mathrm{E}\{n[X_{(1)} - Y_{(1)}]\} = (\alpha - \beta).$$

Also,

$$\mathrm{V}\{n[X_{(1)} - Y_{(1)}]\} = n^2\left[\frac{\alpha^2}{n^2} + \frac{\beta^2}{n^2}\right] = (\alpha^2 + \beta^2).$$

Since $\lim_{n\to\infty} \mathrm{V}\{n[X_{(1)} - Y_{(1)}]\} = (\alpha^2 + \beta^2) \neq 0$, $n[X_{(1)} - Y_{(1)}]$ is *not* a consistent estimator for θ. Also, this estimator is *not* a function of $S_x = \sum_{i=1}^n X_i$ and $S_y = \sum_{i=1}^n Y_i$, the sufficient statistics for α and β.

(b) Now, $(S_x - S_y)/n = (\bar{X} - \bar{Y})$ is an unbiased estimator of θ that is a function of the sufficient statistics for α and β; in fact, $(\bar{X} - \bar{Y})$ is the best linear unbiased estimator of θ. Also, $\mathrm{V}(\bar{X} - \bar{Y}) = (\alpha^2 + \beta^2)/n$. Now, consider the function:

$$\left[\frac{\sqrt{\alpha^2 + \beta^2}}{\sqrt{\bar{X}^2 + \bar{Y}^2}}\right]\left[\frac{(\bar{X} - \bar{Y}) - (\alpha - \beta)}{\sqrt{\frac{(\alpha^2 + \beta^2)}{n}}}\right].$$

Since the term on the left converges in probability to 1 as $n \to \infty$, and since the term on the right converges in distribution to an N(0,1) variate as $n \to \infty$, it follows from Slutsky's Theorem that the statistic

$$\frac{(\bar{X} - \bar{Y}) - (\alpha - \beta)}{\sqrt{\frac{(\bar{X}^2 + \bar{Y}^2)}{n}}} \mathrel{\dot\sim} \mathrm{N}(0,1) \text{ for large } n.$$

Hence, an approximate $100\%(1 - \alpha)$ large sample (n large) confidence interval for θ is:

$$(\bar{X} - \bar{Y}) \pm Z_{1-\frac{\alpha}{2}}\sqrt{\frac{(\bar{X}^2 + \bar{Y}^2)}{n}}.$$

Solution 5.43. Let X_i denote the number of windshields breaking for the i-th breakage category, $i = 1, 2, 3, 4$, where X_1 = number of windshields breaking on the first strike, X_2 = number of windshields breaking on the second strike, X_3 = number of windshields breaking on the third strike, and X_4 = number of windshields requiring more than three strikes to break. Clearly, $n = X_1 + X_2 + X_3 + X_4$. Now, for any particular windshield,

$$\begin{aligned}
\text{pr}(X = 1) &= (1-\theta), \\
\text{pr}(X = 2) &= (1-\theta)\theta, \\
\text{pr}(X = 3) &= (1-\theta)\theta^2, \\
\text{pr}(X \geq 4) &= 1 - \text{pr}(X \leq 3) = 1 - (1-\theta) - (1-\theta)\theta - (1-\theta)\theta^2 = \theta^3.
\end{aligned}$$

So, $\mathcal{L}(x_1, x_2, x_3, x_4) \equiv \mathcal{L}$ is a multinomial distribution, namely,

$$\begin{aligned}
\mathcal{L} &= \frac{n}{x_1!x_2!x_3!x_4!}(1-\theta)^{x_1}[(1-\theta)\theta]^{x_2}[(1-\theta)\theta^2]^{x_3}(\theta^3)^{x_4} \\
&\propto (1-\theta)^{(x_1+x_2+x_3)}\theta^{(x_2+2x_3+3x_4)}.
\end{aligned}$$

So,

$$\ln\mathcal{L} \propto (x_1 + x_2 + x_3)\ln(1-\theta) + (x_2 + 2x_3 + 3x_4)\ln\theta.$$

So,

$$\frac{\partial \ln\mathcal{L}}{\partial\theta} = \frac{-(x_1 + x_2 + x_3)}{(1-\theta)} + \frac{(x_2 + 2x_3 + 3x_4)}{\theta},$$

and the equation $\frac{\partial \ln\mathcal{L}}{\partial\theta} = 0$ gives the MLE

$$\hat{\theta} = \frac{(x_2 + 2x_3 + 3x_4)}{(x_1 + 2x_2 + 3x_3 + 3x_4)}.$$

For the available data

$$\hat{\theta} = \frac{(36 + 44 + 90)}{(112 + 72 + 66 + 90)} = \frac{1}{2}.$$

Now,

$$\frac{\partial^2 \ln\mathcal{L}}{\partial\theta^2} = -\frac{(x_1 + x_2 + x_3)}{(1-\theta)^2} - \frac{(x_2 + 2x_3 + 3x_4)}{\theta^2},$$

so that the estimated *observed* information value is

$$-\frac{\partial^2 \ln\mathcal{L}}{\partial\theta^2}_{|\theta=\hat{\theta}=0.50} = \frac{(112 + 36 + 22)}{\left(1 - \frac{1}{2}\right)^2} + \frac{(36 + 44 + 90)}{\left(\frac{1}{2}\right)^2} = 1,360.$$

So, the large-sample 95% confidence interval for θ based on *observed* information is

$$\frac{1}{2} \pm 1.96\sqrt{\frac{1}{1360}} = (0.4469, 0.5531).$$

Using *expected* information, since

$$-\frac{\partial^2 \ln\mathcal{L}}{\partial\theta^2} = \frac{(n-x_4)}{(1-\theta)^2} + \frac{(x_2+2x_3+3x_4)}{\theta^2},$$

we have

$$\begin{aligned}
\mathrm{E}\left(-\frac{\partial^2 \ln\mathcal{L}}{\partial\theta^2}\right) &= \frac{n-n\theta^3}{(1-\theta)^2} + \frac{n[(1-\theta)\theta + 2(1-\theta)\theta^2 + 3\theta^3]}{\theta^2} \\
&= \frac{n(1-\theta^3)}{\theta(1-\theta)^2}.
\end{aligned}$$

So, the estimated *expected* information is

$$\frac{(200)[1-(\frac{1}{2})^3]}{(\frac{1}{2})(1-\frac{1}{2})^2} = 1400.$$

Thus, the large-sample 95% confidence interval based on *expected* information is

$$\frac{1}{2} \pm 1.96\sqrt{\frac{1}{1400}} = (0.4476, 0.5524).$$

Solution 5.45. Since we only have available $\{S_1, S_2, \ldots, S_n\}$, we need to find the distribution of S. Consider a 1-to-1 transformation from (X,Y) to (S,T), where $S=(X+Y)$ and $T=Y$. For this transformation, $X=(S-T)$, $Y=T$, $|J|=1$, so that $\mathrm{f}_{S,T}(s,t;\theta) = 3\theta^{-3}s$, $0<t<s<\theta$. So,

$$\mathrm{f}_S(s;\theta) = \int_0^s 3\theta^{-3}s\,\mathrm{d}t = 3\theta^{-3}s^2,\ 0<s<\theta;$$

and

$$F_S(s;\theta) = \theta^{-3}s^3, 0<s<\theta.$$

Thus,

$$\begin{aligned}
\mathrm{f}_{S_{(n)}}(s_{(n)};\theta) &= n\left[\frac{s_{(n)}^3}{\theta^3}\right]^{n-1} 3\theta^{-3}s_{(n)}^2 \\
&= 3n\theta^{-3n}s_{(n)}^{3n-1},\ 0<s_{(n)}<\theta.
\end{aligned}$$

Now, since $S_{(n)}<\theta$, let U have the general structure $U=cS_{(n)}$, where $c>1$. Then,

$$\begin{aligned}
\mathrm{pr}\left[cS_{(n)}>\theta\right] &= \mathrm{pr}\left[S_{(n)}>\frac{\theta}{c}\right] \\
&= \int_{\theta/c}^{\theta} 3n\theta^{-3n}s_{(n)}^{3n-1}\mathrm{d}s_{(n)} = \theta^{-3n}\left[s_{(n)}^{3n}\right]_{\theta/c}^{\theta} \\
&= 1-c^{-3n} = (1-\alpha),
\end{aligned}$$

so that $c = \alpha^{-1/3n}$. So, the exact $100(1-\alpha)\%$ confidence interval for θ is

$$\left(0, \alpha^{-1/3n} S_{(n)}\right), \text{ where } S_{(n)} = \max\{S_1, S_2, \ldots, S_n\}.$$

For the given data, the computed exact upper one-sided 95% confidence interval for θ is

$$\left[0, (0.05)^{-1/75}(8.20)\right] = (0, 8.53).$$

Solution 5.47. (a)

$$\begin{aligned} \text{pr}(X > k) &= \sum_{x=k+1}^{\infty} (1-\theta)\theta^{x-1} = \frac{(1-\theta)}{\theta} \sum_{x=k+1}^{\infty} \theta^{x} \\ &= \frac{(1-\theta)}{\theta}\left[\frac{\theta^{k+1}}{(1-\theta)}\right] = \theta^{k}. \end{aligned}$$

(b) First, we need to find a sufficient statistic for θ^k. Now,

$$\text{p}_{X_1,X_2,\ldots,X_n}(x_1, x_2, \ldots, x_n; \theta^k) = [1 - (\theta^k)^{1/k}]^n (\theta^k)^{\frac{1}{k}(u-n)}$$

where $u = \sum_{i=1}^{n} x_i$; clearly, $U = \sum_{i=1}^{n} X_i$ is a sufficient statistic for θ^k by the Factorization Theorem. And, from exponential family theory, U is a *complete* sufficient statistic for θ^k.

The Rao-Blackwell Theorem should be used to find the MVUE of θ^k. Let

$$U^* = \begin{cases} 1 \text{ if } X_1 > k, \\ 0 \text{ if } X_1 \leq k \end{cases}.$$

Then,

$$\text{E}(U^*) = (1)\text{pr}(X_1 > k) + (0)\text{pr}(X_1 \leq k) = \text{pr}(X_1 > k) = \theta^k.$$

Also, we know that U has a negative binomial distribution, namely,

$$\text{p}_U(u) = C_{n-1}^{u-1}(1-\theta)^n \theta^{u-n}, \quad u = n, n+1, \ldots, \infty.$$

So, by the Rao-Blackwell Theorem, the MVUE of θ^k is

$$\text{E}(U^*|U = u) = \text{pr}(X_1 > k|U = u) = \frac{\text{pr}[(X_1 > k) \cap (U = u)]}{\text{pr}(U = u)}.$$

Note that the numerator is 0 if $u < (n + k)$ since we need $X_1 \geq (k + 1)$ and

$\sum_{i=2}^{n} X_i \geq (n-1)$. So, for $u \geq (n+k)$,

$$\begin{aligned}
\text{pr}(X_1 > k | U = u) &= \frac{\text{pr}[(X_1 > k) \cap (U = u)]}{\text{pr}(U = u)} \\
&= \frac{[\sum_{j=k+1}^{u-(n-1)} \text{pr}(X_1 = j)] \{\text{pr}[\sum_{i=2}^{n} X_i = (u-j)]\}}{\text{pr}(U = u)} \\
&= \frac{[\sum_{k+1}^{u-(n-1)} (1-\theta)\theta^{j-1}][C_{n-2}^{(u-j)-1}(1-\theta)^{n-1}\theta^{(u-j-n+1)}]}{C_{n-1}^{u-1}(1-\theta)^n \theta^{u-n}} \\
&= \frac{\sum_{j=k+1}^{u-(n-1)} C_{n-2}^{(u-j)-1}}{C_{n-1}^{u-1}}.
\end{aligned}$$

So, the MVUE of θ^k is:

$$\widehat{\theta^k} = \begin{cases} \sum_{j=k+1}^{U-(n-1)} C_{n-2}^{(U-j)-1} / C_{n-1}^{U-1} & \text{if } U \geq (n+k) \\ 0 & \text{if } U < (n+k) \end{cases},$$

where $U = \sum_{i=1}^{n} X_i$. For the given data, $n = 5$, $k = 6$, and $u = \sum_{i=1}^{5} x_i = 15$, so that

$$\begin{aligned}
\widehat{\theta^k} &= \left(\frac{1}{C_{5-1}^{15-1}}\right) \sum_{j=7}^{11} C_{(5-2)}^{(14-j)} = \frac{C_3^7 + C_3^6 + C_3^5 + C_3^4 + C_3^3}{C_4^{14}} \\
&= \frac{(35 + 20 + 10 + 4 + 1)}{1001} = \frac{70}{1001} \doteq 0.07.
\end{aligned}$$

Solution 5.49.

(a) Since $S_n = \sum_{i=1}^{n} X_i = n\overline{X} \sim \text{POI}(n\lambda)$, it follows that $S_3 \sim \text{POI}(0.75)$. So,

$$\begin{aligned}
\text{pr}\left[\frac{1}{3}\sum_{i=1}^{3} X_i > 0.50\right] &= \text{pr}\left[S_3 > 1.50\right] = 1 - \text{pr}[S_3 \leq 1] \\
&= 1 - \frac{(0.75)^0 e^{-0.75}}{0!} - \frac{(0.75)^1 e^{-0.75}}{1!} = 0.1733.
\end{aligned}$$

(b) First, again note that $S = n\overline{X} \sim \text{POI}(n\lambda)$. For any fixed x,

$$\begin{aligned}
E(\widehat{\pi}_x) &= E\left[\frac{\left(\frac{S}{n}\right)^x e^{-\left(\frac{S}{n}\right)}}{x!}\right] = \sum_{s=0}^{\infty} \left[\frac{\left(\frac{s}{n}\right)^x e^{-\left(\frac{s}{n}\right)}}{x!}\right] \frac{(n\lambda)^s e^{-n\lambda}}{s!} \\
&= \frac{e^{-n\lambda}}{n^x x!} \sum_{s=0}^{\infty} s^x \frac{(n\lambda e^{-\frac{1}{n}})^s}{s!} = \frac{e^{-n\lambda} e^{n\lambda e^{-\frac{1}{n}}}}{n^x x!} \sum_{s=0}^{\infty} s^x \frac{(n\lambda e^{-\frac{1}{n}})^s e^{-(n\lambda e^{-\frac{1}{n}})}}{s!} \\
&= \frac{e^{-n\lambda(1-e^{-1/n})}}{n^x x!} E[Y^x]
\end{aligned}$$

where $Y \sim \text{POI}(n\lambda e^{-1/n})$.

(c) For $x = 0$,

$$\begin{aligned} E(\widehat{\pi}_0) &= n^{-0}(0!)^{-1}e^{-n\lambda(1-e^{-1/n})}E(Y^0) = e^{-n\lambda(1-e^{-1/n})} \\ &= e^{-n\lambda\left[1-\sum_{i=0}^{\infty}\frac{(-1/n)^i}{i!}\right]} = e^{-n\lambda\left[1-1+\frac{1}{n}-\sum_{i=2}^{\infty}\frac{(-1/n)^i}{i!}\right]} \\ &= e^{-\lambda+\lambda\sum_{i=2}^{\infty}\frac{(-1)^i}{n^{i-1}(i!)}}. \end{aligned}$$

So,

$$\lim_{n\to\infty} E(\widehat{\pi}_0) = \lim_{n\to\infty} e^{-\lambda+\lambda\sum_{i=2}^{\infty}\frac{(-1)^i}{n^{i-1}(i!)}} = e^{-\lambda} = \pi_0.$$

In other words, $\widehat{\pi}_0$ is an "asymptotically unbiased" estimator of π_0. In general, since $\overline{X}$ is the MLE of λ, it follows that $\frac{(\bar{X})^x \mathrm{e}^{-\bar{X}}}{x!}$ is the MLE of $\frac{\lambda^x \mathrm{e}^{-\lambda}}{x!}$, and so

$$\lim_{n\to\infty} E(\widehat{\pi}_x) = \pi_x.$$

Solution 5.51.

(a) With $\boldsymbol{y}' = (y_1, y_2, \ldots, y_n), \boldsymbol{X}' = (X_1, X_2, \ldots, X_n)$, and $\boldsymbol{x}' = (x_1, x_2, \ldots, x_n)$, we have

$$\begin{aligned} \mathcal{L}(\boldsymbol{y};\beta|\boldsymbol{X}=\boldsymbol{x}) \equiv \mathcal{L} &= \prod_{i=1}^{n}\left\{(\beta x_i)^{-1}\mathrm{e}^{-y_i/\beta x_i}\right\} \\ &= \beta^{-n}\left(\prod_{i=1}^{n} x_i\right)^{-1}\mathrm{e}^{-\beta^{-1}\sum_{i=1}^{n}x_i^{-1}y_i}. \end{aligned}$$

So,

$$\begin{aligned} \ln\mathcal{L} &= -n\ln\beta - \sum_{i=1}^{n}\ln x_i - \beta^{-1}\sum_{i=1}^{n}x_i^{-1}y_i \\ \Rightarrow \frac{\mathrm{d}\ln\mathcal{L}}{\mathrm{d}\beta} &= \frac{-n}{\beta} + \beta^{-2}\sum_{i=1}^{n}x_i^{-1}y_i = 0 \\ \Rightarrow \hat{\beta}_{\mathrm{ML}} &= \frac{\sum_{i=1}^{n}x_i^{-1}Y_i}{n}. \end{aligned}$$

Since $\mathrm{E}(Y_i|X_i = x_i) = \beta x_i$ and $\mathrm{V}(Y_i|X_i = x_i) = \beta^2 x_i^2$, it follows that

$$\begin{aligned} \mathrm{E}(\hat{\beta}_{\mathrm{ML}}|\boldsymbol{X}=\boldsymbol{x}) &= n^{-1}\sum_{i=1}^{n}x_i^{-1}(\beta x_i) = \beta, \text{ and} \\ \mathrm{V}(\hat{\beta}_{\mathrm{ML}}|\boldsymbol{X}=\boldsymbol{x}) &= n^{-2}\sum_{i=1}^{n}x_i^{-2}(\beta^2 x_i^2) = \beta^2/n. \end{aligned}$$

(b) We need to choose $\hat{\beta}_{\text{LS}}$ to minimize $Q_u = \sum_{i=1}^{n}[Y_i - \text{E}(Y_i|X_i = x_i)]^2 = \sum_{i=1}^{n}(Y_i - \beta x_i)^2$. So,

$$\frac{\text{d}Q_u}{\text{d}\beta} = 2\sum_{i=1}^{n}(Y_i - \beta x_i)(-x_i) = 0 \Rightarrow \hat{\beta}_{\text{LS}} = \sum_{i=1}^{n} x_i Y_i / \sum_{i=1}^{n} x_i^2.$$

So,

$$\begin{aligned}
\text{E}(\hat{\beta}_{\text{LS}}|\boldsymbol{X} = \boldsymbol{x}) &= \sum_{i=1}^{n} x_i(x_i\beta) / \sum_{i=1}^{n} x_i^2 = \beta, \text{ and} \\
\text{V}(\hat{\beta}_{\text{LS}}|\boldsymbol{X} = \boldsymbol{x}) &= \sum_{i=1}^{n} x_i^2(\beta^2 x_i^2) / \left(\sum_{i=1}^{n} x_i^2\right)^2 = \beta^2 \sum_{i=1}^{n} x_i^4 / \left(\sum_{i=1}^{n} x_i^2\right)^2.
\end{aligned}$$

(c) To find $\hat{\beta}_{\text{MM}}$, we equate $\bar{Y} = n^{-1}\sum_{i=1}^{n} Y_i$ to $\text{E}(\bar{Y}|\boldsymbol{X} = \boldsymbol{x})$, and solve for $\hat{\beta}_{\text{MM}}$.
Now,

$$\text{E}(\bar{Y}|\boldsymbol{X} = \boldsymbol{x}) = n^{-1}\sum_{i=1}^{n} \text{E}(Y_i|X_i = x_i) = n^{-1}\sum_{i=1}^{n} \beta x_i = \beta\bar{x},$$

where $\bar{x} = n^{-1}\sum_{i=1}^{n} x_i$.
So,

$$\begin{aligned}
\bar{Y} &= \beta\bar{x} \Rightarrow \hat{\beta}_{\text{MM}} = \bar{Y}/\bar{x}; \text{ so,} \\
\text{E}(\hat{\beta}_{\text{MM}}) &= \frac{\text{E}(\bar{Y})}{\bar{x}} = \frac{\beta\bar{x}}{\bar{x}} = \beta, \text{ and} \\
\text{V}(\hat{\beta}_{\text{MM}}) &= \frac{\text{V}(\bar{Y})}{(\boldsymbol{x})^2} = n^{-2}\sum_{i=1}^{n} \beta^2 x_i^2 \Big/ n^{-2}\left(\sum_{i=1}^{n} x_i\right)^2 \\
&= \beta^2 \sum_{i=1}^{n} x_i^2 / \left(\sum_{i=1}^{n} x_i\right)^2.
\end{aligned}$$

(d) Since the three estimators $\hat{\beta}_{\text{ML}}$, $\hat{\beta}_{\text{LS}}$, and $\hat{\beta}_{\text{MM}}$ are all unbiased estimators of β, it makes sense to use the one with the smallest variance. Since

$$\begin{aligned}
\frac{\text{d}^2\mathcal{L}}{\text{d}\beta^2} &= \frac{n}{\beta^2} - \frac{2}{\beta^3}\sum_{i=1}^{n} x_i^{-1} y_i, \text{ so that} \\
-\text{E}\left(\frac{\text{d}^2\mathcal{L}}{\text{d}\beta^2}|\boldsymbol{X} = \boldsymbol{x}\right) &= \frac{-n}{\beta^2} + \frac{2}{\beta^3}\sum_{i=1}^{n} x_i^{-1}(\beta x_i) = \frac{n}{\beta^2},
\end{aligned}$$

the Cramér-Rao lower bound for the variance of any unbiased estimator of β is $\frac{1}{(n/\beta^2)} = \beta^2/n$, which is the variance of the ML estimator.

Also, note that $\hat{\beta}_{\text{ML}}$ is the only estimator involving the *sufficient statistic* $\sum_{i=1}^{n} x_i^{-1}Y_i$. Finally, the use of $\hat{\beta}_{\text{ML}}$ means that we can utilize the well-established large-sample distributional properties of ML estimators. For all of these reasons, $\hat{\beta}_{\text{ML}}$ is the estimator to use.
Since

$$\frac{\hat{\beta}_{\text{ML}} - \beta}{\hat{\beta}_{\text{ML}}/\sqrt{n}} \mathrel{\dot{\sim}} \text{N}(0,1)$$

for large n based on ML theory, an appropriate large-sample 95% confidence interval for β is:

$$\hat{\beta}_{\text{ML}} \pm 1.96\left(\frac{\hat{\beta}_{\text{ML}}}{\sqrt{n}}\right).$$

Solution 5.53.

(a) In general,

$$\text{f}_{X_{(n)}}(x_{(n)}) = n\left[\text{F}_X(x_{(n)})\right]^{n-1}\text{f}_X(x_{(n)}).$$

Since $\text{F}_X(x;\theta_1) = \frac{x}{\theta_1}, 0 < x < \theta_1$, and $\text{F}_Y(y;\theta_2) = \frac{y}{\theta_2}, 0 < y < \theta_2$, it follows that $\text{f}_{X_{(n)}}(x_{(n)}) = n\theta_1^{-n}{x_{(n)}}^{n-1}$, $\text{f}_{Y_{(n)}}(y_{(n)}) = n\theta_2^{-n}{y_{(n)}}^{n-1}$, and that $\text{f}_{X_{(n)},Y_{(n)}}(x_{(n)},y_{(n)}) = \text{f}_{X_{(n)}}(x_{(n)})\cdot\text{f}_{Y_{(n)}}(y_{(n)}) = n^2(\theta_1\theta_2)^{-n}[x_{(n)}y_{(n)}]^{n-1}$, $0 < x_{(n)} < \theta_1$, $0 < y_{(n)} < \theta_2$. Let

$$R = \frac{X_{(n)}}{Y_{(n)}}, \quad V = Y_{(n)};$$

then

$$X_{(n)} = RV, \quad Y_{(n)} = V.$$

So, the Jacobian is

$$J = \begin{vmatrix} \frac{\partial X_{(n)}}{\partial R} & \frac{\partial X_{(n)}}{\partial V} \\ \frac{\partial Y_{(n)}}{\partial R} & \frac{\partial Y_{(n)}}{\partial V} \end{vmatrix} = \begin{vmatrix} V & R \\ 0 & 1 \end{vmatrix} = V,$$

and $|J| = V$.
Hence,

$$\begin{aligned}\text{f}_{R,V}(r,v) &= n^2(\theta_1\theta_2)^{-n}(rv^2)^{n-1}v \\ &= n^2(\theta_1\theta_2)^{-n}r^{n-1}v^{2n-1}, 0 < v < \theta_2, 0 < r < \theta_1/v.\end{aligned}$$

Thus,

$$\begin{aligned}\text{f}_R(r;\rho) &= n^2(\theta_1\theta_2)^{-n}r^{n-1}\int_0^{\theta_2} v^{2n-1}dv = n^2(\theta_1\theta_2)^{-n}r^{n-1}\left(\frac{v^{2n}}{2n}\right)_0^{\theta_2} \\ &= \frac{n}{2}\left(\frac{\theta_1}{\theta_2}\right)^{-n}r^{n-1} = \frac{n}{2}\rho^{-n}r^{n-1}, 0 < r < \rho;\end{aligned}$$

and

$$\begin{aligned} \mathrm{f}_R(r;\rho) &= n^2(\theta_1\theta_2)^{-n}r^{n-1}\int_0^{\theta_1/r} v^{2n-1}\mathrm{d}v = n^2(\theta_1\theta_2)^{-n}r^{n-1}\left(\frac{v^{2n}}{2n}\right)_0^{\theta_1/r} \\ &= \frac{n}{2}\rho^n r^{-(n+1)}, \rho < r < +\infty. \end{aligned}$$

So,

$$\mathrm{f}_R(r;\rho) = \begin{cases} \frac{n}{2}\rho^{-n}r^{n-1}, 0 < r < \rho; \\ \frac{n}{2}\rho^n r^{-(n+1)}, \rho < r < +\infty. \end{cases}$$

(b) Now,

$$\begin{aligned} \mathrm{E}(R) &= \frac{n}{2}\rho^{-n}\int_0^{\rho} r^n\mathrm{d}r + \frac{n}{2}\rho^n\int_{\rho}^{\infty} r^{-n}\mathrm{d}r \\ &= \frac{n}{2}\rho^{-n}\left[\frac{r^{n+1}}{(n+1)}\right]_0^{\rho} + \frac{n}{2}\rho^n\left[\frac{-r^{-(n-1)}}{(n-1)}\right]_{\rho}^{\infty} \\ &= \frac{n}{2(n+1)}\rho + \frac{n}{2(n-1)}\rho \\ &= \frac{n^2}{(n^2-1)}\rho. \end{aligned}$$

And

$$\begin{aligned} \mathrm{E}(R^2) &= \frac{n}{2}\rho^{-n}\int_0^{\rho} r^{n+1}\mathrm{d}r + \frac{n}{2}\rho^n\int_{\rho}^{\infty} r^{-(n-1)}\mathrm{d}r \\ &= \frac{n}{2}\rho^{-n}\left[\frac{r^{n+2}}{(n+2)}\right]_0^{\rho} + \frac{n}{2}\rho^n\left[\frac{-r^{-(n-2)}}{(n-2)}\right]_{\rho}^{\infty} \\ &= \frac{n}{2(n+2)}\rho^2 + \frac{n}{2(n-2)}\rho^2 \\ &= \frac{n^2}{(n^2-4)}\rho^2. \end{aligned}$$

So,

$$\begin{aligned} \mathrm{V}(R) &= \frac{n^2}{(n^2-4)}\rho^2 - \frac{n^4}{(n^2-1)^2}\rho^2 \\ &= \frac{n^2(2n^2+1)}{(n^2-4)(n^2-1)^2}\rho^2. \end{aligned}$$

Note that $\lim_{n\to\infty}\mathrm{E}(R) = \rho$ and $\lim_{n\to\infty}\mathrm{V}(R) = 0$, so that R is consistent for ρ. As long as n is large, the statistician's suggestion to use R as a point estimator of ρ seems reasonable.

Solution 5.55. The appropriate likelihood function $\mathcal{L}$ is

$$\mathcal{L} = \prod_{i=1}^{n} \mathrm{p}_{Y_i}(y_i) = \prod_{i=1}^{n} \left[\frac{1 + y_i(2 - y_i)}{2}\right] \frac{\theta^{y_i(2-y_i)}}{(1+\theta)},$$

so that

$$\ln\mathcal{L} \propto \left[\sum_{i=1}^{n} y_i(2 - y_i)\right] \ln\theta - n\ln(1+\theta).$$

Now, the equation

$$\frac{\partial \ln\mathcal{L}}{\partial\theta} = \frac{\sum_{i=1}^{n} y_i(2-y_i)}{\theta} - \frac{n}{(1+\theta)} = 0$$

gives

$$\hat{\theta} = \frac{\sum_{i=1}^{n} y_i(2-y_i)}{n - \sum_{i=1}^{n} y_i(2-y_i)}$$

as the maximum likelihood estimate of θ.

Now, using *observed* information, the large-sample variance $\mathrm{V}_o(\hat{\theta})$ of $\hat{\theta}$ is

$$\begin{aligned} \mathrm{V}_o(\hat{\theta}) &= \left[-\frac{\partial^2 \ln\mathcal{L}}{\partial\theta^2}\right]^{-1} \\ &= \left[\frac{\sum_{i=1}^{n} y_i(2-y_i)}{\theta^2} - \frac{n}{(1+\theta)^2}\right]^{-1}. \end{aligned}$$

And, since, for $i = 1, 2, \ldots, n$,

$$\mathrm{E}(Y_i) = 1 \text{ and } \mathrm{E}(Y_i^2) = \left(\frac{2+\theta}{1+\theta}\right),$$

it follows that the large-sample variance $\mathrm{V}_e(\hat{\theta})$ of $\hat{\theta}$ using *expected* information is

$$\begin{aligned} \mathrm{V}_e(\hat{\theta}) &= \left[-\mathrm{E}\left(\frac{\partial^2 \ln\mathcal{L}}{\partial\theta^2}\right)\right]^{-1} \\ &= \left\{\frac{\sum_{i=1}^{n}[2\mathrm{E}(Y_i) - \mathrm{E}(Y_i^2)]}{\theta^2} - \frac{n}{(1+\theta)^2}\right\}^{-1} \\ &= \left\{\theta^{-2}\left[2n(1) - n\left(\frac{2+\theta}{1+\theta}\right)\right] - \frac{n}{(1+\theta)^2}\right\}^{-1} \\ &= \frac{\theta(1+\theta)^2}{n}. \end{aligned}$$

So, for the given set of data, it follows that

$$\sum_{i=1}^{100} = 50(0) + 20(1) + 30(0) = 20 \text{ and that } \hat{\theta} = \frac{20}{(100-20)} = 0.25.$$

Thus, the 95% confidence interval for θ using *observed* information is equal to

$$\begin{aligned}
\hat{\theta} \pm \sqrt{\hat{\mathrm{V}}_o(\hat{\theta})} &= 0.25 \pm \left[\frac{20}{(0.25)^2} - \frac{100}{(1.25)^2}\right]^{-1/2} \\
&= 0.25 \pm 0.1225, \text{ or } (0.1275, 0.3725).
\end{aligned}$$

And, the 95% confidence interval for θ using *expected* information is equal to

$$\begin{aligned}
\hat{\theta} \pm \sqrt{\hat{\mathrm{V}}_e(\hat{\theta})} &= 0.25 \pm \left[\frac{(0.25)(1.25)^2}{100}\right]^{1/2} \\
&= 0.25 \pm 0.1225, \text{ or } (0.1275, 0.3725).
\end{aligned}$$

Although these two confidence intervals are numerically exactly the same in this particular example, this will not typically happen for more complicated data analysis scenarios.

Finally, note that, under the proposed statistical model for $\mathrm{p}_{Y_i}(y_i)$, it follows that

$$\mathrm{p}_{Y_i}(0) = \mathrm{p}_{Y_i}(2) = \frac{1}{2(1+\theta)} \text{ and } \mathrm{p}_{Y_i}(1) = \frac{\theta}{(1+\theta)}.$$

Hence, since $0 < \frac{\theta}{1+\theta)} < \frac{1}{3}$ when $0 < \theta < \frac{1}{2}$, it follows that this data analysis provides statistical evidence in favor of the theory that monozygotic twins separated at birth tend, as adults, to be more alike than different with regard to their exercise habits.

Solution 5.57.

(a) Based on the proposed sampling plan, it is reasonable to assume that $X_i \sim \text{NEGBIN}(k, \pi)$, so that the likelihood function $\mathcal{L}$ has the structure

$$\begin{aligned}
\mathcal{L} &= \prod_{i=1}^{n} \mathrm{p}_{X_i}(x_i) = \prod_{i=1}^{n} \mathrm{C}_{k-1}^{x_i-1} \pi^k (1-\pi)^{x_i-k} \\
&= \left[\prod_{i=1}^{n} \mathrm{C}_{k-1}^{x_i-1}\right] \pi^{nk} (1-\pi)^{s-nk}
\end{aligned}$$

where $s = \sum_{i=1}^{n} x_i$.

So,

$$\frac{\partial \ln \mathcal{L}}{\partial \pi} = \frac{nk}{\pi} - \frac{(s-nk)}{(1-\pi)} = 0$$

gives

$$\hat{\pi} = \frac{nk}{S} = \frac{nk}{(\sum_{i=1}^{n} X_i)}$$

as the MLE of π.

Now,

$$\frac{\partial^2 \ln \mathcal{L}}{\partial \pi^2} = \frac{-nk}{\pi^2} - \frac{(s-nk)}{(1-\pi)^2};$$

and, since $\mathrm{E}(S) = nk/\pi$, it follows that

$$\begin{aligned} -E\left(\frac{\partial^2 \ln \mathcal{L}}{\partial \pi^2}\right) &= \frac{nk}{\pi^2} + \frac{\left(\frac{nk}{\pi} - nk\right)}{(1-\pi)^2} \\ &= \frac{nk}{\pi^2(1-\pi)}. \end{aligned}$$

So, an appropriate large-sample 95% confidence interval for π is

$$\hat{\pi} \pm 1.96\sqrt{\frac{\hat{\pi}^2(1-\hat{\pi})}{nk}}.$$

When $n = 25, k = 5$, and $\hat{\pi}$=0.40, the computed large-sample 95% confidence interval for π is equal to

$$0.40 \pm 1.96\sqrt{\frac{(0.40)^2(1-0.40)}{(25)(5)}} = 0.40 \pm 0.0543, \text{ or } (0.3457, 0.4543).$$

(b) For $0 < \pi \leq 0.50$, $\pi^2(1-\pi)$ is a monotonically increasing function of π, with a maximum value of $(0.50)^2(1-0.50) = 0.125$. So, for $n = 25$, the width W of the large-sample 95% confidence interval developed in part (a) satisfies the inequality

$$W \leq 2(1.96)\sqrt{\frac{0.125}{25k}} = \frac{0.2772}{\sqrt{k}}.$$

So, the inequality $0.2772/\sqrt{k} \leq 0.05$ will be satisfied if $k^* = 31$.

Solution 5.59.

(a) Note that X can take the values 0 and 1, with

$$\begin{aligned} \mathrm{pr}(X=1) &= \mathrm{pr}\left[(W=1) \cap (U_1=1)\right] + \mathrm{pr}\left[(W=0) \cap (U_2=1)\right] \\ &= \mathrm{pr}(W=1)\mathrm{pr}(U_1=1) + \mathrm{pr}(W=0)\mathrm{pr}(U_2=1) \\ &= \theta\pi + (1-\theta)\pi = \pi, \end{aligned}$$

so that X has a Bernoulli distribution with $\mathrm{E}(X) = \pi$ and $\mathrm{V}(X) = \pi(1-\pi)$. Analogously, it follows that Y also has a Bernoulli distribution with $\mathrm{E}(Y) = \pi$ and $\mathrm{V}(Y) = \pi(1-\pi)$.
Now,

$$\mathrm{cov}(X, Y) \;=\; \mathrm{E}(XY) - \mathrm{E}(X)\mathrm{E}(Y) = \mathrm{E}(XY) - \pi^2.$$

And, since

$$\begin{aligned}
\mathrm{E}(XY) &= \mathrm{E}\left\{[WU_1 + (1-W)U_2]\,[WU_1 + (1-W)U_3]\right\} \\
&= \mathrm{E}\big[W^2U_1^2 + W(1-W)U_1U_3 + (1-W)WU_2U_1 \\
&\quad + (1-W)^2U_2U_3\big] \\
&= \mathrm{E}(W^2)\mathrm{E}(U_1^2) + \mathrm{E}\left[W(1-W)\right]\mathrm{E}(U_1)\mathrm{E}(U_3) \\
&\quad + \mathrm{E}\left[(1-W)W\right]\mathrm{E}(U_2)\mathrm{E}(U_1) + \mathrm{E}[(1-W)^2]\mathrm{E}(U_2)\mathrm{E}(U_3) \\
&= \theta\pi + 0 + 0 + (1-\theta)\pi^2 = \pi^2 + \theta\pi(1-\pi),
\end{aligned}$$

we have $\mathrm{cov}(X, Y) = \pi^2 + \theta\pi(1-\pi) - \pi^2 = \theta\pi(1-\pi)$.
Clearly, since $\mathrm{E}(X_i) = \mathrm{E}(Y_i) = \pi, i = 1, 2, \ldots, n$, it follows directly that $\mathrm{E}\left(\hat{\pi}_1\right) = \mathrm{E}\left(\hat{\pi}_2\right) = \pi$. So, the estimator with the smaller variance would be preferred.
First, it is clear that $\mathrm{V}\left(\hat{\pi}_1\right) = \mathrm{V}\left(\bar{X}\right) = \pi(1-\pi)/n$. And,

$$\begin{aligned}
\mathrm{V}\left(\hat{\pi}_2\right) &= \frac{1}{4}\mathrm{V}\left(\bar{X} + \bar{Y}\right) \\
&= \frac{1}{4}\mathrm{V}\left(n^{-1}\sum_{i=1}^{n} X_i + n^{-1}\sum_{i=1}^{n} Y_i\right) \\
&= \frac{1}{4n^2}\mathrm{V}\left[\sum_{i=1}^{n}(X_i + Y_i)\right] = \frac{1}{4n}\mathrm{V}\left(X_i + Y_i\right) \\
&= \frac{1}{4n}\left[\mathrm{V}(X_i) + \mathrm{V}(Y_i) + 2\mathrm{cov}(X_i, Y_i)\right] \\
&= \frac{1}{4n}\left[\pi(1-\pi) + \pi(1-\pi) + 2\theta\pi(1-\pi)\right] \\
&= \frac{\pi(1-\pi)(1+\theta)}{2n},
\end{aligned}$$

which is clearly less than $\mathrm{V}(\hat{\pi}_1)$ since $(1+\theta) < 2$ for $0 < \theta < 1$. So, $\hat{\pi}_2$ is the preferred estimator.

(b) Let $D_i = (X_i + Y_i), i = 1, 2, \ldots, n$. Then, with $\bar{D} = n^{-1}\sum_{i=1}^{n} D_i$, the random variable

$$S_d^2 = \frac{1}{(n-1)}\sum_{i=1}^{n}(D_i - \bar{D})^2$$

has an expected value equal to $\mathrm{V}(D_i) = 2\pi(1-\pi)(1+\theta)$, so that

$$\hat{\theta} = \frac{S_d^2}{2\pi(1-\pi)} - 1$$

is an unbiased estimator of θ assuming that the value of π is known.

Solution 5.61. An appropriate likelihood function $\mathcal{L}$ for these data is the multinomial distribution, namely,

$$\mathcal{L} = \frac{n!}{\prod_{i=1}^{4} n_i!}\left(\frac{2+\theta}{4}\right)^{n_1}\left(\frac{1-\theta}{4}\right)^{n_2}\left(\frac{1-\theta}{4}\right)^{n_3}\left(\frac{\theta}{4}\right)^{n_4},$$

where $0 \le n_i \le n$ for $i = 1, 2, 3, 4$, and where $\sum_{i=1}^{4} n_i = n$.

So, since

$$\ln\mathcal{L} \propto n_1\ln(2+\theta) + (n_2+n_3)\ln(1-\theta) + n_4\ln\theta,$$

we have

$$\frac{\partial\ln\mathcal{L}}{\partial\theta} = \frac{n_1}{(2+\theta)} - \frac{(n_2+n_3)}{(1-\theta)} + \frac{n_4}{\theta} = 0,$$

which, with some algebra, can be shown to be equivalent to the quadratic equation $A\theta^2 + B\theta + C = 0$, where

$$A = n, B = -\left[n_1 - 2(n_2+n_3) - n_4\right], \text{ and } C = -2n_4.$$

For the given data, $A = 100, -B = [70 - 2(10+15) - 5] = 15$, and $C = -2(5) = -10$, so that the two roots are

$$\frac{15 \pm \sqrt{(-15)^2 - 4(100)(-10)}}{2(100)} = \frac{15 \pm 65}{200}.$$

Since we require $\hat{\theta} > 0$, we choose the positive root, so that $\hat{\theta} = 0.40$.

Now,

$$\frac{\partial^2\ln\mathcal{L}}{\partial\theta^2} = -\frac{n_1}{(2+\theta)^2} - \frac{(n_2+n_3)}{(1-\theta)^2} - \frac{n_4}{\theta^2},$$

so that

$$\begin{aligned}
-\mathrm{E}\left(\frac{\partial^2\ln\mathcal{L}}{\partial\theta^2}\right) &= \frac{\mathrm{E}(n_1)}{(2+\theta)^2} + \frac{\mathrm{E}(n_2+n_3)}{(1-\theta)^2} + \frac{\mathrm{E}(n_4)}{\theta^2} \\
&= \frac{n\left(\frac{2+\theta}{4}\right)}{(2+\theta)^2} + \frac{n\left[\frac{(1-\theta)}{4} + \frac{(1-\theta)}{4}\right]}{(1-\theta)^2} + \frac{n\left(\frac{\theta}{4}\right)}{\theta^2} \\
&= \frac{n(1+2\theta)}{2\theta(1-\theta)(2+\theta)}.
\end{aligned}$$

Thus, for large n,

$$\mathrm{V}(\hat{\theta}) \approx \frac{2\theta(1-\theta)(2+\theta)}{n(1+2\theta)};$$

so, since $\hat{\theta} = 0.40$, the estimated variance of $\hat{\theta}$ equals

$$\hat{\mathrm{V}}(\hat{\theta}) = \frac{2(0.40)(1-0.40)(2+0.40)}{100[1+2(0.40)]} = 0.0064.$$

Hence, the computed large-sample 95% confidence interval for θ is equal to

$$\hat{\theta} \pm 1.96\sqrt{\hat{\mathrm{V}}(\hat{\theta})} = 0.40 \pm 1.96\sqrt{0.0064} = 0.40 \pm 1.96(0.08) = 0.40 \pm 0.1568,$$

or (0.2432, 0.5568).

Solution 5.63. First, note that the random variable Y is unobservable, and that we only have observations on the random variable X. So, in order to proceed, we need to find the marginal distribution of the random variable X. Now, under the stated assumptions, it follows that

$$\mathrm{p}_X(x|Y=y) = \mathrm{C}_x^y \pi^x (1-\pi)^{y-x}, x = 0, 1, \ldots, y \text{ and } 0 < \pi < 1.$$

So,

$$\begin{aligned}
\mathrm{p}_X(x) &= \sum_{y=x}^{\infty} \mathrm{p}_X(x|Y=y)\mathrm{p}_Y(y) = \sum_{y=x}^{\infty} \mathrm{C}_x^y \pi^x (1-\pi)^{y-x}\left(\frac{\lambda^y \mathrm{e}^{-\lambda}}{y!}\right) \\
&= \sum_{y=x}^{\infty} \left(\frac{y!}{x!(y-x)!}\right) \pi^x (1-\pi)^{y-x}\left(\frac{\lambda^y \mathrm{e}^{-\lambda}}{y!}\right) \\
&= \sum_{u=0}^{\infty} \left(\frac{1}{x!u!}\right) \pi^x (1-\pi)^u \lambda^{u+x} \mathrm{e}^{-\lambda} \\
&= \frac{(\pi\lambda)^x \mathrm{e}^{-\lambda}}{x!} \sum_{u=0}^{\infty} \frac{[\lambda(1-\pi)]^u}{u!} \\
&= \frac{(\pi\lambda)^x \mathrm{e}^{-\lambda} \mathrm{e}^{\lambda(1-\pi)}}{x!} \\
&= \frac{(\pi\lambda)^x \mathrm{e}^{-\pi\lambda}}{x!}, x = 0, 1, \ldots, \infty,
\end{aligned}$$

so that $X \sim \mathrm{POI}(\pi\lambda)$.

Thus, the appropriate likelihood function $\mathcal{L}$ for the available data takes the form

$$\mathcal{L} = \prod_{i=1}^{n} \left[\frac{(\pi\lambda)^{x_i} \mathrm{e}^{-\pi\lambda}}{x_i!}\right] = \frac{(\pi\lambda)^{n\bar{x}} \mathrm{e}^{-n\pi\lambda}}{\prod_{i=1}^{n} x_i!}.$$

So,

$$\ln\mathcal{L} \propto (n\bar{x})\ln(\pi\lambda) - n\pi\lambda,$$

so that

$$\frac{\partial \mathrm{ln}\mathcal{L}}{\partial \lambda} = \frac{n\bar{x}}{\lambda} - n\pi = 0$$

gives $\hat{\lambda} = \bar{x}/\pi$ as the MLE of λ.

And, since

$$\frac{\partial^2 \mathrm{ln}\mathcal{L}}{\partial \lambda^2} = -\frac{n\bar{x}}{\lambda^2},$$

so that

$$-\mathrm{E}\left(\frac{\partial^2 \mathrm{ln}\mathcal{L}}{\partial \lambda^2}\right) = \frac{n\pi\lambda}{\lambda^2} = \frac{n\pi}{\lambda},$$

it follows that

$$\mathrm{V}(\hat{\lambda}) \approx \left(\frac{n\pi}{\lambda}\right)^{-1} = \frac{\lambda}{n\pi}$$

for large n.

Thus, the large-sample 95% confidence interval for λ is

$$\hat{\lambda} \pm 1.96\sqrt{\hat{V}(\hat{\lambda})} = \hat{\lambda} \pm 1.96\sqrt{\frac{\hat{\lambda}}{n\pi}}.$$

For the available data, $\hat{\lambda} = 24.80/0.75 = 33.07$, so that the computed large-sample 95% confidence interval for λ is equal to

$$33.07 \pm 1.96\sqrt{\frac{(33.07)}{(50)(0.75)}} = 33.07 \pm 1.84,$$

or (31.23, 34.91).

Solution 5.65.

(a) First,

$$\begin{aligned}
\mathrm{MSE}(k\hat{\theta}, \theta) &= \mathrm{E}\left\{\left[\left(k\hat{\theta} - \mathrm{E}(k\hat{\theta})\right) + \left(\mathrm{E}(k\hat{\theta}) - \theta\right)\right]^2\right\} \\
&= \mathrm{V}(k\hat{\theta}) + \left[\mathrm{E}(k\hat{\theta}) - \theta\right]^2 \\
&= k^2\mathrm{V}(\hat{\theta}) + \theta^2(k-1)^2.
\end{aligned}$$

So,

$$\frac{\partial \mathrm{MSE}(k\hat{\theta}, \theta)}{\partial k} = 2k\mathrm{V}(\hat{\theta}) + 2\theta^2(k-1) = 0$$

gives

$$k^* = \frac{\theta^2}{\theta^2 + \mathrm{V}(\hat{\theta})}, \text{ so that } 0 < k^* < 1.$$

Since $\frac{\partial^2 \text{MSE}(k\hat{\theta},\theta)}{\partial k^2} > 0, k^*$ minimizes $\text{MSE}(k\hat{\theta},\theta)$.
The estimator $k^*\hat{\theta}$ cannot be computed without knowledge of parameter values. Note that k^* can be written in the form

$$k^* = \frac{1}{1+\frac{\text{V}(\hat{\theta})}{\theta^2}} = \frac{1}{1+\xi^2},$$

where ξ is the *coefficient of variation* of the estimator $\hat{\theta}$. So, if the value of ξ is known, then the estimator $k^*\hat{\theta}$ can be computed.

(b) Clearly, $\text{MSE}(\hat{\theta}) = \text{V}(\hat{\theta})$. And, from part (a), we have

$$\begin{aligned} \text{MSE}(k^*\hat{\theta},\theta) &= \left[\frac{\theta^2}{\theta^2+\text{V}(\hat{\theta})}\right]\text{V}(\hat{\theta}) + \theta^2\left[\left(\frac{\theta^2}{\theta^2+\text{V}(\hat{\theta})}\right)-1\right]^2 \\ &= \frac{\theta^2\text{V}(\hat{\theta})}{\theta^2+\text{V}(\hat{\theta})} = k^*\text{V}(\hat{\theta}) < \text{V}(\hat{\theta}) = \text{MSE}(\hat{\theta}). \end{aligned}$$

Since $|k^*\hat{\theta}| < |\hat{\theta}|$, this result suggests that an estimator can be improved, in terms of mean-squared error, by *shrinking* it toward the value 0 (or, more generally, to some other known constant); the estimator $k^*\hat{\theta}$ is thus known as a *shrinkage estimator.*

(c) Note that

$$\hat{\theta} = \frac{\theta}{n}\sum_{i=1}^{n}\left(\frac{X_i}{\sqrt{\theta}}\right)^2 = \frac{\theta}{n}\sum_{i=1}^{n}Z_i^2,$$

where $Z_i \sim \text{N}(0,1)$ and where $Z_1, Z_2 \ldots, Z_n$ are mutually independent. Since $Z_i^2 \sim \chi_1^2$, so that $\text{E}(Z_i^2) = 1$ and $\text{V}(Z_i^2) = 2$, we have $\text{E}(\hat{\theta}) = \theta$ and $\text{V}(\hat{\theta}) = 2\theta^2/n$.
So, in this special case, we have

$$k^* = \frac{\theta^2}{\theta^2+\left(\frac{2\theta^2}{n}\right)} = \frac{n}{(n+2)}.$$

Solution 5.67. First, since $\text{E}(Y_{ij}) = \alpha + \beta x_{ij} + \text{E}(U_i) + \text{E}(E_{ij}) = \alpha + \beta x_{ij}$, and since $\sum_{j=1}^{n}(x_{ij} - \bar{x}_i) = 0$, it follows that

$$
\begin{aligned}
\mathrm{E}(\hat{\beta}) &= \frac{\sum_{i=1}^{k}\sum_{j=1}^{n}(x_{ij} - \bar{x}_i)\mathrm{E}(Y_{ij})}{\sum_{i=1}^{k}\sum_{j=1}^{n}(x_{ij} - \bar{x}_i)^2} \\
&= \frac{\sum_{i=1}^{k}\sum_{j=1}^{n}(x_{ij} - \bar{x}_i)(\alpha + \beta x_{ij})}{\sum_{i=1}^{k}\sum_{j=1}^{n}(x_{ij} - \bar{x}_i)^2} \\
&= \alpha\left[\frac{\sum_{i=1}^{k}\sum_{j=1}^{n}(x_{ij} - \bar{x}_i)}{\sum_{i=1}^{k}\sum_{j=1}^{n}(x_{ij} - \bar{x}_i)^2}\right] + \beta\left[\frac{\sum_{i=1}^{k}\sum_{j=1}^{n}(x_{ij} - \bar{x}_i)x_{ij}}{\sum_{i=1}^{k}\sum_{j=1}^{n}(x_{ij} - \bar{x}_i)^2}\right] \\
&= 0 + \beta\left[\frac{\sum_{i=1}^{k}\sum_{j=1}^{n}(x_{ij} - \bar{x}_i)(x_{ij} - \bar{x}_i)}{\sum_{i=1}^{k}\sum_{j=1}^{n}(x_{ij} - \bar{x}_i)^2}\right] \\
&= \beta,
\end{aligned}
$$

so that $\hat{\beta}$ is an unbiased estimator of β.

To develop an explicit expression for $\mathrm{V}(\hat{\beta})$, first note that $\hat{\beta}$ can be written in the form

$$
\left[\sum_{i=1}^{k}\sum_{j=1}^{n}(x_{ij} - \bar{x}_i)^2\right]^{-1}\sum_{i=1}^{k}L_i,
$$

where $L_i = \sum_{j=1}^{n}(x_{ij} - \bar{x}_i)Y_{ij}$ and where $L_1, L_2, \ldots, L_k$ constitute a set of k mutually independent random variables.

Thus,

$$
\mathrm{V}(\hat{\beta}) = \left[\sum_{i=1}^{k}\sum_{j=1}^{n}(x_{ij} - \bar{x}_i)^2\right]^{-2}\sum_{i=1}^{k}\mathrm{V}(L_i).
$$

Now, using conditional expectation theory, it follows that

$$\begin{aligned}
\mathrm{V}(L_i) &= \mathrm{E}\left[\mathrm{V}(L_i|U_i=u_i)\right] + \mathrm{V}\left[\mathrm{E}(L_i|U_i=u_i)\right] \\
&= \mathrm{E}_{u_i}\left[\sum_{j=1}^{n}(x_{ij}-\bar{x}_i)^2\mathrm{V}(Y_{ij}|U_i=u_i)\right] \\
&\quad + \mathrm{V}_{u_i}\left[\sum_{j=1}^{n}(x_{ij}-\bar{x}_i)\mathrm{E}(Y_{ij}|U_i=u_i)\right] \\
&= \mathrm{E}_{u_i}\left[\sum_{j=1}^{n}(x_{ij}-\bar{x}_i)^2(\sigma_e^2)\right] + \mathrm{V}_{u_i}\left[\sum_{j=1}^{n}(x_{ij}-\bar{x}_i)(\alpha+\beta x_{ij}+u_i)\right] \\
&= \sigma_e^2\sum_{j=1}^{n}(x_{ij}-\bar{x}_i)^2 + \mathrm{V}_{u_i}\left[0+\beta\sum_{j=1}^{n}(x_{ij}-\bar{x}_i)x_{ij}+0\right] \\
&= \sigma_e^2\sum_{j=1}^{n}(x_{ij}-\bar{x}_i)^2 + 0 = \sigma_e^2\sum_{j=1}^{n}(x_{ij}-\bar{x}_i)^2.
\end{aligned}$$

Finally,

$$\begin{aligned}
\mathrm{V}(\hat{\beta}) &= \left[\sum_{i=1}^{k}\sum_{j=1}^{n}(x_{ij}-\bar{x}_i)^2\right]^{-2}\sum_{i=1}^{k}\left[\sigma_e^2\sum_{j=1}^{n}(x_{ij}-\bar{x}_i)^2\right] \\
&= \frac{\sigma_e^2}{\sum_{i=1}^{k}\sum_{j=1}^{n}(x_{ij}-\bar{x}_i)^2}.
\end{aligned}$$

One could also obtain this same result less directly by first noting that

$$\begin{aligned}
\mathrm{V}(L_i) &= \sum_{j=1}^{n}(x_{ij}-\bar{x}_i)^2\mathrm{V}(Y_{ij}) \\
&\quad + 2\sum_{j=1}^{n-1}\sum_{j'=j+1}^{n}(x_{ij}-\bar{x}_i)(x_{ij'}-\bar{x}_i)\mathrm{cov}(Y_{ij},Y_{ij'}) \\
&= \sum_{j=1}^{n}(x_{ij}-\bar{x}_i)^2(\sigma_u^2+\sigma_e^2) + 2\sum_{j=1}^{n-1}\sum_{j'=j+1}^{n}(x_{ij}-\bar{x}_i)(x_{ij'}-\bar{x}_i)(\sigma_u^2),
\end{aligned}$$

and then performing further algebraic manipulations.

Solution 5.69*. If n_x is the realized value of N_x for $x = 0, 1, \ldots, \infty$, then the multinomial-type likelihood function is

$$\mathcal{L} = \frac{n!}{\prod_{x=0}^{\infty} n_x!}(1-\pi)^{n_0}\prod_{x=1}^{\infty}\left[\frac{\pi\lambda^x}{x!(\mathrm{e}^{\lambda}-1)}\right]^{n_x},$$

where

$$\sum_{x=0}^{\infty} n_x = n$$

and

$$(1-\pi) + \pi \sum_{x=1}^{\infty} \frac{\lambda^x}{x!(\mathrm{e}^{\lambda}-1)} = (1-\pi) + \pi = 1.$$

So,

$$\begin{aligned}
\ln\mathcal{L} &\sim n_0\ln(1-\pi) + \sum_{x=1}^{\infty} \left\{ n_x\ln\pi + xn_x\ln\lambda - n_x\ln x! - n_x\ln(\mathrm{e}^{\lambda}-1) \right\} \\
&= n_0\ln(1-\pi) + (n-n_0)\ln\pi + \left(\sum_{x=1}^{\infty} xn_x \right) \ln\lambda \\
&\quad - \sum_{x=1}^{\infty} n_x\ln x! - (n-n_0)\ln(\mathrm{e}^{\lambda}-1).
\end{aligned}$$

So,

$$\frac{\partial\ln\mathcal{L}}{\partial\pi} = -\frac{n_0}{(1-\pi)} + \frac{(n-n_0)}{\pi} = 0 \implies \hat{\pi} = \frac{(n-n_0)}{n} = 1 - \frac{n_0}{n}.$$

Now,

$$\begin{aligned}
\frac{\partial\ln\mathcal{L}}{\partial\lambda} &= \frac{\sum_{x=1}^{\infty} xn_x}{\lambda} - \frac{(n-n_0)\mathrm{e}^{\lambda}}{(\mathrm{e}^{\lambda}-1)} = 0 \\
&\implies (\mathrm{e}^{\lambda}-1)\sum_{x=1}^{\infty} xn_x - (n-n_0)\lambda\mathrm{e}^{\lambda} = 0 \\
&\implies (n-n_0)\frac{\lambda\mathrm{e}^{\lambda}}{(\mathrm{e}^{\lambda}-1)} = \sum_{x=1}^{\infty} xn_x \\
&\implies \frac{\lambda\mathrm{e}^{\lambda}}{(\mathrm{e}^{\lambda}-1)} = \frac{\sum_{x=1}^{\infty} xn_x}{(n-n_0)},
\end{aligned}$$

which can be solved iteratively.

Now,

$$\frac{\partial^2\ln\mathcal{L}}{\partial\pi^2} = -\frac{n_0}{(1-\pi)^2} - \frac{(n-n_0)}{\pi^2},$$

so that

$$-\mathrm{E}\left(\frac{\partial^2\ln\mathcal{L}}{\partial\pi^2} \right) = \frac{n(1-\pi)}{(1-\pi)^2} + \frac{n[1-(1-\pi)]}{\pi^2} = \frac{n}{\pi(1-\pi)}.$$

Also, note that

$$\frac{\partial^2 \ln \mathcal{L}}{\partial \pi \partial \lambda} = \frac{\partial^2 \ln \mathcal{L}}{\partial \lambda \partial \pi} = 0.$$

And,

$$\begin{aligned}
\frac{\partial^2 \ln \mathcal{L}}{\partial \lambda^2} &= \frac{-\sum_{x=1}^{\infty} x n_x}{\lambda^2} - (n - n_0)\left[\frac{e^{\lambda}(e^{\lambda}-1) - e^{2\lambda}}{(e^{\lambda}-1)^2}\right] \\
&= -\frac{\sum_{x=1}^{\infty} x n_x}{\lambda^2} + \frac{(n - n_0)e^{\lambda}}{(e^{\lambda}-1)^2}.
\end{aligned}$$

Now,

$$\begin{aligned}
\sum_{x=1}^{\infty} x\mathrm{E}(N_x) &= \sum_{x=1}^{\infty} x\left[\frac{n\pi\lambda^x}{x!(e^{\lambda}-1)}\right] \\
&= \frac{n\pi}{(e^{\lambda}-1)}\sum_{x=1}^{\infty} x\frac{\lambda^x}{x!} = \frac{n\pi e^{\lambda}}{(e^{\lambda}-1)}\sum_{x=0}^{\infty} x\frac{\lambda^x e^{-\lambda}}{x!} = \frac{n\pi\lambda e^{\lambda}}{(e^{\lambda}-1)}.
\end{aligned}$$

So,

$$-\mathrm{E}\left(\frac{\partial^2 \ln \mathcal{L}}{\partial \lambda^2}\right) = \frac{n\pi e^{\lambda}}{\lambda(e^{\lambda}-1)} - \frac{n[1-(1-\pi)]e^{\lambda}}{(e^{\lambda}-1)^2} = \frac{n\pi e^{\lambda}}{\lambda(e^{\lambda}-1)^2}(e^{\lambda}-1-\lambda).$$

So, the expected information matrix is

$$\boldsymbol{\mathcal{I}} = \begin{bmatrix} \frac{n}{\pi(1-\pi)} & 0 \\ 0 & \frac{n\pi e^{\lambda}(e^{\lambda}-1-\lambda)}{\lambda(e^{\lambda}-1)^2} \end{bmatrix}$$

and hence

$$\hat{\boldsymbol{\mathcal{I}}}^{-1} = \begin{bmatrix} \frac{\hat{\pi}(1-\hat{\pi})}{n} & 0 \\ 0 & \frac{\hat{\lambda}(e^{\hat{\lambda}}-1)^2}{n\hat{\pi}e^{\hat{\lambda}}(e^{\hat{\lambda}}-1-\hat{\lambda})} \end{bmatrix}.$$

For the given data, the appropriate large-sample 95% confidence intervals for π and λ are:

$$\begin{aligned}
\hat{\pi} &\pm 1.96\sqrt{\hat{\pi}(1-\hat{\pi})/n} \\
&= 0.30 \pm 1.96\sqrt{0.30(0.70)/50} \\
&= (0.173, 0.427),
\end{aligned}$$

and

$$\begin{aligned}
\hat{\lambda} &\pm 1.96\sqrt{\frac{\hat{\lambda}(e^{\hat{\lambda}}-1)^2}{n\hat{\pi}e^{\hat{\lambda}}(e^{\hat{\lambda}}-1-\hat{\lambda})}} \\
&= 2.75 \pm 1.96\sqrt{\frac{(2.75)(e^{2.75}-1)^2}{(50)(0.30)e^{2.75}(e^{2.75}-1-2.75)}} \\
&= (1.849, 3.651).
\end{aligned}$$

Solution 5.71*.

(a) We have pr($X_i = 1$)=pr($X_i = 1$|spinner points to A) pr(spinner points to A) + pr($X_i = 1$|spinner points to $\overline{A}$) pr(spinner points to $\overline{A}$)= $\pi\theta + (1-\pi)(1-\theta) = \pi(2\theta - 1) + (1-\theta)$.

(b) Since the maximum likelihood estimator (MLE) of pr($X_i = 1$) is S/n, where $S = \sum_{i=1}^{n} X_i$, and since this MLE satisfies the equation

$$\frac{S}{n} = \hat{\pi}(2\theta - 1) + (1 - \theta),$$

we have

$$\hat{\pi} = \frac{\frac{S}{n} - (1-\theta)}{(2\theta - 1)} = \frac{(\theta - 1)}{2\theta - 1)} + \frac{S}{n(2\theta - 1)}, \theta \neq \frac{1}{2}.$$

One would also obtain this result by maximizing with respect to π the likelihood function

$$\begin{aligned} \mathcal{L} &= \prod_{i=1}^{n} \left\{ [\pi\theta + (1-\pi)(1-\theta)]^{x_i} [\theta(1-\pi) + \pi(1-\theta)]^{(1-x_i)} \right\} \\ &= [\pi\theta + (1-\pi)(1-\theta)]^{s} [\theta(1-\pi) + \pi(1-\theta)]^{(n-s)}. \end{aligned}$$

Also,

$$\mathrm{E}(\hat{\pi}) = \frac{(\theta - 1)}{(2\theta - 1)} + \frac{\mathrm{E}(S)}{n(2\theta - 1)} = \frac{(\theta - 1)}{(2\theta - 1)} + \frac{n[\pi\theta + (1-\pi)(1-\theta)]}{n(2\theta - 1)} = \pi,$$

so that $\hat{\pi}$ is an unbiased estimator of the unknown parameter π.

(c) The exact variance of $\hat{\pi}$ is

$$\mathrm{V}(\hat{\pi}) = \frac{\mathrm{V}(S)}{n^2(2\theta - 1)^2} = \frac{[\pi\theta + (1-\pi)(1-\theta)][\pi(1-\theta) + \theta(1-\pi)]}{n(2\theta - 1)^2},$$

which is the same expression that is obtained by evaluating $-\mathrm{E}\left[\frac{\partial^2 \ln(\mathcal{L})}{\partial \pi^2}\right]$. For $n = 100, \theta = 0.20$, and $\hat{\pi} = 0.10$, the computed 95% confidence interval for π is $\hat{\pi} \pm 1.96\sqrt{\hat{\mathrm{V}}(\hat{\pi})} = 0.10 \pm 1.96\sqrt{0.0053} = 0.10 \pm 0.1433$, or (0, 0.2433).

(d) We want to choose n^* so that

$$\mathrm{pr}(|\hat{\pi} - \pi| < \delta) = \mathrm{pr}[-\delta < (\hat{\pi} - \pi) < \delta]$$

$$= \mathrm{pr}\left[\frac{-\delta}{\sqrt{\mathrm{V}(\hat{\pi})}} < \frac{(\hat{\pi} - \pi)}{\sqrt{\mathrm{V}(\hat{\pi})}} < \frac{\delta}{\sqrt{\mathrm{V}(\hat{\pi})}}\right]$$

$$\approx \text{pr}\left[\frac{-\delta}{\sqrt{\text{V}(\hat{\pi})}} < Z < \frac{\delta}{\sqrt{\text{V}(\hat{\pi})}}\right] \geq 0.95,$$

where

$$Z = \frac{(\hat{\pi} - \pi)}{\sqrt{\text{V}(\hat{\pi})}} \dot{\sim} \text{N}(0, 1) \text{ for large } n.$$

So, n^* is the smallest positive integer such that $\frac{\delta}{\sqrt{\text{V}(\hat{\pi})}} \geq 1.96$ when $\theta = 0.20$, so that n^* must satisfy the inequality

$$n^* \geq \frac{(1.96)^2}{(0.36)\delta^2}(0.80 - 0.60\pi)(0.20 + 0.60\pi).$$

The right-hand side of the above inequality is a maximum when $\pi = \frac{1}{2}$, so that n^* takes its maximum value when $\pi = \frac{1}{2}$.

Solution 5.73*. Since $\hat{\beta} = \sum_{i=1}^{n}(x_i - \bar{x})Y_i/(n-1)s_x^2$, it follows that

$$\begin{aligned} \text{E}(\hat{\beta}) &= \frac{1}{(n-1)s_x^2}\sum_{i=1}^{n}(x_i - \bar{x})(\alpha + \beta x_i) \\ &= \frac{1}{(n-1)s_x^2}\left\{\alpha\sum_{i=1}^{n}(x_i - \bar{x}) + \beta\sum_{i=1}^{n}(x_i - \bar{x})x_i\right\} \\ &= \frac{1}{(n-1)s_x^2}\left\{0 + \beta(n-1)s_x^2\right\} \\ &= \beta \end{aligned}$$

So, the expected value of $\hat{\beta}$ is unaffected by correlations among the Y_is. Now,

$$\begin{aligned} \text{V}(\hat{\beta}) &= \text{V}\left[\frac{\sum_{i=1}^{n}(x_i - \bar{x})(Y_i - \bar{Y})}{(n-1)s_x^2}\right] = \text{V}\left[\frac{\sum_{i=1}^{n}(x_i - \bar{x})Y_i}{(n-1)s_x^2}\right] \\ &= \frac{1}{[(n-1)s_x^2]^2}\left\{\sum_{i=1}^{n}(x_i - \bar{x})^2\sigma^2 + 2\sum_{i=2}^{n-1}\sum_{j=i+1}^{n}(x_i - \bar{x})(x_{i'} - \bar{x})\rho\sigma^2\right\} \\ &= \frac{\sigma^2}{(n-1)s_x^2} + \frac{2\rho\sigma^2\sum_{i=2}^{n-1}\sum_{j=i+1}^{n}(x_i - \bar{x})(x_{i'} - \bar{x})}{[(n-1)s_x^2]^2}. \end{aligned}$$

Now, since

$$\left[\sum_{i=1}^{n}(x_i - \bar{x})\right]^2 = 0 = \sum_{i=1}^{n}(x_i - \bar{x})^2 + 2\sum_{i=2}^{n-1}\sum_{j=i+1}^{n}(x_i - \bar{x})(x_{i'} - \bar{x}),$$

it follows that

$$\text{V}(\hat{\beta}) = \frac{\sigma^2}{(n-1)s_x^2} - \frac{\rho\sigma^2(n-1)s_x^2}{[(n-1)s_x^2]^2} = \frac{\sigma^2(1-\rho)}{(n-1)s_x^2}.$$

When $\rho = 0$, we get the usual expression for $\mathrm{V}(\hat{\beta})$ under the standard assumption of uncorrelated Y_is. If $-1 < \rho < 0$, then

$$\mathrm{V}(\hat{\beta}) > \frac{\sigma^2}{(n-1)s_x^2}.$$

And, if $0 < \rho < 1$, then

$$\mathrm{V}(\hat{\beta}) < \frac{\sigma^2}{(n-1)s_x^2}.$$

Solution 5.75*.

(a) Since $Y_i \sim \mathrm{BIN}\left[g_i, 1-(1-\pi)^{n_i}\right], i = 1, 2, \ldots, k$, the likelihood function $\mathcal{L}$ has the form $\mathcal{L} = \prod_{i=1}^{k} \mathcal{L}_i$, where

$$\mathcal{L}_i = \mathrm{C}_{y_i}^{g_i}\left[1-(1-\pi)^{n_i}\right]^{y_i}\left[(1-\pi)^{n_i}\right]^{g_i-y_i},$$

so that

$$\mathrm{ln}\mathcal{L}_i = \mathrm{lnC}_{y_i}^{g_i} + y_i\mathrm{ln}\left[1-(1-\pi)^{n_i}\right] + (g_i - y_i)\left[n_i\mathrm{ln}(1-\pi)\right].$$

Thus,

$$\frac{\partial \mathrm{ln}\mathcal{L}_i}{\partial \pi} = \frac{n_i y_i (1-\pi)^{n_i-1}}{1-(1-\pi)^{n_i}} - \frac{n_i(g_i-y_i)}{(1-\pi)};$$

so, the maximum estimator $\hat{\pi}$ can be found iteratively via the equation

$$\sum_{i=1}^{k} \frac{n_i y_i (1-\pi)^{n_i}}{1-(1-\pi)^{n_i}} - \sum_{i=1}^{k} n_i(g_i - y_i) = 0,$$

or, equivalently, the equation

$$\sum_{i=1}^{k} \frac{n_i y_i}{1-(1-\pi)^{n_i}} = \sum_{i=1}^{k} n_i g_i.$$

(b) Now, since $\frac{\partial^2 \mathrm{ln}\mathcal{L}_i}{\partial \pi^2}$ is equal to

$$n_i y_i \left\{ \frac{-(n_i-1)(1-\pi)^{n_i-2}\left[1-(1-\pi)^{n_i}\right] - (1-\pi)^{n_i-1} n_i (1-\pi)^{n_i-1}}{\left[1-(1-\pi)^{n_i}\right]^2} \right\}$$

$$-\frac{n_i(g_i-y_i)}{(1-\pi)^2},$$

and since

$$\mathrm{E}(Y_i) = g_i\left[1-(1-\pi)^{n_i}\right],$$

it can be shown (with some algebraic manipulations) that

$$-\mathrm{E}\left(\frac{\partial^2 \ln\mathcal{L}}{\partial\pi^2}\right) = -\sum_{i=1}^{k}\mathrm{E}\left(\frac{\partial^2 \ln\mathcal{L}_i}{\partial\pi^2}\right) = \sum_{i=1}^{k}\frac{n_i^2 g_i(1-\pi)^{n_i-2}}{1-(1-\pi)^{n_i}}.$$

Thus, the large-sample variance $\mathrm{V}(\hat{\pi})$ of $\hat{\pi}$ is equal to

$$\mathrm{V}(\hat{\pi}) = \left[-\mathrm{E}\left(\frac{\partial^2 \ln\mathcal{L}}{\partial\pi^2}\right)\right]^{-1} = \left[\sum_{i=1}^{k}\frac{n_i^2 g_i(1-\pi)^{n_i-2}}{1-(1-\pi)^{n_i}}\right]^{-1}.$$

(c) Now, for the given data, we have $k = 1, g_1 = 10, n_1 = 20$, and $y_1 = 6$. So, using the result from part (a), we have

$$\frac{n_1 y_1}{1-(1-\pi)^{n_1}} = n_1 g_1,$$

which gives

$$\hat{\pi} = 1 - \left(1 - \frac{y_1}{g_1}\right)^{1/n_1}.$$

For the available data, it follows that

$$\hat{\pi} = 1 - \left(1 - \frac{6}{10}\right)^{1/20} = 0.0448.$$

Since the estimated large-sample variance of $\hat{\pi}$ equals

$$\hat{\mathrm{V}}(\hat{\pi}) = \left[\frac{n_1^2 g_1(1-\hat{\pi})^{n_1-2}}{1-(1-\hat{\pi})^{n_1}}\right]^{-1} = \left[\frac{(20)^2(10)(0.9552)^{18}}{1-(0.9552)^{20}}\right]^{-1} = (2,920.53)^{-1},$$

the computed 95% confidence interval for π is equal to

$$\hat{\pi} \pm 1.96\sqrt{\hat{\mathrm{V}}(\hat{\pi})} = 0.0448 \pm 1.96\sqrt{(2,920.53)^{-1}} = 0.0448 \pm 0.0363,$$

or (0.0085, 0.0811).

Solution 5.77*.

(a) The likelihood function $\mathcal{L}$ is

$$\mathcal{L} = \left[\frac{m}{\Gamma\left(\frac{1}{2m}\right)}\right]^n \mathrm{e}^{-\sum_{i=1}^{n}(x_i-\theta)^{2m}},$$

so that

$$\ln\mathcal{L} \propto -\sum_{i=1}^{n}(x_i-\theta)^{2m}.$$

So,

$$\frac{\partial \ln \mathcal{L}}{\partial \theta} = 2m \sum_{i=1}^{n} (x_i - \theta)^{2m-1},$$

$$\frac{\partial^2 \ln \mathcal{L}}{\partial \theta^2} = -2m(2m-1) \sum_{i=1}^{n} (x_i - \theta)^{2(m-1)},$$

and

$$-\mathrm{E}\left[\frac{\partial^2 \ln \mathcal{L}}{\partial \theta^2}\right] = 2m(2m-1)n\mathrm{E}\left[(X-\theta)^{2(m-1)}\right].$$

Now, for r a non-negative integer, and with $u = (x-\theta)^{2m}$, so that $(x-\theta) = u^{1/2m}$ and $dx = \frac{1}{2m}u^{\frac{1}{2m}-1}$, it follows (appealing to properties of the gamma distribution) that

$$\begin{aligned}
\mathrm{E}\left[(X-\theta)^{2r}\right] &= \frac{m}{\Gamma\left(\frac{1}{2m}\right)} \int_{-\infty}^{\infty} (x-\theta)^{2r} \mathrm{e}^{-(x-\theta)^{2m}} dx \\
&= \frac{2m}{\Gamma\left(\frac{1}{2m}\right)} \int_{0}^{\infty} (x-\theta)^{2r} \mathrm{e}^{-(x-\theta)^{2m}} dx \\
&= \frac{2m}{\Gamma\left(\frac{1}{2m}\right)} \int_{0}^{\infty} u^{r/m} \mathrm{e}^{-u} \frac{1}{2m} u^{\frac{1}{2m}-1} du \\
&= \frac{1}{\Gamma\left(\frac{1}{2m}\right)} \int_{0}^{\infty} u^{\left(\frac{2r+1}{2m}\right)-1} \mathrm{e}^{-u} du \\
&= \frac{\Gamma\left(\frac{2r+1}{2m}\right)}{\Gamma\left(\frac{1}{2m}\right)}.
\end{aligned}$$

Using this result with $r = (m-1)$, we have

$$-\mathrm{E}\left(\frac{\partial^2 \ln \mathcal{L}}{\partial \theta^2}\right) = 2m(2m-1)n \frac{\Gamma\left[\frac{2(m-1)+1}{2m}\right]}{\Gamma\left(\frac{1}{2m}\right)},$$

so that

$$\mathrm{CRLB} = \frac{\Gamma\left(\frac{1}{2m}\right)}{2m(2m-1)n\Gamma\left(\frac{2m-1}{2m}\right)}.$$

And, since $r = 1$ gives $\mathrm{E}\left[(X-\theta)^2\right] = \Gamma\left(\frac{3}{2m}\right)/\Gamma\left(\frac{1}{2m}\right)$, it follows that

$$\begin{aligned}
\mathrm{EFF}(\bar{X}, \theta) &= \frac{\mathrm{CRLB}}{\mathrm{V}(\bar{X})} \\
&= \frac{\Gamma\left(\frac{1}{2m}\right)/2m(2m-1)n\Gamma\left(\frac{2m-1}{2m}\right)}{\Gamma\left(\frac{3}{2m}\right)/n\Gamma\left(\frac{1}{2m}\right)} \\
&= \frac{\left[\Gamma\left(\frac{1}{2m}\right)\right]^2}{2m(2m-1)\Gamma\left(\frac{3}{2m}\right)\Gamma\left(\frac{2m-1}{2m}\right)} = \eta_m, \text{ say.}
\end{aligned}$$

(b) When $m = 1, \eta_1 = 1$. Since $X \sim \mathrm{N}(\theta, 1/2)$ when $m = 1$, this answer makes sense because $\mathrm{V}(\bar{X})$ equals the CRLB for every n when random sampling from a normal population with unknown mean and known variance. Also, when $m = 2, \eta_2 = 0.7295$; when $m = 3, \eta_3 = 0.5160$; and when $m = 4, \eta_4 = 0.3924$. So, η_m apparently decreases as m increases.

Using the fact that $\Gamma(y) = (y-1)\Gamma(y-1), y > 1$, it follows that

$$\Gamma\left(\frac{2m+1}{2m}\right) = \left(\frac{1}{2m}\right)\Gamma\left(\frac{1}{2m}\right),$$

that

$$\Gamma\left(\frac{3+2m}{2m}\right) = \left(\frac{3}{2m}\right)\Gamma\left(\frac{3}{2m}\right),$$

and that

$$\Gamma\left(\frac{4m-1}{2m}\right) = \left(\frac{2m-1}{2m}\right)\Gamma\left(\frac{2m-1}{2m}\right).$$

Then, using the above results in the expression for η_m, we obtain

$$\begin{aligned}
\eta_m &= \frac{\left[(2m)\Gamma\left(\frac{2m+1}{2m}\right)\right]^2}{2m(2m-1)\left(\frac{2m}{3}\right)\left(\frac{2m}{2m-1}\right)\Gamma\left(\frac{3+2m}{2m}\right)\Gamma\left(\frac{4m-1}{2m}\right)} \\
&= \left(\frac{3}{2m}\right)\frac{\left[\Gamma\left(\frac{2+\frac{1}{m}}{2}\right)\right]^2}{\Gamma\left(\frac{\frac{3}{m}+2}{2}\right)\Gamma\left(\frac{4-\frac{1}{m}}{2}\right)}.
\end{aligned}$$

Thus, it follows directly that $\lim_{m\to\infty}\eta_m = 0$.

Solution 5.79*.

(a) Since $\ln f_X(x) = \ln\left[2(1-\theta)x + 2\theta(1-x)\right]$, it follows that

$$\frac{\partial \ln f_X(x)}{\partial\theta} = \frac{-2x + 2(1-x)}{2(1-\theta)x + 2\theta(1-x)} = \frac{(1-2x)}{(1-\theta)x + \theta(1-x)}$$

and that

$$\frac{\partial^2 \ln f_X(x)}{\partial\theta^2} = -(1-2x)^2\left[(1-\theta)x + \theta(1-x)\right]^{-2}.$$

So,

$$\begin{aligned}
-\mathrm{E}\left[\frac{\partial^2 \ln f_X(x)}{\partial^2\theta^2}\right] &= \int_0^1 \frac{(1-2x)^2}{[(1-\theta)x + \theta(1-x)]^2}\left[2(1-\theta)x + 2\theta(1-x)\right]\mathrm{d}x \\
&= 2\int_0^1 \frac{(1-2x)^2}{[(1-\theta)x + \theta(1-x)]}\mathrm{d}x.
\end{aligned}$$

Now, for $\theta \neq \frac{1}{2}$, let $u = (1-\theta)x + \theta(1-x)$, so that $\mathrm{d}u = (1-2\theta)\mathrm{d}x$, $x = (u-\theta)/(1-2\theta)$, and

$$(1-2x)^2 = \left[1 - 2\left(\frac{u-\theta}{1-2\theta}\right)\right]^2 = \frac{(1-2u)^2}{(1-2\theta)^2}.$$

Thus, for $\theta \neq \frac{1}{2}$, we have

$$\begin{aligned} -\mathrm{E}\left[\frac{\partial^2 \ln f_X(x)}{\partial\theta^2}\right] &= 2\int_{\theta}^{1-\theta} \frac{(1-2u)^2}{(1-2\theta)^2}\left(u^{-1}\right)(1-2\theta)^{-1}\mathrm{d}u \\ &= \frac{2}{(1-2\theta)^3}\int_{\theta}^{1-\theta}\left(\frac{1}{u} - 4 + 4u\right)\mathrm{d}u. \end{aligned}$$

So, for $0 < \theta < \frac{1}{2}$, we have

$$\begin{aligned} -\mathrm{E}\left[\frac{\partial^2 \ln f_X(x)}{\partial\theta^2}\right] &= \frac{2}{(1-2\theta)^3}\left[\ln u - 4u + 2u^2\right]_{\theta}^{1-\theta} \\ &= \frac{2}{(1-2\theta)^3}\left[\ln\left(\frac{1-\theta}{\theta}\right) - 2(1-2\theta)\right]. \end{aligned}$$

And, for $\frac{1}{2} < \theta < 1$, we have

$$\begin{aligned} -\mathrm{E}\left[\frac{\partial^2 \ln f_X(x)}{\partial\theta^2}\right] &= -\frac{2}{(1-2\theta)^3}\left[\ln u - 4u + 2u^2\right]_{1-\theta}^{\theta} \\ &= \frac{2}{(1-2\theta)^3}\left[\ln\left(\frac{1-\theta}{\theta}\right) - 2(1-2\theta)\right]. \end{aligned}$$

So, for $\theta \neq \frac{1}{2}$, it follows that the CRLB is equal to

$$n^{-1}\left\{-\mathrm{E}\left[\frac{\partial^2 \ln f_X(x)}{\partial\theta^2}\right]\right\}^{-1} = \frac{(1-2\theta)^3}{2n\left[\ln\left(\frac{1-\theta}{\theta}\right) - 2(1-2\theta)\right]}.$$

When $\theta = \frac{1}{2}$, we have

$$\begin{aligned} -\mathrm{E}\left[\frac{\partial^2 \ln f_X(x)}{\partial\theta^2}\right] &= 2\int_0^1 \frac{(1-2x)^2}{\left[\left(1-\frac{1}{2}\right)x + \frac{1}{2}(1-x)\right]}\mathrm{d}x \\ &= 4\int_0^1 (1-2x)^2\mathrm{d}x = 4\left[\frac{-(1-2x)^3}{6}\right]_0^1 \\ &= 4\left(\frac{1}{6} + \frac{1}{6}\right) = \frac{4}{3}. \end{aligned}$$

Hence, when $\theta = \frac{1}{2}$, the CRLB is equal to 3/4n. In general, the CRLB takes its maximum value of 3/4n when $\theta = \frac{1}{2}$, and it is symmetric about the value $\theta = \frac{1}{2}$. The CRLB decreases to zero as $\theta \to 0$ and as $\theta \to 1$.

(b) For r a non-negative integer,

$$\begin{aligned}
\mathrm{E}(X^r) &= \int_0^1 x^r\left[2(1-\theta)x + 2\theta(1-x)\right]\mathrm{d}x \\
&= 2\int_0^1 x^r(x+\theta-2\theta x)\,\mathrm{d}x = 2\int_0^1\left[(1-2\theta)x^{r+1}+\theta x^r\right]\mathrm{d}x \\
&= 2\left[\left(\frac{1-2\theta}{r+2}\right)x^{r+2} + \left(\frac{\theta}{r+1}\right)x^{r+1}\right]_0^1 \\
&= 2\left[\left(\frac{1-2\theta}{r+2}\right) + \left(\frac{\theta}{r+1}\right)\right] \\
&= \frac{2\left[1+r(1-\theta)\right]}{(r+1)(r+2)}.
\end{aligned}$$

So, it follows directly that $\mathrm{E}(X) = (2-\theta)/3$, that $\mathrm{E}\left(X^2\right) = (3-2\theta)/6$, and that

$$\mathrm{V}(X) = \mathrm{E}\left(X^2\right) - \left[\mathrm{E}(X)\right]^2 = \frac{1+2\theta(1-\theta)}{18}.$$

Now, since $\mathrm{E}(\bar{X}) = (2-\theta)/3$, it follows directly that $\hat{\theta} = (2-3\bar{X})$, and that

$$\mathrm{V}(\hat{\theta}) = (-3)^2\mathrm{V}(\bar{X}) = 9n^{-1}\left[\frac{1+2\theta(1-\theta)}{18}\right] = \frac{1+2\theta(1-\theta)}{2n}.$$

Clearly, when $\theta = \frac{1}{2}, \mathrm{V}(\hat{\theta}) = \frac{3}{4n}$ =CRLB, so that $\hat{\theta}$ is fully efficient when $\theta = \frac{1}{2}$. However, when $\theta \neq \frac{1}{2}$, then

$$\mathrm{EFF}(\hat{\theta},\theta) = \frac{(1-2\theta)^3}{\left[\ln\left(\frac{1-\theta}{\theta}\right) - 2(1-2\theta)\right]\left[1+2\theta(1-\theta)\right]}$$

decreases monotonically as $\theta \rightarrow 0$ and as $\theta \rightarrow 1$. In particular, $\lim_{\theta\rightarrow 0}\mathrm{EFF}(\hat{\theta},\theta) = \lim_{\theta\rightarrow 1}\mathrm{EFF}(\hat{\theta},\theta) = 0$.

Solution 5.81*.

(a) The likelihood function $\mathcal{L}$ is equal to

$$\begin{aligned}
\mathcal{L} &= \prod_{i=1}^{n}\mathrm{p}_{X_i}(x_i) = \prod_{i=1}^{n}(1-\pi)^{\frac{1}{2}(2-x_i)(1-x_i)}\left[\pi(1-\theta)\right]^{x_i(2-x_i)}(\pi\theta)^{\frac{1}{2}x_i(x_i-1)} \\
&= (1-\pi)^{n_0}\left[\pi(1-\theta)\right]^{n_1}(\pi\theta)^{n_2},
\end{aligned}$$

so that

$$\ln\mathcal{L} = n_0\ln(1-\pi) + n_1\left[\ln\pi + \ln(1-\theta)\right] + n_2\left[\ln\pi + \ln\theta\right].$$

Hence,

$$\frac{\partial \ln \mathcal{L}}{\partial \pi} = -\frac{n_0}{(1-\pi)} + \frac{n_1}{\pi} + \frac{n_2}{\pi} = 0$$

gives $\hat{\pi} = (n_1 + n_2)/n$. And,

$$\frac{\partial \ln \mathcal{L}}{\partial \theta} = -\frac{n_1}{(1-\theta)} + \frac{n_2}{\theta} = 0$$

gives $\hat{\theta} = n_2/(n_1 + n_2)$.

First, note that the random variables N_0, N_1, and N_2 (with respective realizations n_0, n_1, and n_2) follow a multinomial distribution, namely,

$$(N_0, N_1, N_2) \sim \text{MULT}\left[n; (1-\pi), \pi(1-\theta), \pi\theta\right].$$

Now,

$$\frac{\partial^2 \ln \mathcal{L}}{\partial \pi^2} = -\frac{n_0}{(1-\pi)^2} - \frac{n_1}{\pi^2} - \frac{n_2}{\pi^2},$$

so that

$$\begin{aligned} -\text{E}\left(\frac{\partial^2 \ln \mathcal{L}}{\partial \pi^2}\right) &= \frac{n(1-\pi)}{(1-\pi)^2} + \frac{n\pi(1-\theta)}{\pi^2} + \frac{n\pi\theta}{\pi^2} \\ &= \frac{n}{\pi(1-\pi)}. \end{aligned}$$

And,

$$\frac{\partial^2 \ln \mathcal{L}}{\partial \theta^2} = -\frac{n_1}{(1-\theta)^2} - \frac{n_2}{\theta^2},$$

so that

$$\begin{aligned} -\text{E}\left(\frac{\partial^2 \ln \mathcal{L}}{\partial \theta^2}\right) &= \frac{n\pi(1-\theta)}{(1-\theta)^2} + \frac{n\pi\theta}{\theta^2} \\ &= \frac{n\pi}{\theta(1-\theta)}. \end{aligned}$$

Since $\partial^2 \ln \mathcal{L}/\partial\pi\partial\theta = 0$, so that the expected information matrix is a diagonal matrix, it follows that the large-sample variances of $\hat{\pi}$ and $\hat{\theta}$ are

$$\text{V}(\hat{\pi}) = \frac{\pi(1-\pi)}{n} \text{ and } \text{V}(\hat{\theta}) = \frac{\theta(1-\theta)}{n\pi}.$$

Using the available data, we have

$$\hat{\pi} = \frac{(30+50)}{100} = 0.800 \text{ and } \hat{\theta} = \frac{50}{(30+50)} = 0.625.$$

Thus, the computed large-sample 95% confidence interval for π is

$$\begin{aligned}\hat{\pi} \pm 1.96\sqrt{\frac{\hat{\pi}(1-\hat{\pi})}{n}} &= 0.800 \pm 1.96\sqrt{\frac{(0.800)(0.200)}{(100)}} \\ &= 0.800 \pm 0.078, \text{ or } (0.722, 0.878).\end{aligned}$$

And, the computed large-sample 95% confidence interval for θ is

$$\begin{aligned}\hat{\theta} \pm 1.96\sqrt{\frac{\hat{\theta}(1-\hat{\theta})}{n\hat{\pi}}} &= 0.625 \pm 1.96\sqrt{\frac{(0.625)(0.375)}{(100)(0.800)}} \\ &= 0.625 \pm 0.106, \text{ or } (0.519, 0.731).\end{aligned}$$

Based on the available data, the prevalence of children with ear infections in this U.S. area is high, and it is apparently more likely than not that an infant will have both ears infected once that infant develops an ear infection.

(b) A reasonable maximum likelihood estimator of γ is $\hat{\gamma} = (N_1 + 2N_2)/2n$, namely, the proportion of infected ears. Since

$$\begin{aligned}\text{V}(N_1 + 2N_2) &= (1)^2\text{V}(N_1) + (2)^2\text{V}(N_2) + (2)(1)(2)\text{cov}(N_1, N_2) \\ &= n\pi(1-\theta)[1-\pi(1-\theta)] + 4n(\pi\theta)(1-\pi\theta) \\ &\quad + 4[-n\pi(1-\theta)(\pi\theta)] \\ &= n\pi\left[1 + 3\theta - \pi(1+\theta)^2\right],\end{aligned}$$

it follows that

$$\text{V}(\hat{\gamma}) = \frac{\pi\left[1 + 3\theta - \pi(1+\theta)^2\right]}{4n}.$$

So, for the available data, the large-sample 95% confidence interval for γ is

$$\begin{aligned}\hat{\gamma} \pm 1.96\sqrt{\hat{\text{V}}(\hat{\gamma})} &= \left[\frac{(n_1 + 2n_2)}{2n}\right] \pm 1.96\sqrt{\frac{\hat{\pi}\left[1 + 3\hat{\theta} - \hat{\pi}(1+\hat{\theta})^2\right]}{4n}} \\ &= 0.650 \pm 0.077, \text{ or } (0.573, 0.727).\end{aligned}$$

Solution 5.83*. For $i = 1, 2, \ldots, n$, since

$$\begin{aligned}\text{pr}(Y_i = 1) &= \text{pr}(Y_i = 1|D_i = 1)\text{pr}(D_i = 1) + \text{pr}(Y_i = 1|D_i = 0)\text{pr}(D_i = 0) \\ &= \gamma\pi + (1-\delta)(1-\pi) = \theta, \text{ say,}\end{aligned}$$

the likelihood function $\mathcal{L}$ for the observed data $y_1, y_2, \ldots, y_n$ takes the form

$$\mathcal{L} = \prod_{i=1}^{n} \theta^{y_i}(1-\theta)^{1-y_i} = \theta^s(1-\theta)^{n-s}, \text{ where } s = \sum_{i=1}^{n} y_i.$$

So,

$$\ln\mathcal{L} = s\ln\theta + (n-s)\ln(1-\theta);$$

and

$$\frac{\partial \ln\mathcal{L}}{\partial\theta} = \frac{s}{\theta} - \frac{(n-s)}{(1-\theta)} = 0 \text{ gives } \hat{\theta} = \frac{s}{n} = \bar{y}$$

as the MLE of θ, so that the equation

$$\gamma\hat{\pi} + (1-\delta)(1-\hat{\pi}) = \hat{\theta} \text{ gives } \hat{\pi} = \frac{(\hat{\theta}+\delta-1)}{(\gamma+\delta-1)}$$

as the MLE of π.

Since

$$\begin{aligned} -\mathrm{E}\left(\frac{\partial^2\ln\mathcal{L}}{\partial\theta^2}\right) &= -\mathrm{E}\left[\frac{-s}{\theta^2} - \frac{(n-s)}{(1-\theta)^2}\right] \\ &= \frac{n\theta}{\theta^2} + \frac{n(1-\theta)}{(1-\theta)^2} = \frac{n}{\theta(1-\theta)} = \left[\mathrm{V}(\hat{\theta})\right]^{-1}, \end{aligned}$$

it follows that

$$\mathrm{V}(\hat{\pi}) = \frac{\mathrm{V}(\hat{\theta})}{(\gamma+\delta-1)^2} = \frac{\theta(1-\theta)}{n(\gamma+\delta-1)^2}.$$

So, an appropriate large-sample 95% confidence interval for π is

$$\hat{\pi} \pm 1.96\sqrt{\frac{\hat{\theta}(1-\hat{\theta})}{n(\gamma+\delta-1)^2}}.$$

For the given data, the computed 95% confidence interval for π is

$$\frac{\left(\frac{20}{100}+0.85-1\right)}{(0.90+0.85-1)} \pm 1.96\sqrt{\frac{(0.20)(1-0.20)}{(100)(0.90+0.85-1)^2}},$$

giving 0.0667 ± 0.1045, or (0, 0.1712).

Note that this is a wide, and not very informative, confidence interval. Possibilities for improving precision include increasing the size of n and using multiple diagnostic tests.

Solution 5.85*.

(a) First, from properties of the multinomial distribution, it follows that $S =$

$(Y_{10}+Y_{01}) \sim \text{BIN}(n; \pi_{10}+\pi_{01})$, so that

$$\begin{aligned}
\text{p}_{Y_{10}}(y_{10}|S=s) &= \frac{\text{pr}\left[(Y_{10}=y_{10})\cap(S=s)\right]}{\text{pr}(S=s)} \\
&= \frac{\text{pr}\left[(Y_{10}=y_{10})\cap(Y_{01}=s-y_{10})\right]}{\text{C}_s^n(\pi_{10}+\pi_{01})^s(\pi_{11}+\pi_{00})^{n-s}} \\
&= \frac{\left[\frac{n!}{y_{10}!y_{01}!(y_{11}+y_{00})!}\right]\pi_{10}^{y_{10}}\pi_{01}^{y_{01}}(\pi_{11}+\pi_{00})^{y_{11}+y_{00}}}{\left[\frac{n!}{s!(n-s)!}\right](\pi_{10}+\pi_{01})^s(\pi_{11}+\pi_{00})^{n-s}} \\
&= \text{C}_{y_{10}}^s\left(\frac{\pi_{10}}{\pi_{10}+\pi_{01}}\right)^{y_{10}}\left(\frac{\pi_{01}}{\pi_{10}+\pi_{01}}\right)^{s-y_{10}} \\
&= \text{C}_{y_{10}}^s\left(\frac{\psi}{\psi+1}\right)^{y_{10}}\left(\frac{1}{\psi+1}\right)^{y_{01}}, y_{10}=0,1,\ldots,s;
\end{aligned}$$

that is, given $S=s, Y_{10} \sim \text{BIN}\left(s; \frac{\psi}{\psi+1}\right)$.

(b) Now,

$$\begin{aligned}
\ln\mathcal{L} &\propto y_{10}[\ln\psi - \ln(\psi+1)] - y_{01}\ln(\psi+1) \\
&= y_{10}\ln\psi - (y_{10}+y_{01})\ln(\psi+1).
\end{aligned}$$

So, the equation

$$\frac{\partial\ln\mathcal{L}}{\partial\psi} = \frac{y_{10}}{\psi} - \frac{(y_{10}+y_{01})}{(\psi+1)} = 0$$

gives $\hat{\psi} = y_{10}/y_{01}$ as the maximum likelihood estimate (MLE) of ψ. This same result follows by noting that y_{10}/s is the MLE of $\psi/(\psi+1)$.
And, since

$$\frac{\partial^2\ln\mathcal{L}}{\partial\psi^2} = -\frac{y_{10}}{\psi^2} + \frac{(y_{10}+y_{01})}{(\psi+1)^2},$$

it follows that

$$\begin{aligned}
-\text{E}\left(\frac{\partial^2\ln\mathcal{L}}{\partial\psi^2}\bigg|S=s\right) &= \frac{(y_{10}+y_{01})\left(\frac{\psi}{\psi+1}\right)}{\psi^2} - \frac{(y_{10}+y_{01})}{(\psi+1)^2} \\
&= \frac{(y_{10}+y_{01})}{\psi(\psi+1)^2};
\end{aligned}$$

hence, for large values of $(y_{10}+y_{01})$,

$$\text{V}(\hat{\psi}) \approx \frac{\psi(\psi+1)^2}{(y_{10}+y_{01})},$$

so that

$$\hat{\mathrm{V}}(\hat{\psi}) \approx \frac{\hat{\psi}(\hat{\psi}+1)^2}{(y_{10}+y_{01})} = \frac{\left(\frac{y_{10}}{y_{01}}\right)\left(\frac{y_{10}}{y_{01}}+1\right)^2}{(y_{10}+y_{01})} = \frac{y_{10}(y_{10}+y_{01})}{y_{01}^3}.$$

Finally, an appropriate large-sample 95% confidence interval for ψ is

$$\hat{\psi} \pm 1.96\sqrt{\hat{\mathrm{V}}(\hat{\psi})} = \left(\frac{y_{10}}{y_{01}}\right) \pm 1.96\sqrt{\frac{y_{10}(y_{10}+y_{01})}{y_{01}^3}}.$$

(c) For the available data, the computed large-sample 95% confidence interval for ψ is equal to

$$\left(\frac{26}{10}\right) \pm 1.96\sqrt{\frac{(26)(26+10)}{(10)^3}} = 2.60 \pm 1.90, \text{ or } (0.70, 4.50).$$

Since the value 1 is not contained in this computed confidence interval, these data provide evidence that there is an exposure-disease association.

(d) Using the delta method, we have

$$\mathrm{V}(\ln\hat{\psi}|S=s) \approx \left(\frac{\mathrm{d}\ln\psi}{\mathrm{d}\psi}\right)^2 \mathrm{V}(\hat{\psi}) = \left(\frac{1}{\psi}\right)^2\left[\frac{\psi(\psi+1)^2}{(y_{10}+y_{01})}\right] = \frac{(\psi+1)^2}{\psi(y_{10}+y_{01})}.$$

Thus,

$$\hat{\mathrm{V}}(\ln\hat{\psi}|S=s) \approx \frac{\left(\frac{y_{10}}{y_{01}}+1\right)^2}{\left(\frac{y_{10}}{y_{01}}\right)(y_{10}+y_{01})} = \frac{1}{y_{10}} + \frac{1}{y_{01}}.$$

Hence, an appropriate large-sample 95% confidence interval for $\ln\psi$ is

$$\ln\hat{\psi} \pm 1.96\sqrt{\hat{\mathrm{V}}(\ln\hat{\psi}|S=s)},$$

so that the corresponding large-sample 95% confidence interval for ψ is

$$\exp\left[\ln\hat{\psi} \pm 1.96\sqrt{\hat{\mathrm{V}}(\ln\hat{\psi}|S=s)}\right] = \left(\frac{y_{10}}{y_{01}}\right)\mathrm{e}^{\pm 1.96\sqrt{\frac{1}{y_{10}}+\frac{1}{y_{01}}}}.$$

For the available data, the computed large-sample 95% confidence interval for ψ is equal to

$$\left(\frac{26}{10}\right)\mathrm{e}^{\pm 1.96\sqrt{\frac{1}{26}+\frac{1}{10}}} = (2.60)\mathrm{e}^{\pm 0.73}, \text{ or } (1.25, 5.40).$$

Although neither computed confidence interval includes the value 1, the two computed confidence intervals are somewhat different, which will typically be the case.

Research suggests that the confidence interval computed in part (d) is to be preferred to the one computed in part (c).

Solution 5.87*. Since

$$\bar{X}_i = \mu + \beta_i + \bar{\epsilon}_i, \text{ where } \bar{\epsilon}_i = k^{-1}\sum_{j=1}^{k} \epsilon_{ij},$$

it follows that $\bar{X}_i \sim \mathrm{N}\left(\mu, \sigma_\beta^2 + \frac{\sigma_\epsilon^2}{k}\right)$, and that

$$\mathrm{cov}\left(\bar{X}_i, \mu_i\right) = \mathrm{cov}\left(\mu + \beta_i + \bar{\epsilon}_i, \mu + \beta_i\right) = \mathrm{V}(\beta_i) = \sigma_\beta^2.$$

So, since $\mu_i \sim \mathrm{N}(\mu, \sigma_\beta^2)$, it follows from properties of the assumed bivariate normal distribution for $(\bar{X}_i, \mu_i)$ that

$$\begin{aligned} \mathrm{E}(\mu_i|\bar{X}_i = \bar{x}_i) &= \mu + \mathrm{corr}(\bar{X}_i, \mu_i)\sqrt{\frac{\mathrm{V}(\mu_i)}{\mathrm{V}(\bar{X}_i)}}(\bar{x}_i - \mu) \\ &= \mu + \left[\frac{\sigma_\beta^2}{\sqrt{\left(\sigma_\beta^2 + \frac{\sigma_\epsilon^2}{k}\right)(\sigma_\beta^2)}}\right]\sqrt{\frac{\sigma_\beta^2}{\left(\sigma_\beta^2 + \frac{\sigma_\epsilon^2}{k}\right)}}(\bar{x}_i - \mu) \\ &= \mu + \left(\frac{1}{1 + \frac{\lambda}{k}}\right)(\bar{x}_i - \mu), \text{ where } \lambda = \frac{\sigma_\epsilon^2}{\sigma_\beta^2}. \end{aligned}$$

Now, since

$$\hat{\theta}_1 = \sum_{i=1}^{n} c_i Y_i, \text{ where } c_i = \frac{(\bar{X}_i - \bar{X})}{\sum_{i=1}^{n}(\bar{X}_i - \bar{X})^2},$$

it follows that

$$\mathrm{E}(\hat{\theta}_1|\bar{X}_i = \bar{x}_i, i = 1, 2, \ldots, n) = \sum_{i=1}^{n} c_i^* \mathrm{E}(Y_i|\bar{X}_i = \bar{x}_i),$$

where $c_i^* = (\bar{x}_i - \overline{x})/\sum_{i=1}^{n}(\bar{x}_i - \overline{x})^2$.

Now, using the nondifferential error assumption, we have

$$\begin{aligned}
\mathrm{E}(Y_i|\bar{X}_i = \bar{x}_i) &= \mathrm{E}_{\mu_i|\bar{X}_i=\bar{x}_i}\left[\mathrm{E}(Y_i|\mu_i, \bar{X}_i = \bar{x}_i)\right] \\
&= \mathrm{E}_{\mu_i|\bar{X}_i=\bar{x}_i}\left[\mathrm{E}(Y_i|\mu_i)\right] \\
&= \mathrm{E}_{\mu_i|\bar{X}_i=\bar{x}_i}\left(\theta_0 + \theta_1\mu_i\right) \\
&= \theta_0 + \theta_1 \mathrm{E}\left(\mu_i|\bar{X}_i = \bar{x}_i\right) \\
&= \theta_0 + \theta_1\left[\mu + \left(\frac{1}{1+\frac{\lambda}{k}}\right)(\bar{x}_i - \mu)\right] \\
&= \theta_0^* + \theta_1^*\bar{x}_i,
\end{aligned}$$

where

$$\theta_0^* = \theta_0 + \mu\left[\theta_1 - \left(\frac{1}{1+\frac{\lambda}{k}}\right)\right] \text{ and } \theta_1^* = \left(\frac{1}{1+\frac{\lambda}{k}}\right)\theta_1.$$

Thus, since $\sum_{i=1}^{n} c_i^* = 0$, we have

$$\begin{aligned}
\mathrm{E}(\hat{\theta}_1|\bar{X}_i = \bar{x}_i, i = 1, 2, \ldots, n) &= \sum_{i=1}^{n} c_i^*(\theta_0^* + \theta_1^*\bar{x}_i) \\
&= \theta_0^*\sum_{i=1}^{n} c_i^* + \theta_1^*\sum_{i=1}^{n} c_i^*\bar{x}_i \\
&= \theta_1^* = \left(\frac{1}{1+\frac{\lambda}{k}}\right)\theta_1 = \mathrm{E}(\hat{\theta}_1),
\end{aligned}$$

so that

$$\gamma = \left(1 + \frac{\lambda}{k}\right)^{-1}, \text{ with } \lambda = \frac{\sigma_\epsilon^2}{\sigma_\beta^2}.$$

Interestingly, γ does not depend on n. Also, the degree of attenuation increases as λ increases and decreases as k increases. Thus, it is advantageous to have σ_ϵ^2 be much smaller than σ_β^2, and to have a large value for k.

Solution 5.89*.

(a) Since $\bar{Y}_i \sim \mathrm{N}\left(\mu_i, \frac{\sigma_\epsilon^2}{2n}\right), i = 1, 2$, and since $\bar{Y}_1$ and $\overline{Y}_2$ are independent random variables, it follows that

$$\frac{(\bar{Y}_1 - \overline{Y}_2) - (\mu_1 - \mu_2)}{\sqrt{\frac{\sigma_\epsilon^2}{2n} + \frac{\sigma_\epsilon^2}{2n}}} \sim \mathrm{N}(0, 1),$$

so that

$$(\bar{Y}_1 - \overline{Y}_2) \pm 1.96\sqrt{\frac{\sigma_\epsilon^2}{n}}$$

is the exact 95% confidence interval for $(\mu_1 - \mu_2)$.

For the available data, the computed 95% confidence interval for $(\mu_1 - \mu_2)$ is equal to

$$(90 - 87) \pm 1.96\sqrt{\frac{21}{10}} = 3 \pm 2.84, \text{ or } (0.16, 5.84).$$

Since the value 0 is not contained in this interval, this data analysis suggests that there is a statistically significant difference between these two experimental drugs with regard to their abilities to lower DBP.

(b) Since $\mathrm{V}(Y_{ij1}) = \mathrm{V}(Y_{ij2}) = (\sigma_\beta^2 + \sigma_\epsilon^2)$ and since $\mathrm{cov}(Y_{ij1}, Y_{ij2}) = \mathrm{V}(\beta_{ij}) = \sigma_\beta^2$, it follows that

$$\mathrm{corr}(Y_{ij1}, Y_{ij2}) = \frac{\sigma_\beta^2}{(\sigma_\beta^2 + \sigma_\epsilon^2)} = \frac{7}{(7 + 21)} = 0.25.$$

Now, since $\overline{Y}_{ij}$ has a normal distribution with $\mathrm{E}(\overline{Y}_{ij}) = \mu_i$ and with

$$\begin{aligned} \mathrm{V}(\overline{Y}_{ij}) &= \frac{1}{4}\mathrm{V}(Y_{ij1} + Y_{ij2}) \\ &= \frac{1}{4}\left[\mathrm{V}(Y_{ij1}) + \mathrm{V}(Y_{ij2}) + 2\mathrm{cov}(Y_{ij1}, Y_{ij2})\right] \\ &= \frac{1}{4}\left[(\sigma_\beta^2 + \sigma_\epsilon^2) + (\sigma_\beta^2 + \sigma_\epsilon^2) + 2(\sigma_\beta^2)\right] \\ &= \sigma_\beta^2 + \frac{\sigma_\epsilon^2}{2}. \end{aligned}$$

Thus, since the $\{\overline{Y}_{ij}\}$ constitute a set of mutually independent random variables, it follows that

$$\bar{Y}_i = \frac{1}{n}\sum_{j=1}^{n} \overline{Y}_{ij} \sim \mathrm{N}\left(\mu_i, \frac{\sigma_\beta^2 + \frac{\sigma_\epsilon^2}{2}}{n}\right).$$

Hence, we have

$$\frac{(\bar{Y}_1 - \overline{Y}_2) - (\mu_1 - \mu_2)}{\sqrt{\frac{2\left(\sigma_\beta^2 + \frac{\sigma_\epsilon^2}{2}\right)}{n}}} \sim \mathrm{N}(0, 1),$$

so that

$$(\bar{Y}_1 - \overline{Y}_2) \pm 1.96\sqrt{\frac{(2\sigma_\beta^2 + \sigma_\epsilon^2)}{n}}$$

is the exact 95% confidence interval for $(\mu_1 - \mu_2)$.
For the available data, the computed 95% confidence interval for $(\mu_1 - \mu_2)$ is

$$(90 - 87) \pm 1.96\sqrt{\frac{[2(7) + 21]}{10}} = 3 \pm 3.67, \text{ or } (-0.67, 6.67).$$

Since the value 0 is included in this interval, this data analysis, which correctly takes into account the positive correlation between the two DBP measurements on each subject, provides evidence that there is no statistically significant difference between these two experimental drugs with regard to their abilities to lower DBP.

More generally, this simple example illustrates that ignoring intra-subject response correlation can lead to invalid statistical conclusions.

Chapter 6

Hypothesis Testing Theory

6.1 Exercises

Exercise 6.1. Let $Y_1, Y_2, \ldots, Y_n$ constitute a random sample of size n from the continuous parent population

$$\mathrm{f}_Y(y;\theta) = (1+\theta y)/2, \quad -1 < y < +1; \quad -1 < \theta < +1.$$

(a) Derive the *general structure* of the rejection region $\mathcal{R}$ of the *most powerful* (MP) test of size α for testing $\mathrm{H}_0 : \theta = 0$ versus $\mathrm{H}_1 : \theta = 0.50$. Is this MP test also the *uniformly most powerful* (UMP) test of $\mathrm{H}_0 : \theta = 0$ versus $\mathrm{H}_1 : \theta > 0$?

(b) When $n = 1$, find the critical value of the MP test statistic so that pr(Type I error) $= 0.05$ when using the MP test developed in part (a) to test $\mathrm{H}_0 : \theta = 0$ versus $\mathrm{H}_1 : \theta = 0.50$.

(c) When $n = 1$, what is the numerical value of the power of the MP test of size $\alpha = 0.05$ developed in part (b) to test $\mathrm{H}_0 : \theta = 0$ versus $\mathrm{H}_1 : \theta = 0.50$ when, in fact, $\theta = 0.50$?

Exercise 6.2. Suppose that the distribution of sulfur dioxide (SO_2) concentration measurements in city i ($i = 1, 2$) is assumed to be of the form

$$\mathrm{f}_Y(y;\theta_i) = \theta_i^{-2} y \mathrm{e}^{-y/\theta_i}, \quad y > 0,\ \theta_i > 0.$$

For $i = 1, 2$, let $Y_{i1}, Y_{i2}, \ldots, Y_{in_i}$ constitute a random sample of n_i SO_2 measurements from city i.

To compare City 1 to City 2 with regard to average SO_2 concentration level, it is proposed to test $\mathrm{H}_0 : \theta_1 = \theta_2\ (= \theta$, say) versus $\mathrm{H}_\mathrm{A} : \theta_1 \neq \theta_2$ using the $(n_1 + n_2)$ observations Y_{ij}, $j = 1, 2, \ldots, n_i$ and $i = 1, 2$.

If $n_1 = n_2 = 25$, $\bar{y}_1 = n_1^{-1} \sum_{j=1}^{n_1} y_{1j} = 10$ and $\bar{y}_2 = n_2^{-1} \sum_{j=1}^{n_2} y_{2j} = 8$, perform an appropriate likelihood ratio test of H_0 versus H_A, and compute the P-value of your test. Would you reject H_0 at the $\alpha = 0.10$ level?

Exercise 6.3. The survival time T (in years) for patients who have had quadruple bypass surgery (QBPS) is assumed to have the distribution

$$\mathrm{f}_T(t;\theta) = \theta^{-2} t \mathrm{e}^{-t/\theta},\ 0 < t < +\infty,\ \theta > 0.$$

Suppose that survival times $t_1, t_2, \ldots, t_n$ are recorded for n randomly selected patients who have had QBPS. In other words, $t_1, t_2, \ldots, t_n$ are the realizations of a random sample $T_1, T_2, \ldots, T_n$ of size n from $\mathrm{f}_T(t;\theta)$.

If $\bar{t} = n^{-1} \sum_{i=1}^{n} t_i = 2.40$ when $n = 40$, what is the P-value for the *likelihood ratio test* of $\mathrm{H}_0 : \theta = 1$ versus $\mathrm{H}_1 : \theta \neq 1$?

Exercise 6.4. For $i = 1, 2$, let $X_{i1}, X_{i2}, \ldots, X_{in}$ constitute a random sample of size $n(\geq 2)$ from an $\mathrm{N}(\mu_i, \sigma^2)$ population, where μ_1 and μ_2 are unknown parameters and where σ^2 has a known value.

It is of interest to test the null hypothesis $\mathrm{H}_0 : \mu_1 = \mu_2$ versus the one-sided alternative hypothesis $\mathrm{H}_1 : \mu_1 > \mu_2$ using the following decision rule. For $i = 1, 2$, and with $\bar{X}_i = n^{-1} \sum_{j=1}^{n} X_{ij}$, compute the confidence interval (L, U), where

$$L = (\bar{X}_1 - \bar{X}_2) - 1.96\sqrt{\frac{2\sigma^2}{n}} \text{ and } U = (\bar{X}_1 - \bar{X}_2) + 1.96\sqrt{\frac{2\sigma^2}{n}}.$$

Then, if the computed value of L is greater than zero, reject H_0 in favor of H_1; otherwise, do not reject H_0.

If $\mu_1 - \mu_2 = 10$ and $\sigma = 20$, what is the smallest value of n, say n^*, required so that the power of this decision rule is at least equal to 0.90?

Exercise 6.5. For $i = 1, 2$, let $X_{i1}, X_{i2}, \ldots, X_{in}$ constitute a random sample of size $n(\geq 2)$ from an $\mathrm{N}(\mu_i, \sigma^2)$ population, where μ_1 and μ_2 are unknown parameters and where σ^2 has a known value.

It is of interest to test the null hypothesis $\mathrm{H}_0 : \mu_1 = \mu_2$ versus the alternative hypothesis $\mathrm{H}_1 : \mu_1 \neq \mu_2$ using the following decision rule. For $i = 1, 2$, and with $\bar{X}_i = n^{-1} \sum_{j=1}^{n} X_{ij}$, compute the two confidence intervals

$$\left(\bar{X}_1 - z_{1-\frac{\alpha}{2}}\frac{\sigma}{\sqrt{n}}, \bar{X}_1 + z_{1-\frac{\alpha}{2}}\frac{\sigma}{\sqrt{n}}\right) \text{ and } \left(\bar{X}_2 - z_{1-\frac{\alpha}{2}}\frac{\sigma}{\sqrt{n}}, \bar{X}_2 + z_{1-\frac{\alpha}{2}}\frac{\sigma}{\sqrt{n}}\right),$$

where $\mathrm{pr}(Z \leq z_{1-\frac{\alpha}{2}}) = \left(1 - \frac{\alpha}{2}\right)$ when $Z \sim \mathrm{N}(0, 1)$ and $0 < \alpha < 1$. Then, if these two intervals have at least one value in common (i.e., the two intervals overlap), then the decision is to *not* reject H_0; and, if these two intervals do not overlap, then the decision is to reject H_0 in favor of H_1.

(a) If H_0 is true, show that the probability θ that these two intervals overlap can be expressed as a function of the CDF $F_Z(z) = \text{pr}(Z \leq z), Z \sim N(0,1)$. If $\alpha = 0.05$, find the numerical value of θ.

(b) For this decision rule, what should be the value of α so that the probability of a Type I error is equal to 0.05?

Exercise 6.6. Let Y_1 and Y_2 constitute a random sample of size $n = 2$ from the density function

$$f_Y(y;\theta) = \theta y^{\theta-1}, \quad 0 < y < 1,\ \theta > 0.$$

(a) Show that the rejection region $\mathcal{R}$ for the most powerful (MP) test of $H_0 : \theta = 1$ versus $H_A : \theta = 2$ has the structure

$$\mathcal{R} = \{(y_1, y_2) : y_1 y_2 \geq k\}.$$

Is this a uniformly most powerful (UMP) rejection region for all $\theta > 1$?

(b) Prove that the density function of $U = Y_1 Y_2$ is

$$f_U(u;\theta) = \theta^2 u^{\theta-1} \ln\left(u^{-1}\right), \quad 0 < u < 1.$$

(c) Based on parts (a) and (b), set up the appropriate integral needed to determine that value of k, say k_α, so that the probability of a Type I error equals α.

(d) Based on parts (a), (b), and (c), set up the appropriate integral needed to find the *power* of the MP test of $H_0 : \theta = 1$ versus $H_A : \theta = 2$ based on the use of Y_1 and Y_2.

Exercise 6.7. Let $Y_1, Y_2, \ldots, Y_n$ constitute a random sample of size $n\,(\geq 2)$ from an $N(\mu, \sigma^2)$ population.

Consider using $\bar{Y} = n^{-1}\sum_{i=1}^{n} Y_i$ to test $H_0 : \mu = 0$ versus $H_A : \mu > 0$ when $\alpha = 0.025$. If σ^2 is *known* to be equal to 1, what is the *minimum* sample size n^* required so that β, the probability of a Type II error, is no more than 0.16 when $\mu = 0.50$?

Exercise 6.8. For the i-th of two very large counties ($i = 1, 2$), the distribution of the number Y_{ij} of deaths per year in city j ($j = 1, 2, \ldots, n$) due to rabies is assumed to be of the geometric form

$$p_{Y_{ij}}(y_{ij};\theta_i) = \theta_i(1-\theta_i)^{y_{ij}}, y_{ij} = 0, 1, \ldots, \infty; 0 < \theta_i < 1.$$

It is being assumed here that the probability parameter θ_i takes the same value for all n cities within the i-the county; in other words, θ_i does not vary with j. Also, assume that $Y_{i1}, Y_{i2}, \ldots, Y_{in}$ are mutually independent random variables, and that the set of random variables $\{Y_{11}, Y_{12}, \ldots, Y_{1n}\}$ for County #1 ($i = 1$) is mutually independent of the set of random variables $\{Y_{21}, Y_{22}, \ldots, Y_{2n}\}$ for County #2 ($i = 2$).

Now, suppose that health scientists representing the county health departments for these two counties provide the following data summarization information:

$$n = 50; \ \bar{y}_1 = (50)^{-1}\sum_{j=1}^{50} y_{1j} = 5.20; \text{ and, } \bar{y}_2 = (50)^{-1}\sum_{j=1}^{50} y_{2j} = 4.80.$$

Further, these health scientists are interested in using the above numerical information to address the question of whether there is statistical evidence that the true average number of deaths due to rabies per year in County #1 is different from the true average number of rabies deaths per year in County #2.

Use the available numerical information to carry out an appropriate likelihood ratio test to address the question of interest to these health scientists.

Exercise 6.9. For residents in a certain city in the United States, suppose that it is reasonable to assume that the distribution of the *proportion* X of a certain protein in a cubic centimeter of blood taken from a randomly chosen resident follows the density function

$$\mathrm{f}_X(x;\theta) = \theta x^{\theta-1}, \quad 0 < x < 1, \ \theta > 0.$$

Let $X_1, X_2, \ldots, X_n$ constitute a random sample of size $n\,(> 2)$ from $\mathrm{f}_X(x;\theta)$.

For testing $\mathrm{H}_0 : \theta = 2$ versus $\mathrm{H}_1 : \theta \neq 2$ at the $\alpha = 0.05$ level, determine the minimum sample size n^* required so that a *Wald test* has power at least equal to 0.80 when, in fact, $\theta \geq 3$.

Exercise 6.10. Suppose that a discrete random variable X has the distribution

$$\mathrm{p}_X(x;\theta) = (1-\theta)\theta^{x-1}, x = 1, 2, \ldots, \infty \text{ and } 0 < \theta < 1.$$

If $X_1, X_2, \ldots, X_n$ constitute a random sample from $\mathrm{p}_X(x;\theta)$, what is the approximate power of the uniformly most powerful (UMP) test of size $\alpha = 0.05$ for testing the null hypothesis $H_0 : \theta = \frac{1}{3}$ versus the alternative hypothesis $H_1 : \theta > \frac{1}{3}$ when $n = 50$ and when the true value of θ equals $\frac{1}{2}$?

Exercise 6.11. For $i = 1, 2, \ldots, k$ and $j = 1, 2, \ldots, n_i$, with $N = \sum_{i=1}^{k} n_i$, suppose that $Y_{ij} \sim \mathrm{N}(\mu_i, \sigma^2)$ and that the Y_{ij}s constitute a set of N mutually independent random variables.

Using all N observations, develop a likelihood ratio test statistic that can be used to test the null hypothesis $\mathrm{H}_0 : \mu_1 = \mu_2 = \cdots = \mu_k$ ($= \mu$, say) versus the alternative hypothesis H_1 : "no restrictions on $\mu_1, \mu_2, \ldots, \mu_k$." Suppose that $k = 3$, $n_1 = 10$, $n_2 = 15$, $n_3 = 12$, $\bar{y}_1 = 4.00$, $\bar{y}_2 = 9.00$, $\bar{y}_3 = 14.00$, $s_1^2 = 4.00$, $s_2^2 = 4.25$, and $s_3^2 = 4.50$, where, in general, $\bar{y}_i = n_i^{-1} \sum_{j=1}^{n_i} y_{ij}$ and $s_i^2 = (n_i - 1)^{-1} \sum_{j=1}^{n_i} (y_{ij} - \bar{y}_i)^2$ for $i = 1, 2, 3$. Based on this numerical information, what is the P-value associated with your likelihood ratio test of H_0 versus H_1?

Exercise 6.12*. The concentration X (in parts per million or ppm) of styrene in the air in a certain styrene manufacturing plant has a lognormal distribution; in particular, $Y = \ln X \sim \mathrm{N}(\mu, \sigma^2)$. Suppose that $x_1, x_2, \ldots, x_n$ represent n measurements of the airborne styrene concentration in this plant; these n measured concentration values can be considered to be realized values of a random sample $X_1, X_2, \ldots, X_n$ of size n from the lognormal distribution for X. With $y_i = \ln x_i, i = 1, 2, \ldots, n$, suppose that $n = 30, \bar{y} = n^{-1} \sum_{i=1}^{n} y_i = 3.00$ and $s_y^2 = (n-1)^{-1} \sum_{i=1}^{n} (y_i - \bar{y})^2 = 2.50$. Using these data, perform an ML-based large-sample test of $\mathrm{H}_0 : \mathrm{E}(X) \leq 30$ versus $\mathrm{H}_1 : \mathrm{E}(X) > 30$ at the $\alpha \approx 0.025$ level, and then comment on your findings.

Exercise 6.13*. The sulfur dioxide concentration (in parts per billion or ppb) in the ambient air near a certain industrial plant is assumed to follow the lognormal distribution

$$\mathrm{f}_X(x) = (2\pi\theta x^2)^{-1/2} \mathrm{e}^{-(\ln x)^2/2\theta}, 0 < x < \infty \text{ and } 0 < \theta < \infty.$$

The Environmental Protection Agency (EPA) is interested in determining whether the true mean concentration of sulfur dioxide in the ambient air near this plant exceeds the EPA standard of 75 ppb. To make this determination, $n(> 1)$ independently selected measurements $x_1, x_2, \ldots, x_n$ of the sulfur dioxide concentration are made; these n measurements can be considered to be the realized values of a random sample $X_1, X_2, \ldots, X_n$ of size n from $\mathrm{f}_X(x)$.

(a) Consider testing $\mathrm{H}_0 : \theta = \theta_0$ versus $\mathrm{H}_1 : \theta > \theta_0$. Find the appropriate value of θ_0 that should be used to assess whether this industrial plant is in violation of the EPA standard of 75 ppb. Using this particular value of θ_0, show that the rejection region $\mathcal{R}$ for a uniformly most powerful (UMP) test of size α for testing $\mathrm{H}_0 : \theta = \theta_0$ versus $\mathrm{H}_1 : \theta > \theta_0$ can be expressed as an explicit function of the random variable $U = \sum_{i=1}^{n} Y_i^2$, where $Y_i = \ln X_i, i = 1, 2, \ldots, n$.

(b) If $n = 10$ and $u = \sum_{i=1}^{10} y_i^2 = 100$, use a testing procedure based on the chi-squared distribution to assess whether this small data set provides statistical evidence that the plant is in violation of the EPA standard.

(c) For *large* n, for $\alpha \approx 0.05$, and for the value of θ_0 determined in part (a), find the minimum sample size, say n^*, required so that the UMP test of $\mathrm{H}_0 : \theta = \theta_0$ versus $\mathrm{H}_1 : \theta > \theta_0$ has power at least equal to 0.90 when $\theta \geq 10$.

Exercise 6.14*. A certain rare cancer can be classified as being one of four types. Based on genetic models, researchers who study the causes of this rare cancer have determined that a subject with this rare cancer has probability $(2 + \theta)/4$ of having type 1, has probability $(1 - \theta)/2$ of having type 2, has probability $(1 - \theta)/2$ of having type 3, and has probability $\theta/4$ of having type 4, where $\theta(0 < \theta < 1)$ is an unknown parameter.

Find a reasonable value for the minimum number, say n^*, of randomly chosen subjects having this rare cancer who need to be examined so that the power for rejecting $\mathrm{H}_0 : \theta = 0.40$ in favor of $\mathrm{H}_1 : \theta > 0.40$ at the $\alpha = 0.05$ level is at least 0.80 in value when using a score-type test statistic and when the true value of θ is equal to 0.50.

Exercise 6.15*. When testing a certain null hypothesis $\mathrm{H}_0 : \theta = \theta_0$ versus a certain alternative hypothesis $\mathrm{H}_1 : \theta > \theta_0$, suppose that H_0 is rejected for large values of a test statistic T_0, where T_0 has the distribution $\mathrm{f}_T(t; \theta_0)$ under H_0. Also, if H_0 is false and H_1 is true, assume that T_1 has the distribution $\mathrm{f}_T(t; \theta_1)$, where θ_1 is a specific value of θ satisfying $\theta_1 > \theta_0$. In addition, assume that T_0 and T_1 are independent random variables.

(a) Show that the expected value ψ of the P-value in this situation is equal to $\psi = \mathrm{pr}(T_0 \geq T_1)$. What is the value of ψ when $\theta_0 = \theta_1$?

(b) Let $X_1, X_2, \ldots, X_n$ constitute a random sample of size n from a $\mathrm{N}(\mu, \sigma^2)$ population. Consider testing $\mathrm{H}_0 : \mu = \mu_0$ versus $\mathrm{H}_1 : \mu > \mu_0$ using the sample mean $\bar{X} = n^{-1} \sum_{i=1}^{n} X_i$ as the test statistic. Show that

$$\psi = \mathrm{F}_Z \left[\frac{\sqrt{n}(\mu_0 - \mu_1)}{\sqrt{2}\sigma} \right],$$

where $\mathrm{F}_Z(z) = \mathrm{pr}(Z \leq z), Z \sim \mathrm{N}(0, 1)$. Comment on this finding.

For further details about the statistical properties of the P-value, see Hung et al. (1997) and Sackrowitz and Samuel-Cahn (1999).

Exercise 6.16*. The distribution of weight Y (in pounds) for adults weighing at least c pounds (c a known constant) is assumed to be adequately described by the Pareto density function.

$$\mathrm{f}_Y(y; \theta) = \theta c^{\theta} y^{-(\theta+1)}, 0 < c < y < +\infty; 1 < \theta < +\infty.$$

(a) Using a random sample $Y_1, Y_2, \ldots, Y_n$ of size n from $\mathrm{f}_Y(y;\theta)$, it is of interest to test the null hypothesis H_0: $\theta = 2$ versus the alternative hypothesis H_A: $\theta > 2$ when $c = 100$. Using an appropriate test statistic involving the maximum likelihood estimator (MLE) $\hat{\theta}$ of θ, find the minimum sample size n^* required so that the approximate power of a large-sample test of H_0 versus H_A involving this test statistic is at least 0.80 when the Type I error rate is to be approximately 0.05 and when the true value of θ is equal to 3.

(b) Using the random sample $Y_1, Y_2, \ldots, Y_n$ from $\mathrm{f}_Y(y;\theta)$, develop the structure of the most powerful (MP) rejection region of size $\alpha (0 < \alpha < 0.50)$ for testing H_0: $\theta = 2$ versus H_A: $\theta = 3$. Is this MP rejection region also a uniformly most powerful (UMP) rejection region of size α for testing H_0: $\theta = 2$ versus H_A: $\theta > 2$?

(c) If $n = 1$ (so that only Y_1 is available) and $c = 100$, what is the numerical value of the power of the most powerful (MP) test of size $\alpha = 0.10$ for rejecting H_0: $\theta = 2$ when θ is actually equal to 3?

Exercise 6.17*. In many important practical data analysis situations, the statistical models being used involve several parameters, only a few of which are relevant for directly addressing the research questions of interest. The irrelevant parameters, generally referred to as "nuisance parameters," are typically employed to ensure that the statistical models make scientific sense, but are generally unimportant otherwise. One method for eliminating the need to estimate these nuisance parameters, and hence to improve both statistical validity and precision, is to employ a *conditional inference* approach, whereby a conditioning argument is used to produce a conditional likelihood function that involves only the relevant parameters. For an excellent discussion of methods of conditional inference, see McCullagh and Nelder (1989).

As an example, consider the matched-pairs case-control study design often used in epidemiologic research to examine the association between a potentially harmful exposure and a particular disease. In such a design, a case (i.e., a diseased person, denoted D) is matched (on covariates such as age, race, and sex) to a control (i.e., a non-diseased person, denoted $\overline{\mathrm{D}}$). Each member of the pair is then categorized with regard to the presence (E) or absence ($\overline{\mathrm{E}}$) of a history of exposure to some potentially harmful substance (e.g., cigarette smoke, asbestos, benzene, etc.). For further details, see the books by Breslow and Day (1980) and by Kleinbaum, Kupper, and Morgenstern (1982).

The data from a case-control study involving n case-control pairs can be presented in tabular form, as follows:

		$\overline{\text{D}}$: E	$\overline{\text{D}}$: $\overline{\text{E}}$	
D	E	Y_{11}	Y_{10}	
D	$\overline{\text{E}}$	Y_{01}	Y_{00}	
				n

Here, Y_{11} is the number of pairs for which *both* the case and the control are exposed (i.e., both have a history of exposure to the potentially harmful agent under study), Y_{10} is the number of pairs for which the case is exposed but the control is not, and so on. Clearly, $\sum_{i=0}^{1}\sum_{j=0}^{1} Y_{ij} = n$.

In what follows, assume that the $\{Y_{ij}\}$ have a multinomial distribution with sample size n and associated cell probabilities $\{\pi_{ij}\}$, where $\sum_{i=0}^{1}\sum_{j=0}^{1} \pi_{ij} = 1$. For example, π_{10} is the probability of obtaining a pair in which the case is exposed and its matched control is not, and π_{01} is the probability of obtaining a pair in which the case is not exposed but the control is.

Now, let $\alpha = \text{pr}(\text{E}|\text{D})$ and let $\beta = \text{pr}(\text{E}|\overline{\text{D}})$, so that $\pi_{10} = \alpha(1-\beta)$ and $\pi_{01} = (1-\alpha)\beta$. A parameter used to quantify the association between exposure status and disease status in a matched-pairs case-control study is the *exposure odds ratio* ψ, namely,

$$\begin{aligned} \psi &= \frac{\text{pr}(\text{E}|\text{D})/\text{pr}(\overline{\text{E}}|\text{D})}{\text{pr}(\text{E}|\overline{\text{D}})/\text{pr}(\overline{\text{E}}|\overline{\text{D}})} \\ &= \frac{\alpha/(1-\alpha)}{\beta/(1-\beta)} = \frac{\alpha(1-\beta)}{(1-\alpha)\beta} \\ &= \frac{\pi_{10}}{\pi_{01}}. \end{aligned}$$

(a) Let $S = (Y_{10} + Y_{01})$ and $s = (y_{10} + y_{01})$. Show that the *conditional* distribution $\text{p}_{Y_{10}}(y_{10}|S = s)$ of Y_{10} given $S = s$ can be expressed as a function of the exposure odds ratio ψ, and not of the parameters α and β.

(b) If $\text{p}_{Y_{10}}(y_{10}|S = s) = \mathcal{L}$ is taken as the *conditional likelihood function*, use $\mathcal{L}$ to develop a score test statistic $\hat{S}$ based on *expected* information for testing $\text{H}_0 : \psi = 1$ versus $\text{H}_1 : \psi \neq 1$. Note that conditioning eliminates the need to consider the two probability parameters α and β.

(c) For a particular matched-pairs case-control study, suppose that the observed value of Y_{10} equals $y_{10} = 26$ and the observed value of Y_{01} equals $y_{01} = 10$. Use the test statistic $\hat{S}$ developed in part (b) to test $\text{H}_0 : \psi = 1$ versus $\text{H}_1 : \psi \neq 1$. Do these data provide evidence of an exposure-disease association?

Exercise 6.18*. It is well-documented that U.S. office workers spend a significant amount of time each workday using the Internet for non-work-related purposes. Suppose that the proportion X of an 8-hour workday that a typical U.S. office worker spends using the Internet for non-work-related purposes is assumed to have the distribution

$$f_X(x) = 2(1-\theta)x + 2\theta(1-x), 0 < x < 1 \text{ and } 0 < \theta < 1.$$

Suppose that a large number n of randomly selected U.S. office workers complete a questionnaire, with the i-th worker providing a value x_i of the random variable $X_i, i = 1, 2, \ldots, n$. The values $x_1, x_2, \ldots, x_n$ can be considered to be realizations of a random sample $X_1, X_2, \ldots, X_n$ of size n from $f_X(x)$.

(a) Suppose that $n = 50$ and that $\bar{x} = n^{-1}\sum_{i=1}^{n} x_i$=0.45. Does this information provide statistical evidence at the $\alpha = 0.05$ level that a typical U.S. office worker spends, on average, more than 40% of an 8-hour workday using the Internet for non-work-related purposes?

(b) If $E(X) \geq 0.42$, provide a reasonable value for the smallest sample size (say, n^*) required so that, at the $\alpha = 0.025$ level, the power will be at least 0.90 for rejecting $H_0 : E(X) \leq 0.40$ in favor of $H_1 : E(X) > 0.40$. Comment on your findings.

Exercise 6.19*. For $i = 1, 2$, let $X_{i1}, X_{i2}, \ldots, X_{in_i}$ constitute a random sample of size n_i from a $N(\mu_i, \sigma_i^2)$ population, where μ_1 and μ_2 are *unknown* parameters and where σ_1^2 and σ_2^2 are *known* parameters.

Due to logistical constraints, suppose that it is only possible to select a total sample size of N from these two normal populations, so that the constraint $(n_1 + n_2) = N$ holds. Subject to this sample size constraint, find expressions (as a function of σ_1, σ_2, and N) for the optimal values n_1^* and n_2^* of n_1 and n_2 that *maximize* the power of a size α test of the null hypothesis $H_0 : \mu_1 = \mu_2$ versus the alternative hypothesis $H_1 : \mu_1 > \mu_2$ using a test statistic that is a function of $\bar{X}_1 = n_1^{-1}\sum_{j=1}^{n_1} X_{1j}$ and $\bar{X}_2 = n_2^{-1}\sum_{j=1}^{n_2} X_{2j}$. Provide an interpretation for your findings. If $N = 100, \sigma_1^2 = 4$, and $\sigma_2^2 = 9$, find the numerical values of n_1^* and n_2^*.

Exercise 6.20*. Let $X_1, X_2, \ldots, X_n$ constitute a random sample of size n from the parent population

$$f_X(x;\theta) = (2\theta)^{-1}e^{-|x|/\theta}, \quad -\infty < x < +\infty, \ \theta > 0.$$

(a) Prove that the *uniformly most powerful* (UMP) test of $H_0 : \theta = \theta_0$ versus $H_1 : \theta > \theta_0$ has the rejection region of size α of the form

$$\mathcal{R} = \{S : S > c_\alpha\},$$

where $S = \sum_{i=1}^{n} |X_i|$ and where c_α is chosen so that

$$\mathrm{pr}\left\{S > c_\alpha \middle| \mathrm{H}_0 : \theta = \theta_0\right\} = \alpha.$$

(b) Prove that the distribution of $Y_i = |X_i|$ is

$$\mathrm{f}_Y(y;\theta) = \theta^{-1}\mathrm{e}^{-y_i/\theta}, \quad y_i > 0.$$

(c) Assuming that n is large, use the results in parts (a) and (b) to obtain a large-sample normal approximation for c_α as a function of θ_0, n, and $Z_{1-\alpha}$, where $\mathrm{pr}(Z > Z_{1-\alpha}) = \alpha$ when $Z \sim \mathrm{N}(0,1)$.

(d) Based on the normal approximation used in part (c), what is the approximate power of this UMP test when testing $\mathrm{H}_0 : \theta = 1$ versus $\mathrm{H}_1 : \theta > 1$ if, in fact, $\theta = 1.2$, $n = 100$, and $\alpha = 0.025$?

Exercise 6.21*. Researchers have theorized that monozygotic twins separated at birth will tend, as adults, to be more alike than different with regard to their exercise habits. To examine this theory, a random sample of n sets of such adult monozygotic twins are interviewed regarding their current exercise habits. For $i = 1, 2, \ldots, n$, suppose that the random variable Y_i takes the value 0 if neither member of the i-th set of twins exercises on a regular basis, that Y_i takes the value 1 if one twin in the i-th set exercises on a regular basis and the other does not, and that Y_i takes the value 2 if both twins in the i-th set exercise on a regular basis.

Further, for $i = 1, 2, \ldots, n$, assume that the random variable Y_i has the probability distribution

$$\mathrm{p}_{Y_i}(y_i) = \left[\frac{1 + y_i(2 - y_i)}{2}\right] \frac{\theta^{y_i(2-y_i)}}{(1+\theta)}, y_i = 0, 1, 2 \text{ and } \theta > 0.$$

(a) For a data set involving $n = 50$ sets of monozygotic twins, suppose that there are no regular exercisers for each of 25 sets of these twins, that there is one regular exerciser and one non-regular exerciser for each of 15 sets of these twins, and that there are two regular exercisers for each of 10 sets of these twins. Using a Wald statistic (based on *expected* information), perform an appropriate statistical test to determine whether these data supply evidence in favor of the proposed theory.

(b) If, in fact, $\theta = 0.40$, find a reasonable value for the smallest number n^* of sets of adult monozygotic twins needed so that the power for rejecting $\mathrm{H}_0 : \theta = 0.50$ in favor of $\mathrm{H}_1 : \theta < 0.50$ at the $\alpha = 0.05$ level is at least 0.80.

Exercise 6.22*. Let $X_1, X_2, \ldots, X_n$ constitute a random sample of size n from a normal population with mean zero and variance θ_1. Further let

$Y_1, Y_2, \ldots, Y_n$ constitute a random sample of size n from a different normal population also with mean zero but with a different variance θ_2^{-1}.

(a) Show that the likelihood ratio statistic $-2\ln\hat{\lambda}$ for testing the null hypothesis $\text{H}_0 : \theta_1 = \theta_2$ ($= \theta$, say) versus the alternative hypothesis $\text{H}_1 : \theta_1 \neq \theta_2$ can be expressed as an explicit function of $\hat{\theta}_1$ and $\hat{\theta}_2$, the maximum likelihood estimators of θ_1 and θ_2, respectively. For a particular set of data where $n = 30$, $\sum_{i=1}^{n} x_i^2 = 60$, and $\sum_{i=1}^{n} y_i^2 = 18$, show that the available data provide no statistical evidence for rejecting H_0 in favor of H_1.

(b) Based on the results of part (a), use all $2n$ observations to derive a general expression for an *exact* $100(1-\alpha)\%$ confidence interval (based on the f-distribution) for the unknown parameter θ, and then use the available data given in part (a) to compute an exact 95% confidence interval for θ.

Exercise 6.23*. Recent research findings support the proposition that human beings who carry a certain gene have increased sensitivity to alcohol. To provide further evidence supporting these research findings, a large random sample of n adult human subjects is selected for study. Each study subject consumes the same fixed amount of alcohol over a 30-minute time period, and then the time (in minutes) needed for that subject to complete a certain complicated manual dexterity test is recorded. Also, each study subject is genetically tested for the presence of the gene under study.

As a proposed statistical model, for $i = 1, 2, \ldots, n$, let the dichotomous random variable X_i take the value 1 with probability $\pi(0 < \pi < 1)$ if the i-th subject carries the gene, and let X_i take the value 0 with probability $(1 - \pi)$ if not. The parameter π represents the *prevalence* of the gene. Further, given that $X_i = x_i$, suppose that the time Y_i (in minutes) required to complete the manual dexterity test has the negative exponential distribution

$$\text{f}_{Y_i}(y_i|X_i = x_i) = \frac{1}{\theta_i}\text{e}^{-y_i/\theta_i}, 0 < y_i < \infty \text{ and } 0 < \theta_i < \infty,$$

where $\ln\text{E}(Y_i|X_i = x_i) = \ln\theta_i = (\alpha + \beta x_i), -\infty < \alpha < \infty$ and $-\infty < \beta < \infty$.

(a) Develop an explicit expresson for $\text{cov}(X_i, Y_i)$, and then comment on your finding with regard to the parameter β and its connection to the proposition under study.

(b) Without loss of generality, suppose that $(0, y_1), (0, y_2), \ldots, (0, y_{n_0})$, $(1, y_{n_0+1}), (1, y_{n_0+2}), \ldots, (1, y_n)$ constitute the observed data for the n randomly selected human subjects, so that the first n_0 data pairs pertain to the n_0 subjects who are not carriers of the gene and the last $(n - n_0) = n_1$ data pairs pertain to the n_1 subjects who are carriers of the gene. These data can be considered to be realizations of the mutually independent pairs of random variables $(X_1, Y_1), (X_2, Y_2), \ldots, (X_n, Y_n)$. Further, let $\bar{y}_0 = n_0^{-1} \sum_{i=1}^{n_0} y_i$ and let $\bar{y}_1 = n_1^{-1} \sum_{i=n_0+1}^{n} y_i$. Using this notation, show that the maximum likelihood (ML) estimates $\hat{\pi}, \hat{\alpha}$, and $\hat{\beta}$ of the parameters π, α, and β are

$$\hat{\pi} = \frac{n_1}{n}, \hat{\alpha} = \ln \bar{y}_0, \text{ and } \hat{\beta} = \ln\left(\frac{\bar{y}_1}{\bar{y}_0}\right).$$

(c) Among $n = 100$ randomly selected subjects, suppose that 20 of these 100 subjects are found to be carriers of the gene. Also, suppose that the average time to complete the manual dexterity test is $\bar{y}_1 = 32$ minutes for the 20 carriers of the gene, and that the average time to complete the manual dexterity test is $\bar{y}_0 = 19$ minutes for the 80 non-carriers of the gene. Use these data to conduct Wald and score tests of the null hypothesis $\text{H}_0 : \beta = 0$ versus the alternative hypothesis $\text{H}_1 : \beta \neq 0$ at the $\alpha = 0.05$ level. Do the results of these tests provide statistical evidence in favor of the proposition?

Exercise 6.24*. Suppose that two equally competent radiologists (designated radiologist #1 and radiologist #2) examine the same set of n x-rays for the presence or absence of breast cancer. It is of interest to assess the level of agreement between these two radiologists. The following statistical model is to be used for this purpose. For the i-th of n x-rays, $i = 1, 2, \ldots, n$, let $X_{i1} = 1$ if radiologist #1 detects the presence of breast cancer, and let $X_{i1} = 0$ otherwise. Also, let $X_{i2} = 1$ if radiologist #2 detects the presence of breast cancer on the i-th x-ray, and let $X_{i2} = 0$ otherwise. Then, the joint probability distribution of X_{i1} and X_{i2} is assumed to have the structure

$$\text{p}_{X_{i1}, X_{i2}}(x_{i1}, x_{i2}) = \pi_{11}^{x_{i1}x_{i2}} \pi_{10}^{x_{i1}(1-x_{i2})} \pi_{01}^{(1-x_{i1})x_{i2}} \pi_{00}^{(1-x_{i1})(1-x_{i2})},$$

where $\pi_{11} = \theta^2 + \rho\theta(1-\theta), \pi_{10} = \pi_{01} = (1-\rho)\theta(1-\theta), \pi_{00} = (1-\theta)^2 + \rho\theta(1-\theta), 0 < \theta < 1$, and $-1 < \rho < 1$. For this statistical model, the parameter $\rho = \text{corr}(X_{i1}, X_{i2})$ is defined to be the *agreement coefficient*, and it is of interest to make statistical inferences concerning this unknown parameter.

Now, suppose that these two radiologists each examine the same $n = 40$ x-rays. Further, suppose that there are $n_{11} = 2$ x-rays for which both radiologists detect the presence of breast cancer, that there are $n_{00} = 30$ x-rays for which neither radiologist detects the presence of breast cancer, that there are $n_{10} = 5$ x-rays for which only radiologist #1 detects the presence of breast cancer, and that there are $n_{01} = 3$ x-rays for which only radiologist #2 detects the presence of breast cancer.

(a) Show that the maximum likelihood estimates $\hat{\theta}$ and $\hat{\rho}$ of θ and ρ can be written as

$$\hat{\theta} = \frac{(2n_{11} + n_{10} + n_{01})}{2n} \text{ and } \hat{\rho} = \frac{4n_{11}n_{00} - (n_{10} + n_{01})^2}{(2n_{11} + n_{10} + n_{01})(2n_{00} + n_{10} + n_{01})},$$

and then use the available data to compute the numerical values of these two maximum likelihood estimates.

(b) Using expected information, develop an explicit expression for a score test statistic $\hat{S}$ for testing $\text{H}_0 : \rho = 0$ versus $\text{H}_1 : \rho \neq 0$, where $\hat{S} = (\hat{\rho} - 0)^2/\hat{\text{V}}_0(\hat{\rho})$ and where $\hat{\text{V}}_0(\hat{\rho})$ is the estimated variance of $\hat{\rho}$ under the null hypothesis $\text{H}_0 : \rho = 0$. Do the available data provide statistical evidence to reject $\text{H}_0 : \rho = 0$ in favor of $\text{H}_1 : \rho \neq 0$ at the $\alpha = 0.05$ level of significance?

For further information concerning measures of agreement among raters, see Bloch and Kraemer (1989).

Exercise 6.25*. Ear infections are quite common in infants. To assess whether ear infections in infants tend to occur in both ears rather than in just one ear in a certain U.S. area, the following statistical model is proposed.

For a random sample of n infants whose parents reside in this U.S. area, suppose, for $i = 1, 2, \ldots, n$, that the random variable $X_i = 0$ with probability $(1 - \pi)$ if the i-th infant does not have an ear infection, that $X_i{=}1$ with probability $\pi(1 - \theta)$ if the i-th infant has an ear infection in only one ear, and that $X_i = 2$ with probability $\pi\theta$ if the i-th infant has ear infections in both ears. Here, $\pi(0 < \pi < 1)$ is the probability that an infant has an infection in at least one ear; that is, π is the *prevalence* in this U.S. area of children with an infection in at least one ear. And, since

$$\begin{aligned} \text{pr}(X_i = 2|X_i \geq 1) &= \frac{\text{pr}\left[(X_i = 2) \cap (X_i \geq 1)\right]}{\text{pr}(X_i \geq 1)} \\ &= \frac{\text{pr}(X_i = 2)}{\text{pr}(X_i \geq 1)} = \frac{\pi\theta}{\pi} = \theta, \end{aligned}$$

it follows that $\theta(0 < \theta < 1)$ is the conditional probability that an infant has ear infections in both ears given that this infant has at least one ear that is infected.

(a) Show that a score test statistic $\hat{S}$ for testing $\text{H}_0 : \theta = \theta_0$ versus $\text{H}_1 : \theta \neq \theta_0$ can be written in the form

$$\hat{S} = \frac{(\hat{\theta} - \theta_0)^2}{\hat{V}_0(\hat{\theta})},$$

where $\hat{V}_0(\hat{\theta})$ is the estimated variance of $\hat{\theta}$ under the null hypothesis $\text{H}_0 : \theta = \theta_0$.

(b) Suppose that $n = 100$, that $n_0 = 20$ is the number of infants with no ear infections, that $n_1 = 35$ is the number of infants with an ear infection in only one ear, and that $n_2 = 45$ is the number of infants with ear infections in both ears. Use these data and the score test developed in part (a) to test $\text{H}_0 : \theta = 0.50$ versus $\text{H}_1 : \theta \neq 0.50$ at the $\alpha = 0.025$ significance level. Do these data provide statistical evidence that it is more likely than not that an infant in this U.S. region will have both ears infected once that infant develops an ear infection?

(c) Assuming that $\pi = 0.80$, provide a reasonable value for the smallest value of n, say n^*, required so that the power of a one-sided score test of $\text{H}_0 : \theta = 0.50$ versus $\text{H}_1 : \theta > 0.50$ at the $\alpha = 0.025$ level is at least 0.90 when, in fact, $\theta = 0.60$.

Exercise 6.26*. Suppose that a randomized clinical trial is conducted to compare two new experimental drugs (denoted drug 1 and drug 2) designed to prolong the lives of patients with metastatic colorectal cancer. For $i = 1, 2$, it is assumed that the survival time X_i (in years) for a patient using drug i can be described by the CDF

$$\text{F}_{X_i}(x_i) = 1 - \text{e}^{-(x_i/\theta_i)^m}, 0 < x_i < \infty, 0 < \theta_i < \infty,$$

where $m(\geq 1)$ is a *known* positive constant.

Suppose that n patients are randomly allocated to each of these two drug therapies. For $i = 1, 2$, let $x_{i1}, x_{i2}, \ldots, x_{in}$ be the n observed survival times (in years) for the n patients receiving drug i. For $i = 1, 2$, these n observed survival times can be considered to be the realizations of a random sample $X_{i1}, X_{i2}, \ldots, X_{in}$ of size n from the CDF $\text{F}_{X_i}(x_i)$.

Suppose that $n = 30, m = 3, \sum_{j=1}^{n} x_{1j}^3 = 210$, and $\sum_{j=1}^{n} x_{2j}^3 = 300$. Use these data to conduct an appropriate likelihood ratio test to assess whether these two drugs perform differently with regard to prolonging the lives of patients with metastatic colon cancer, and then comment on your findings. (HINT: For $i = 1, 2$, consider the random variable $Y_i = X_i^m$.)

¿¿ chapter6.tex

6.2 Solutions to Odd-Numbered Exercises

Solution 6.1.

(a) In general,

$$\mathcal{L}(\boldsymbol{y}; \theta) = \prod_{i=1}^{n} \left[\frac{1}{2}(1 + \theta y_i) \right] = 2^{-n} \prod_{i=1}^{n} (1 + \theta y_i).$$

So,

$$\frac{\mathcal{L}(\boldsymbol{y};0)}{\mathcal{L}(\boldsymbol{y};0.50)} = \frac{2^{-n}}{2^{-n}\prod_{i=1}^{n}\left(1+\frac{y_i}{2}\right)} \leq k,$$

$$\text{or } \prod_{i=1}^{n}\left(1+\frac{y_i}{2}\right) \geq k^{-1}.$$

So,

$$\mathcal{R} = \left\{(y_1, y_2, \ldots, y_n) : \prod_{i=1}^{n}\left(1+\frac{y_i}{2}\right) \geq k_\alpha\right\},$$

where k_α is chosen so that $\text{pr}\{(Y_1, Y_2, \ldots, Y_n) \in \mathcal{R}|\text{H}_0 : \theta = 0\} = \alpha$. Since, for any fixed $\theta_1 > 0$, we have a region of the form

$$\prod_{i=1}^{n}(1+\theta_1 y_i) \geq k^{-1},$$

which depends on θ_1, this is *not* a UMP region and test.

(b) When $n = 1$,

$$\mathcal{R} = \left\{y_1 : \left(1+\frac{y_1}{2}\right) \geq k_{0.05}\right\}.$$

Now,

$$\left(1+\frac{y_1}{2}\right) \geq k_{0.05} \Leftrightarrow y_1 \geq 2(k_{0.05}-1) = k^*, \text{ say.}$$

So, under $\text{H}_0 : \theta = 0$,

$$\begin{aligned}\text{pr}(Y_1 \geq k^*|\theta = 0) &= \int_{k^*}^{1}\left(\frac{1}{2}\right)\text{d}y_1 = \frac{(1-k^*)}{2} = 0.05,\\ \text{or } k^* &= 0.90.\\ \text{So, } k_{0.05} &= 1+\frac{k^*}{2} = 1.45.\end{aligned}$$

Thus, for a size $\alpha = 0.05$ test when $n = 1$, we reject $\text{H}_0 : \theta = 0$ in favor of $\text{H}_1 : \theta = 0.50$ when $Y_1 \geq 0.90$, or equivalently, when

$$\left(1+\frac{Y_1}{2}\right) \geq 1.45.$$

(c) When the true value of $\theta = 0.50$, then

$$\begin{aligned}\text{POWER} &= \text{pr}\{Y_1 \geq 0.90|\theta = 0.50\}\\ &= \int_{0.90}^{1}\frac{1}{2}\left(1+\frac{y_1}{2}\right)\text{d}y_1 \doteq 0.074.\end{aligned}$$

The power is small because $n = 1$.

Solution 6.3. The unrestricted likelihood $\mathcal{L}_\Omega$ is

$$\begin{aligned} \mathcal{L}_\Omega &= \prod_{i=1}^{n}\left\{\theta^{-2}t_i\mathrm{e}^{-t_i/\theta}\right\} = \theta^{-2n}\left(\prod_{i=1}^{n}t_i\right)\mathrm{e}^{-\sum_{i=1}^{n}t_i/\theta}, \\ \text{so that } \ln\mathcal{L}_\Omega &= -2n\ln\theta + \sum_{i=1}^{n}\ln t_i - \theta^{-1}\sum_{i=1}^{n}t_i. \end{aligned}$$

So,

$$\frac{\mathrm{d}\ln\mathcal{L}_\Omega}{\mathrm{d}\theta} = \frac{-2n}{\theta} + \frac{\sum_{i=1}^{n}t_i}{\theta^2} = 0 \Rightarrow \hat{\theta} = \frac{\sum_{i=1}^{n}t_i}{2n} = \frac{\bar{t}}{2}.$$

Now, since the restricted likelihood $\mathcal{L}_\omega$ is obtained by setting $\theta = 1$ in the expression for $\mathcal{L}_\Omega$, we obtain

$$\hat{\lambda} = \frac{\mathcal{L}_\omega}{\hat{\mathcal{L}}_\Omega} = \frac{\left(\prod_{i=1}^{n}t_i\right)\mathrm{e}^{-\sum_{i=1}^{n}t_i}}{\hat{\theta}^{-2n}\left(\prod_{i=1}^{n}t_i\right)\mathrm{e}^{-\sum_{i=1}^{n}t_i/\hat{\theta}}} = \left(\frac{\bar{t}}{2}\right)^{2n}\mathrm{e}^{-n\bar{t}}\mathrm{e}^{2n}.$$

So,

$$-2\ln\hat{\lambda} = -2\ln\left[\left(\frac{\bar{t}}{2}\right)^{2n}\mathrm{e}^{-n\bar{t}}\mathrm{e}^{2n}\right] = 2n\left[\ln\left(\frac{2}{\bar{t}}\right)^2 + (\bar{t}-2)\right].$$

Note that $\bar{t} = 2$, or $\hat{\theta} = 1$, gives $-2\ln\hat{\lambda} = 0$, as desired.

When $n = 40$ and $\bar{t} = 2.40$,

$$-2\ln\hat{\lambda} = 2(40)\left[\ln\left(\frac{2}{2.40}\right)^2 + (2.40-2)\right] = 2.83.$$

Since $-2\ln\hat{\lambda}\dot{\sim}\chi_1^2$ under $\mathrm{H_0}: \theta = 1$, the P-value $\doteq \mathrm{pr}[\chi_1^2 > 2.83] \doteq 0.09$, so that we would not reject $\mathrm{H_0}: \theta = 1$ at the $\alpha = 0.05$ level based on the observed data.

Solution 6.5.

(a) Since $Z = (\bar{X}_1 - \bar{X}_2)/(\sqrt{2}\sigma/\sqrt{n}) \sim \mathrm{N}(0,1)$, we have

$$\begin{aligned}\theta &= \mathrm{pr}\left[|\bar{X}_1 - \bar{X}_2| \leq 2z_{1-\frac{\alpha}{2}}\frac{\sigma}{\sqrt{n}}\right] \\ &= \mathrm{pr}\left[-2z_{1-\frac{\alpha}{2}}\frac{\sigma}{\sqrt{n}} \leq (\bar{X}_1 - \bar{X}_2) \leq 2z_{1-\frac{\alpha}{2}}\frac{\sigma}{\sqrt{n}}\right] \\ &= \mathrm{pr}\left[-\sqrt{2}z_{1-\frac{\alpha}{2}} \leq Z \leq \sqrt{2}z_{1-\frac{\alpha}{2}}\right] \\ &= \mathrm{F}_Z(\sqrt{2}z_{1-\frac{\alpha}{2}}) - \mathrm{F}_Z(-\sqrt{2}z_{1-\frac{\alpha}{2}}) \\ &= 2\mathrm{F}_Z(\sqrt{2}z_{1-\frac{\alpha}{2}}) - 1.\end{aligned}$$

When $\alpha = 0.05$, then $\theta = 2\mathrm{F}_Z\left[\sqrt{2}(1.96)\right] - 1 = 2\mathrm{F}_Z(2.772) - 1$ $= 2(0.997) - 1 = 0.994$.

(b) Since $2\mathrm{F}_Z(1.96) - 1 = 2(0.975) - 1 = 0.95$, we require $\sqrt{2}z_{1-\frac{\alpha}{2}} = 1.96$, or $z_{1-\frac{\alpha}{2}} = 1.386$, so that $\alpha = 0.166$.

Solution 6.7. Using the test statistic

$$\frac{\bar{Y} - 0}{1/\sqrt{n}},$$

we want the smallest n such that

$$\beta = \mathrm{pr}\left\{\frac{\bar{Y} - 0}{1/\sqrt{n}} < 1.96 \middle| \mu = \frac{1}{2}\right\} \leq 0.16.$$

Thus, we must choose n^* such that

$$\mathrm{pr}\left\{\frac{\bar{Y} - 1/2}{1/\sqrt{n}} < 1.96 - \frac{1/2}{1/\sqrt{n}} \middle| \mu = \frac{1}{2}\right\} = \mathrm{pr}\left\{Z < 1.96 - \frac{\sqrt{n}}{2}\right\} \leq 0.16,$$

where, given $\mu = 1/2$, $Z = \frac{\bar{Y} - 1/2}{1/\sqrt{n}} \sim \mathrm{N}(0,1)$. Hence, we want to pick n so that

$$\begin{aligned}\left(1.96 - \frac{\sqrt{n}}{2}\right) &\leq -1 \\ &\Rightarrow \frac{\sqrt{n}}{2} \geq 2.96 \\ &\Rightarrow n \geq [2(2.96)]^2 = 35.05 \\ &\Rightarrow n^* = 36.\end{aligned}$$

Solution 6.9. First,

$$\mathcal{L}(\boldsymbol{x};\theta) \equiv \mathcal{L} = \prod_{i=1}^{n} \theta x_i^{\theta-1}, \ \ln\mathcal{L} = n\ln\theta + (\theta - 1)\sum_{i=1}^{n} \ln x_i, \text{ so that}$$

$\mathcal{I}(\theta) = -\frac{\mathrm{d}^2\ln\mathcal{L}}{\mathrm{d}\theta^2} = n/\theta^2$. So,

$$\begin{aligned}
\hat{W} &= (\hat{\theta}-\theta_0)\mathcal{I}(\hat{\theta})(\hat{\theta}-\theta_0) = \frac{(\hat{\theta}-\theta_0)^2}{\hat{\theta}^2/n} = \left[\frac{\hat{\theta}-\theta_0}{\hat{\theta}/\sqrt{n}}\right]^2, \text{ and} \\
\frac{\hat{\theta}-\theta_0}{\hat{\theta}/\sqrt{n}} &\dot{\sim} \mathrm{N}(0,1) \text{ for large } n \text{ under } \mathrm{H}_0: \theta = \theta_0. \\
\text{So, POWER} &= \mathrm{pr}[\hat{W} > 3.84|\theta \geq 3] \geq \mathrm{pr}[\hat{W} > 3.84|\theta = 3] \\
&= \mathrm{pr}\left[\left(\frac{\hat{\theta}-2}{\hat{\theta}/\sqrt{n}}\right)^2 > 3.84|\theta = 3\right] \\
&= \mathrm{pr}\left\{\left[\frac{\hat{\theta}-2}{\hat{\theta}/\sqrt{n}} < -1.96\right] \cup \left[\frac{\hat{\theta}-2}{\hat{\theta}/\sqrt{n}} > 1.96\right] |\theta = 3\right\} \\
&= \mathrm{pr}\left[\frac{\hat{\theta}-2}{\hat{\theta}/\sqrt{n}} < -1.96|\theta = 3\right] + \mathrm{pr}\left[\frac{\hat{\theta}-2}{\hat{\theta}/\sqrt{n}} > 1.96|\theta = 3\right] \\
&\doteq \mathrm{pr}\left[\frac{\hat{\theta}-2}{\hat{\theta}/\sqrt{n}} > 1.96|\theta = 3\right], \\
&\qquad \text{since pr}\left[\frac{\hat{\theta}-2}{\hat{\theta}/\sqrt{n}} < -1.96|\theta = 3\right] \doteq 0.
\end{aligned}$$

So,

$$\begin{aligned}
\text{POWER} &\doteq \mathrm{pr}\left[\frac{\hat{\theta}-2}{\hat{\theta}/\sqrt{n}} > 1.96|\theta = 3\right] = \mathrm{pr}\left[\hat{\theta} > \frac{1.96\hat{\theta}}{\sqrt{n}} + 2|\theta = 3\right] \\
&= \mathrm{pr}\left[\frac{\hat{\theta}-3}{\hat{\theta}/\sqrt{n}} > 1.96 - \frac{1}{\hat{\theta}/\sqrt{n}}\right] \doteq \mathrm{pr}\left[Z > 1.96 - \frac{\sqrt{n}}{3}\right] \geq 0.80,
\end{aligned}$$

where $Z \dot{\sim} \mathrm{N}(0,1)$ and $\hat{\theta} \approx 3$ for large n.

So, we require $(1.96 - \sqrt{n}/3) \leq -0.842$ or $n \geq 70.6608$. Thus, we require $n^* = 71$.

Solution 6.11. The unrestricted likelihood function is

$$\begin{aligned}
\mathcal{L}_\Omega &= \prod_{i=1}^{k}\prod_{j=1}^{n_i}\left\{\frac{1}{\sqrt{2\pi}\sigma}e^{-\frac{1}{2\sigma^2}(y_{ij}-\mu_i)^2}\right\} \\
&= (2\pi)^{-N/2}(\sigma^2)^{-N/2}\exp\left\{-\frac{1}{2\sigma^2}\sum_{i=1}^{k}\sum_{j=1}^{n_i}(y_{ij}-\mu_i)^2\right\},
\end{aligned}$$

so that

$$\ln\mathcal{L}_{\Omega} = -\frac{N}{2}\ln(2\pi) - \frac{N}{2}\ln(\sigma^2) - \frac{1}{2\sigma^2}\sum_{i=1}^{k}\sum_{j=1}^{n_i}(y_{ij} - \mu_i)^2$$

The equation

$$\frac{\partial\ln\mathcal{L}_{\Omega}}{\partial\mu_i} = \frac{-1}{2\sigma^2}(-2)\sum_{j=1}^{n_i}(y_{ij} - \mu_i) = 0$$

yields the unrestricted MLE of μ_i, namely,

$$\hat{\mu}_i = \bar{y}_i = \frac{1}{n_i}\sum_{j=1}^{n_i} y_{ij}, \quad i = 1, 2, \ldots, k.$$

Similarly, the equation

$$\frac{\partial\ln\mathcal{L}_{\Omega}}{\partial\sigma^2} = -\frac{N}{2\sigma^2} + \frac{1}{2\sigma^4}\sum_{i=1}^{k}\sum_{j=1}^{n_i}(y_{ij} - \mu_i)^2 = 0,$$

yields the unrestricted MLE of σ^2, namely,

$$\hat{\sigma}^2_{\Omega} = \frac{1}{N}\sum_{i=1}^{k}\sum_{j=1}^{n_i}(y_{ij} - \bar{y}_i)^2.$$

So,

$$\hat{\mathcal{L}}_{\Omega} = (2\pi)^{-N/2}(\hat{\sigma}^2_{\Omega})^{-N/2}e^{-N/2}.$$

Under the restriction $\mu_1 = \mu_2 = \cdots = \mu_k$ ($= \mu$, say), the restricted log-likelihood function is

$$\ln\mathcal{L}_{\omega} = -\frac{N}{2}\ln(2\pi) - \frac{N}{2}\ln(\sigma^2) - \frac{1}{2\sigma^2}\sum_{i=1}^{k}\sum_{j=1}^{n_i}(y_{ij} - \mu)^2.$$

Solving

$$\frac{\partial\ln\mathcal{L}_{\omega}}{\partial\mu} = \frac{-1}{2\sigma^2}(-2)\sum_{i=1}^{k}\sum_{j=1}^{n_i}(y_{ij} - \mu) = 0$$

yields the restricted MLE

$$\hat{\mu} = \frac{1}{N}\sum_{i=1}^{k}\sum_{j=1}^{n_i} y_{ij}.$$

Similarly,

$$\frac{\partial \ln \mathcal{L}_\omega}{\partial \sigma^2} = -\frac{N}{2\sigma^2} + \frac{1}{2\sigma^4}\sum_{i=1}^{k}\sum_{j=1}^{n_i}(y_{ij}-\mu)^2 = 0$$

yields

$$\hat{\sigma}^2_\omega = \frac{1}{N}\sum_{i=1}^{k}\sum_{j=1}^{n_i}(y_{ij}-\hat{\mu})^2.$$

Thus,

$$\hat{\mathcal{L}}_\omega = (2\pi)^{-N/2}(\hat{\sigma}^2_\omega)^{-N/2}e^{-N/2},$$

and so the likelihood ratio statistic is

$$\hat{\lambda} = \frac{\hat{\mathcal{L}}_\omega}{\hat{\mathcal{L}}_\Omega} = \left(\frac{\hat{\sigma}^2_\Omega}{\hat{\sigma}^2_\omega}\right)^{N/2}.$$

Under H_0 and for large samples,

$$-2\ln\hat{\lambda} \,\dot{\sim}\, \chi^2_{k-1}.$$

For the given numerical information,

$$\hat{\sigma}^2_\Omega = \frac{1}{N}\sum_{i=1}^{k}(n_i-1)s_i^2 = \frac{145}{37} = 3.9189.$$

Now,

$$\begin{aligned}
\hat{\sigma}^2_\omega = \frac{1}{N}\sum_{i=1}^{k}\sum_{j=1}^{n_i}(y_{ij}-\hat{\mu})^2 &= \frac{1}{N}\sum_{i=1}^{k}\sum_{j=1}^{n_i}\left[(y_{ij}-\bar{y}_i)+(\bar{y}_i-\hat{\mu})\right]^2 \\
&= \frac{1}{N}\sum_{i=1}^{k}\sum_{j=1}^{n_i}(y_{ij}-\bar{y}_i)^2 + \frac{1}{N}\sum_{i=1}^{k}n_i(\bar{y}_i-\hat{\mu})^2 \\
&= \hat{\sigma}^2_\Omega + \frac{1}{N}\sum_{i=1}^{k}n_i(\bar{y}_i-\hat{\mu})^2.
\end{aligned}$$

Since

$$\hat{\mu} = \sum_{i=1}^{k}n_i\bar{y}_i/N = \frac{1}{37}\left[(10)(4.00)+(15)(9.00)+(12)(14.00)\right] = 9.2703,$$

we have

$$\begin{aligned}
\hat{\sigma}^2_\omega &= 3.9189 + \frac{1}{37}[10(4.00-9.2703)^2 + 15(9.00-9.2703)^2 \\
&\quad + 12(14.00-9.2703)^2] = 18.7107.
\end{aligned}$$

Finally,

$$-2\ln\hat{\lambda} = -N\ln\left(\frac{3.9189}{18.7107}\right) = 57.8415.$$

So,

$$\text{P-value} = \text{pr}\left[\chi_2^2 > 57.8415\right] \dot{\approx} 2.75 \times 10^{-13},$$

and so H_0 is clearly rejected in favor of H_1.

Solution 6.13*.

(a) From the connection between the lognormal and normal distributions, we know that $Y = \ln X \sim N(0, \theta)$ and that $E(X) = e^{\theta/2}$. Hence, the requirement that $E(X) \leq 75$ is equivalent to the requirement that $\theta \leq 8.635$, so that $\theta_0 = 8.635$. Now, with $\mathbf{y} = (y_1, y_2, \ldots, y_n)$, the likelihood function $\mathcal{L}(\mathbf{y}; \theta)$ is

$$\begin{aligned}\mathcal{L}(\mathbf{y}; \theta) &= = \prod_{i=1}^{n} (2\pi\theta)^{-1/2} e^{-y_i^2/2\theta} \\ &= (2\pi\theta)^{-n/2} e^{-u/2\theta},\end{aligned}$$

where $u = \sum_{i=1}^{n} y_i^2$.
Now, using the Neyman-Pearson Lemma with θ_1 being a specific value of θ such that $\theta_1 > 8.635$, we have

$$\frac{\mathcal{L}(\mathbf{y}; 8.635)}{\mathcal{L}(\mathbf{y}; \theta_1)} = \frac{[2\pi(8.635)]^{-n/2} e^{-u/2(8.635)}}{(2\pi\theta_1)^{-n/2} e^{-u/2\theta_1}} \leq k,$$

or equivalently

$$-\frac{n}{2}\ln\left(\frac{8.635}{\theta_1}\right) - u\left(\frac{1}{17.270} - \frac{1}{2\theta_1}\right) \leq \ln k,$$

or equivalently $\mathcal{R} = \{\mathbf{y} : u \geq k_\alpha\}$, where k_α is chosen so that $\text{pr}(U \geq k_\alpha | H_0 : \theta = 8.635) = \alpha$.

Since the same form of rejection region is obtained for every specific value of $\theta > 8.635$, it follows that $\mathcal{R}$ is also a UMP region (and the associated test is a UMP test) for testing $H_0 : \theta = 8.635$ versus $H_1 : \theta > 8.635$.

(b) Since $Y_i/\sqrt{\theta} \sim N(0, 1)$ for $i = 1, 2, \ldots, n$ and since $Y_1, Y_2, \ldots, Y_n$ are mutually independent random variables, it follows that $Y_i^2/\theta \sim \chi_1^2, i = 1, 2, \ldots, n$, so that $\sum_{i=1}^{n} Y_i^2/\theta = U/\theta \sim \chi_n^2$. So, in our particular situation, $U/8.635 \sim \chi_n^2$ under $H_0 : \theta = 8.635$. For the given data set $(n = 10, u = 100)$, the observed value of $U/8.635$ is $u/8.635 = 100/8.635 = 11.581$. Since the P-value$= \text{pr}\left(\chi_{10}^2 > 11.581\right) > 0.30$, this small data set provides no statistical evidence that the plant is in violation of the EPA standard.

(c) For $i = 1, 2, \ldots, n$, since $Y_i^2/\theta \sim \chi_1^2$, it follows that $\mathrm{E}(Y_i^2) = \theta$ and that $\mathrm{V}(Y_i^2) = 2\theta^2$, so that $\mathrm{E}(U) = n\theta$ and $\mathrm{V}(U) = 2n\theta^2$.
Hence, by the Central Limit Theorem, it follows that the random variable $(U - n\theta)/\sqrt{2n\theta^2} \dot{\sim} \mathrm{N}(0,1)$ for large n. Hence, we would reject $\mathrm{H}_0 : \theta = 8.635$ in favor of $\mathrm{H}_1 : \theta > 8.635$ at the $\alpha = 0.05$ level if the random variable $(U - 8.635n)/\sqrt{2(8.635)^2 n}$ is greater than 1.645. So, we have

$$
\begin{aligned}
\text{POWER} &= \mathrm{pr}\left[\frac{U - 8.635n}{\sqrt{2(8.635)^2 n}} > 1.645 \middle| \theta \geq 10\right] \\
&= \mathrm{pr}\left[U > 20.088\sqrt{n} + 8.635n \middle| \theta \geq 10\right] \\
&\geq \mathrm{pr}\left[\frac{U - 10n}{\sqrt{2(10)^2 n}} > \frac{20.088\sqrt{n} + 8.635n - 10n}{\sqrt{2(10)^2 n}}\right] \\
&\doteq \mathrm{pr}\left(Z > 1.4204 - 0.0965\sqrt{n}\right),
\end{aligned}
$$

where $Z \dot{\sim} \mathrm{N}(0,1)$ for large n.
So, for POWER≥ 0.90, we require $(1.4204 - 0.0965\sqrt{n}) \leq -1.282$, which gives $n^* = 785$, which is a surprisingly large required sample size.

Solution 6.15*.

(a) In this situation, the P-value is equal to $g(t_1) = \mathrm{pr}(T_0 \geq t_1 | T_1 = t_1)$. Hence,

$$
\begin{aligned}
\psi &= \mathrm{E}_{t_1}\left[g(t_1)\right] \\
&= \mathrm{E}_{t_1}\left[\mathrm{pr}(T_0 \geq t_1 | T_1 = t_1)\right] \\
&= \mathrm{pr}(T_0 \geq T_1).
\end{aligned}
$$

When $\theta_0 = \theta_1$, T_0 and T_1 are independent random variables having exactly the same distribution, so that $\psi = 1/2$.

(b) Under $\mathrm{H}_0 : \mu = \mu_0, T_0 \sim \mathrm{N}\left(\mu_0, \frac{\sigma^2}{n}\right)$. And, under $\mathrm{H}_1 : \mu > \mu_0$, with μ_1 a specific value of μ satisfying $\mu_1 > \mu_0, T_1 \sim \mathrm{N}\left(\mu_1, \frac{\sigma^2}{n}\right)$. Thus, since $(T_1 - T_0) \sim \mathrm{N}(\mu_1 - \mu_0, 2\sigma^2/n)$, it follows that

$$
\begin{aligned}
\psi &= \mathrm{pr}(T_0 \geq T_1) = \mathrm{pr}(T_1 - T_0 \leq 0) \\
&= \mathrm{pr}\left[\frac{(T_1 - T_0) - (\mu_1 - \mu_0)}{\sqrt{2}\sigma/\sqrt{n}} \leq \frac{-(\mu_1 - \mu_0)}{\sqrt{2}\sigma/\sqrt{n}}\right] \\
&= \mathrm{pr}\left[Z \leq \frac{(\mu_0 - \mu_1)}{\sqrt{2}\sigma/\sqrt{n}}\right] \\
&= \mathrm{F}_Z\left[\frac{\sqrt{n}(\mu_0 - \mu_1)}{\sqrt{2}\sigma}\right], Z \sim \mathrm{N}(0,1).
\end{aligned}
$$

Since $(\mu_0 - \mu_1) < 0$, ψ decreases as n and $(\mu_1 - \mu_0)$ increase, and ψ increases as σ increases. These are anticipated properties that are shared by the power function in this situation.

Solution 6.17*.

(a) First, from properties of the multinomial distribution, it follows that $S = (Y_{10} + Y_{01}) \sim \text{BIN}(n; \pi_{10} + \pi_{01})$, so that

$$\begin{aligned}
\text{p}_{Y_{10}}(y_{10}|S=s) &= \frac{\text{pr}\left[(Y_{10}=y_{10}) \cap (S=s)\right]}{\text{pr}(S=s)} \\
&= \frac{\text{pr}\left[(Y_{10}=y_{10}) \cap (Y_{01}=s-y_{10})\right]}{\text{C}_s^n(\pi_{10}+\pi_{01})^s(\pi_{11}+\pi_{00})^{n-s}} \\
&= \frac{\left[\frac{n!}{y_{10}!y_{01}!(y_{11}+y_{00})!}\right]\pi_{10}^{y_{10}}\pi_{01}^{y_{01}}(\pi_{11}+\pi_{00})^{y_{11}+y_{00}}}{\left[\frac{n!}{s!(n-s)!}\right](\pi_{10}+\pi_{01})^s(\pi_{11}+\pi_{00})^{n-s}} \\
&= \text{C}_{y_{10}}^s\left(\frac{\pi_{10}}{\pi_{10}+\pi_{01}}\right)^{y_{10}}\left(\frac{\pi_{01}}{\pi_{10}+\pi_{01}}\right)^{s-y_{10}} \\
&= \text{C}_{y_{10}}^s\left(\frac{\psi}{\psi+1}\right)^{y_{10}}\left(\frac{1}{\psi+1}\right)^{y_{01}}, y_{10}=0,1,\ldots,s;
\end{aligned}$$

that is, given $S = s, Y_{10} \sim \text{BIN}\left(s; \frac{\psi}{\psi+1}\right)$.

(b) Now,

$$\begin{aligned}
\text{ln}\mathcal{L} &\propto y_{10}[\text{ln}\psi - \text{ln}(\psi+1)] - y_{01}\text{ln}(\psi+1) \\
&= y_{10}\text{ln}\psi - (y_{10}+y_{01})\text{ln}(\psi+1),
\end{aligned}$$

so that

$$S(\psi) = \frac{\partial \text{ln}\mathcal{L}}{\partial \psi} = \frac{y_{10}}{\psi} - \frac{(y_{10}+y_{01})}{(\psi+1)}.$$

And, since

$$\frac{\partial^2 \text{ln}\mathcal{L}}{\partial \psi^2} = -\frac{y_{10}}{\psi^2} + \frac{(y_{10}+y_{01})}{(\psi+1)^2},$$

it follows that

$$\begin{aligned}
-\text{E}\left(\frac{\partial^2 \text{ln}\mathcal{L}}{\partial \psi^2}\middle| S=s\right) &= \frac{(y_{10}+y_{01})\left(\frac{\psi}{\psi+1}\right)}{\psi^2} - \frac{(y_{10}+y_{01})}{(\psi+1)^2} \\
&= \frac{(y_{10}+y_{01})}{\psi(\psi+1)^2},
\end{aligned}$$

so that

$$\mathcal{I}^{-1}(\psi|S=s) = \frac{\psi(\psi+1)^2}{(y_{10}+y_{01})}.$$

Thus,

$$\begin{aligned}\hat{S} &= \left\{[S(\psi)]^2\mathcal{I}^{-1}(\psi|S=s)\right\}\Big|_{\psi=1} \\ &= \left[\frac{(y_{10}-y_{01})}{2}\right]^2\left[\frac{4}{(y_{10}+y_{01})}\right] \\ &= \frac{(y_{10}-y_{01})^2}{(y_{10}+y_{01})}.\end{aligned}$$

Under $\text{H}_0 : \psi = 1, \hat{S}\dot{\sim}\chi_1^2$ for large $(y_{10}+y_{01})$. A test of $\text{H}_0 : \psi = 1$ versus $\text{H}_1 : \psi \neq 1$ using $\hat{S}$ is called *McNemar's test.*

(c) For the available data,

$$\hat{S} = \frac{(26-10)^2}{(26+10)} = 7.11,$$

so that the P-value is less than 0.01. Thus, these data provide fairly strong evidence of an exposure-disease association.

Solution 6.19*. Under $\text{H}_0 : \mu_1 = \mu_2$, the random variable

$$\frac{(\bar{X}_1-\bar{X}_2)-0}{\sqrt{\frac{\sigma_1^2}{n_1}+\frac{\sigma_2^2}{n_2}}} \sim \text{N}(0,1).$$

So, with $\text{pr}(Z > Z_{1-\alpha}) = \alpha$ when $Z \sim \text{N}(0,1)$, it follows that

$$\begin{aligned}\text{POWER} &= \text{pr}\left[\frac{(\bar{X}_1-\bar{X}_2)-0}{\sqrt{\frac{\sigma_1^2}{n_1}+\frac{\sigma_2^2}{n_2}}} > Z_{1-\alpha}\Bigg|\mu_1 > \mu_2\right] \\ &= \text{pr}\left[\frac{(\bar{X}_1-\bar{X}_2)-(\mu_1-\mu_2)}{\sqrt{\frac{\sigma_1^2}{n_1}+\frac{\sigma_2^2}{n_2}}} > Z_{1-\alpha} - \frac{(\mu_1-\mu_2)}{\sqrt{\frac{\sigma_1^2}{n_1}+\frac{\sigma_2^2}{n_2}}}\Bigg|\mu_1 > \mu_2\right] \\ &= \text{pr}\left[Z > Z_{1-\alpha} - \frac{(\mu_1-\mu_2)}{\sqrt{\frac{\sigma_1^2}{n_1}+\frac{\sigma_2^2}{n_2}}}\right].\end{aligned}$$

To maximize POWER, we need to minimize the quantity $\left(\frac{\sigma_1^2}{n_1}+\frac{\sigma_2^2}{n_2}\right)$ subject to the constraint $(n_1+n_2) = N$. So, using the method of Lagrange multipliers, we consider the expression

$$Q = \left(\frac{\sigma_1^2}{n_1}+\frac{\sigma_2^2}{n_2}\right) + \lambda(n_1+n_2-N).$$

Now,

$$\frac{\partial Q}{\partial n_1} = \sigma_1^2 n_1^{-2} + \lambda = 0,$$

$$\frac{\partial Q}{\partial n_2} = \sigma_2^2 n_2^{-2} + \lambda = 0,$$

and

$$\frac{\partial Q}{\partial \lambda} = n_1 + n_2 - N = 0.$$

Solving the first two equations above simultaneously gives

$$\sigma_1^2 n_1^{-2} = \sigma_2^2 n_2^{-2}, \text{ or } \frac{n_1}{n_2} = \frac{\sigma_1}{\sigma_2}.$$

Then, using the third equation, we obtain

$$n_1 = \left(\frac{\sigma_1}{\sigma_2}\right) n_2 = \left(\frac{\sigma_1}{\sigma_2}\right)(N - n_1),$$

so that

$$\left(1 + \frac{\sigma_1}{\sigma_2}\right) n_1 = \left(\frac{\sigma_1}{\sigma_2}\right) N,$$

which gives

$$n_1^* = \left(\frac{\sigma_1}{\sigma_1 + \sigma_2}\right) N \text{ and } n_2^* = (N - n_1^*) = \left(\frac{\sigma_2}{\sigma_1 + \sigma_2}\right) N.$$

Thus, if $\sigma_1 > \sigma_2$, then $n_1^* > n_2^*$, indicating that more observations should be selected from the more variable normal population. Similarly, if $\sigma_1 < \sigma_2$, then $n_1^* < n_2^*$; and, if $\sigma_1 = \sigma_2$, then $n_1^* = n_2^* = N/2$.

Finally, if $N = 100, \sigma_1^2 = 4$, and $\sigma_2^2 = 9$, then

$$n_1^* = \left(\frac{2}{2+3}\right)(100) = 40, \text{ and } n_2^* = (100 - 40) = 60.$$

Solution 6.21*.

(a) First, under the proposed statistical model for $\mathrm{p}_{Y_i}(y_i)$, note that

$$\mathrm{p}_{Y_i}(0) = \mathrm{p}_{Y_i}(2) = \frac{1}{2(1+\theta)} \text{ and } \mathrm{p}_{Y_i}(1) = \frac{\theta}{(1+\theta)}.$$

Hence, since $0 < \frac{\theta}{(1+\theta)} < \frac{1}{3}$ when $0 < \theta < \frac{1}{2}$, one can use a Wald statistic to test $\mathrm{H}_0 : \theta = \frac{1}{2}$ versus $\mathrm{H}_1 : \theta < \frac{1}{2}$. A small P-value (say, P-value < 0.05)

would indicate that these data supply statistical evidence in favor of the proposed theory.

Now, the appropriate likelihood function $\mathcal{L}$ is

$$\mathcal{L} = \prod_{i=1}^{n} \mathrm{p}_{Y_i} y_i = \prod_{i=1}^{n} \left[\frac{1 + y_i(2-y_i)}{2} \right] \frac{\theta^{y_i(2-y_i)}}{(1+\theta)},$$

so that

$$\ln\mathcal{L} \propto \left[\sum_{i=1}^{n} y_i(2-y_i) \right] \ln\theta - n\ln(1+\theta).$$

Thus, the equation

$$\frac{\partial \ln\mathcal{L}}{\partial\theta} = \frac{\sum_{i=1}^{n} y_i(2-y_i)}{\theta} - \frac{n}{(1+\theta)} = 0$$

gives

$$\hat{\theta} = \frac{\sum_{i=1}^{n} y_i(2-y_i)}{n - \sum_{i=1}^{n} y_i(2-y_i)}$$

as the maximum likelihood estimate of θ.

Also,

$$\frac{\partial^2 \ln\mathcal{L}}{\partial\theta^2} = -\frac{\sum_{i=1}^{n} y_i(2-y_i)}{\theta^2} + \frac{n}{(1+\theta)^2}.$$

And, since, for $i = 1, 2, \ldots, n$,

$$\mathrm{E}(Y_i) = 1 \text{ and } \mathrm{E}(Y_i^2) = \left(\frac{2+\theta}{1+\theta} \right),$$

it follows that the large-sample variance $\mathrm{V}_e(\hat{\theta})$ of $\hat{\theta}$ using *expected* information is

$$\begin{aligned}
\mathrm{V}_e(\hat{\theta}) &= \left[-\mathrm{E}\left(\frac{\partial^2 \ln\mathcal{L}}{\partial\theta^2} \right) \right]^{-1} \\
&= \left\{ \frac{\sum_{i=1}^{n} [2\mathrm{E}(Y_i) - \mathrm{E}(Y_i^2)]}{\theta^2} - \frac{n}{(1+\theta)^2} \right\}^{-1} \\
&= \left\{ \theta^{-2} \left[2n(1) - n\left(\frac{2+\theta}{1+\theta} \right) \right] - \frac{n}{(1+\theta)^2} \right\}^{-1} \\
&= \frac{\theta(1+\theta)^2}{n}.
\end{aligned}$$

Finally, the Wald-type statistic $\hat{W}$ for testing $\mathrm{H}_0 : \theta = 1/2$ versus $\mathrm{H}_1 : \theta < 1/2$ has the structure

$$\hat{W} = \frac{\hat{\theta} - \frac{1}{2}}{\sqrt{\frac{\hat{\theta}(1+\hat{\theta})^2}{n}}};$$

under $H_0 : \theta = 1/2, \hat{W} \dot{\sim} N(0,1)$ for large n.
For the given set of data, it follows that

$$\sum_{i=1}^{50} = 25(0) + 15(1) + 10(0) = 15 \text{ and that } \hat{\theta} = \frac{15}{(50-15)} = 0.429.$$

Thus,

$$\hat{W} = \frac{(0.429 - 0.50)}{\left[\frac{0.429(1+0.429)^2}{50}\right]^{1/2}} = -0.536,$$

so that these data provide absolutely no evidence in favor of the proposed theory.

(b) Now, with $\alpha = 0.05$,

$$\begin{aligned}
\text{POWER} &= \text{pr}(\hat{W} < -1.645 | \theta = 0.40) \\
&= \text{pr}\left[\frac{\hat{\theta} - 0.50}{\sqrt{\frac{\hat{\theta}(1+\hat{\theta})^2}{n}}} < -1.645 \middle| \theta = 0.40\right] \\
&= \text{pr}\left[\hat{\theta} < 0.50 - 1.645\sqrt{\frac{\hat{\theta}(1+\hat{\theta})^2}{n}} \middle| \theta = 0.40\right] \\
&= \text{pr}\left[\frac{\hat{\theta} - 0.40}{\sqrt{\frac{0.40(1+0.40)^2}{n}}} < \frac{0.50 - 1.645\sqrt{\frac{\hat{\theta}(1+\hat{\theta})^2}{n}} - 0.40}{\sqrt{\frac{0.40(1+0.40)^2}{n}}}\right]
\end{aligned}$$

Thus, for large n, so that $\hat{\theta} \doteq \theta$, it follows that

$$\text{POWER} \doteq \text{pr}\left[Z < 0.113\sqrt{n} - 1.645\right],$$

where

$$Z = \frac{\hat{\theta} - 0.40}{\sqrt{\frac{0.40(1+0.40)^2}{n}}} \dot{\sim} N(0,1) \text{ for large } n \text{ when } \theta = 0.40.$$

So, for POWER≥ 0.80, we wish to find the smallest integer value of n, say n^*, such that $(0.113\sqrt{n} - 1.645) \geq 0.842$, which gives $n \geq 484.39$, so that $n^* = 485$.

Solution 6.23*.

(a) Clearly, $E(X_i) = \pi$. Also,

$$\begin{aligned}
E(Y_i) &= E_{x_i}\left[E(Y_i | X_i = x_i)\right] = E_{x_i}\left[e^{\alpha + \beta x_i}\right] \\
&= (1 - \pi)e^{\alpha} + \pi e^{\alpha + \beta},
\end{aligned}$$

and

$$\begin{aligned}\mathrm{E}(X_iY_i) &= \mathrm{E}_{x_i}\left[x_i\mathrm{E}(Y_i|X_i = x_i)\right] \\ &= \mathrm{E}_{x_i}\left[x_i\mathrm{e}^{\alpha+\beta x_i}\right] = \pi\mathrm{e}^{\alpha+\beta}.\end{aligned}$$

So, we have

$$\begin{aligned}\mathrm{cov}(X_i, Y_i) &= \mathrm{E}(X_iY_i) - \mathrm{E}(X_i)\mathrm{E}(Y_i) \\ &= \pi\mathrm{e}^{\alpha+\beta} - \pi\left[(1-\pi)\mathrm{e}^{\alpha} + \pi\mathrm{e}^{\alpha+\beta}\right] \\ &= \pi(1-\pi)\mathrm{e}^{\alpha}(\mathrm{e}^{\beta} - 1).\end{aligned}$$

Thus, $\mathrm{cov}(X_i, Y_i) = 0$ if and only if $\beta = 0$, so that statistical evidence that $\beta > 0$, or equivalently that $\mathrm{cov}(X_i, Y_i) > 0$, would support the proposition.

(b) With $\boldsymbol{\theta} = (\pi, \alpha, \beta)$, and with $\sum_{i=1}^{n} x_i = n_1$, the likelihood function $\mathcal{L}(\boldsymbol{\theta}) \equiv \mathcal{L}$ is equal to

$$\begin{aligned}\mathcal{L} &= \prod_{i=1}^{n} \pi^{x_i}(1-\pi)^{1-x_i}\theta_i^{-1}\mathrm{e}^{-y_i/\theta_i} \\ &= \pi^{n_1}(1-\pi)^{n_0}\left(\prod_{i=1}^{n}\theta_i^{-1}\right)\mathrm{e}^{-\sum_{i=1}^{n}\theta_i^{-1}y_i}\end{aligned}$$

so that

$$\begin{aligned}\mathrm{ln}\mathcal{L} &= n_1\mathrm{ln}\pi + n_0\mathrm{ln}(1-\pi) - \sum_{i=1}^{n}\mathrm{ln}\theta_i - \sum_{i=1}^{n}\theta_i^{-1}y_i \\ &= n_1\mathrm{ln}\pi + n_0\mathrm{ln}(1-\pi) - \sum_{i=1}^{n}(\alpha + \beta x_i) - \sum_{i=1}^{n}\mathrm{e}^{-(\alpha+\beta x_i)}y_i.\end{aligned}$$

So,

$$\frac{\partial\mathrm{ln}\mathcal{L}}{\partial\pi} = \frac{n_1}{\pi} - \frac{n_0}{(1-\pi)} = 0 \text{ gives } \hat{\pi} = \frac{n_1}{(n_0 + n_1)} = \frac{n_1}{n} = \bar{x}.$$

And, solving simultaneously the two equations

$$\begin{aligned}\frac{\partial\mathrm{ln}\mathcal{L}}{\partial\alpha} &= -n + \sum_{i=1}^{n}\mathrm{e}^{-(\alpha+\beta x_i)}y_i \\ &= -n + n_0\bar{y}_0\mathrm{e}^{-\alpha} + n_1\bar{y}_1\mathrm{e}^{-(\alpha+\beta)} = 0\end{aligned}$$

and

$$\begin{aligned}\frac{\partial\mathrm{ln}\mathcal{L}}{\partial\beta} &= -n_1 + \sum_{i=1}^{n}\mathrm{e}^{-(\alpha+\beta x_i)}x_iy_i \\ &= -n_1 + n_1\bar{y}_1\mathrm{e}^{-(\alpha+\beta)} = 0\end{aligned}$$

gives $\hat{\alpha} = \mathrm{ln}\bar{y}_0$ and $\hat{\beta} = \ln\left(\bar{y}_1/\bar{y}_0\right)$.

(b) We first need to determine the structure of the expected information matrix $\boldsymbol{\mathcal{I}}(\boldsymbol{\theta})$ and its inverse $\boldsymbol{\mathcal{I}}^{-1}(\boldsymbol{\theta})$. First,

$$\frac{\partial^2 \ln \mathcal{L}}{\partial \pi^2} = -\frac{n_1}{\pi^2} - \frac{n_0}{(1-\pi)^2},$$

so that

$$-\mathrm{E}\left(\frac{\partial^2 \ln \mathcal{L}}{\partial \pi^2}\right) = \frac{n\pi}{\pi^2} + \frac{n(1-\pi)}{(1-\pi)^2} = \frac{n}{\pi(1-\pi)}.$$

And,

$$\frac{\partial^2 \ln \mathcal{L}}{\partial \alpha^2} = -\sum_{i=1}^{n} \mathrm{e}^{-(\alpha+\beta x_i)} y_i,$$

so that, with $\boldsymbol{X} = (X_1, X_2, \ldots, X_n)$ and $\boldsymbol{x} = (x_1, x_2, \ldots, x_n)$, we have

$$\begin{aligned} -\mathrm{E}\left(\frac{\partial^2 \ln \mathcal{L}}{\partial \alpha^2}\right) &= -\mathrm{E}_{\boldsymbol{x}}\left[\mathrm{E}\left(\frac{\partial^2 \ln \mathcal{L}}{\partial \alpha^2}\bigg| \boldsymbol{X} = \boldsymbol{x}\right)\right] \\ &= -\mathrm{E}_{\boldsymbol{x}}\left[-\sum_{i=1}^{n} \mathrm{e}^{-(\alpha+\beta x_i)} \mathrm{e}^{(\alpha+\beta x_i)}\right] = n. \end{aligned}$$

And,

$$\frac{\partial^2 \ln \mathcal{L}}{\partial \alpha \partial \beta} = -\sum_{i=1}^{n} x_i \mathrm{e}^{-(\alpha+\beta x_i)} y_i,$$

so that

$$\begin{aligned} -\mathrm{E}\left(\frac{\partial^2 \ln \mathcal{L}}{\partial \alpha \partial \beta}\right) &= -\mathrm{E}_{\boldsymbol{x}}\left[\mathrm{E}\left(\frac{\partial^2 \ln \mathcal{L}}{\partial \alpha \partial \beta}\bigg| \boldsymbol{X} = \boldsymbol{x}\right)\right] \\ &= -\mathrm{E}_{\boldsymbol{x}}\left[-\sum_{i=1}^{n} x_i \mathrm{e}^{-(\alpha+\beta x_i)} \mathrm{e}^{(\alpha+\beta x_i)}\right] \\ &= \sum_{i=1}^{n} \mathrm{E}(X_i) = n\pi. \end{aligned}$$

Also,

$$\frac{\partial^2 \ln \mathcal{L}}{\partial \beta^2} = -\sum_{i=1}^{n} x_i^2 \mathrm{e}^{-(\alpha+\beta x_i)} y_i,$$

so that clearly

$$-\mathrm{E}\left(\frac{\partial^2 \ln \mathcal{L}}{\partial \beta^2}\right) = \sum_{i=1}^{n} \mathrm{E}(X_i^2) = n\pi.$$

Since

$$\frac{\partial^2 \ln \mathcal{L}}{\partial \pi \partial \alpha} = \frac{\partial^2 \ln \mathcal{L}}{\partial \pi \partial \beta} = 0,$$

it follows that

$$\mathcal{I}(\boldsymbol{\theta}) = \begin{bmatrix} \frac{n}{\pi(1-\pi)} & 0 & 0 \\ 0 & n & n\pi \\ 0 & n\pi & n\pi \end{bmatrix},$$

and that

$$\mathcal{I}^{-1}(\boldsymbol{\theta}) = \begin{bmatrix} \frac{\pi(1-\pi)}{n} & 0 & 0 \\ 0 & \frac{1}{n(1-\pi)} & \frac{-1}{n(1-\pi)} \\ 0 & \frac{-1}{n(1-\pi)} & \frac{1}{n\pi(1-\pi)} \end{bmatrix}.$$

For the Wald test, $\boldsymbol{R}(\boldsymbol{\theta}) = R_1(\boldsymbol{\theta}) = (\beta - 0) = \beta$, so that

$$\boldsymbol{T}(\boldsymbol{\theta}) = \left[\frac{\partial R_1(\boldsymbol{\theta})}{\partial \pi}, \frac{\partial R_1(\boldsymbol{\theta})}{\partial \alpha}, \frac{\partial R_1(\boldsymbol{\theta})}{\partial \beta}\right] = (0, 0, 1).$$

So,

$$\boldsymbol{\Lambda}(\boldsymbol{\theta}) = \boldsymbol{T}(\boldsymbol{\theta})\mathcal{I}^{-1}(\boldsymbol{\theta})\boldsymbol{T}'(\boldsymbol{\theta}) = \frac{1}{n\pi(1-\pi)},$$

which gives

$$\begin{aligned} \hat{W} &= \boldsymbol{R}(\hat{\boldsymbol{\theta}})\boldsymbol{\Lambda}^{-1}(\hat{\boldsymbol{\theta}})\boldsymbol{R}'(\hat{\boldsymbol{\theta}}) \\ &= \frac{(\hat{\beta})^2}{[1/n\hat{\pi}(1-\hat{\pi})]}, \end{aligned}$$

which, as expected, is of the form

$$\hat{W} = \left[\frac{\hat{\beta} - 0}{\sqrt{1/n\hat{\pi}(1-\hat{\pi})}}\right]^2 = \left[\frac{\hat{\beta} - 0}{\sqrt{\hat{\mathrm{V}}(\hat{\beta})}}\right]^2.$$

For the given data, $\hat{\pi} = 20/100 = 0.20$, and $\hat{\beta} = \ln(\bar{y}_1/\bar{y}_0) = \ln(32/19) = 0.5213$, so that

$$\hat{W} = n\hat{\pi}(1-\hat{\pi})(\hat{\beta})^2 = (100)(0.20)(1-0.20)(0.5213)^2 = 4.3481.$$

Under $\mathrm{H}_0 : \beta = 0$ and for large n, $\hat{W} \dot{\sim} \chi_1^2$. Since $\chi^2_{1,0.95} = 3.84$, we reject $\mathrm{H}_0 : \beta = 0$ in favor of $\mathrm{H}_1 : \beta \neq 0$. Also, since

$$\frac{\hat{\beta} - 0}{\sqrt{1/n\hat{\pi}(1-\hat{\pi})}} \dot{\sim} \mathrm{N}(0, 1)$$

for large n and under $\mathrm{H}_0 : \beta = 0$, and since this random variable has the numerical value $+\sqrt{\hat{W}} = +\sqrt{4.3481} = 2.0852$ because $\hat{\beta} > 0$, these data supply statistical evidence in support of the proposition based on the use of the Wald test.

For the score test,

$$\begin{aligned}\boldsymbol{S}(\boldsymbol{\theta}) &= \left(\frac{\partial\ln\mathcal{L}}{\partial\pi}, \frac{\partial\ln\mathcal{L}}{\partial\alpha}, \frac{\partial\ln\mathcal{L}}{\partial\beta}\right)\\ &= \left[\frac{(n_1 - n\pi)}{\pi(1-\pi)}, -n + \sum_{i=1}^{n} \mathrm{e}^{-(\alpha+\beta x_i)} y_i, -n_1 + \sum_{i=1}^{n} \mathrm{e}^{-(\alpha+\beta x_i)} x_i y_i\right].\end{aligned}$$

Now, with $\beta = 0$, we have

$$\ln\mathcal{L}_\omega = n_1\ln\pi + n_0\ln(1-\pi) - n\alpha - \mathrm{e}^{-\alpha}\sum_{i=1}^{n} y_i.$$

Clearly, $\partial\ln\mathcal{L}_\omega/\partial\pi = 0$ gives $\hat{\pi}_\omega = \hat{\pi} = n_1/n$. And,

$$\frac{\partial\ln\mathcal{L}_\omega}{\partial\alpha} = -n + \mathrm{e}^{-\alpha}\sum_{i=1}^{n} y_i$$

gives

$$\hat{\alpha}_\omega = \ln\left(\frac{\sum_{i=1}^{n} y_i}{n}\right) = \ln(\bar{y}) = \ln\left[\frac{n_0\bar{y}_0 + n_1\bar{y}_1}{n}\right].$$

So,

$$\hat{\boldsymbol{\theta}}_\omega = [\hat{\pi}, \ln(\bar{y}), 0],$$

so that

$$\begin{aligned}\boldsymbol{S}(\hat{\boldsymbol{\theta}}_\omega) &= \left[\frac{(n_1 - n\hat{\pi})}{\hat{\pi}(1-\hat{\pi})}, -n + \sum_{i=1}^{n} \mathrm{e}^{-\hat{\alpha}_\omega} y_i, -n_1 + \sum_{i=1}^{n} \mathrm{e}^{-\hat{\alpha}_\omega} x_i y_i\right]\\ &= \left[0, 0, \frac{n_0 n_1(\bar{y}_1 - \bar{y}_0)}{(n_0\bar{y}_0 + n_1\bar{y}_1)}\right].\end{aligned}$$

Finally, the score statistic $\hat{S}$ has the structure

$$\begin{aligned}\hat{S} &= \boldsymbol{S}(\hat{\boldsymbol{\theta}}_\omega)\boldsymbol{\mathcal{I}}^{-1}(\hat{\boldsymbol{\theta}}_\omega)\boldsymbol{S}'(\hat{\boldsymbol{\theta}}_\omega)\\ &= \frac{n_0^2 n_1^2(\bar{y}_1 - \bar{y}_0)^2}{n\hat{\pi}(1-\hat{\pi})(n_0\bar{y}_0 + n_1\bar{y}_1)^2}.\end{aligned}$$

For the available data, the numerical value of $\hat{S}$ is 5.7956, which is a value supporting the rejection of $\mathrm{H}_0: \beta = 0$ in favor of $\mathrm{H}_1: \beta \neq 0$. Also, since $\bar{y}_1 > \bar{y}_0$, the score test results also provide statistical evidence in favor of the proposition.

Solution 6.25*.

(a) The likelihood function $\mathcal{L}$ is equal to

$$\begin{aligned}\mathcal{L} &= \prod_{i=1}^{n} \mathrm{p}_{X_i}(x_i) = \prod_{i=1}^{n} (1-\pi)^{\frac{1}{2}(2-x_i)(1-x_i)} \left[\pi(1-\theta)\right]^{x_i(2-x_i)} (\pi\theta)^{\frac{1}{2}x_i(x_i-1)} \\ &= (1-\pi)^{n_0} \left[\pi(1-\theta)\right]^{n_1} (\pi\theta)^{n_2},\end{aligned}$$

so that

$$\ln\mathcal{L} = n_0 \ln(1-\pi) + n_1 \left[\ln\pi + \ln(1-\theta)\right] + n_2 \left[\ln\pi + \ln\theta\right].$$

Hence,

$$\frac{\partial \ln\mathcal{L}}{\partial \pi} = -\frac{n_0}{(1-\pi)} + \frac{n_1}{\pi} + \frac{n_2}{\pi} = 0$$

gives $\hat{\pi} = (n_1 + n_2)/n$. And,

$$\frac{\partial \ln\mathcal{L}}{\partial \theta} = -\frac{n_1}{(1-\theta)} + \frac{n_2}{\theta} = 0$$

gives $\hat{\theta} = n_2/(n_1 + n_2)$.

First, note that the random variables N_0, N_1, and N_2 (with respective realizations n_0, n_1, and n_2) follow a multinomial distribution, namely,

$$(N_0, N_1, N_2) \sim \text{MULT}\left[n; (1-\pi), \pi(1-\theta), \pi\theta\right].$$

Now,

$$\frac{\partial^2 \ln\mathcal{L}}{\partial \pi^2} = -\frac{n_0}{(1-\pi)^2} - \frac{n_1}{\pi^2} - \frac{n_2}{\pi^2},$$

so that

$$\begin{aligned}-\mathrm{E}\left(\frac{\partial^2 \ln\mathcal{L}}{\partial \pi^2}\right) &= \frac{n(1-\pi)}{(1-\pi)^2} + \frac{n\pi(1-\theta)}{\pi^2} + \frac{n\pi\theta}{\pi^2} \\ &= \frac{n}{\pi(1-\pi)}.\end{aligned}$$

And,

$$\frac{\partial^2 \ln\mathcal{L}}{\partial \theta^2} = -\frac{n_1}{(1-\theta)^2} - \frac{n_2}{\theta^2},$$

so that

$$\begin{aligned}-\mathrm{E}\left(\frac{\partial^2 \ln\mathcal{L}}{\partial \theta^2}\right) &= \frac{n\pi(1-\theta)}{(1-\theta)^2} + \frac{n\pi\theta}{\theta^2} \\ &= \frac{n\pi}{\theta(1-\theta)}.\end{aligned}$$

Since $\partial^2 \ln\mathcal{L}/\partial\pi\partial\theta = 0$, the expected information matrix $\boldsymbol{\mathcal{I}}(\pi,\theta)$ has the structure

$$\boldsymbol{\mathcal{I}}(\pi,\theta) = \begin{bmatrix} \frac{n}{\pi(1-\pi)} & 0 \\ 0 & \frac{n\pi}{\theta(1-\theta)} \end{bmatrix}.$$

Also, since $\partial^2 \ln\mathcal{L}/\partial\pi\partial\theta = 0$, so that $\hat{\pi} = (n_1+n_2)/n$ is the MLE of π under $\mathrm{H}_0 : \theta = \theta_0$, it follows that

$$\begin{aligned} \boldsymbol{S}(\hat{\pi},\theta_0) &= \left[0, -\frac{n_1}{(1-\theta_0)} + \frac{n_2}{\theta_0}\right] \\ &= \left[0, \frac{(n_1+n_2)(\hat{\theta}-\theta_0)}{\theta_0(1-\theta_0)}\right], \end{aligned}$$

where $\hat{\theta} = n_2/(n_1+n_2)$.
So, since $\hat{\pi} = (n_1+n_2)/n$, it follows directly that the score statistic $\hat{S}$ is equal to

$$\begin{aligned} \hat{S} &= \boldsymbol{S}(\hat{\pi},\theta_0)\boldsymbol{\mathcal{I}}^{-1}(\hat{\pi},\theta_0)\boldsymbol{S}'(\hat{\pi},\theta_0) \\ &= \left[\frac{(n_1+n_2)(\hat{\theta}-\theta_0)}{\theta_0(1-\theta_0)}\right]^2 \left[\frac{\theta_0(1-\theta_0)}{n\hat{\pi}}\right] \\ &= \frac{(\hat{\theta}-\theta_0)^2}{\theta_0(1-\theta_0)\frac{(n_1+n_2)}{(n_1+n_2)^2}} = \frac{(\hat{\theta}-\theta_0)^2}{\theta_0(1-\theta_0)/(n_1+n_2)} \\ &= \frac{(\hat{\theta}-\theta_0)^2}{\theta_0(1-\theta_0)/n\hat{\pi}} = \frac{(\hat{\theta}-\theta_0)^2}{\mathrm{V}_0(\hat{\theta})}. \end{aligned}$$

(b) For the available data, $\hat{\pi} = (35+45)/100 = 0.800$ and $\hat{\theta} = 45/(35+45) = 0.563$, so that

$$\hat{S} = \frac{(0.563-0.50)^2}{0.50(1-0.50)/(100)(0.800)} = 1.27.$$

For large n and under $\mathrm{H}_0 : \theta = 0.50$, $\hat{S} \dot{\sim} \chi^2_1$. Since $\chi^2_{1,0.975} = 5.024$, there is not sufficient evidence to reject $\mathrm{H}_0 : \theta = 0.50$ in favor of $\mathrm{H}_1 : \theta \neq 0.50$. These data do *not* provide evidence to support the contention that it is more likely than not that an infant in this U.S. region will have both ears infected once that infant develops an ear infection.

(c) For large n and under $\mathrm{H}_0 : \theta = 0.50$, the random variable

$$\frac{\hat{\theta}-\theta_0}{\sqrt{\mathrm{V}_0(\hat{\theta})}} = \frac{\hat{\theta}-0.50}{\sqrt{(0.50)(1-0.50)/0.80n}} \dot{\sim} \mathrm{N}(0,1).$$

So, for $\alpha = 0.025$,

$$\begin{aligned}
\text{POWER} &= \text{pr}\left[\frac{\hat{\theta} - 0.50}{\sqrt{0.50(1-0.50)/0.80n}} > 1.96 \middle| \theta = 0.60\right] \\
&= \text{pr}\left[\hat{\theta} > 0.50 + 1.96\sqrt{\frac{0.25}{0.80n}} \middle| \theta = 0.60\right] \\
&= \text{pr}\left[\frac{\hat{\theta} - 0.60}{\sqrt{\frac{0.60(1-0.60)}{0.80n}}} > \frac{0.50 - 0.60 + 1.96\sqrt{\frac{0.25}{0.80n}}}{\sqrt{\frac{0.60(1-0.60)}{0.80n}}}\right] \\
&\doteq \text{pr}\left[Z > -0.183\sqrt{n} + 2.000\right],
\end{aligned}$$

where $Z \dot{\sim} \text{N}(0,1)$ for large n.

Thus, for POWER≥ 0.90, we need to find the smallest integer value of n, say n^*, such that $-0.183\sqrt{n} + 2.000 \leq -1.282$, which gives $n \geq 321.64$, so that we need $n^* = 322$.

Appendix: Useful Math Results

1. **Summations**

   (a) *Binomial:* $\sum_{j=0}^{n} C_j^n a^j b^{(n-j)} = (a+b)^n$, where $C_j^n = \frac{n!}{j!(n-j)!}$.

   (b) *Geometric:*

   i. $\sum_{j=0}^{\infty} r^j = \frac{1}{1-r}, |r| < 1.$

   ii. $\sum_{j=1}^{\infty} r^j = \frac{r}{1-r}, |r| < 1.$

   iii. $\sum_{j=0}^{n} r^j = \frac{1-r^{(n+1)}}{1-r}, -\infty < r < +\infty.$

   (c) *Negative Binomial:* $\sum_{j=0}^{\infty} C_k^{j+k} \pi^j = (1-\pi)^{-(k+1)}, 0 < \pi < 1, k$ a positive integer.

   (d) *Exponential:* $\sum_{j=0}^{\infty} \frac{x^j}{j!} = e^x, -\infty < x < +\infty.$

   (e) *Sums of Integers:*

   i. $\sum_{i=1}^{n} i = \frac{n(n+1)}{2}.$

   ii. $\sum_{i=1}^{n} i^2 = \frac{n(n+1)(2n+1)}{6}.$

   iii. $\sum_{i=1}^{n} i^3 = \left[\frac{n(n+1)}{2}\right]^2.$

2. **Limits**

   (a) $\lim_{n\to\infty} \left(1 + \frac{a}{n}\right)^n = e^a, -\infty < a < +\infty.$

3. **Important Calculus-Based Results**

   (a) *L'Hôpital's Rule:* For differentiable functions $f(x)$ and $g(x)$ and an "extended" real number c (i.e., $c \in \Re_1$ or $c = \pm\infty$), suppose that $\lim_{x\to c} f(x) = \lim_{x\to c} g(x) = 0$, or that $\lim_{x\to c} f(x) = \lim_{x\to c} g(x) = \pm\infty$. Suppose also that $\lim_{x\to c} f'(x)/g'(x)$ exists [in particular, $g'(x) \neq 0$

near c, except possibly at c]. Then,

$$\lim_{x \to c} \frac{\mathrm{f}(x)}{\mathrm{g}(x)} = \lim_{x \to c} \frac{\mathrm{f}'(x)}{\mathrm{g}'(x)}.$$

L'Hôpital's Rule is also valid for one-sided limits.

(b) *Integration by Parts:* Let $u = \mathrm{f}(x)$ and $v = \mathrm{g}(x)$, with differentials $\mathrm{d}u = \mathrm{f}'(x)\mathrm{d}x$ and $\mathrm{d}v = \mathrm{g}'(x)\mathrm{d}x$. Then,

$$\int u\,\mathrm{d}v = uv - \int v\,\mathrm{d}u.$$

(c) *Jacobians for One- and Two-Dimensional Change-of-Variable Transformations:* Let X be a scalar variable with support $\mathcal{A} \subseteq \Re_1$. Consider a one-to-one transformation $U = \mathrm{g}(X)$ that maps $\mathcal{A} \to \mathcal{B} \subseteq \Re_1$. Denote the inverse of U as $X = \mathrm{h}(U)$. Then, the corresponding one-dimensional Jacobian of the transformation is defined as

$$J = \frac{\mathrm{d}[\mathrm{h}(U)]}{\mathrm{d}U},$$

so that

$$\int_{\mathcal{A}} \mathrm{f}(X)\mathrm{d}X = \int_{\mathcal{B}} \mathrm{f}[\mathrm{h}(U)]|J|\mathrm{d}U.$$

Similarly, consider scalar variables X and Y defined on a two-dimensional set $\mathcal{A} \subseteq \Re_2$, and let $U = \mathrm{g}_1(X, Y)$ and $V = \mathrm{g}_2(X, Y)$ define a one-to-one transformation that maps $\mathcal{A}$ in the xy-plane to $\mathcal{B} \subseteq \Re_2$ in the uv-plane. Define $X = \mathrm{h}_1(U, V)$ and $Y = \mathrm{h}_2(U, V)$. Then, the Jacobian of the (two-dimensional) transformation is given by the second-order determinant

$$J = \begin{vmatrix} \frac{\partial \mathrm{h}_1(U,V)}{\partial U} & \frac{\partial \mathrm{h}_1(U,V)}{\partial V} \\ \frac{\partial \mathrm{h}_2(U,V)}{\partial U} & \frac{\partial \mathrm{h}_2(U,V)}{\partial V} \end{vmatrix},$$

so that

$$\int\int_{\mathcal{A}} \mathrm{f}(X, Y)\mathrm{d}X\mathrm{d}Y = \int\int_{\mathcal{B}} \mathrm{f}[\mathrm{h}_1(U, V), \mathrm{h}_2(U, V)]|J|\mathrm{d}U\mathrm{d}V.$$

4. **Special Functions**

(a) *Gamma Function:*

i. For any real number $t > 0$, the Gamma function is defined as

$$\Gamma(t) = \int_0^{\infty} y^{t-1}\mathrm{e}^{-y}\mathrm{d}y.$$

ii. For any real number $t > 0$, $\Gamma(t+1) = t\Gamma(t)$.
iii. For any positive integer n, $\Gamma(n) = (n-1)!$
iv. $\Gamma(1/2) = \sqrt{\pi}$; $\Gamma(3/2) = \sqrt{\pi}/2$; $\Gamma(5/2) = (3\sqrt{\pi})/4$.

(b) *Beta Function:*

i. For $\alpha > 0$ and $\beta > 0$, the Beta function is defined as

$$\mathrm{B}(\alpha, \beta) = \int_0^1 y^{\alpha-1}(1-y)^{\beta-1}\mathrm{d}y.$$

ii. $\mathrm{B}(\alpha, \beta) = \frac{\Gamma(\alpha)\Gamma(\beta)}{\Gamma(\alpha+\beta)}$

(c) *Convex and Concave Functions:* A real-valued function $\mathrm{f}(\cdot)$ is said to be *convex* if, for any two points x and y in its domain and any $t \in [0, 1]$, we have

$$\mathrm{f}[tx + (1-t)y] \leq t\mathrm{f}(x) + (1-t)\mathrm{f}(y).$$

Likewise, $\mathrm{f}(\cdot)$ is said to be *concave* if

$$\mathrm{f}[tx + (1-t)y] \geq t\mathrm{f}(x) + (1-t)\mathrm{f}(y).$$

Also, $\mathrm{f}(x)$ is concave on $[a, b]$ if and only if $-\mathrm{f}(x)$ is convex on $[a, b]$.

5. **Approximations**

(a) *Stirling's Approximation:*

For n a large non-negative integer, $n! \approx \sqrt{2\pi n}\left(\frac{n}{\mathrm{e}}\right)^n$.

(b) *Taylor Series Approximations:*

(i) Univariate Taylor Series: If $\mathrm{f}(x)$ is a real-valued function of x that is infinitely differentiable in a neighborhood of a real number a, then a Taylor series expansion of $\mathrm{f}(x)$ around a is equal to

$$\mathrm{f}(x) = \sum_{k=0}^{\infty} \frac{\mathrm{f}^{(k)}(a)}{k!}(x-a)^k,$$

where

$$\mathrm{f}^{(k)}(a) = \left[\frac{\mathrm{d}^k\mathrm{f}(x)}{\mathrm{d}x^k}\right]_{|x=a}, k = 0, 1, \ldots, \infty.$$

When $a = 0$, the infinite series expansion above is called a Maclaurin series.

As examples, a *first-order* (or linear) Taylor series approximation to $\mathrm{f}(x)$ around the real number a is equal to

$$\mathrm{f}(x) \approx \mathrm{f}(a) + \left[\frac{\mathrm{d}\mathrm{f}(x)}{\mathrm{d}x}\right]_{|x=a}(x-a),$$

and a *second-order* Taylor series approximation to $\mathrm{f}(x)$ around the real number a is equal to

$$\mathrm{f}(x) \approx \mathrm{f}(a) + \left[\frac{\mathrm{df}(x)}{\mathrm{d}x}\right]_{|x=a}(x-a) + \frac{1}{2!}\left[\frac{\mathrm{d}^2\mathrm{f}(x)}{\mathrm{d}x^2}\right]_{|x=a}(x-a)^2.$$

(ii) Multivariate Taylor series: For $p \geq 2$, if $\mathrm{f}(x_1, x_2, \ldots, x_p)$ is a real-valued function of $x_1, x_2, \ldots, x_p$ that is infinitely differentiable in a neighborhood of $(a_1, a_2, \ldots, a_p)$, where $a_i, i = 1, 2, \ldots, p$, is a real number, then a multivariate Taylor series expansion of $\mathrm{f}(x_1, x_2, \ldots, x_p)$ around $(a_1, a_2, \ldots, a_p)$ is equal to

$$\begin{aligned}\mathrm{f}(x_1, x_2, \ldots, x_p) &= \sum_{k_1=0}^{\infty}\sum_{k_2=0}^{\infty}\cdots\sum_{k_p=0}^{\infty}\frac{\mathrm{f}^{(k_1+k_2+\cdots+k_p)}(a_1, a_2, \ldots, a_p)}{k_1!k_2!\cdots k_p!} \\ &\quad \times \prod_{i=1}^{p}(x_i - a_i)^{k_i},\end{aligned}$$

where

$$\begin{aligned}&\mathrm{f}^{(k_1+k_2+\cdots+k_p)}(a_1, a_2, \ldots, a_p) \\ &= \left[\frac{\partial^{k_1+k_2+\cdots+k_p}\mathrm{f}(x_1, x_2, \ldots, x_p)}{\partial x_1^{k_1}\partial x_2^{k_2}\cdots\partial x_p^{k_p}}\right]_{|(x_1,x_2,\ldots,x_p)=(a_1,a_2,\ldots,a_p)}.\end{aligned}$$

As examples, when $p = 2$, a *first-order* (or *linear*) multivariate Taylor series approximation to $\mathrm{f}(x_1, x_2)$ around (a_1, a_2) is equal to

$$\mathrm{f}(x_1, x_2) \approx f(a_1, a_2) + \sum_{i=1}^{2}\left[\frac{\partial \mathrm{f}(x_1, x_2)}{\partial x_i}\right]_{|(x_1,x_2)=(a_1,a_2)}(x_i - a_i),$$

and a *second-order* multivariate Taylor series approximation to $\mathrm{f}(x_1, x_2)$ around (a_1, a_2) is equal to

$$\begin{aligned}\mathrm{f}(x_1, x_2) &\approx \mathrm{f}(a_1, a_2) + \sum_{i=1}^{2}\left[\frac{\partial \mathrm{f}(x_1, x_2)}{\partial x_i}\right]_{|(x_1,x_2)=(a_1,a_2)}(x_i - a_i) \\ &+ \frac{1}{2!}\sum_{i=1}^{2}\left[\frac{\partial^2 \mathrm{f}(x_1, x_2)}{\partial x_i^2}\right]_{|(x_1,x_2)=(a_1,a_2)}(x_i - a_i)^2 \\ &+ \left[\frac{\partial^2 \mathrm{f}(x_1, x_2)}{\partial x_1 \partial x_2}\right]_{|(x_1,x_2)=(a_1,a_2)}(x_1 - a_1)(x_2 - a_2).\end{aligned}$$

6. **Lagrange Multipliers:** The method of *Lagrange multipliers* provides a

strategy for finding stationary points $\boldsymbol{x}^*$ of a differentiable function $\mathrm{f}(\boldsymbol{x})$ subject to the constraint $\mathbf{g}(\boldsymbol{x}) = \boldsymbol{c}$, where $\boldsymbol{x} = (x_1, x_2, \ldots, x_p)'$, where $\mathbf{g}(\boldsymbol{x}) = [\mathrm{g}_1(\boldsymbol{x}), \mathrm{g}_2(\boldsymbol{x}), \ldots, \mathrm{g}_m(\boldsymbol{x})]'$ is a set of $m(< p)$ constraining functions, and where $\boldsymbol{c} = (c_1, c_2, \ldots, c_m)'$ is a vector of known constants. The stationary points $\boldsymbol{x}^* = (x_1^*, x_2^*, \ldots, x_p^*)'$ can be (local) maxima, (local) minima, or saddle points. The Lagrange multipler method involves consideration of the Lagrange function

$$\Lambda(\boldsymbol{x}, \boldsymbol{\lambda}) = \mathrm{f}(\boldsymbol{x}) - [\mathbf{g}(\boldsymbol{x}) - \mathbf{c}]' \boldsymbol{\lambda},$$

where $\boldsymbol{\lambda} = (\lambda_1, \lambda_2, \ldots, \lambda_m)'$ is a vector of scalars called "Lagrange multipliers." In particular, the stationary points $\boldsymbol{x}^*$ are obtained as the solutions for $\boldsymbol{x}$ using the $(p + m)$ equations

$$\begin{aligned} \frac{\partial \Lambda(\boldsymbol{x}, \boldsymbol{\lambda})}{\partial \boldsymbol{x}} &= \frac{\partial \mathrm{f}(\boldsymbol{x})}{\partial \boldsymbol{x}} - \left\{ \frac{\partial \, [\mathbf{g}(\boldsymbol{x}) - \mathbf{c}]'}{\partial \boldsymbol{x}} \right\} \boldsymbol{\lambda} = \mathbf{0} \text{ and} \\ \frac{\partial \Lambda(\boldsymbol{x}, \boldsymbol{\lambda})}{\partial \boldsymbol{\lambda}} &= - [\mathbf{g}(\boldsymbol{x}) - \mathbf{c}] = \mathbf{0}, \end{aligned}$$

where $\partial \mathrm{f}(\boldsymbol{x})/\partial \boldsymbol{x}$ is a $(p\mathrm{x}1)$ column vector with i-th element equal to $\partial \mathrm{f}(\boldsymbol{x})/\partial x_i, i = 1, 2, \ldots, p$, where $\partial[\mathbf{g}(\boldsymbol{x}) - \mathbf{c}]'/\partial \boldsymbol{x}$ is a $(p\mathrm{x}m)$ matrix with (i, j)-th element equal to $\partial g_j(\boldsymbol{x})/\partial x_i, i = 1, 2, \ldots, p$ and $j = 1, 2, \ldots, m$, and where $\mathbf{0}$ denotes a column vector of zeros.

Note that the second matrix equation gives $\mathbf{g}(\boldsymbol{x}) = \mathbf{c}$.

As an example, consider the problem of finding the stationary points (x^*, y^*) of the function $\mathrm{f}(x, y) = (x^2 + y^2)$ subject to the constraint $\mathbf{g}(x, y) = \mathrm{g}_1(x, y) = (x + y) = 1$. Here, $p = 2, m = 1$, and the Lagrange multiplier function is given by

$$\Lambda(x, y, \lambda) = (x^2 + y^2) - \lambda(x + y - 1).$$

The stationary points (x^*, y^*) are obtained by solving the system of equations

$$\begin{aligned} \frac{\partial \Lambda(x, y, \lambda)}{\partial x} &= 2x - \lambda = 0, \\ \frac{\partial \Lambda(x, y, \lambda)}{\partial y} &= 2y - \lambda = 0, \\ \frac{\partial \Lambda(x, y, \lambda)}{\partial \lambda} &= x + y - 1 = 0. \end{aligned}$$

Solving these three equations yields the solution $x^* = y^* = 1/2$. Since $\frac{\partial \Lambda^2(x,y,\lambda)}{\partial x^2} = \frac{\partial \Lambda^2(x,y,\lambda)}{\partial y^2} > 0$ and $\frac{\partial \Lambda^2(x,y,\lambda)}{\partial x \partial y} = 0$, this solution yields a *minimum* subject to the constraint $x + y = 1$.

References

Agresti A. 2012. *Categorical Data Analysis*, Third Edition, John Wiley & Sons, Inc., Hoboken, NJ.

Behboodian J. 1990. "Examples of uncorrelated dependent variables using a bivariate mixture," *The American Statistician*, 44(3), 218.

Blackwell D. 1947. "Conditional expectation and unbiased sequential estimation," *Annals of Mathematical Statistics*, 18, 105-110.

Bloch DA and Kraemer HC. 1989. "2x2 kappa coefficients: Measures of agreement or association," *Biometrics*, 45, 269-287.

Breslow NE and Day NE. 1980. *Statistical Methods in Cancer Research, Volume 1: The Analysis of Case-Control Studies*, International Agency for Research on Cancer (IARC) Scientific Publications.

Bryson MC. 1973. "Craps with crooked dice," *The American Statistician*, 27(4), 167-168.

Casella G and Berger RL. 2002. *Statistical Inference*, Second Edition, Duxbury, Thomson Learning, Belmont, CA.

Copas JB. 1983. "Regression, prediction, and shrinkage," *Journal of the Royal Statistical Society, Series B (Methodological)*, 45(3), 311-354.

Cornfield J, Haenszel W, Hammond E, Lilienfeld A, Shimkin M, and Wynder E. 1959. "Smoking and lung cancer: Recent evidence and a discussion of some questions," *Journal of the National Cancer Institute*, 22, 173-203.

Cortina JM. 1993. "What is coefficient alpha? An examination of theory and applications," *Journal of Applied Psychology*, 78(1), 98-104.

Coyle CA and Wang C. 1993. "Wanna bet: On gambling strategies that may or may not work in a casino," *The American Statistician*, 47(2), 108-111.

Cramér H. 1946. *Mathematical Methods of Statistics*, Princeton University Press, Princeton, NJ.

Cronbach LJ. 1951. "Coefficient alpha and the internal structure of tests," *Psychometrika*, 16, 297-334.

Diggle PJ, Liang K-Y, and Zeger SL. 1994. *Analysis of Longitudinal Data*, Oxford University Press, New York.

Gelman A, Carlin JB, Stern HS, and Rubin DB. 2004. *Bayesian Data Analysis*, Second Edition, Chapman & Hall, Boca Raton, FL.

Goldstein H. 1995. *Multilevel Statistical Models*, Second Edition, Oxford University Press, New York.

Greenland S. 1996. "A lower bound for the correlation of exponentiated bivariate normal pairs," *The American Statistician*, 50(2), 163-164.

Griffiths M. 2011. "Trying to pull out of the drive," *Significance*, 8(2), 89-91.

Gross D, Shortle JF, Thompson JM, and Harris CM. 2008. *Fundamentals of Queueing Theory*, Fourth Edition, John Wiley & Sons, Inc., Hoboken, NJ.

Halmos PR and Savage LJ. 1949. "Applications of the Radon-Nikodym theorem to the theory of sufficient statistics," *Annals of Mathematical Statistics*, 20, 225-241.

Hasselblad V and Hedges LV. 1995. "Meta-analysis of screening and diagnostic tests," *Psychological Bulletin*, 117, 167-168.

Hastie T, Tibshirani R, and Friedman J. 2009. *The Elements of Statistical Learning*, Second Edition, Springer, New York.

Hepworth G. 2005. "Confidence intervals for proportions estimated by group testing with groups of unequal size," *Journal of Agricultural, Biological, and Environmental Statistics*, 10(4), 478-497.

Hoff PD. 2009. *A First Course in Bayesian Statistical Methods*, Springer, New York.

Hogg RV, Craig AT, and McKean JW. 2005. *Introduction to Mathematical Statistics*, Sixth Edition, Prentice-Hall, Upper Saddle River, NJ.

Hung HMJ, O'Neill RT, Bauer P, and Köhne K. 1997. "The behavior of the P-value when the alternative hypothesis is true," *Biometrics*, 53, 11-22.

Kalbfleisch JG. 1985. *Probability and Statistical Inference, Volume 1: Probability*, Second Edition, Springer, New York.

Kalbfleisch JG. 1985. *Probability and Statistical Inference, Volume 2: Statistical Inference*, Second Edition, Springer, New York.

Kleinbaum DG, Kupper LL, and Morgenstern H. 1982. *Epidemiologic Research: Principles and Quantitative Methods*, John Wiley & Sons, Inc., Hoboken, NJ.

Kleinbaum DG, Kupper LL, Nizam A, and Muller KE. 2008. *Applied Regression Analysis and Other Multivariable Methods*, Fourth Edition, Duxbury Press, Belmont, CA.

Lefebvre M. 2007. *Applied Stochastic Processes*, Springer, New York.

Lehmann EL. 1983. *Theory of Point Estimation*, Springer, New York.

Likert R. 1931. "A technique for the measurement of attitudes," *Archives of Psychology*, Columbia University Press, New York.

McCullagh P and Nelder JA. 1989. *Generalized Linear Models*, Second Edition, Chapman & Hall/CRC Press, London, UK.

Morris MD and Ebey SF. 1984. "An interesting property of the sample mean under a first-order autoregressive model," *The American Statistician*, 38(2), 127-129.

Neyman J and Pearson ES. 1928. "On the use and interpretation of certain test criteria for purposes of statistical inference," *Biometrika*, 20A, 175-240 and 263-294.

Neyman J and Pearson ES. 1933. "On the problem of the most efficient tests of statistical hypotheses," *Philosophical Transactions, Series A*, 231, 289-337.

Pepe MS and Janes H. 2007. "Insights into latent class analysis of diagnostic test performance," *Biostatistics*, 8(2), 474-484.

Press WH. 2009. "Strong profiling is not mathematically optimal for discovering rare malfeasors," *Proceedings of the National Academy of Sciences USA*, 106, 1716-1719.

Press WH. 2010. "To catch a terrorist: Can ethnic profiling work?" *Significance*, 7(4), 164-167.

Rao CR. 1945. "Information and accuracy attainable in the estimation of statistical parameters," *Bulletin of the Calcutta Mathematical Society*, 37, 81-91.

Rao CR. 1947. "Large sample tests of statistical hypotheses concerning several parameters with applications to problems of estimation," *Proceedings of the Cambridge Philosophical Society*, 44, 50-57.

Rao CR. 1973. *Linear Statistical Inference and Its Applications*, Second Edition, John Wiley & Sons, Inc., Hoboken, NJ.

Rappaport SM and Kupper LL. 2008. *Quantitative Exposure Assessment*, Lulu Press, Raleigh, NC.

Rappaport SM, Symanski E, Yager JW, and Kupper LL. 1995. "The relationship between environmental monitoring and biological markers in exposure assessment," *Environmental Health Perspectives*, 103(Supplement 3), 49-53.

Ross S. 2006. *A First Course in Probability*, Seventh Edition, Prentice-Hall, Inc., Upper Saddle River, NJ.

Sackrowitz H and Samuel-Cahn E. 1999. "P values as random variables - expected P values," *The American Statistician*, 53(4), 326-331.

Serfling RJ. 2002. *Approximation Theorems of Mathematical Statistics*, John Wiley & Sons, Inc., Hoboken, NJ.

Stefanski LA. 1992. "Monotone likelihood ratio of a faulty inspection distribution," *The American Statistician*, 46(2), 110-114.

Wackerly DD, Mendenhall III W, and Scheaffer RL. 2008. *Mathematical Statistics With Applications*, Seventh Edition, Duxbury, Thomson Learning, Belmont, CA.

Wald A. 1943. "Tests of statistical hypotheses concerning several parameters when the number of observations is large," *Transactions of the American Mathematical Society*, 54, 426-482.

Warner SL. 1965. "Randomized response: a survey technique for eliminating evasive answer bias," *Journal of the American Statistical Association*, 60(309), 63-69.

Webb RY, Smith PJ, and Firag A. 2010. "On the probability of improved accuracy with increased sample size," *The American Statistician*, 64(3), 257-262.

Index

N

O

P

Q

R

S

T

U

Printed and bound by CPI Group (UK) Ltd, Croydon, CR0 4YY
23/10/2024
01777672-0018